T0214614

Lecture Notes in Computer Science 11417

Commenced Publication in 1973
Founding and Former Series Editors:
Gerhard Goos, Juris Hartmanis, and Jan van Leeuwen

Carlos Martín-Vide · Alexander Okhotin
Dana Shapira (Eds.)

Language
and Automata Theory
and Applications

13th International Conference, LATA 2019
St. Petersburg, Russia, March 26–29, 2019
Proceedings

Springer

Editors
Carlos Martín-Vide (iD)
Rovira i Virgili University
Tarragona, Spain

Dana Shapira (iD)
Ariel University
Ariel, Israel

Alexander Okhotin (iD)
Saint Petersburg State University
St. Petersburg, Russia

ISSN 0302-9743   ISSN 1611-3349 (electronic)
Lecture Notes in Computer Science
ISBN 978-3-030-13434-1   ISBN 978-3-030-13435-8 (eBook)
https://doi.org/10.1007/978-3-030-13435-8

Library of Congress Control Number: 2019931952

LNCS Sublibrary: SL1 – Theoretical Computer Science and General Issues

This Springer imprint is published by the registered company Springer Nature Switzerland AG
The registered company address is: Gewerbestrasse 11, 6330 Cham, Switzerland

Preface

These proceedings contain the papers that were presented at the 13th International Conference on Language and Automata Theory and Applications (LATA 2019), held in Saint Petersburg, Russia, during March 26–29, 2019.

The scope of LATA is rather broad, including: algebraic language theory, algorithms for semi-structured data mining, algorithms on automata and words, automata and logic, automata for system analysis and program verification, automata networks, automatic structures, codes, combinatorics on words, computational complexity, concurrency and Petri nets, data and image compression, descriptional complexity, foundations of finite state technology, foundations of XML, grammars (Chomsky hierarchy, contextual, unification, categorial, etc.), grammatical inference and algorithmic learning, graphs and graph transformation, language varieties and semigroups, language-based cryptography, mathematical and logical foundations of programming methodologies, parallel and regulated rewriting, parsing, patterns, power series, string processing algorithms, symbolic dynamics, term rewriting, transducers, trees, tree languages and tree automata, and weighted automata.

LATA 2019 received 98 submissions. Every paper was reviewed by three Programme Committee members. There were also some external experts consulted. After a thorough and lively discussion phase, the committee decided to accept 31 papers (which represents a competitive acceptance rate of about 32%). The conference program included five invited talks as well.

The excellent facilities provided by the EasyChair conference management system allowed us to deal with the submissions successfully and handle the preparation of these proceedings in time.

We would like to thank all invited speakers and authors for their contributions, the Program Committee and the external reviewers for their cooperation, and Springer for its very professional publishing work.

January 2019

Carlos Martín-Vide
Alexander Okhotin
Dana Shapira

Organization

Program Committee

Krishnendu Chatterjee	Institute of Science and Technology, Austria
Bruno Courcelle	University of Bordeaux, France
Manfred Droste	University of Leipzig, Germany
Travis Gagie	Diego Portales University, Chile
Peter Habermehl	Paris Diderot University, France
Tero Harju	University of Turku, Finland
Markus Holzer	University of Giessen, Germany
Radu Iosif	Verimag, France
Kazuo Iwama	Kyoto University, Japan
Juhani Karhumäki	University of Turku, Finland
Lila Kari	University of Waterloo, Canada
Juha Kärkkäinen	University of Helsinki, Finland
Bakhadyr Khoussainov	The University of Auckland, New Zealand
Sergey Kitaev	University of Strathclyde, UK
Shmuel Tomi Klein	Bar-Ilan University, Israel
Olga Kouchnarenko	University of Franche-Comté, France
Thierry Lecroq	University of Rouen, France
Markus Lohrey	University of Siegen, Germany
Sebastian Maneth	University of Bremen, Germany
Carlos Martín-Vide (Chair)	Rovira i Virgili University, Spain
Giancarlo Mauri	University of Milano-Bicocca, Italy
Filippo Mignosi	University of L'Aquila, Italy
Victor Mitrana	Polytechnic University of Madrid, Spain
Joachim Niehren	Inria Lille, France
Alexander Okhotin	Saint Petersburg State University, Russian Federation
Dominique Perrin	University of Paris-Est, France
Matteo Pradella	Polytechnic University of Milan, Italy
Jean-François Raskin	Université Libre de Bruxelles, Belgium
Marco Roveri	Bruno Kessler Foundation, Italy
Karen Rudie	Queen's University, Canada
Wojciech Rytter	University of Warsaw, Poland
Kai Salomaa	Queen's University, Canada
Sven Schewe	University of Liverpool, UK
Helmut Seidl	Technical University of Munich, Germany
Ayumi Shinohara	Tohoku University, Japan
Hans Ulrich Simon	Ruhr-University of Bochum, Germany
William F. Smyth	McMaster University, Canada
Frank Stephan	National University of Singapore, Singapore

Martin Sulzmann	Karlsruhe University of Applied Sciences, Germany
Jorma Tarhio	Aalto University, Finland
Stefano Tonetta	Bruno Kessler Foundation, Italy
Rob van Glabbeek	Data61, CSIRO, Australia
Margus Veanes	Microsoft Research, USA
Mahesh Viswanathan	University of Illinois, Urbana-Champaign, USA
Mikhail Volkov	Ural Federal University, Russian Federation
Fang Yu	National Chengchi University, Taiwan
Hans Zantema	Eindhoven University of Technology, New Zealand

Additional Reviewers

Abdeddaïm, Saïd
Baeza-Yates, Ricardo
Balogh, Jozsef
Beier, Simon
Bernardinello, Luca
Bersani, Marcello M.
Bollig, Benedikt
Bride, Hadrien
Cadilhac, Michaël
Capelli, Florent
Chiari, Michele
Choudhury, Salimur
Ciobanu, Laura
Courcelle, Bruno
Crespi Reghizzi, Stefano
Crosetti, Nicolas
De La Higuera, Colin
Delecroix, Vincent
Dennunzio, Alberto
Dose, Titus
Dück, Stefan
Ferretti, Claudio
Fici, Gabriele
Fülöp, Zoltan
Geffert, Viliam
Gilroy, Sorcha

Giorgetti, Alain
Godin, Thibault
Groote, Jan Friso
Groschwitz, Jonas
Gutiérrez, Martín
Guyeux, Christophe
Heam, Pierre-Cyrille
Hellouin de Menibus,
    Benjamin
Hendrian, Diptarama
Hugot, Vincent
Jeandel, Emmanuel
Kari, Jarkko
Kociumaka, Tomasz
Kopczynski, Eryk
Kuske, Dietrich
Lemay, Aurelien
Lisitsa, Alexei
Lonati, Violetta
Luttik, Bas
Maletti, Andreas
Mandrioli, Dino
Mens, Irini
Mhaskar, Neerja
Morawska, Barbara
Morzenti, Angelo

Nakamura, Katsuhiko
Nehaniv, Chrystofer
Nugues, Pierre
Paul, Erik
Penelle, Vincent
Persiano, Giuseppe
Plandowski, Wojciech
Popa, Alexandru
Praveen, M.
Prunescu, Mihai
Pyatkin, Artem
Quaas, Karin
Radoszewski, Jakub
Rossi, Matteo
Sakho, Momar
Sangnier, Arnaud
Schmidt-Schauss,
    Manfred
Starikovskaya, Tatiana
Talbot, Jean-Marc
Vaszil, György
Wendlandt, Matthias
Willemse, Tim
Yakovets, Nikolay
Yoshinaka, Ryo
Zetzsche, Georg

Abstracts of Invited Papers

Searching and Indexing Compressed Text

Paweł Gawrychowski

Institute of Computer Science, University of Wrocław, Wrocław, Poland
gawry@cs.uni.wroc.pl

Abstract. Two basic problems considered in algorithms on strings are pattern matching and text indexing. An instance of the pattern matching problem comprises two strings, usually called the pattern and the text, and the goal is to locate an occurrence of the former in the latter. In the more general text indexing problem we wish to preprocess a long text for multiple queries with (probably short) patterns. Efficient solutions to both problems have been already designed in the 1970s, and by now we know several linear-time pattern matching algorithms as well as linear-space indexes answering queries in linear (or almost linear) time that are simple enough to be taught in a basic algorithms course. However, with the ever-increasing amount of data being generated and stored, it is not clear if the seemingly optimal linear complexity is actually good enough. This is because linear in the length of the pattern or the text might be larger than the size of its description, for example when we are working with strings over a small alphabet and are able to store multiple characters in a single machine word. The difference might be even more dramatic if we store the data in a compressed form. For some compression schemes, such as the Lempel-Ziv family of compression algorithms, it may as well be the case that the length of the original string is exponential in the size of its compressed representation. In such a case, we would like to design a solution running in time and space proportional to the size of the compressed representation, or at least close to it. This brings the question of developing compressed pattern matching algorithms and designing compressed text indexes.

I will survey the landscape of searching and indexing compressed text, focusing on the Lempel-Ziv family of compression algorithms and the related grammar-based compression. For compressed pattern matching, I will mostly assume that only the text is compressed, but will also briefly describe the recent progress on the more general case of fully compressed pattern matching, where both the text and the pattern are compressed, and on the approximate version of the problem, in which we seek fragments of the text within small Hamming or edit distance to the pattern. For compressed text indexing, I will discuss the known trade-offs between the size of the structure and the query time and highlight the remaining open questions.

Pattern Discovery in Biological Sequences

Esko Ukkonen ⓘ

Department of Computer Science, University of Helsinki,
P. O. Box 68 (Gustaf Hällströmin katu 2b), 00014 Helsinki, Finland
esko.ukkonen@helsinki.fi

Abstract. Sequence motifs are patterns of symbols that have recurrent occurrences in biological sequences such as DNA and are presumed to have biological function. Modeling and identification of regulatory DNA motifs such as the binding sites of so-called transcription factors (TFs) is in the core of the attempts to understand gene regulation and functioning of the genome as a whole. Transcription factors are proteins that may bind to DNA, close to transcription start site of a gene. Binding activates or inhibits the transcription machinery (expression) of the associated gene. As the regulated gene may itself be a transcription factor, such pairwise regulatory relation between genes induces a genome-wide network model for gene regulation.

The possible binding sites of a transcription factor T are short DNA segments (DNA words). Different sites of T are close variants of an underlying consensus word specific to T. As for most transcription factors an accurate biophysical modeling of this variation is currently infeasible, simplified combinatorial and probabilistic models of binding motifs are used. The parameters of the models are learned from training DNA sequences that contain plenty of instances of the motif but their exact location within the sequences may not be known a priori.

Applying concepts of formal languages and automata, motifs are modeled with words in generalized alphabets and with other regular-expression-like structures representing the language of possible words of the motif. Such motifs can be extracted from training data using string processing algorithms that find repetitions in sequences.

On probabilistic side, sequence motifs can be modeled with inhomogeneous Markov chains of order 0 or higher and also with more general Markov models. Markov chains of order 0, usually called Position Weight Matrices (PWMs) and visualized with so-called sequence logos, is the motif class commonly used in motif databases. PWM assumes that the motif positions are independent. For some TFs this is too weak, and then Markov models of order higher than 0 capable of representing dependencies between two or more positions suit better. Given training DNA sequences that contain occurrences of motif instances proportionally to the target distribution, machine learning methods can estimate the distribution by learning a probabilistic model that fits best the data.

The talk surveys representations and corresponding discovery algorithms for transcription factor binding motifs. We will discuss suffix-tree based methods for discovery of combinatorial models as well as expectation maximization (EM) algorithm based learning of probabilistic models. We consider basic motifs for single factors (monomers) as well as composite motifs for pairs of factors (dimers) and for chains of factors. Such chains model regulatory modules that are built of clusters of several factors making together a regulatory complex. Regulatory modules can be discovered from alignments of genomes of related species. Alignment-based method is possible as regulatory modules are conserved in evolution.

References

1. Bailey, T.L., Elkan, C.: The value of prior knowledge in discovering motifs with MEME. In: Proceedings of Third International Conference on Intelligent Systems for Molecular Biology, pp. 21–29. AAAI Press (1995)
2. Jolma, A., Yan, J., Whitington, T., Toivonen, J., et al.: DNA-binding specifities of human transcription factors. Cell, **152**(1–2), 327–339 (2013)
3. Khan, A., et al.: JASPAR 2018: update of the open-access database of transcription factor binding profiles and its web framework. Nucleic Acids Res. **46**, D260–D266 (2018)
4. Marsan, L., Sagot, M.: Algorithms for extracting structured motifs using a suffix tree with application to promoter and regulatory site consensus identification. J. Comput. Biol. **7**, 345–360 (2000)
5. Omidi, S., Zavolan, M., Pachkov, M., Breda, J., Berger, S., van Nimwegen, E.: Automated incorporation of pairwise dependency in transcription factor binding site prediction using dinucleotide weight tensors. PLoS Comput. Biol. **13**(7), e1005176 (2017)
6. Palin, K., Taipale, J., Ukkonen, E.: Locating potential enhancer elements by comparative genomics using the EEL software. Nat. Protoc. **1**(1), 368 (2006)
7. Pavesi, G., Mauri, G., Pesole, G.: An algorithm for finding signals of unknown length in DNA sequences. Bioinf. **17**(Suppl. 1), S207–S214 (2001)
8. Ruan, S., Stormo, G.D.: Inherent limitations of probabilistic models for protein-DNA binding specificity. PLoS Comput. Biol. **13**(7), e1005638 (2017)
9. Stormo, G.D., Schneider, T.D., Gold, L., Ehrenfeucht, A.: Use of the 'Perceptron' algorithm to distinguish translational initiation sites in E. coli. Nucleic Acids Res. **10**, 2997–3011 (1982)
10. Siebert, M., Söding, J.: Bayesian markov models consistently outperform pwms at predicting motifs in nucleotide sequences. Nucleic Acids Res. **44**(13), 6055–6069 (2016)
11. Toivonen, J., Kivioja, T., Jolma, A., Yin, Y., Taipale, J., Ukkonen, E.: Modular discovery of monomeric and dimeric transcription factor binding motifs for large data sets. Nucleic Acids Res. **46**(8), e44 (2018)

Contents

Graphs, Trees and Rewriting

Words and Codes

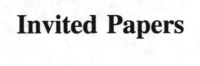

Invited Papers

Invited Papers

Modern Aspects of Complexity Within Formal Languages

Henning Fernau[✉][iD]

Fachbereich 4 – Abteilung Informatikwissenschaften, CIRT, Universität Trier,
54286 Trier, Germany
fernau@uni-trier.de

Abstract. We give a survey on some recent developments and achievements of modern complexity-theoretic investigations of questions in Formal Languages (FL). We will put a certain focus on multivariate complexity analysis, because this seems to be particularly suited for questions concerning typical questions in FL.

Keywords: String problems · Finite automata ·
Context-free grammars · Multivariate analysis ·
Fixed-parameter tractability · Fine-grained complexity

1 Introduction

Formal Languages and Complexity Theory have a long (common) history. Both fields can be seen as two of the major backbones of Theoretical Computer Science. Both fields are often taught together in undergraduate courses, mostly obligatory in Computer Science curricula. This is also testified by looking at classical textbooks like [50]. Yet, modern developments in complexity and algorithmics are barely mirrored in Formal Languages. We want to argue in this paper that this is a fact that need to be changed.

We will work through six case studies to explain several findings in recent years. We will also expose a number of open problems in each of these cases. This should motivate researchers working in Formal Languages to look deeper into recent developments in Complexity Theory, but also researchers more oriented towards these modern aspects of Complexity Theory to look into (possibly apparently rather old) problems in Formal Languages to see if they could offer solutions or at least new approaches to these problems.

In most cases, the problems we discuss in the area of Formal Languages can be easily understood with the already mentioned background knowledge each computer scientists gets already during the corresponding bachelor courses. Therefore, we will only fix notations but assume that no further explanations are necessary. By way of contrast, we will spend some more time explaining at least some ideas of the concepts discussed nowadays in Complexity Theory and Algorithms, so that readers with a background rooted in Formal Languages

C. Martín-Vide et al. (Eds.): LATA 2019, LNCS 11417, pp. 3–30, 2019.
https://doi.org/10.1007/978-3-030-13435-8_1

could easily follow this introductory exposition. In any case, we want to make clear why these modern approaches are particularly suited for attacking decision problems in Formal Languages.

2   Some Modern Concepts of Complexity Theory

As most of our case studies deal with multivariate analysis, let us first delve into the general scheme behind this idea. This is intimately linked to the basic ideas of Parameterized Complexity, which could be paraphrased as follows. If we can show that some decision problem is computationally complicated, which is typically formalized by proving that it is NP-hard or PSPACE-hard, what can we do about it then, assuming that we still like to solve it with the help of computers? More traditionally, help was expected from approximation algorithms or, more practically, from heuristics, and in fact the proof of computational hardness often gave a sort of excuse of using these rules of thumb called heuristics. However, in particular using heuristics with no success guarantees is not very satisfying from a more theoretical perspective at least. What can be done about such a situation?

It is exactly here where Parameterized Complexity and also the more practical side of it, namely Parameterized Algorithms, steps in. The basic idea is to not merely look at the bitstring length as the only source of information that an instance of a computational problem might give as defining or measuring its complexity, but to also look at other aspects of the instance, formalized in terms of a so-called parameter. What happens if they are always small in the concrete instances that practitioners look at? Does this still mean that the problem is computationally hard? Or can we possibly solve these practically relevant instances efficiently, although the problem is NP-hard when considering all possible instances? Interestingly, one of the most successful approaches within Parameterized Algorithms can be viewed as an analysis of natural heuristics, which means, in more formal terms, the use of reduction rules to provide means to (repeatedly) move from an instance to an equivalent instance of smaller size.

What could these parameter be? Let us look at various examples.

- The most classical NP-hard problem is arguably SATISFIABILITY, or SAT for short. Given a Boolean formula φ in conjunctive normal form (CNF), decide if φ is satisfiable. A possibly natural choice of a parameter could be a bound k on the number of literals that may appear in clauses. When fixing this upper-bound to k, we arrive at problems like k-SAT. This parameter choice is not that helpful, because it is well-known that even 3-SAT remains NP-hard, and in fact there are quite a number of much more restricted variants of SAT that cannot be solved in polynomial time, assuming that P does not equal NP, see [57,61,93] for several such restrictions. Only 2-SAT is still solvable in polynomial time.
- Looking back at the proof of the theorem of Cook, phrased in (nowadays) non-standard terminology in [21], one can argue that the more basic NP-complete problem is in fact the following one, which can be considered as a

problem arising in the field of Formal Languages: Given a nondeterministic Turing machine M, an input string x and an integer k, does M halt within k steps accepting x? Now, if k is fixed to a small constant, this problem looks more amenable than the previous one. More precisely, let us focus on one-tape machines for now. Assume that we have ℓ symbols in the work alphabet of M and assume that M has t states. Then at each time step, being in a specific state, upon reading a symbol, M has at most $3\ell t$ choices to continue its computation, the factor three entering because of the different choices for moves. Moreover, after k steps, at most $\mathcal{O}(tk\ell^k)$ many configurations are possible, simply because at most k tape cells could have been visited within k steps, and also the current state and head position has to be memorized. The difference to Turing machines with a dedicated read-only input tape is not of importance here. It can be easily checked if an accepting configuration can be reached from the initial configuration within the directed graph implicitly given by the configuration and the reachability relation. Instead of writing ℓ and t, we can also consider the size of M, with reasonable ways to measure this; the size of the state transition table should be always encaptured here, and this also bounds the number of choices. So, we might write the running time of this algorithm like $\mathcal{O}(|M|^k)$, which is polynomial when k is really fixed, but if k is considered as part of the input, this algorithm is clearly exponential. Similar considerations are valid for multi-tape nondeterministic Turing machines. For future reference, let us call these problems SHORT NTM ACCEPTANCE (when referring to single tapes) and SHORT MULTI-TAPE NTM ACCEPTANCE.

- Recall that a hypergraph can be specified as $G = (V, E)$ with $E \subseteq 2^V$. Elements of V are called vertices, while elements of E are called (hyper-)edges. In different terminology, V is the ground-set or universe, and E is a set system. $C \subseteq V$ is called a *hitting set* if $e \cap C \neq \emptyset$ for all $e \in E$. In the decision problem HITTING SET, we are given a hypergraph $G = (V, E)$ and an integer k, and the question is if one can find a hitting set of cardinality at most k for G. This question is well-known to be NP-hard. There is a simple relation to SAT: One can view the hyperedges as clauses. Now, a hitting set corresponds to the set of variables set to true. It is also evident why we need the bound k here: setting all variables to true corresponds to selecting V as a hitting set, and this is clearly a satisfying assignment, because none of the clauses contains any negation. This relation also motivates to study d-HITTING SET, where d upper-bounds the number of vertices in each hyperedge. In contrast to SAT, this restriction looks quite promising, considering the following algorithm: As long as there are hyperedges in G and as long as the integer k is positive, do the following: pick such a hyperedge e and branch according to the at most d possibilities for hitting e. In each case, delete all hyperedges from G that contain e, decrement k and continue. If and only if this loop is ever exited with $E = \emptyset$, the original instance was a YES-instance. As at most d^k many possibilities are tested, the running time of this algorithm can be estimated as $\mathcal{O}(d^k p(|G|))$, where p is some polynomial and $|G|$ gives the size of the original input. Without this additional accounting of d, it is not clear how to solve this problem better than $\mathcal{O}(|G|^k)$.

- Observe that hypergraph instances of 2-HITTING SET can be also viewed as undirected simple graphs, because loops (i.e., singletons in the set of hyper-edges) can be easily removed, as the constituent elements must be put into the hitting set. 2-HITTING SET is also known as VERTEX COVER. Then, an alternative parameterization might be $k_d = |V| - k$. This problem is also known as INDEPENDENT SET, and it can be rephrased as asking for a set I of k_d many vertices such that no edge contains two vertices from I. Such a set is also known as an *independent set*. A related question asks, given a graph G and an integer ℓ, if there is a *clique* of size ℓ in G, i.e., a set K of ℓ vertices such that each pair of vertices from K is adjacent. All these decision problems are also known to be NP-complete.

With these problems from different fields in mind, it might make sense to consider a decision problem $\mathcal{P}$ equipped with a computable parameterization function κ that maps instances to integers. Two ways in which algorithms can behave nicely on instances x might be considered. (a) There is an algorithm that solves instances x of $\mathcal{P}$ in time $\mathcal{O}(|x|^{\kappa(x)})$. (b) There is an algorithm that solves instances x of $\mathcal{P}$ in time $\mathcal{O}(|x|^d f(\kappa(x)))$ for some constant degree d and some (computable) function f. Both definitions imply that $\mathcal{P}$ can be solved in polynomial time, when restricted to instances whose parameter is smaller than some constant c. Yet, possibility (b) means that, assuming d to be reasonably small and f not behaving too badly, then $\mathcal{P}$ can be solved not only for instances where the parameter $\kappa(x)$ is bounded by a constant but it may grow moderately, i.e., this is a far more desirable property. (Parameterized) problems $(\mathcal{P}, \kappa)$ that satisfy (a) are also said to belong to XP, while if $(\mathcal{P}, \kappa)$ satisfies (b), it belongs to FPT (fixed parameter tractable).

Let us look at our example problems again.

- Let κ_1 map a Boolean formula φ to the number of variables that occur in φ. Then, (SAT, κ_1) is in FPT, because there are only $2^{\kappa_1(\varphi)}$ many assignments one has to check to see if φ is satisfiable. Let κ_2 map a Boolean formula φ in CNF to the maximum number of literals appearing in any clause of φ. Then, unless P equals NP, (SAT, κ_2) does not belong to XP.
- For an instance $I = (M, x, k)$ of SHORT NTM ACCEPTANCE, let $\kappa_3(I) = k$. As argued above, (SHORT NTM ACCEPTANCE, κ_3) belongs to XP. However, it seems to be hard to put it in FPT. If we consider the size of M as an additional parameter, i.e., $\kappa_4(I) = |M| + k$, then we have also seen that (SHORT NTM ACCEPTANCE, κ_4) belongs to FPT. Similar considerations hold true for the multi-tape case. Possibly more interestingly, our considerations show that also with the parameterization κ_5 that adds k and the size of the overall alphabet, we end up in FPT. A reader knowledgeable about the early days of Descriptional Complexity (within FL) might ponder for a moment if there might be a possibility to put (SHORT NTM ACCEPTANCE, κ_3) into FPT by resorting to a theorem of Shannon [84] that states that Turing machines with an arbitrary number of working tape symbols can be simulated by Turing

machines with binary tapes. However, this idea is problematic at least for two reasons: (a) the simulating machine (with binary working tape) needs a considerable amount of time for the simulation, this way changing the upper-bound k on the number of steps; (b) in Shannon's simulation, the order of magnitude of the product of alphabet size and number of states stays the same. In combination, both effects counter-act this idea.

- Let us study various parameterizations for HITTING SET. Let $G = (V, E)$ with $E \subseteq 2^V$ and k form an instance and define $\kappa_6 : (G, k) \mapsto k$, $\kappa_7 : (G, k) \mapsto |V|$, $\kappa_8 : (G, k) \mapsto |E|$ and $\kappa_9 : (G, k) \mapsto \max\{|e| : e \in E\}$. As $\binom{n}{k} \in \mathcal{O}(n^k)$, in roughly $\mathcal{O}(|V|^k)$ steps, the instance (G, k) can be solved by testing all k-element subsets if they form a hitting set, putting (HITTING SET, κ_6) in XP. With the same idea, (HITTING SET, κ_7) is in FPT. By using dynamic programming as explained in [30, 41], also (HITTING SET, κ_8) is in FPT. Due to the NP-hardness of VERTEX COVER, if (HITTING SET, κ_9) is in XP, then P equals NP. When combining parameters, $\kappa_{10} := \kappa_6 + \kappa_9$, we conclude with the reasoning given above that (HITTING SET, κ_{10}) is in FPT. This is quite instructive, because it shows that even combining relatively weak parameters, one can obtain relatively nice algorithmic results. More details, also for special cases, can be found in [30–32, 99].
- Reconsider $\kappa_6 : (G, k) \mapsto k$ for VERTEX COVER. Then, by the equivalence to 2-HITTING SET, (VERTEX COVER, κ_6) belongs to FPT. However, reparameterizing by $\kappa_{11} : (G, k) \mapsto |V| - k$, we only know membership in XP for (VERTEX COVER, κ_{11}). Recall that this is equivalent to considering (INDEPENDENT SET, κ_6). By moving over from G to the graph complement $\overline{G}$, with $\overline{(V, E)} = (V, \binom{V}{2} \setminus E)$, one understands that also (CLIQUE, κ_6) has the same complexity status as (VERTEX COVER, κ_{11}).

So far, we introduced the classes FPT and XP of parameterized problems. Clearly, FPT $\subseteq$ XP. As often in Complexity Theory, it is unknown if this inclusion is strict, but it is generally assumed that this is the case. Moreover, we have seen examples of parameterized problems that are not in XP, assuming that P is not equal to NP. In order to have a more refined picture of the world of parameterized problems, it is a good idea to define appropriate reductions. It should be clear what properties such a *many-one* FPT *reduction* relating problem $(\mathcal{P}, \kappa)$ to $(\mathcal{P}', \kappa')$ should satisfy: (a) it should translate an instance I of $\mathcal{P}$ to an instance I' of $\mathcal{P}'$ within time $f(\kappa(I))|I|^{\mathcal{O}(1)}$ for some computable function f; (b) it should preserve the parameterization in the sense there is a function g such that $\kappa'(I') \leq g(\kappa(I))$; (c) I is a YES-instance of $\mathcal{P}$ if and only if I' is a YES-instance of $\mathcal{P}'$. Such a notion allows us to define further complexity classes, based on the idea of being interreducibility with respect to FPT reductions, or FPT-*equivalent*. Observe that the classes FPT and XP studied so far are closed under FPT reductions. Let W[1] be the class of parameterized problems that are FPT-equivalent to (SHORT NTM ACCEPTANCE, κ_3), and let W[2] be the class of parameterized problems that are FPT-equivalent to (SHORT MULTITAPE NTM ACCEPTANCE, κ_3). In fact, there is a whole (presumably infinite) hierarchy of complexity classes captured in the following inclusion chain:

$$\mathsf{FPT} \subseteq \mathsf{W}[1] \subseteq \mathsf{W}[2] \subseteq \cdots \subseteq \mathsf{XP}.$$

Looking back at our examples, it is known that (INDEPENDENT SET, κ_6) and (CLIQUE, κ_6) are complete for $\mathsf{W}[1]$, while (HITTING SET, κ_6) is complete for $\mathsf{W}[2]$. As we will see in the following sections, many natural parameterizations of computational problems stemming from Formal Languages lead to $\mathsf{W}[1]$-hardness results. If we still want to employ the idea of getting FPT algorithms, we need to find different, often multiple parameterizations. This approach is also called *mutlivariate analysis*. We refer to [12, 28, 71] for further discussions. At this point, we only recall that we also used this idea when looking at (HITTING SET, κ_{10}).

There is a nice characterization of FPT based on the idea of the existence of a polynomial-time many-one reduction termed *kernelization* that maps instances I of $(\mathcal{P}, \kappa)$ to instances I', also of $(\mathcal{P}, \kappa)$, satisfying $|I'| \leq f(\kappa(I))$ for some computable function f, where $|I|$ yields the size of instances I in the classical sense.[1] I' is also called the *kernel* of I. The existence of kernelizations is often shown by providing a collection of so-called reduction rules that should be applied exhaustively in order to produce the kernel. As an example, the two reduction rules together provide a kernelization for (VERTEX COVER, κ_6). (a) Delete vertices of degree zero, or, more formally, $((V, E), k) \mapsto ((V \setminus \{v \mid v \notin \bigcup_{e \in E} e\}), k)$, and (b) $(G, k) \mapsto (G - v, k - 1)$ if there is some $v \in V(G)$ with more than k neighbors. In fact, it is not hard to see that the resulting kernels (G', k') even satisfy a polynomial bound on the size of G' (with some reasonable size measure) with respect to $\kappa_6(G', k') = k'$. We also say that (VERTEX COVER, κ_6) admits a polynomial kernel.

With this notion at hand, the question is if one can always produce kernels of polynomial size for parameterized problems in FPT. This is not the case unless the polynomial-time hierarchy collapses to the third level, which is considered to be unlikely. For instance, under this condition, it can be shown that both (HITTING SET, κ_7) and (HITTING SET, κ_8) and hence also (SAT, κ_1) have no polynomial-size kernels.

Another venue that one could follow is refining the questions about optimality of existing algorithms further beyond the question if the decision problem at hand belongs to P or if it is NP-hard. This line of research is nowadays captured under the umbrella of *Fine-Grained Complexity*.

For instance, as discussed, there is a trivial algorithm to solve a SAT instance by testing all assignments. Neglecting polynomial factors, as standard by the very definition of FPT, we can also state that SAT instances on n variables can be solved in time $O^*(2^n)$, where the O^*-notation was invented just to suppress polynomial factors. Now, one can ask if there is any algorithm that solves SAT instances in time $O^*((2 - \varepsilon)^n)$ for any $\varepsilon > 0$. No such algorithm is known today.

---

[1] In the literature, is often required that $|I'| + \kappa(I') \leq f(\kappa(I))$, but this is equivalent to the present requirement, because the parameterization can be computed from I', i.e., $\kappa(I')$ is also bounded by a function in $\kappa(I)$ if $|I'|$ is.

The hypothesis that no such algorithm exists is also called *Strong Exponential-Time Hypothesis*, or SETH for short. A weaker assumption is to believe that there is no function $f(n) \in o(n)$ such that SAT, or in this case equivalently also 3-SAT, can be solved in time $O^*(2^{f(n)})$. This hypothesis is also known as *Exponential-Time Hypothesis*, or ETH for short.[2] The outcome of the famous sparsification lemma is sometimes good to know: Under ETH, there is also no $O^*(2^{o(n)})$-time algorithm for 3-SAT on instances that have $\mathcal{O}(n)$ many clauses. For instance, it is known that under ETH, no $2^{o(n)}$ algorithm exists for solving HITTING SET on instances with n vertices. We also refer to [22]. Further consequences of this approach to the (non-)existence of certain types of FPT algorithms are also discussed in the survey paper [62]. For instance, while there are quite a number of algorithms that solve (VERTEX COVER, κ_6) in time $O^*(2^{\mathcal{O}(\kappa_6(G))})$, under ETH there is no algorithm for doing this in time $O^*(2^{o(\kappa_6(G))})$. It should be clear that for obtaining such results, another form of reduction is needed. We do not explain any details here, but just observe that many well-known reductions suffice for showing some basic ETH-based results. For instance, the typical text-book reductions for showing NP-hardness of VERTEX COVER start from 3-SAT and then introduce gadgets with two or three vertices for variables or clauses, respectively. Hence, by the outcome of the sparsification lemma, under ETH there is no $O^*(2^{o(|V|)})$-time algorithm for computing a minimum vertex cover for $G = (V, E)$. However, not all lower bounds of this type that one assumes to hold can be shown to be rooted in ETH. For instance, the *Set Cover Conjecture* claims (using the vocabulary of this paper) that (HITTING SET, κ_7) cannot be solved in time $O^*(2^{o(\kappa_7(G))})$. As discussed in [22], it is not clear how this (plausible) conjecture relates to ETH.

Finally, one might wonder how to attack XP-problems, trying to understand how the parameter(s) influence the running time. For instance, consider the problem of finding a clique of size k in a graph with n vertices. Using brute-force, this problem can be solved in time $O^*(\binom{n}{k}) = \mathcal{O}(n^{k+c})$ for some small constant c. Nešetřil and Poljak showed in [66] that this can be improved to $\mathcal{O}(n^{k\omega/3+c})$, where ω is the exponent such that $n \times n$ matrices can be multiplied in time $\mathcal{O}(n^\omega)$. The underlying idea is that triangles can be found by multiplying adjacency matrices. Nowadays, it is believed that this is indeed the correct bound for detecting k-cliques. The hypothesis that no better algorithms are possible than those intimately linked to matrix multiplication is known as *k-Clique Conjecture*. This is one of the various examples of conjectures within the realm of polynomial-time algorithms on which several hardness assertions are based. In this context, it is also worth mentioning that there is a common belief that $\omega > 2$; this also links to interesting combinatorial conjectures as exhibited in [5]. We will re-encounter this conjecture in our last case study. Virginia Vassilevska Williams wrote several surveys on Fine-Grained Complexity, the most recent published one being [104].[3] When dealing with distinguishing problems within P, also adapted

---

[2] The definitions in [52] are a bit different, but this can be neglected in the current discussion.

[3] A new survey is announced to appear in [85].

notions of reductions have to be introduced. For the sake of space and because this is not that central to this paper, we are not going to present them here but refer to the mentioned survey papers. Also due to the nature of these reductions, in this part of the complexity world, polylogarithmic factors in the running time are often ignored, leading to notations like $\tilde{O}(n^2)$ for denoting quadratic running times up to terms like $(\log(n))^3$. Vassilevska Williams put the central question of Fine-Grained Complexity as follows in the survey to appear in [85]: *For each of the problems of interest with textbook running time $\mathcal{O}(t(n))$ and nothing much better known, is there a barrier to obtaining an $\mathcal{O}(t(n)^{1-\varepsilon})$ time algorithm for $\varepsilon > 0$?* Notice this formulation ignores polylogarithmic factors. Also, SETH perfectly fits into this line of questions.

Many more details can be found in the textbooks that have appeared in the meantime in the context of Parameterized Complexity, often also capturing aspects of ETH and also of SETH. We refer to [23–25,40,42,70].

In the following, we present six case studies, focusing on typical effects that show up when dealing with computational problems in Formal Languages. We start with a problem dealing with strings only, continuing with problems involving grammars and automata. There are some common themes throughout all these studies, for instance, the (sometimes surprising) role played by the size of the alphabet concerning the complexity status of the problems. Another recurring theme is that rather simple algorithms can be shown to be optimal.

3   First Case Study: String-to-String Correction S2S

The *edit distance* is a measure introduced to tell the distance between two strings $S, T \subset \Sigma^*$, where S is the source string and T is the target string, by counting the number of elementary edit operations that are necessary to turn S into T. The complexity of this problem depends on the permitted operations. Let O be the set of permitted operations, $O \subseteq \{\mathbf{C}, \mathbf{D}, \mathbf{I}, \mathbf{S}\}$, with:

C Change: replace/substitute a letter
D Delete a letter
I Insert a letter
S Swap: transpose neighboring letters

For each O, define the problem O-STRING-TO-STRING CORRECTION, or O-S2S for short, with input Σ, $S, T \in \Sigma^*$, $k \in \mathbb{N}$, to be the following question: Is it possible to turn S into T with a sequence of at most k operations from O?

Wagner [97] obtained a by now classical dichotomy result that can be stated as follows.[4]

Theorem 1. *Consider $O \subseteq \{\boldsymbol{C}, \boldsymbol{D}, \boldsymbol{I}, \boldsymbol{S}\}$.*

– If $O \in \{\{\boldsymbol{S}, \boldsymbol{D}\}, \{\boldsymbol{S}, \boldsymbol{I}\}\}$, then O-S2S is NP-complete.

---

[4] The result was phrased in different terminology back in 1975. Wagner actually proved stronger results in the sense that weights on the operations are permitted.

– *If $O \notin \{\{S, D\}, \{S, I\}\}$, then O-S2S is solvable in polynomial time.*

Note: $\{S, D\}$-S2S is equivalent to $\{S, I\}$-S2S.

How is this dichotomy result obtained? What is the *source of* NP-*hardness*? Conversely, how do the algorithms work? Here, dynamic programming is the key, and the corresponding algorithms (or variants thereof) have made their way into textbooks on algorithms.

Let us first study the NP-hard variant, focusing on $\{S, D\}$-S2S. What are natural parameters of an instance I defined by Σ, $S, T \in \Sigma^*$, $k \in \mathbb{N}$? From the viewpoint of now traditional parameterized complexity, $\kappa_1(I) = k$ is a first pick. This has been considered in [3]. Its main result can be stated as follows.

Theorem 2. $\{S, D\}$-*S2S with parameter* $\kappa_1(\Sigma, S, T \in \Sigma^*, k) = k$ *is in* FPT. *More precisely, an instance* $I = (\Sigma, S, T \in \Sigma^*, k)$ *can be solved in time* $\mathcal{O}(\varphi^k(|S|)\log(|\Sigma|))$, *where* $\varphi < 1.62$ *is the golden ratio number.*

In addition, a polynomial kernel was obtained in [101]. One of the important observations is that we can assume to always execute $k_1 = |S| - |T|$ deletions prior to swaps. Moreover, the at most $k - k_1$ swaps can be described by one position in the string. Hence, k is upper-bounded by a function in $|S|$ and we can use the previously mentioned algorithm to prove:

Proposition 3. $\{S, D\}$-*S2S with parameter* $\kappa_2(\Sigma, S, T, k) = |S|$ *is in* FPT.

With quite a similar reasoning, one can obtain the next result.

Proposition 4. $\{S, D\}$-*S2S with parameter* $\kappa_3(\Sigma, S, T, k) = |T|$ *is in* FPT.

The previous two results seem to be a bit boring, because $|S|$ and $|T|$ appear to be the natural choice to describe the overall size of the input. Also, these string lengths would be rather big in practice, while one could assume k to be rather small if one compares strings that are somehow similar to each other.

There is one last choice of a parameter that one might tend to overlook when first being confronted with this problem, namely, the size of the alphabet over which the strings S and T are formed. Yet, studying the proof of NP-hardness of Wagner, one is led to the conclusion that $|\Sigma|$ is crucial for this proof, which simply does not work if $|\Sigma|$ is bounded, for instance, if we consider binary alphabets only. This might look a bit surprising, as it might be hard to imagine that the alphabet size itself could carry such an importance, given the fact that the alphabet carries no visible or obvious structure. Yet, the consideration of the alphabet size will be a recurring theme in this paper, and we will see various situations where this is in fact a crucial parameter. This problem was first resolved in [36] for binary alphabets, showing the following result.

Proposition 5. $\{S, D\}$-*S2S on binary alphabets can be solved in cubic time.*

This was soon superseded by a more general result by Meister [64], showing that indeed the size of the alphabet was the crucial source of hardness for this problem. This was later improved by Barbay and Pérez-Lantero [10] concerning the dependence of the degree of the polynomial describing the running time on the alphabet size parameter.

Theorem 6. $\{S, D\}$-*S2S with parameter* $\kappa_4(\Sigma, S, T, k) = |\Sigma|$ *is in* XP. *More precisely, it can be solved in time* $\mathcal{O}(|S|^{|\Sigma|+1})$.

Still, this result is likely only practical only for very small alphabet sizes. Also, it is still open if one can put the problem in FPT. From a more practical perspective, as the parameters k and Σ are rather unrelated and also because the algorithmic approaches leading to Theorems 2 and 6 are quite different, it would be interesting to see if one can combine both ideas to produce an algorithm that is really useful for instances with a small alphabet size and a moderate number of permitted edit operations. Of course, this should be also checked by computational experiments that seem to be lacking so far.

It would also be interesting to study further parameters for this problem. In [10], several such suggestions have been considered. For instance, Barbay and Pérez-Lantero [10] have shown the following consequence for their algorithm. This relates to the number of deletions ($|T| - |S|$) already studied above.

Proposition 7. $\{S, D\}$-*S2S with parameter* $\kappa_5(\Sigma, S, T, k) = |\Sigma| + (|T| - |S|)$ *is in* FPT. *More precisely, it can be solved in time* $\mathcal{O}^*((|T| - |S|)^{|\Sigma|})$.

Let us also discuss some fine-grained complexity results for $\{C, D, I\}$-S2S. This problem is also known as computing the edit distance (in a more restricted sense) or as computing the Levenshtein-distance between two strings. The already mentioned textbook algorithms, often based on [98], take quadratic time. Whether or not these are optimal was actually a question already investigated 40 years ago. In those days, however, lower bound results were dependent on particular models of computation, while more modern approaches to lower bounds are independent of such a choice. For instance, Wong and Chandra [106] showed a quadratic lower bound assuming that only equality tests of single symbols are permitted as the basic operation of comparison. Masek and Paterson [63] managed to shave off a logarithmic factor by making some clever use of matrix multiplication tricks. Yet, whether essentially better algorithms are possible remained an open question up to recently. Backurs and Indyk [8] finally proved that assuming SETH, no $\mathcal{O}(n^{2-\varepsilon})$-time algorithm can be expected for any $\varepsilon > 0$. Interestingly enough, their proof was working only for alphabet sizes at least seven. This result has then been improved by Bringmann and Künnemann [14] to binary alphabets. Let us summarized these results in the following statement.

Theorem 8. $\{S, D, I\}$-*S2S can be solved in quadratic time. Assuming SETH, no* $\mathcal{O}(|S|^{2-\varepsilon})$-*time algorithm exists even on binary alphabets, for any* $\varepsilon > 0$.

Suggestions for Further Studies. (a) Does {**S, D**}-S2S with parameterization $\kappa_4(\Sigma, S, T, k) = |\Sigma|$ belong to FPT, or is it hard or even complete for some level of the W-hierarchy? (b) Assuming that {**S, D**}-S2S with parameter $\kappa_4(\Sigma, S, T, k) = |\Sigma|$ belongs to FPT, is it possible to give, e.g., SETH-based lower bounds for showing that existing algorithms are (close to) optimal? (c) Assuming that existing algorithms for {**S, D**}-S2S with parameter $\kappa_4(\Sigma, S, T, k) = |\Sigma|$ are optimal, it might make sense to study FPT-approaches even for fixed alphabets, because the running times that are obtainable at present are impractical for, say, the ASCII alphabet, not to speak about Unicode. (D) We are not aware of any studies concerning the polynomial-time approximability of the minimization problem related to {**S, D**}-S2S.

4   Second Case Study: Grammar-Based Compression

One of the main ideas behind data compression algorithms is to use regularities found in an input to find representations that are much smaller than the original data. Although data compression algorithms usually come with their own special data structures, there are some common schemes to be found in the algorithms of Lempel, Ziv, Storer, Szymanski and Welch from 1970s and 1980s [90, 103, 108] that easily generalize to the idea to use context-free grammars producing singleton languages for data compression purposes. Such context-free grammars are also called straight-line programs in the literature. Another perspective on this question is offered by Grammatical Inference, a perspective that can be traced back to the work of Nevill-Manning and Witten [67–69] and Kieffer and Yang [55]. We refer to [86] for quite a number of other papers that link to Grammatical Inference and applications thereof.

This leads us to consider the following decision problem, called GRAMMAR-BASED COMPRESSION, or GBC for short: Given a word w over some alphabet Σ and an integer k, decide if there exists a context-free grammar G of size at most k such that $L(G) = \{w\}$.

In a sense, this question is still ill-posed, because we did not explain how to measure the size of a grammar. In fact, there are several natural ways to do this. The main results that we are citing in the following do not really depend on the choice. Hence, we follow the common definition that the size of a context-free grammar G is computed by summing up the lengths of all right-hand sides of rules of G.

Based on reductions due to Storer [89], Charikar *et al.* [19] showed the following complexity result.

Theorem 9. *GBC is* NP-*complete (on unbounded terminal alphabets).*

Although occasionally there were claims in the literature that such a hardness result would be also true for bounded alphabet sizes (see [7]), this question was in fact open until recently. In the journal version of [17], the authors showed the following result.

Theorem 10. *GBC is* NP*-complete (on terminal alphabets of size at least 17).*

Let us now again study this problem with a multivariate analysis. With grammars, we have some natural choices of parameters, given as input I an alphabet Σ, a word $w \in \Sigma^*$ and an integer k: $\kappa_1(I) = \Sigma$, $\kappa_2(I) = |w|$, $\kappa_3(I) = k$. Observe that we can assume (after some straightforward reductions) that $\kappa_1(I) \leq \kappa_3(I) \leq \kappa_2(I)$. This indicates that κ_1 is the most challenging parameterization, while finding FPT-results should be easiest for κ_2. We now look at these parameters in the chosen order. Our intuition on the strength of the parameters will be certified.

From Theorem 10, we can conclude:

Corollary 11. *GBC with parameterization κ_1 is not in* XP, *unless* P = NP.

Theorem 12 [17]. *GBC with parameterization κ_2 belongs to* FPT. *More precisely, instance $I = (\Sigma, w, k)$ can be solved in time $\mathcal{O}^*(3^{\kappa_2(I)})$.*

We now demonstrate that such an FPT-result also holds for κ_3.

Theorem 13. *GBC with parameterization κ_3 belongs to* FPT.

Proof. A context-free grammar $G = (N, \Sigma, R, S)$ generating a singleton is called an F-grammar if $F = \{u \in \Sigma^+ \mid \exists A \in N : A \Rightarrow^* u\}$. As exhibited in [17], a related INDEPENDENT DOMINATING SET problem can be used to compute, for a given finite set F with $w \in F$, the smallest F-grammar that generates $\{w\}$ in polynomial time. For each F with $F \subseteq \Sigma^+$, $\sum_{u \in F} |u| \leq \kappa_3(\Sigma, w, k) = k$ we can find the smallest F-grammar for w in polynomial time. As $|\Sigma| \leq k$, only $f(k)$ many sets F have to be considered. $\quad\square$

In [17], the idea of a multivariate analysis has been taken further by considering further properties of the grammars we are looking for. For instance, is there a shortest grammar for w that uses $\leq r$ rules? The paper shows that this question is NP-hard. Considering r as a parameter, this problem is in XP and W[1]-hard.

There are also some studies on the approximability of the related minimization problem MIN-GBC, see [7,17,19,51,59,81]. We now summarize the main results in this direction. Notice the strange role that the size of the alphabet plays again.

Theorem 14. *If $m^*(w)$ denotes the size of the smallest context-free grammar for string w, then* MIN-GBC *can be approximated in polynomial time up to a factor of $\mathcal{O}\left(\log\left(\frac{|w|}{m^*(w)}\right)\right)$. Conversely, there is no polynomial-time algorithm achieving an approximation ratio better than $\frac{8569}{8568}$ unless* P = NP *(unbounded terminal alphabets). Furthermore, if there would be a polynomial-time constant-factor approximation algorithm for* MIN-GBC *on binary alphabets, there would be also some polynomial-time constant-factor approximation algorithm for* MIN-GBC *on unbounded alphabets.* MIN-GBC *is* APX*-hard on bounded terminal alphabets.*

Charikar *et al.* [19] also proved an interesting connection to a long standing open problem on approximating so-called addition chains. This approach might be interesting from a Fine-Grained Complexity perspective.

Suggestions for Further Studies. (A) The most natural complexity task is to further reduce the size of the terminal alphabet in Theorem 10. More specifically: Is GBC still NP-hard for binary terminal alphabets? From a practical point of view, i.e., when applying this technique to compressing data files, this is crucial to know. (B) Is there a polynomial-time constant-factor approximation algorithm for the smallest grammar problem? (C) Storer and Szymanski [89, 90] studied macro schemes that can be viewed as extensions of context-free grammars. No multivariate analysis of the related NP-hard compression problems has been undertaken so far. Notice that recent experimental studies [60] show the potential of these ideas. (D) We also mention generalizations of GBC to finite languages described by context-free grammars (and use them to encode specific words) as proposed in [86] which have not yet been studied from a more theoretical perspective.

5   Third Case Study: Synchronizing Words

A word $x \in \Sigma^*$ is called *synchronizing* for a deterministic finite automaton A, or DFA for short, with $A = (S, \Sigma, \delta, s_0, F)$ if there is a state s_f, such that for all states s, $\delta^*(s, x) = s_f$. An automaton is called *synchronizable* if it possesses a synchronizing word. It is known that A is synchronizable iff for every pair (s, s') of states, there exists a word $x_{s,s'}$ such that $\delta^*(s, x_{s,s'}) = \delta^*(s', x_{s,s'})$. This notion relates to the best known open combinatorial problem in Formal Languages, namely Černý's Conjecture: Any synchronizable DFA with t states has a synchronizing word of length $\leq (t-1)^2$. We are not going to give further details on this famous combinatorial problem, but only refer to the original paper by Černý [18], to two survey articles [82, 95] that also describe a couple of applications, to one very recent paper [92] that describes the best upper bound of $(85059t^3 + 90024t^2 + 196504t - 10648)/511104$ on the length of a synchronizing word for a t-state synchronizable DFA.

Rather, we will now turn to the related decision problem SYNCHRONIZING WORDS, or DFA-SW for short. The input consists of a DFA A and an integer k. The question is if there is a synchronizing word w for A with $|w| \leq k$. In [26], Eppstein has shown the following complexity result:

Theorem 15. DFA-SW *is* NP-*complete.*

How could a multivariate analysis of this problem look like? Natural parameterizations of an instance $I = (A, k)$ with $A = (S, \Sigma, \delta, s_0, F)$ include: $\kappa_1(I) = |\Sigma|$, $\kappa_2(I) = |S|$, and $\kappa_3(I) = k$. Clearly, one could also study combined parameters, like $\kappa_4(I) = |I| + k$. Also, notice that $\kappa_5(I) = |\delta|$ corresponds to $|S^{\Sigma \times S}|$, which would therefore be again a combined parameter. We are going to report on results from [33, 96].

Theorem 16. DFA-SW *with parameterization* $\kappa_1(I) = |\Sigma|$ *does not belong to* XP, *unless* P = NP.

In fact, the reduction from [26] can be used to show the previous results, as it shows NP-hardness for binary input alphabets.

Theorem 17. DFA-SW *with parameterization* $\kappa_2(I) = |S|$ *lies in* FPT. *More precisely, it can be solved in time* $\mathcal{O}^*(2^{\kappa_2(I)})$. *Yet, it does not admit polynomial kernels unless the polynomial-time hierarchy collapses to the third level.*

The FPT-algorithm is actually quite simple. It is based on the well-known subset construction, reducing the problem to a path-finding problem in an exponentially large graph. Yet, this algorithm is close to optimal in the following sense.

Proposition 18. *Assuming ETH, DFA-SW is not solvable in time* $\mathcal{O}^*(2^{o(\kappa_2(I))})$.

This result can be easily obtained by re-analyzing the reduction leading to Theorem 17.

Theorem 19. DFA-SW *with parameterization* $\kappa_3(I) = |k|$ *is* W*[2]-hard.*

We provide a sketch of this hardness result in Fig. 1, also to give an example of a parameterized reduction from HITTING SET introduced above. Observe that now the size of the input alphabet of the resulting DFA is unbounded, as it is the vertex set of the given hypergraph. Recently, Montoya and Nolasco [65] showed that (even) for planar automata, DFA-SW with parameterization $\kappa_3(I) = |k|$ is complete for the parameterized space complexity class NWL that embraces the whole W-hierarchy. Also, strong inapproximability results are known for the corresponding minimization problem; see [43].

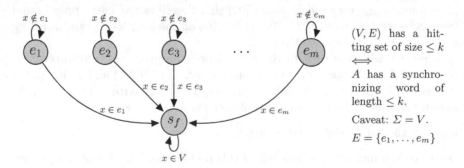

(V, E) has a hitting set of size $\leq k$

$\Longleftrightarrow$

A has a synchronizing word of length $\leq k$.

Caveat: $\Sigma = V$.

$E = \{e_1, \ldots, e_m\}$

Fig. 1. An example showing how an FPT reduction works.

For the combined parameterization κ_4, we can state:

Theorem 20. DFA-SW *with parameterization* $\kappa_4(I) = |\Sigma| + k$ *lies in* FPT. *More precisely, it can be solved in time* $\mathcal{O}^*(|\Sigma|^k)$. *Yet, it does not admit polynomial kernels unless the polynomial-time hierarchy collapses to the third level. Moreover, there is no* $\mathcal{O}^*((|\Sigma| - \varepsilon)^k)$-*time algorithm, unless SETH fails.*

Suggestions for Further Studies. (A) Although the proof sketch in Fig. 1 indicates that DFA-SW remains NP-hard for rather restricted forms of automata, it might be interesting to study classes of subregular languages regarding the question if DFA-SW might become simpler when restricted to these classes. (B) There are quite a number of notions similar to synchronizing words that have been introduced over the years, also due to the practical motivation of this notion, see [82]. No systematic study of computational complexity aspects has been undertaken for all these notions. (C) In view of the FPT result concerning parameterization κ_2, the number of states, the parameterization κ_5 might not look that interesting. Yet, as κ_2 does not allow for polynomial kernels, this question could be of interest for the variation $\kappa_5'(I) = |S| + |\Sigma|$.

6   Fourth Case Study: Consistency Problem for DFAs

The problem DFA-CONSISTENCY takes as input an alphabet Σ, two disjoint finite sets of words $X^+, X^- \subseteq \Sigma^*$, and some integer t. The question is if there is a DFA A with $\leq t$ states that is *consistent* with X^+, X^-, i.e., $L(A) \supseteq X^+$ and $L(A) \cap X^- = \emptyset$. This problem arises in various contexts, for instance, also in connection with Grammatical Inference; see [47]. Its classical complexity status was settled four decades ago.

Theorem 21 [6,44]. DFA-CONSISTENCY *is* NP-*complete.*

Let us explore the possible natural choices for parameterizations for instance $I = (\Sigma, X^+, X^+, t)$. We could look at $\kappa_1(I) = |\Sigma|$, $\kappa_2 = |X^+ \cup X^-|$, $\kappa_3(I) = \max\{|w| \mid w \in X^+ \cup X^-\}$, $\kappa_4(I) = t$, and there are quite a number of further ways to parameterize with respect to the sets X^+ and X^-. The NP-hardness results (with different constructions) extend to situations when $\kappa_1(I) = 2$ or when $\kappa_3(I) = 2$ or when $\kappa_4(I) = 2$. This immediately entails the following results.

Theorem 22. DFA-CONSISTENCY *with parameterization* $\kappa_1(I) = |\Sigma|$ *does not belong to* XP, *unless* P = NP.

Theorem 23. DFA-CONSISTENCY *with parameterization* $\kappa_3(I) = \max\{|w| \mid w \in X^+ \cup X^-\}$ *does not belong to* XP, *unless* P = NP.

Theorem 24. DFA-CONSISTENCY *with parameterization* $\kappa_4(I) = t$ *does not belong to* XP, *unless* P = NP.

Intuitively, this last result might appear most surprising, because there are only four ways how a letter can act with respect to two states. The literature situation was also a bit weird for some time. The result was mentioned in Sect. 1.2 of [75], as well as in [47], referring to an unpublished work of Angluin. However, no proof was given in these two references. Therefore, in the process of writing [33], we contacted Angluin, also to see how our solution compared to hers. We were quite impressed to receceive an email from Angluin within a couple of days, sending us a scanned copy of her proof, dating back to August 2nd, 1988. An NP-hardness proof can now be found in [33].

Interestingly, it is open if DFA-CONSISTENCY is NP-hard for any constant value of $\kappa_2(I)$. In the sense of multivariate analysis, we should continue to look into combined parameters. Let $\kappa_{i,j}$ for $1 \leq i < j \leq 4$ the parameterization given by $\kappa_{i,j}(I) = \kappa_i(I) + \kappa_j(I)$.

It was shown in [33] that DFA-CONSISTENCY is NP-hard even for 3-state DFAs with word lengths at most two in $X^+ \cup X^-$. This implies:

Theorem 25. DFA-CONSISTENCY *with parameterization* $\kappa_{3,4}(I)$ *does not belong to* XP, *unless* P = NP.

Conversely, by a trivial algorithm one can show the following (positive) result. The ETH hardness follows from a construction in [33].

Theorem 26. DFA-CONSISTENCY *with parameterization* $\kappa_{1,4}(I) = |\Sigma| + t$ *belongs to* FPT, *namely in time* $\mathcal{O}^*(t^{|\Sigma|t})$. *Assuming ETH, there is no* $\mathcal{O}^*(t^{o(|\Sigma|t)})$-*time algorithm for* DFA-CONSISTENCY.

The remaining parameter combinations seem to be open. Only one three-parameter combination was found in [33] that admitted a further FPT-result, combining κ_2, κ_3 and κ_4.

Suggestions for Further Studies. (A) Quite a number of parameter combinations are still open regarding their complexity status. Also, there are more parameters that could be related to X^+ and X^-. One potentially interesting scenario (pondering practical applications) would be to see what happens if there are much less negative than positive samples. (B) It might be an idea to look into classes of subregular languages and find some that allow for efficient consistency checks. (C) There are related questions that have not yet been studied from a multivariate perspective, for instance, what about REGULAR EXPRESSION CONSISTENCY?

7 Fifth Case Study: Lower Bounds for UNIVERSALITY

Possibly, the reader would have expected that we focus on problems like DFA INTERSECTION EMPTINESS and similar problems traditionally studied (with respect to their complexity) in textbooks on Formal Languages. This will be partially rectified in this section. We will mainly concentrate on UNIVERSALITY, which is the following problem. Given a finite automaton A with input alphabet Σ, is $L(A) = \Sigma^*$? Clearly, this makes only sense for nondeterministic automata, as the problem can be solved in linear time for DFAs. Natural parameters are the number t of states of A and the size of Σ. $|\Sigma|$ plays again an important role. Notice that the problem is PSPACE-complete in general, but co-NP-complete for unary input alphabets; see [56,87,88] and possibly more explicit in [34]. Natural parameterizations are $\kappa_1(A) = |\Sigma|$ and $\kappa_2(A) = |S|$, where $A = (S, \Sigma, \delta, s_0, F)$ is an NFA. By the classical hardness results, we see:

Proposition 27. UNIVERSALITY. *parameterized by κ_1, is not in* XP, *unless* P = NP.

Conversely, by the classical subset construction to produce a DFA, followed by final state complementation and a simple emptiness check, one sees:

Proposition 28. UNIVERSALITY. *parameterized by κ_2, belongs to* FPT.

We are now going to study complexity aspects under the ETH perspective.

Theorem 29 [34]. *Unless ETH fails, there is no $\mathcal{O}^*(2^{o(t^{1/3})})$ -time algorithm for deciding* UNIVERSALITY *on t-state NFAs with unary inputs.*

There is a slight gap to the known upper bound by Chrobak [20] who showed:

Theorem 30. UNIVERSALITY *on t-state NFAs with unary inputs can be solved in time $2^{\Theta((t \log t)^{1/2})}$.*

For larger alphabets, the situation looks a bit different for UNIVERSALITY.

Theorem 31 [34]. *Unless ETH fails, there is no $\mathcal{O}^*(2^{o(t)})$ -time algorithm for deciding* UNIVERSALITY *on t-state NFAs with binary inputs, or larger alphabets.*

The results is obtained by a parsimonous reduction from 3-COLORABILITY. This is the correct bound, because the power-set construction gives that UNIVERSALITY on t-state NFAs can be solved in time $\mathcal{O}^*(2^t)$.

Also to overcome the fact that UNIVERSALITY is PSPACE-complete, a length-bounded variant has been introduced. LB-UNIVERSALITY: Given NFA A and length bound ℓ, does A accept all words up to length ℓ? This length bound puts the problem into NP. In fact, it is NP-complete. From a multivariate perspective, this introduces a natural third parameter, $\kappa_3(A, \ell) = \ell$.

Theorem 32. LB-UNIVERSALITY, *parameterized by κ_3, is* W[2]*-hard.*

As there is no formal proof of this result in the literature, we provide an explicit construction. In fact, it is quite similar to the construction illustrated in Fig. 1 which the reader might want to consult.

Proof. We show how to solve any instance $G = (V, E)$ and k of HITTING SET, parameterized by the size k of the solution, with the help of an instance of LB-UNIVERSALITY, parameterized by κ_3. Set $\Sigma = V$, $S = \{s_0, s_f\} \cup E$. Let s_0 be the initial and $E \cup \{s_0\}$ be the set of final states. We include the following transitions in the transition relation.

- (s_0, a, e) for any $a \in \Sigma$ and any $e \in E$;
- (e, a, e) for any $e \in E$ and $a \notin e$;
- (e, a, s_f) for any $e \in E$ and $a \in e$;
- (s_f, a, s_f) for any $a \in \Sigma$.

Furthermore, we set $\ell = k + 1$. Now, we claim that there is a hitting set of size at most k in G if and only if there is a word of length at most $k + 1$ that is not accepted by the constructed automaton. Namely, the only way not to accept a word by the automaton would be a word ending in s_f irrespectively what state $e \in E$ was entered inbetween. This shows that the encoded set of vertices of the hypergraph indeed hits all hyperedges.   □

Membership in W[2] is unknown. As the parameterized complexity results for the other two parameters transfer, we get a rather diverse picture of what could happen in a multivariate analysis. As parameters κ_1 and κ_3 yield intractability results, the following (straightforward) result is interesting for the combined parameter $\kappa_1 + \kappa_3$.

Proposition 33. LB-UNIVERSALITY *can be solved in time* $\mathcal{O}^*(|\Sigma|^\ell)$.

Namely, just enumerate and test all strings up to length ℓ. This has been complemented by the following result that proves conditional optimality of this simple algorithm.

Theorem 34 [34]. *There is no algorithm that solves* LB-UNIVERSALITY *in time* $\mathcal{O}^*\left((|\Sigma| - \varepsilon)^\ell\right)$ *for any* $\varepsilon > 0$, *unless SETH fails.*

Let us finally discuss the issue of kernelization for this problem. The size of an instance is vastly dominated by the size of the transition table. Measured in terms of number of states and input alphabet size, this size can be as large as $\mathcal{O}(2^{|\Sigma||S|^2})$. Is there any hope to bring this down to a size only polynomial in $|S|$, a result that would complement Proposition 28? Interestingly, this question seems to be open, while it is possible to show the non-existence of polynomial kernels for the length-bounded variation. We can even show this result for the combined parameter $\kappa_2 + \kappa_3$.

Theorem 35. LB-UNIVERSALITY, *parameterized by* $\kappa_2 + \kappa_3$, *does not admit polynomial kernels, unless the polynomial-time hierarchy collapses to the third level.*

Proof. It is known that under the stated complexity assumptions, HITTING SET, parameterized by the number of hyperedges plus an upper-bound k on the size of the solution, does not admit polynomial kernels; see [23]. Now, assume that LB-UNIVERSALITY, parameterized by $\kappa_2 + \kappa_3$, would have a kernelization algorithm A that produces polynomial kernels. Now, start with an instance (G, k) of HITTING SET, where $G = (V, E)$, and first translate it to an equivalent instance (A, ℓ) of LB-UNIVERSALITY, using the construction from Theorem 32. Observe that $\kappa_2(A, \ell) = |E|$ and $\kappa_3(A, \ell) = k + 1$. Next, run algorithm A, yielding an instance (A', ℓ') of size polynomial in $\kappa_2(A, \ell) + \kappa_3(A, \ell)$ and hence polynomial in $|E| + k$. Finally, observe that as LB-UNIVERSALITY is in NP, while HITTING SET is NP-hard, there is a polynomial-time transformation of (A', ℓ') into an equivalent instance (G', k') of HITTING SET. Clearly, also (G', k') would be of polynomial size, measured in $|E| + k$, which contradicts the non-existence of polynomial kernels.   □

Let us mention one further exploit of the construction of Theorem 32.

Theorem 36. *Under the Set Cover Conjecture, there is no $\mathcal{O}^*(2^{o(\kappa_2(A))})$-time algorithm for solving instances A of* UNIVERSALITY.

Suggestions for Further Studies. (A) For the simple FPT-results for this (and similar) automata problems, polynomial kernel questions have barely been studied. This is also true for all the related classical automata problems. (B) There are slight but noticeable gaps between lower and upper bounds on running times (assuming ETH). More gaps can be found in related automata problems, as discussed in [34]. Is it possible to close these gaps, possibly using hypotheses different from ETH? (C) Unlike this section might suggest, most work has been put into studying automata intersection problems (among the classical algorithmic questions about finite automata); see [34, 73, 91, 100, 102]. Relatively few efforts have been put into related questions or into other automata models; we only mention here [37, 38] and the references given in these papers.

8   Sixth Case Study: Parsing Theory

Coming from FL theory courses, where the Chomsky hierarchy is often taught with indicating a certain relevance to areas like Compiler Construction or also to (Computational) Linguistics, one might get disappointed when actually encountering these two mentioned areas. In the former case, the regular languages seem to be relevant and also some parts of the context-free languages, but not much more. In the second case, the situation is even more disillusioning: there, formal language classes more expressive than context-free but being not much more complex with respect to parsing are most interesting.

Even the typical parsing algorithms like CYK or Earley's mostly taught at FL undergraduate courses are not really relevant, as their cubic complexity is too much for typical applications in Compiler Construction. Rather, one resorts to *deterministic* context-free languages, also because they allow for giving unambiguous interpretations in the sense of unique parse trees. But is really necessary to spend cubic time for parsing context-free languages?

In a positive (algorithmic) sense, this question was answered already by Valiant [94] in a paper entitled *Parsing (general context-free recognition) in time $\mathcal{O}(n^\omega)$*. Here, n is the length of the string to be parsed, and ω is the exponent of multiplying two square matrices. At the time of that paper, this was still Strassen's multiplication, i.e., $\omega \approx 2.81$. If we want to use this method in a practical algorithm, this might be still a method of choice. Yet, in theory, ω has improved to 2.3727, as shown by Vassilevska Williams in [105]. Whether it can be further improved or not, as well as relations to other problems, is discussed in a recent FOCS paper of Alman and Vassilevska Williams [4]. Alternatives to Valiant's original algorithm are discussed in [45, 80]. Actually, Rytter called Valiant's algorithm *probably the most interesting algorithm related to formal languages*. This is a good reason to study it further here.

A natural question in our context is: Can we parse context-free grammars faster than multiplying matrices? This question was first addressed in a paper of Lee [58] with the title *Fast context-free parsing requires fast Boolean matrix multiplication*. The drawback of the underlying construction is that this is only true for grammars whose size grows with n^6, where n is the length of the string to be parsed. This is not a very realistic scenario. Abboud, Backurs, and Vassilevska Williams have fixed this issue in [1]. This fine-grained reduction works for a specific CF grammar, so that the previous dependence between grammar size and string length no longer holds. To get an idea how these results look like on a more technical level, we cite the following theorem.

Theorem 37. *There is context-free grammar G_{fix} of constant size such that if we can determine if a string of length n belongs to $L(G_{fix})$ in $T(n)$ time, then k-CLIQUE on n-vertex graphs can be solved in $\mathcal{O}(T(n^{k/3+1}))$ time, for any $k \geq 3$.*

Hence, under the mentioned k-Clique Conjecture, context-free parsing cannot be faster than $\mathcal{O}(n^\omega)$.

We remark that there are extensions of context-free grammars, like Boolean grammars [72], that admit parsers like Valiant's; therefore, the lower bounds transfer to them immediately. For several related problems in computational biology, we refer to [13, 74, 107].

We are now reporting on one more problem directly related to parsing, namely to parsing tree-adjoining grammars. Notice that these are quite important for computational linguistics. We are not going to give a detailed introduction into tree-adjoining grammars, but rather refer to the textbook [54] that covers this and similar mechanisms from a linguistic yet mathematically profound perspective. Tree-adjoining grammars extend context-free ones in a way that allows for representing several linguistically relevant features beyond context-free languages. They yield one of the basic examples for mildly context-sensitive languages. The parsing is still possible in polynomial time, more precisely, the textbook algorithm will be in $\mathcal{O}(n^6)$, where n is the length of the string to be parsed. Yet, Rajasekaran and Yooseph's parser [76, 77] runs in $\mathcal{O}(n^{2\omega})$. While this solved a previously well-known open problem in Computational Linguistics, it is interesting that a negative result pre-dated this algorithmic one. Satta [83] showed a reverse relation, actually inspiring Lee's work. But, not surprisingly, it comes with a similar drawback: This lower-bound is only true for grammars whose size grows with n^6. Bringmann and Wellnitz [15] have improved this result as follows.

Theorem 38. *There is a tree-adjoining grammar G_{fix} of constant size such that if we can decide in time $T(n)$ whether a given string of length n can be generated from G_{fix}, then $6k$-CLIQUE can be solved in time $\mathcal{O}(T(n^{k+1} \log n))$, for any fixed $k \geq 1$.*

A consequence would be: An $\mathcal{O}(n^{2\omega - \varepsilon})$-algorithm for TAL recognition would prove that $6k$-CLIQUE can be solved in time $\widetilde{\mathcal{O}}(n^{(2\omega - \varepsilon)(k+1)}) \subseteq \mathcal{O}(n^{(\omega/3 - \delta)6k})$, contradicting the k-Clique Conjecture.

Suggestions for Further Studies. (A) Tree-adjoining grammars (TAGs) have been quite popular in Computational Linguistics in the 1990s, but this has calmed down a bit due to various shortcomings, both regarding parsing complexity (as discussed above) and the expressiveness of this formalism. Possibly, Formal Languages could help in the second issue by coming up with grammatical mechanisms that are more powerful than TAGs but do not need more computational resources for parsing. For instance, can the ideas underlying Boolean grammars be extended towards TAGs? (B) The whole topic of parsing has been a bit neglected in the Formal Language community. This is something that should change, in the best interest of the FL community. Whoever likes to start working in this direction should not overlook the rich annotated bibliography with nearly 2000 entries by Grune and Jacobs [46], available at https://dickgrune. com/Books/PTAPG_2nd_Edition/Additional.html. (C) Since four decades, it is open if EDT0L systems can be parsed in polynomial time [79, Page 315]. Weakening this question, one could also ask [29] if there is some $\mathcal{O}^*(f(|N|))$-time algorithm for parsing, where N is the set of nonterminal symbols.

9 Conclusions

With this survey, we could only highlight some of the many results that have been obtained in the meantime regarding multivariate analysis, but also regarding fine-grained complexity results. Yet, there are some common themes, as the role of the alphabet size, or also the richness of natural parameter choices. Another typical observation is that often simple parameterized algorithms cannot be improved under certain complexity assumptions. All this gives these problem a flavor different from, say, graph problems.

We preferred to focus on six problems, rather than trying to discuss all of them. Yet, in these conclusions, we are going to mention at least some further papers.

For instance, there is a vast body of literature on string problems. In fact, string problems were among the first ones where a true multivariate analysis was undertaken (without naming it such); see [27]. For a survey on these types of analyses for string problems, we refer to [16]. String problems have been also further investigated from the viewpoint of fine-grained complexity; see [2].

The related area of pattern matching would have also deserved a closer look. Let us suggest [35, 39] and the literature cited therein for further reading. To the readers otherwise more interested in graph-theoretic problems, it might be interesting to learn that the parameter treewidth well-known from graph algorithms has been also introduced in the context of patterns in [78].

String problems have also tight connections to several problems arising in computational biology. We refrain from giving any further references here, as this would finally surpass any reasonable length of the list of citations, but it should be clear that there are scores of papers on the parameterized and also on the fine-grained complexity of such problems.

In the context of stochastic automata, the Viterbi algorithm is central; its optimality is considered in [9].

Finally, let us discuss possible connections to Descriptional Complexity (within FL). One question of this sort is about smallest representations (within certain formalisms). One such example is also grammar-based compression, another one the minimization of automata or expressions, see [11,53]. Further on, one could consider questions as *Given an automaton, is there an equivalent representation with certain additional restrictions?* which are typical for this area, but have not yet been considered from a multivariate or fine-grained angle. We only refer to two survey papers of Holzer and Kutrib [48,49].

Acknowledgements. We are grateful to many people giving feedback to the ideas presented in this paper. In particular, Anne-Sophie Himmel, Ulrike Stege, and Petra Wolf commented on earlier versions of the manuscript.

References

1. Abboud, A., Backurs, A., Williams, V.V.: If the current clique algorithms are optimal, so is Valiant's parser. In: Guruswami, V. (ed.) IEEE 56th Annual Symposium on Foundations of Computer Science, FOCS, pp. 98–117. IEEE Computer Society (2015)
2. Abboud, A., Williams, V.V., Weimann, O.: Consequences of faster alignment of sequences. In: Esparza, J., Fraigniaud, P., Husfeldt, T., Koutsoupias, E. (eds.) ICALP 2014. LNCS, vol. 8572, pp. 39–51. Springer, Heidelberg (2014). https://doi.org/10.1007/978-3-662-43948-7_4
3. Abu-Khzam, F.N., Fernau, H., Langston, M.A., Lee-Cultura, S., Stege, U.: A fixed-parameter algorithm for string-to-string correction. Discrete Optim. **8**, 41–49 (2011)
4. Alman, J., Williams, V.V.: Limits on all known (and some unknown) approaches to matrix multiplication. In: Thorup, M. (ed.) 59th IEEE Annual Symposium on Foundations of Computer Science, FOCS, pp. 580–591. IEEE Computer Society (2018)
5. Alon, N., Shpilka, A., Umans, C.: On sunflowers and matrix multiplication. Comput. Complex. **22**(2), 219–243 (2013)
6. Angluin, D.: On the complexity of minimum inference of regular sets. Inf. Control (Now Inf. Comput.) **39**, 337–350 (1978)
7. Arpe, J., Reischuk, R.: On the complexity of optimal grammar-based compression. In: 2006 Data Compression Conference (DCC), pp. 173–182. IEEE Computer Society (2006)

8. Backurs, A., Indyk, P.: Edit distance cannot be computed in strongly sub-quadratic time (unless SETH is false). SIAM J. Comput. **47**(3), 1087–1097 (2018)
9. Backurs, A., Tzamos, C.: Improving Viterbi is hard: better runtimes imply faster clique algorithms. In: Precup, D., Teh, Y.W. (eds.) Proceedings of the 34th International Conference on Machine Learning, ICML, Proceedings of Machine Learning Research, vol. 70, pp. 311–321. PMLR (2017)
10. Barbay, J., Pérez-Lantero, P.: Adaptive computation of the swap-insert correction distance. ACM Trans. Algorithms **14**(4), 49:1–49:16 (2018)
11. Björklund, H., Martens, W.: The tractability frontier for NFA minimization. J. Comput. Syst. Sci. **78**(1), 198–210 (2012)
12. Bodlaender, H.L., Downey, R., Fomin, F.V., Marx, D. (eds.): The Multivariate Algorithmic Revolution and Beyond - Essays Dedicated to Michael R. Fellows on the Occasion of His 60th Birthday. LNCS, vol. 7370. Springer, Heidelberg (2012). https://doi.org/10.1007/978-3-642-30891-8
13. Bringmann, K., Grandoni, F., Saha, B., Williams, V.V.: Truly sub-cubic algorithms for language edit distance and RNA-folding via fast bounded-difference min-plus product. In: Dinur, I. (ed.) IEEE 57th Annual Symposium on Foundations of Computer Science, FOCS, pp. 375–384. IEEE Computer Society (2016)
14. Bringmann, K., Künnemann, M.: Quadratic conditional lower bounds for string problems and dynamic time warping. In: Guruswami, V. (ed.) IEEE 56th Annual Symposium on Foundations of Computer Science, FOCS, pp. 79–97. IEEE Computer Society (2015)
15. Bringmann, K., Wellnitz, P.: Clique-based lower bounds for parsing tree-adjoining grammars. In: Kärkkäinen, J., Radoszewski, J., Rytter, W. (eds.) 28th Annual Symposium on Combinatorial Pattern Matching, CPM. LIPIcs, vol. 78, pp. 12:1–12:14. Schloss Dagstuhl - Leibniz-Zentrum für Informatik (2017)
16. Bulteau, L., Hüffner, F., Komusiewicz, C., Niedermeier, R.: Multivariate algorithmics for NP-hard string problems. EATCS Bull. **114** (2014). http://bulletin.eatcs.org/index.php/beatcs/article/view/310/292
17. Casel, K., Fernau, H., Gaspers, S., Gras, B., Schmid, M.L.: On the complexity of grammar-based compression over fixed alphabets. In: Chatzigiannakis, I., Mitzenmacher, M., Rabani, Y., Sangiorgi, D. (eds.) International Colloquium on Automata, Languages and Programming, ICALP, Leibniz International Proceedings in Informatics (LIPIcs), vol. 55, pp. 122:1–122:14. Schloss Dagstuhl-Leibniz-Zentrum für Informatik (2016)
18. Černý, J.: Poznámka k homogénnym experimentom s konečnými automatmi. Matematicko-fyzikálny časopis **14**(3), 208–216 (1964)
19. Charikar, M., et al.: The smallest grammar problem. IEEE Trans. Inf. Theory **51**(7), 2554–2576 (2005)
20. Chrobak, M.: Finite automata and unary languages. Theor. Comput. Sci. **47**, 149–158 (1986)
21. Cook, S.A.: The complexity of theorem-proving procedures. In: Proceedings of the Third Annual ACM Symposium on Theory of Computing, STOC, pp. 151–158. ACM (1971)
22. Cygan, M., et al.: On problems as hard as CNF-SAT. ACM Trans. Algorithms **12**(3), 41:1–41:24 (2016)
23. Cygan, M., et al.: Parameterized Algorithms. Springer, Heidelberg (2015). https://doi.org/10.1007/978-3-319-21275-3
24. Downey, R.G., Fellows, M.R.: Parameterized Complexity. Springer, Heidelberg (1999). https://doi.org/10.1007/978-1-4612-0515-9

25. Downey, R.G., Fellows, M.R.: Fundamentals of Parameterized Complexity. Texts in Computer Science. Springer, Heidelberg (2013)
26. Eppstein, D.: Reset sequences for monotonic automata. SIAM J. Comput. 19(3), 500–510 (1990)
27. Fellows, M.R., Gramm, J., Niedermeier, R.: On the parameterized intractability of CLOSEST SUBSTRING and related problems. In: Alt, H., Ferreira, A. (eds.) STACS 2002. LNCS, vol. 2285, pp. 262–273. Springer, Heidelberg (2002). https://doi.org/10.1007/3-540-45841-7_21
28. Fellows, M.R., Jansen, B.M.P., Rosamond, F.A.: Towards fully multivariate algorithmics: parameter ecology and the deconstruction of computational complexity. Eur. J. Combin. 34(3), 541–566 (2013)
29. Fernau, H.: Parallel grammars: a phenomenology. GRAMMARS 6, 25–87 (2003)
30. Fernau, H.: Parameterized Algorithmics: A Graph-Theoretic Approach. Universität Tübingen, Germany, Habilitationsschrift (2005)
31. Fernau, H.: Parameterized algorithmics for d-hitting set. Int. J. Comput. Math. 87(14), 3157–3174 (2010)
32. Fernau, H.: A top-down approach to search-trees: improved algorithmics for 3-hitting set. Algorithmica 57, 97–118 (2010)
33. Fernau, H., Heggernes, P., Villanger, Y.: A multi-parameter analysis of hard problems on deterministic finite automata. J. Comput. Syst. Sci. 81(4), 747–765 (2015)
34. Fernau, H., Krebs, A.: Problems on finite automata and the exponential time hypothesis. Algorithms 10(24), 1–25 (2017)
35. Fernau, H., Manea, F., Mercaş, R., Schmid, M.L.: Pattern matching with variables: fast algorithms and new hardness results. In: Mayr, E.W., Ollinger, N. (eds.) 32nd International Symposium on Theoretical Aspects of Computer Science (STACS 2015), Leibniz International Proceedings in Informatics (LIPIcs), vol. 30, pp. 302–315. Schloss Dagstuhl-Leibniz-Zentrum für Informatik (2015)
36. Fernau, H., Meister, D., Schmid, M.L., Stege, U.: Editing with swaps and inserts on binary strings (2014). Manuscript
37. Fernau, H., Paramasivan, M., Schmid, M.L., Thomas, D.G.: Simple picture processing based on finite automata and regular grammars. J. Comput. Syst. Sci. 95, 232–258 (2018)
38. Fernau, H., Paramasivan, M., Schmid, M.L., Vorel, V.: Characterization and complexity results on jumping finite automata. Theor. Comput. Sci. 679, 31–52 (2017)
39. Fernau, H., Schmid, M.L., Villanger, Y.: On the parameterised complexity of string morphism problems. Theory Comput. Syst. 59(1), 24–51 (2016)
40. Flum, J., Grohe, M.: Parameterized Complexity Theory. Springer, Heidelberg (2006). https://doi.org/10.1007/3-540-29953-X
41. Fomin, F.V., Kratsch, D., Woeginger, G.J.: Exact (exponential) algorithms for the dominating set problem. In: Hromkovič, J., Nagl, M., Westfechtel, B. (eds.) WG 2004. LNCS, vol. 3353, pp. 245–256. Springer, Heidelberg (2004). https://doi.org/10.1007/978-3-540-30559-0_21
42. Fomin, F.V., Kratsch, D.: Exact Exponential Algorithms. Texts in Theoretical Computer Science. Springer, Heidelberg (2010). https://doi.org/10.1007/978-3-642-16533-7
43. Gawrychowski, P., Straszak, D.: Strong inapproximability of the shortest reset word. In: Italiano, G.F., Pighizzini, G., Sannella, D.T. (eds.) MFCS 2015. LNCS, vol. 9234, pp. 243–255. Springer, Heidelberg (2015). https://doi.org/10.1007/978-3-662-48057-1_19
44. Gold, E.M.: Complexity of automaton identification from given data. Inf. Control (Now Inf. Comput.) 37, 302–320 (1978)

45. Graham, S.L., Harrison, M.A., Ruzzo, W.L.: An improved context-free recognizer. ACM Trans. Program. Lang. Syst. **2**(3), 415–462 (1980)
46. Grune, D., Jacobs, C.J.H.: Parsing Techniques - A Practical Guide. Monographs in Computer Science. Springer, Heidelberg (2008). https://doi.org/10.1007/978-0-387-68954-8
47. Higuera, C.: Grammatical inference. Learning automata and grammars. Cambridge University Press, Cambridge (2010)
48. Holzer, M., Kutrib, M.: The complexity of regular(-like) expressions. Int. J. Found. Comput. Sci. **22**(7), 1533–1548 (2011)
49. Holzer, M., Kutrib, M.: Descriptional and computational complexity of finite automata - a survey. Inf. Comput. **209**(3), 456–470 (2011)
50. Hopcroft, J.E., Ullman, J.D.: Introduction to Automata Theory, Languages, and Computation. Addison-Wesley, Reading (1979)
51. Hucke, D., Lohrey, M., Reh, C.P.: The smallest grammar problem revisited. In: Inenaga, S., Sadakane, K., Sakai, T. (eds.) SPIRE 2016. LNCS, vol. 9954, pp. 35–49. Springer, Cham (2016). https://doi.org/10.1007/978-3-319-46049-9_4
52. Impagliazzo, R., Paturi, R., Zane, F.: Which problems have strongly exponential complexity? J. Comput. Syst. Sci. **63**(4), 512–530 (2001)
53. Jiang, T., Ravikumar, B.: Minimal NFA problems are hard. SIAM J. Comput. **22**(6), 1117–1141 (1993)
54. Kallmeyer, L.: Parsing Beyond Context-Free Grammars. Springer, Heidelberg (2010). https://doi.org/10.1007/978-3-642-14846-0
55. Kieffer, J.C., Yang, E.: Grammar-based codes: a new class of universal lossless source codes. IEEE Trans. Inf. Theory **46**(3), 737–754 (2000)
56. Kozen, D.: Lower bounds for natural proof systems. In: 18th Annual Symposium on Foundations of Computer Science, FOCS, pp. 254–266. IEEE Computer Society (1977)
57. Kratochvíl, J.: A special planar satisfiability problem and a consequence of its NP-completeness. Discrete Appl. Math. **52**, 233–252 (1994)
58. Lee, L.: Fast context-free grammar parsing requires fast boolean matrix multiplication. J. ACM **49**(1), 1–15 (2002)
59. Lehman, E., Shelat, A.: Approximations algorithms for grammar-based compression. In: Thirteenth Annual Symposium on Discrete Algorithms SODA. ACM Press (2002)
60. Liao, K., Petri, M., Moffat, A., Wirth, A.: Effective construction of relative lempel-ziv dictionaries. In: Bourdeau, J., Hendler, J., Nkambou, R., Horrocks, I., Zhao, B.Y. (eds.) Proceedings of the 25th International Conference on World Wide Web, WWW, pp. 807–816. ACM (2016)
61. Lichtenstein, D.: Planar formulae and their uses. SIAM J. Comput. **11**, 329–343 (1982)
62. Lokshtanov, D., Marx, D., Saurabh, S.: Lower bounds based on the exponential time hypothesis. EATCS Bull. **105**, 41–72 (2011)
63. Masek, W.J., Paterson, M.: A faster algorithm computing string edit distances. J. Comput. Syst. Sci. **20**(1), 18–31 (1980)
64. Meister, D.: Using swaps and deletes to make strings match. Theor. Comput. Sci. **562**, 606–620 (2015)
65. Andres Montoya, J., Nolasco, C.: On the synchronization of planar automata. In: Klein, S.T., Martín-Vide, C., Shapira, D. (eds.) LATA 2018. LNCS, vol. 10792, pp. 93–104. Springer, Cham (2018). https://doi.org/10.1007/978-3-319-77313-1_7
66. Nešetřil, J., Poljak, S.: On the complexity of the subgraph problem. Comment. Math. Univ. Carolinae **26**(2), 415–419 (1985)

67. Nevill-Manning, C.G.: Inferring sequential structure. Ph.D. thesis, University of Waikato, New Zealand (1996)
68. Nevill-Manning, C.G., Witten, I.H.: Identifying hierarchical structure in sequences: a linear-time algorithm. J. Artif. Intell. Res. **7**, 67–82 (1997)
69. Nevill-Manning, C.G., Witten, I.H.: On-line and off-line heuristics for inferring hierarchies of repetitions in sequences. Proc. IEEE **88**, 1745–1755 (2000)
70. Niedermeier, R.: Invitation to Fixed-Parameter Algorithms. Oxford University Press, Oxford (2006)
71. Niedermeier, R.: Reflections on multivariate algorithmics and problem parameterization. In: Marion, J.Y., Schwentick, T. (eds.) 27th International Symposium on Theoretical Aspects of Computer Science (STACS 2010), Leibniz International Proceedings in Informatics (LIPIcs), vol. 5, pp. 17–32. Schloss Dagstuhl-Leibniz-Zentrum für Informatik (2010)
72. Okhotin, A.: Parsing by matrix multiplication generalized to Boolean grammars. Theor. Comput. Sci. **516**, 101–120 (2014)
73. de Oliveira Oliveira, M., Wehar, M.: Intersection non-emptiness and hardness within polynomial time. In: Hoshi, M., Seki, S. (eds.) DLT 2018. LNCS, vol. 11088, pp. 282–290. Springer, Cham (2018). https://doi.org/10.1007/978-3-319-98654-8_23
74. Pinhas, T., Zakov, S., Tsur, D., Ziv-Ukelson, M.: Efficient edit distance with duplications and contractions. Algorithms Mole. Biol. **8**, 27 (2013)
75. Pitt, L., Warmuth, M.K.: The minimum consistent DFA problem cannot be approximated within any polynomial. J. ACM **40**, 95–142 (1993)
76. Rajasekaran, S.: Tree-adjoining language parsing in $O(n^6)$ time. SIAM J. Comput. **25**(4), 862–873 (1996)
77. Rajasekaran, S., Yooseph, S.: TAL recognition in $O(M(n^2))$ time. J. Comput. Syst. Sci. **56**(1), 83–89 (1998)
78. Reidenbach, D., Schmid, M.L.: Patterns with bounded treewidth. Inf. Comput. **239**, 87–99 (2014)
79. Rozenberg, G., Salomaa, A.K.: The Mathematical Theory of L Systems. Academic Press, Cambridge (1980)
80. Rytter, W.: Context-free recognition via shortest paths computation: a version of Valiant's algorithm. Theor. Comput. Sci. **143**(2), 343–352 (1995)
81. Rytter, W.: Application of Lempel-Ziv factorization to the approximation of grammar-based compression. Theor. Comput. Sci. **302**, 211–222 (2003)
82. Sandberg, S.: 1 homing and synchronizing sequences. In: Broy, M., Jonsson, B., Katoen, J.-P., Leucker, M., Pretschner, A. (eds.) Model-Based Testing of Reactive Systems. LNCS, vol. 3472, pp. 5–33. Springer, Heidelberg (2005). https://doi.org/10.1007/11498490_2
83. Satta, G.: Tree-adjoining grammar parsing and Boolean matrix multiplication. J. Comput. Linguist. **20**(2), 173–191 (1994)
84. Shannon, C.E.: A universal Turing machine with two internal states. In: Shannon, C.E., McCarthy, J. (eds.) Automata Studies, Annals of Mathematics Studies, vol. 34, pp. 157–165. Princeton University Press (1956)
85. Sirakov, B., de Souza, P.N., Viana, M. (eds.): Proceedings of the International Congress of Mathematicians 2018 (ICM 2018). World Scientific (2019)
86. Siyari, P., Gallé, M.: The generalized smallest grammar problem. In: Verwer, S., van Zaanen, M., Smetsers, R. (eds.) Proceedings of the 13th International Conference on Grammatical Inference, ICGI 2016, JMLR Workshop and Conference Proceedings, vol. 57, pp. 79–92. JMLR.org (2017)

87. Stockmeyer, L.J.: The complexity of decision problems in automata theory and logic. Ph.D. thesis, Massachusetts Institute of Technology, Department of Electrical Engineering (1974)

88. Stockmeyer, L.J., Meyer, A.R.: Word problems requiring exponential time: preliminary report. In: Aho, A.V., et al. (eds.) Proceedings of the 5th Annual ACM Symposium on Theory of Computing, STOC, pp. 1–9. ACM (1973)

89. Storer, J.A.: NP-completeness results concerning data compression. Technical report 234, Department of Electrical Engineering and Computer Science, Princeton University, USA, November 1977

90. Storer, J.A., Szymanski, T.G.: Data compression via textual substitution. J. ACM **29**(4), 928–951 (1982)

91. Swernofsky, J., Wehar, M.: On the complexity of intersecting regular, context-free, and tree languages. In: Halldórsson, M.M., Iwama, K., Kobayashi, N., Speckmann, B. (eds.) ICALP 2015. LNCS, vol. 9135, pp. 414–426. Springer, Heidelberg (2015). https://doi.org/10.1007/978-3-662-47666-6_33

92. Szykuła, M.: Improving the upper bound on the length of the shortest reset word. In: Niedermeier, R., Vallée, B. (eds.) 35th Symposium on Theoretical Aspects of Computer Science (STACS 2018), Leibniz International Proceedings in Informatics (LIPIcs), vol. 96, pp. 56:1–56:13. Schloss Dagstuhl-Leibniz-Zentrum für Informatik (2018)

93. Tovey, C.A.: A simplified NP-complete satisfiability problem. Discrete Appl. Math. **8**, 85–89 (1984)

94. Valiant, L.G.: General context-free recognition in less than cubic time. J. Comput. Syst. Sci. **10**(2), 308–315 (1975)

95. Volkov, M.V.: Synchronizing automata and the Černý conjecture. In: Martín-Vide, C., Otto, F., Fernau, H. (eds.) LATA 2008. LNCS, vol. 5196, pp. 11–27. Springer, Heidelberg (2008). https://doi.org/10.1007/978-3-540-88282-4_4

96. Vorel, V., Roman, A.: Parameterized complexity of synchronization and road coloring. Discrete Math. Theor. Comput. Sci. **17**, 283–306 (2015)

97. Wagner, R.A.: On the complexity of the extended string-to-string correction problem. In: Proceedings of seventh Annual ACM Symposium on Theory of Computing, STOC 1975, pp. 218–223. ACM Press (1975)

98. Wagner, R.A., Fischer, M.J.: The string-to-string correction problem. J. ACM **21**(1), 168–173 (1974)

99. Wahlström, M.: Algorithms, measures and upper bounds for satisfiability and related problems. Ph.D. thesis, Department of Computer and Information Science, Linköpings universitet, Sweden (2007)

100. Todd Wareham, H.: The parameterized complexity of intersection and composition operations on sets of finite-state automata. In: Yu, S., Păun, A. (eds.) CIAA 2000. LNCS, vol. 2088, pp. 302–310. Springer, Heidelberg (2001). https://doi.org/10.1007/3-540-44674-5_26

101. Watt, N.: String to string correction kernelization. Master's thesis, University of Victoria, Canada (2013)

102. Wehar, M.: Hardness results for intersection non-emptiness. In: Esparza, J., Fraigniaud, P., Husfeldt, T., Koutsoupias, E. (eds.) ICALP 2014. LNCS, vol. 8573, pp. 354–362. Springer, Heidelberg (2014). https://doi.org/10.1007/978-3-662-43951-7_30

103. Welch, T.A.: A technique for high-performance data compression. IEEE Comput. **17**, 8–19 (1984)

104. Williams, V.V.: Hardness of easy problems: basing hardness on popular conjectures such as the strong exponential time hypothesis (invited talk). In: Husfeldt, T., Kanj, I.A. (eds.) 10th International Symposium on Parameterized and Exact Computation, IPEC, LIPIcs, vol. 43, pp. 17–29. Schloss Dagstuhl - Leibniz-Zentrum für Informatik (2015)
105. Williams, V.V.: Multiplying matrices faster than Coppersmith-Winograd. In: Karloff, H.J., Pitassi, T. (eds.) Proceedings of the 44th Symposium on Theory of Computing Conference, STOC, pp. 887–898. ACM (2012)
106. Wong, C.K., Chandra, A.K.: Bounds for the string editing problem. J. ACM **23**(1), 13–16 (1976)
107. Zakov, S., Tsur, D., Ziv-Ukelson, M.: Reducing the worst case running times of a family of RNA and CFG problems, using Valiant's approach. Algorithms Mole. Biol. **6**, 20 (2011)
108. Ziv, J., Lempel, A.: Compression of individual sequences via variable-rate coding. IEEE Trans. Inf. Theory **24**, 530–536 (1978)

Observation and Interaction
Invited Paper

Edward A. Lee[✉] [iD]

UC Berkeley, Berkeley, CA 94720, USA
eal@berkeley.edu
https://ptolemy.berkeley.edu/~eal/

Abstract. This paper connects three concepts in computer science, zero-knowledge proofs, causal reasoning, and bisimulation, to show that interaction is more powerful than observation. Observation is the use of input data plus, possibly, tractable computation, in such a way that the observer has no effect on the source of the data. Interaction is observation plus action that affects the source of the data. Observation lets the data "speak for itself" and is objective, whereas interaction is first-person and subjective. Zero-knowledge proofs are a strategy for building confidence in some fact while acquiring no additional information other than that the fact is likely to be true. They fall short of absolute certainty and they require interaction. This paper shows that absolutely certainty for such scenarios can be modeled by a bisimulation relation. Causal reasoning has also been shown to require subjective involvement. It is not possible by observation alone, and like zero-knowledge proofs, requires first-person involvement and interaction. This paper shows that bisimulation relations can reveal flaws in causal reasoning.

Keywords: Zero-knowledge proof · Causal reasoning · Bisimulation · Randomized controlled trials

1 Interaction vs. Observation

A number of researchers have argued that interaction is more powerful than observation [1,4,10,16,18]. What I mean by "observation" here is the use of input data plus, possibly, tractable computation. What I mean by "interaction" is observation plus action that affects the input data. Interaction combines observation with action in a closed feedback loop.

In this paper, which is largely an extract from my forthcoming book [8], I will connect three Turing-Award-winning concepts that I believe have never before been connected in this way. Specifically, I will connect zero-knowledge proofs (Goldwasser and Micali, 2012 Turing Award), bisimulation (Milner, 1991

This work was supported in part by the National Science Foundation, NSF award #1446619 (Mathematical Theory of CPS).

C. Martín-Vide et al. (Eds.): LATA 2019, LNCS 11417, pp. 31–42, 2019.
https://doi.org/10.1007/978-3-030-13435-8_2

Turing Award), and causal reasoning (Pearl, 2011 Turing Award) with each other and with the notion that interaction is more powerful than observation. I will assume in this paper that the reader is familiar with the oldest of these concepts, bisimulation, but I assume no prior knowledge of the two newer ones. For a gentle introduction to bisimulation, see [7], Chap. 14. I will also boldly (and perhaps foolishly) relate these concepts to a treacherous quagmire in philosophy, the notion of free will.

Interaction as a tool is closely related to the concept of feedback, which has a long history. In the 1920s, at Bell Labs, Harold Black found that negative feedback could compensate for the deficiencies in amplifiers of the day [3]. His feedback circuits push on their environment, measure the extent to which its reaction deviates from the desired reaction, and adjust the pushing to get closer to a desired objective.

Wiener, during World War II, also used feedback for the automatic aiming and firing of anti-aircraft guns. Wiener coined the term "cybernetics" for the conjunction of physical processes, computation that governs the actions of those physical processes, and communication between the parts [19]. He derived the term from the Greek word for helmsman, governor, pilot, or rudder.

Feedback, which is used in many engineered systems today, is a tight interaction between a system and its environment. Turing-Church computation can be used as building blocks, for example to calculate adjustments, but fundamentally, they are just components in a bigger picture. Interactive systems go well beyond what Turing-Church computation alone can accomplish.

To make a connection with the concept of free will, I will rely on current trends in psychology, specifically the thesis of *embodied cognition*, where the mind "simply does not exist as something decoupled from the body and the environment in which it resides" [17, p. 7]. The mind does not just interact with its environment, but rather the mind *is* an interaction of the brain with its environment. A cognitive being is not an *observer* of its environment, but rather a collection of feedback loops that include the body and its environment, an *interactive* system.

Zero Knowledge Proofs

Zero-knowledge proofs were first developed by Goldwasser and Micali [5,6]. They were a first instance of a more general idea, interactive proofs, which bring randomness and interaction together. An interactive proof, developed independently by Babai [2], can be thought of as a game with two players, a prover (named Merlin by Babai) and a verifier (named Arthur by Babai). The verifier, Arthur, has limited ability to compute. Specifically, Arthur is assumed to be able to perform only computations that can be completed in a reasonable amount of time on a modern sequential computer. The prover, Merlin, is allowed to perform more difficult computations, but I will not make use of that feature in this paper.

Zero-knowledge proofs are easy to understand using a story developed by Quisquater and Guillou [14]. Assume that Merlin knows something important, like a password, and wants to prove to Arthur that he knows this. Merlin is a very private person, so while he wants to convince Arthur that he knows the password, he does not want Arthur to be able to convincingly tell anyone else that he knows the password. His objective is only to convince Arthur and give him exactly zero additional information. Note that Merlin's objective cannot be accomplished by simply telling Arthur the password because then Arthur will then also know the password.

In this story, there is an oddly shaped cave (see Fig. 1), where the entrance tunnel forks into two tunnels labeled A and B. Both tunnels are dead ends, but there is door connecting the two ends. The door can only be opened with a password that only Merlin knows.

One way that Merlin could prove to Arthur that he knows the password is to enter the cave together with Arthur, and while Arthur waits at the mouth of the cave, go down tunnel A and come back out through tunnel B. Arthur will be convinced that Merlin knows the password, and Arthur will not himself know the password. But if Arthur surreptitiously records the event with a video camera, then Arthur would be able to convince anyone else that Merlin knows the password. This makes the information that Merlin knows the password available to a third-person observer. The goal is that the information be available only to the first-person interactor, Arthur.

So, instead, Arthur waits outside the cave while Merlin goes in and picks one of the tunnels to go down. Suppose he picks tunnel B and goes as far as the door. Then Arthur comes into the cave as far as the fork and randomly calls out either A or B. He cannot see which tunnel Merlin went down. If he calls A, then Merlin has to use his password, open the door, and come out through tunnel A. Arthur is not yet *sure* that Merlin knows the password, but he can conclude that it is equally likely that he knows it as that he doesn't know it.

Arthur and Merlin then repeat the experiment. If Merlin successfully comes out of the tunnel that Arthur identifies a second time, then Arthur can conclude that the probability that he knows the password is now 3/4. It would have

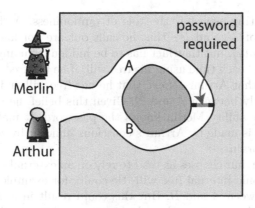

Fig. 1. Ali Baba's cave, illustrating zero-knowledge proofs.

required quite a bit of luck for him to not have to use the password twice in a row. Repeating the experiment again will raise the probability to 7/8. After 10 repeats, the likelihood that he didn't need the password drops to about 1 in 1000. By repeating the experiment, Merlin can convince Arthur to any level he demands short of absolute certainty.

Unlike the previous experiment, where Merlin just went in one tunnel and came out the other, this new experiment does not give Arthur the power to convince a third party, say Sarah, that Merlin knows the password. Arthur could videotape the whole experiment, but Sarah is a savvy third party, and she suspects that Arthur and Merlin colluded and agreed ahead of time on the sequence of A's and B's that Arthur would call out. Only Arthur and Merlin can know whether collusion occurred. So Sarah is not convinced that Merlin knows the password the way Arthur is convinced. Merlin retains plausible deniability, and only Arthur knows for sure (almost for sure) that Merlin knows the password.

There are several fascinating aspects to this story. First, for Merlin to prove to Arthur that he knows the password while not giving Arthur the power to pass on that knowledge, interaction is required. If Arthur simply watches Merlin, observing but not interacting, then anything Merlin does to convince Arthur that he knows the password gives Arthur the power to pass on that knowledge, for example by making a video. He can then convince Sarah that Merlin knows the password by simply showing her the video. But by interacting, Merlin is able to convince Arthur and only Arthur. No third party observer will be convinced. You have to actively participate to be convinced. Interaction is more powerful than observation, but for interaction to work, you have to be a first-person participant in the interaction. This is what interaction means! Arthur's first-person action, choosing A or B at random, is necessarily subjective. Only he knows that no collusion was involved.

Another fascinating aspect of this story is the role of uncertainty. Using this scheme, it is not possible to give Arthur absolute certainty without giving Arthur more than Merlin wants to. The residual uncertainty that Arthur retains can be made as small as we like, but it cannot be reduced to zero, at least not by this technique.

A third fascinating aspect is the role of randomness. Arthur has to know that the sequence of A's and B's that he calls out are not knowable to Merlin (with high probability), but that fact has to be hidden from anyone else. Arthur could choose A or B each time using his free will, if he has free will. Actually, all that is required is that Arthur *believe* that he has free will and believes that he has chosen randomly between A and B. Given this belief, he will be convinced that with high probability Merlin knows the password. It makes no difference whether the choice is made by Arthur's conscious mind or by some unconscious mechanism in his brain.

Suppose that Arthur chooses instead to rely on an external source of randomness rather than some internal free will. He could, for example, flip a coin each time to choose between A and B. But this could result in leaking information because now he could videotape the coin flipping, and the resulting video would convince Sarah and any other third party as much as it convinces Arthur. It will be evident to any observer that Arthur is not colluding with Merlin. Observers could easily imagine themselves flipping the coin, so Arthur is just a proxy for their own first-person interaction. I will later leverage this strategy to explain why randomized controlled trials work to determine causal relationships. But for the goal of preserving Merlin's privacy, Arthur has to generate the choices between A and B in a hidden way, and by hiding this, he gives up the ability to convince any third party.

Even Arthur's knowledge, however, is not certainty. Some background assumptions are needed. Arthur has to believe in his free will, and dismiss ideas like that Merlin is somehow manipulating his subconscious brain to make colluding choices. Ultimately, a little bit of trust is required to get past all the conspiracy theories. Once we open the door to trust, we have to admit that a third person may decide to trust Arthur and assume that he is not colluding with Merlin, in which case, despite Merlin's wishes, his secret will be out.

Merlin and Arthur Bisimulate

Merlin and Arthur's interaction in Fig. 1 can be modeled using automata, as shown in Fig. 2. The model for Arthur is shown at the top. It shows that Arthur enters the cave in the first time instant, then nondeterministically calls out A or B, ending in one of two possible states, $endA$ or $endB$. The second model shows Merlin under the assumption that he does not know the password. He also enters the cave in the first instant, but nondeterministically goes to one of two locations, $insideA$ or $insideB$. Once he is one of these locations, he has no choice but to come out the same way he went in.

The third model in the figure shows Merlin under the assumption that he does know the password. One way to understand the difference between the second and third models is that, in the second, the decision about which tunnel to exit from is made earlier than in the third model. To make the decision later, in the second reaction of the state machine, Merlin needs to know the password. To make it earlier, in the first reaction, there is no need to know the password.

Here is where I will boldly make a connection with the concept of free will. Arthur has to make one decision, which tunnel to call out, A or B. Merlin has to make two decisions, which tunnel to enter, and which to exit. If Merlin does not know the password, the first decision determines the second, and, once the first decision is made, Merlin has no free will to make the second. On the other hand, if Merlin does know the password, then the second decision remains free, and Merlin is free to exit from the tunnel called out by Arthur. Here, "knowing the password" is a proxy for an ability to exercise the choice to pass through the door. If Merlin does not know the password, there is no such choice, and the tunnel by which he exits has been preordained. This lack of free will illustrates the incompatibilist interpretation in philosophy, where free will is incompatible with determinism. On the other hand, Arthur's free will in choosing to call out A or B illustrates the compatibilist interpretation, where it doesn't really matter whether the resolution of alternatives is predetermined or not. If Arthur's brain internally uses a deterministic pseudo-random sequence generator, the outcome is the same as long as he believes the choice was free.

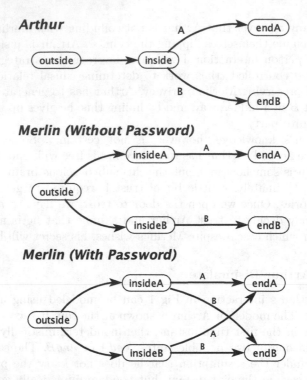

Fig. 2. Automata models of Arthur and Merlin, with and without the password.

Given the many trajectories that the game can follow, we can ask *why* one trajectory occurs over another, *how* the determination of a trajectory is made, or *when* the determination between alternatives is made. If, for example, the determination between alternative trajectories is made early, then according to the incompatibilist interpretation, later in the game, there is no free will. If on the other hand, the determination between alternative trajectories is made as late as possible, say, just before the selection of alternatives has any effect on anything else, then there remains at least a possibility of free will. In any case, the questions of *how* and *why* a determination is made can only really make sense after we answer the question of *when*.

How can we determine whether the selection between alternatives is made early or late? I will leverage insights first exposed by Robin Milner, who showed how to compare automata using simulation and bisimulation relations. In automata theory, a passive observer of a system *cannot tell* whether selection is made early or late. In order to be able to tell, an observer must *interact* with the system. It is not sufficient to just observe the system. Interaction is required to determine whether there is free will, and first-person interaction yields more than observation. This theory, in fact, helps to explain why first-person interaction is so different from third-person observation. It may even help us understand what we mean by "first person."

Merlin (Guessing the Password)

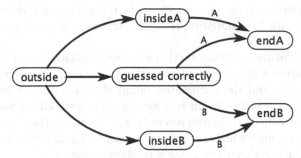

Fig. 3. Automata model Merlin where he guesses the password.

Notice that all three automata in Fig. 2 are language equivalent. Each is capable of producing the output A or B and nothing more. But language equivalence is not enough. Milner's notion of simulation captures the difference between Merlin (without password) and Merlin (with password). Specifically, Merlin (with password) simulates Arthur, but Merlin (without password) does not. Merlin is unable to make some of the moves that Arthur may demand.

The fact that Arthur simulates Merlin is what makes it possible for Arthur to collude with Merlin. Arthur can match the decisions Merlin has already made. Equivalently, Merlin can anticipate whether Arthur will call out A or B. If Merlin *does* know the password, then Merlin is bisimilar to Arthur. They can perfectly match each other's moves regardless of who moves first at each time instant. No collusion is needed.

Simulation relations, however, are not quite enough. Suppose instead that Merlin does not know the password but rather guesses it each time he needs it. This can be represented by the automaton in Fig. 3. Here, if Merlin correctly guesses the password, he is able to fool Arthur no matter how many times they perform the experiment. This gives Merlin's automaton the ability to simulate Arthur's automaton. So Merlin (with guessing) simulates Arthur, and Arthur simulates Merlin (with guessing). But Merlin (with guessing) is still not fundamentally equivalent to Arthur. The possibility of guessing incorrectly remains.

A bit of history may be helpful here. In the 1970s, Milner had introduced the idea of simulation relations between automata. In 1980, David Park found a gap in Milner's prior notion of simulation. He noticed that even if two automata simulate each other, they can nevertheless exhibit significant differences in behavior when they *interact*. Milner's prior notion of simulation was unable to distinguish Merlin (with password) from Merlin (with guessing).

Milner and Park together came up with a stronger notion of modeling that they decided to call "bisimulation" [9,11]. Milner then fully developed and popularized the idea.[1] He showed that the difference between Merlin (with password)

---

[1] Sangiorgi gives a nice overview of the historical development of this idea [15]. He notes that essentially the same concept of bisimulation had also been developed in the fields of philosophical logic and set theory.

and Merlin (with guessing) becomes evident only if the two automata interact with one another. It is not enough to just observe each other, as he had done previously with his simulation relations. Interaction is more powerful than observation.

How is bisimulation about interaction whereas simulation is only about observation? To construct a simulation relation, the automaton being simulated moves first in each round, and the automaton doing the simulating must match the move. To construct a bisimulation relation, in each turn, either automaton can move first and the other automaton has to be able to match the move. The ability in the game to alternate which automaton moves first makes this fundamentally an interactive game rather than a one-way observation.

It is easy to verify that there is no bisimulation relation between Merlin (with guessing) and Arthur, nor between Merlin (with guessing) and Merlin (with password). The lack of a bisimulation relation reveals the mismatch. But there is a subtlety. To know that there is no bisimulation relation, we need to know the structure of the automata. If we know that Merlin's automaton has the structure shown in Fig. 3, then we know that he does not know the password, even if the possibility of a lucky guess remains.

This subtlety lends insight into why zero-knowledge proofs do not yield certainty. Arthur is never absolutely certain that Merlin knows the password, though by repeating the trial, be can reach any level of certainty he desires short of absolute certainty. If Arthur were instead given the bisimulation relation, he would have a proof that Merlin knows the password. No uncertainty would remain. But constructing that proof requires knowing the structure of Merlin's automaton, or equivalently, knowing that Merlin knows the password.

What really does bisimulation mean in this case? The two automata, Arthur and Merlin (with password), have different structure, but they are fundamentally indistinguishable. Arthur's automaton represents what he demands from someone who knows the password. Merlin's automaton represents the capabilities he acquires by knowing the password. The fact that these two automata are bisimilar shows conclusively what Arthur is able to conclude with repeated experiments, that Merlin knows the password. Hence, the repeated experiments may provide evidence of bisimilarity that does not require knowing the detailed structure of the automata. Such evidence will only be provided if the repeated experiments are fair in the sense that all of the possible nondeterministic transitions occur in at least some of the trials (or infinitely often in an infinite experiment).

Causal Reasoning

Pearl has argued that interacting with a system enables drawing conclusions about causal relationships between pieces of that system, conclusions that are much harder to defend without interaction [12,13]. Specifically, consider the classic problem of determining whether administering an experimental drug causes a patient's condition to improve. The gold standard for making such a determination is a double-blind randomized controlled trial (RCT), where a subset of patients from a population is chosen at random to receive the drug, and the other patients in the trial receive a placebo. "Double blind" means that neither the medical staff administering the drug nor the patients know whether they are using the real drug or a placebo. "Randomized" means that the selection of patients to receive treatment is not causally affected by anything other than chance.

Why does an RCT work so well? Pearl explains this using causal diagrams, which represent the ability one variable in a system has to cause perturbations in another. Consider the causal diagram on the left in Fig. 4. The solid arrows represent an assumed causal relationship between a "confounding factor" and both the treatment and health of the patient. For example, suppose that some treatment, when made available to a population, is more likely to be taken by males than by females, and males are also more likely to recover than females. In this case, a statistician will tell you that it is necessary to control for the sex of the patient. Otherwise, you may derive erroneous conclusions from the data. But the challenge, in many practical cases, is that the confounding factors are not known or data about them are not available.

Consider instead the scenario in the center of Fig. 4. Suppose for example that the "colliding factor" is whether a patient ends up in the hospital. Taking the treatment, because of side effects, may cause the patient to end up in the hospital. Poor health, where the treatment has been ineffective, could also result in the patient ending up in the hospital. In this case, it would be a statistical error to control for whether the patient ends up in the hospital. An effective treatment could be rejected because, among patients that end up in the hospital, whether they got the treatment and whether their health improved could be uncorrelated, and also among patients who do not end up in the hospital, while in the general population, there is a correlation between patients who receive treatment and those whose health improves.

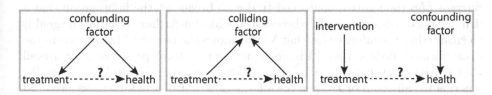

Fig. 4. A causal diagram on the left guiding the evaluation of a treatment's effectiveness that requires controlling for a confounder on the left and *not* controlling for collider in the center. On the right, intervention removes the effect of a confounder.

At the right of Fig. 4 is a causal diagram representing an intervention, a form of interaction that Pearl calls a "do operator." The intervention in a randomized controlled trial (RCT) breaks any causal dependencies on whether the treatment is taken by forcing the treatment to be taken or not taken according to a random outcome. This removes the need to control for any other factors, known or unknown.

Randomized Controlled Trial

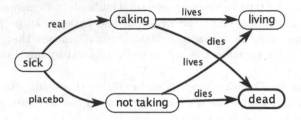

Flawed Trial

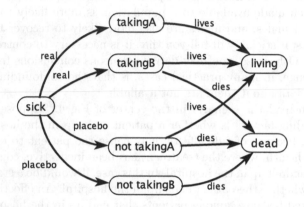

Fig. 5. Randomized controlled trial model and model of a flawed trial.

The intervention is analogous to Arthur's calling out of A or B to specify the tunnel from which Merlin should exit. But there is an interesting twist here. The purpose of a randomized controlled trial is to broadcast the information that a drug works or does not work, whereas in the Merlin-Arthur scenario, the goal is to ensure that the information that Merlin knows the password (analogous to the drug works or does not work) is *not* available to a third party observer. Recall that if Arthur visibly flips a coin, as opposed to using free will, to determine whether to call out A or B, then the information that Merlin knows the password becomes available to a third party observer. Analogous, in an RCT, the decision of whether to administer the drug or a placebo should be made by a verifiably random choice, not secretly by someone's free will, in order for the outcome of the trial to be trusted by an outside observer.

A properly constructed RCT can be represented by the automaton at the top of Fig. 5. The important feature of this automaton is the determination of whether the patient lives or dies is made *after* the determination of whether to administer a placebo or the real drug. In an incorrectly constructed trial, shown at the bottom of the figure, it is possible for a patient who is doomed to die will get assigned a placebo and one that is destined to live will be given the real drug. An unscrupulous researcher could, for example, assign the real drug to younger and healthier patients and the placebo to older and sicker patients, thereby skewing the results of the trial.

The two automata in Fig. 5 simulate each other, but they are not bisimilar. These automata say nothing about the probabilities of outcomes. They only express possibilities. Hence, it is still possible to construct an invalid trial that is bisimilar to the top automaton. For example, adding transitions from *takingA* to *dead* and *not takingB* to *living* would make the lower automaton bisimilar to the upper one, but the trial could still be skewed. But any automaton that is *not* bisimilar to the upper one will *surely* be invalid.

Humanity Requires Interaction

Interacting components can observe *and be observed* and can affect and *be affected*. Such interaction can accomplish things that are not possible with observation alone. The implications of this are profound. It reinforces Milner's observation that machines that look identical to an observer are not identical if you can interact with them. It reinforces Goldwasser and Micali's observation that interaction can do things that are not possible without interaction. It reinforces Pearl's observation that reasoning about causality requires interaction. It also reinforces the hypothesis of embodied cognition from psychology. If our sense of self depends on bidirectional interaction, the kind of dialog of Milner's model, where either party can observe or be observed, then our sense of self cannot be separated from our social interactions. Our minds cannot exist as an observer of the universe alone. And indeed, our interaction with the world around us has this bidirectional character. Sometimes we react to stimulus in ways that affect those around us, and sometimes we produce stimulus and watch the reactions of those around us. Such dialog seems to be an essential part of being human and may even form the foundations for language and even thought.

Moreover, such dialog has deep roots in physics. Quantum physics has taught us that no observation of a physical system is possible without disrupting the system in some way. In fact, quantum physics has real problems with any attempt to separate the observer from the observed. The observed automaton necessarily observes the observer. Passive observation in the form of unidirectional simulation is impossible in our natural universe. This suggests that simulation relations alone are not a reasonable model of modeling (a "metamodel," if you will permit me). Bisimulation is a better choice.

In an objectivist approach to science, we are often taught to let the data "speak for itself," to avoid subjective bias, where our actions may affect the data. I have collected in this paper several powerful arguments that being so objective has serious limitations. Subjectivity, first-person involvement, and interaction with the sources of data are sometimes essential.

References

1. Agha, G.A.: Abstracting interaction patterns: a programming paradigm for open distributed systems. In: Najm, E., Stefani, J.B. (eds.) Formal Methods for Open Object-based Distributed Systems. IFIPAICT, pp. 135–153. Springer, Heidelberg (1997). https://doi.org/10.1007/978-0-387-35082-0_10
2. Babai, L.: Trading group theory for randomness. In: Symposium on Theory of Computing (STOC), pp. 421–429. ACM (1985). https://doi.org/10.1145/22145.22192
3. Black, H.S.: Stabilized feed-back amplifiers. Electr. Eng. **53**, 114–120 (1934)
4. Goldin, D., Smolka, S., Attie, P., Sonderegger, E.: Turing machines, transition systems, and interaction. Inf. Comput. **194**(2), 101–128 (2004)
5. Goldwasser, S., Micali, S., Rackoff, C.: The knowledge complexity of interactive proof systems (extended abstract). In: Symposium on Theory of Computing (STOC), pp. 291–304. ACM (1985)
6. Goldwasser, S., Micali, S., Rackoff, C.: The knowledge complexity of interactive proof systems. SIAM J. Comput. **18**(1), 186–208 (1989). https://doi.org/10.1137/0218012
7. Lee, E.A., Seshia, S.A.: Introduction to Embedded Systems - A Cyber-Physical Systems Approach, Second edn. MIT Press, Cambridge (2017). http://LeeSeshia.org
8. Lee, E.A.: Living Digital Beings – A New Life Form on Our Planet? MIT Press, Cambridge (2020, to appear)
9. Milner, R.: Communication and Concurrency. Prentice Hall, Englewood Cliffs (1989)
10. Milner, R.: Elements of interaction. Commun. ACM **36**, 78–89 (1993)
11. Park, D.: Concurrency and automata on infinite sequences. In: Deussen, P. (ed.) GI-TCS 1981. LNCS, vol. 104, pp. 167–183. Springer, Heidelberg (1981). https://doi.org/10.1007/BFb0017309
12. Pearl, J.: Causality: Models, Reasoning, and Inference, 2nd edn. Cambridge University Press, Cambridge (2000). (2009)
13. Pearl, J., Mackenzie, D.: The Book of Why: The New Science of Cause and Effect. Basic Books, New York (2018)
14. Quisquater, J.-J., et al.: How to explain zero-knowledge protocols to your children. In: Brassard, G. (ed.) CRYPTO 1989. LNCS, vol. 435, pp. 628–631. Springer, New York (1990). https://doi.org/10.1007/0-387-34805-0_60
15. Sangiorgi, D.: On the origins of bisimulation and coinduction. ACM Trans. Program. Lang. Syst. **31**(4), 15:1–15:41 (2009). https://doi.org/10.1145/1516507.1516510. Article 15, Pub. date: May 2009
16. Talcott, C.L.: Interaction semantics for components of distributed systems. In: Najm, E., Stefani, J.B. (eds.) Formal Methods for Open Object-Based Distributed Systems (FMOODS). IFIPAICT, pp. 154–169. Springer, Heidelberg (1996). https://doi.org/10.1007/978-0-387-35082-0_11
17. Thelen, E.: Grounded in the world: developmental origins of the embodied mind. Infancy **1**(1), 3–28 (2000)
18. Wegner, P.: Why interaction is more powerful than algorithms. Commun. ACM **40**(5), 80–91 (1997). https://doi.org/10.1145/253769.253801
19. Wiener, N.: Cybernetics: Or Control and Communication in the Animal and the Machine. Librairie Hermann & Cie, Paris, and MIT Press, Cambridge (1948)

From Words to Graphs, and Back

Vadim Lozin[(⊠)]

Mathematics Institute, University of Warwick, Coventry CV4 7AL, UK
V.Lozin@warwick.ac.uk

Abstract. In 1918, Heinz Prüfer discovered a fascinating relationship between labelled trees with n vertices and words of length $n-2$ over the alphabet $\{1, 2, \ldots, n\}$. Since the discovery of the Prüfer code for trees, the interplay between words and graphs has repeatedly been explored and exploited in both directions. In the present paper, we review some of the many results in this area and discuss a number of open problems related to this topic.

1 Introduction

In the beginning was the word. Graphs have appeared much later. Since then, these two notions frequently interact and cooperate. Graphs help reveal structure in words and (not necessarily formal) languages (see e.g. [34]). Words are used to represent graphs, which became a important issue with the advent of the computer era. We discuss this issue in Sect. 3 of the paper. Section 2 is devoted to the interplay between words and graphs around the notion of well-quasi-ordering. In the rest of the present section, we introduce basic terminology and notation used in the paper.

A binary relation is a *quasi-order* (also known as *pre-order*) if it is reflexive and transitive. A set of pairwise comparable elements in a quasi-ordered set is called a *chain* and a set of pairwise incomparable elements is called an *antichain*. A quasi-ordered set is *well-quasi-ordered* if it contains neither infinite strictly decreasing chains, nor infinite antichains.

Given a finite set B (an alphabet), we denote by B^* the set of all words over B. For a word $\alpha \in B^*$, $|\alpha|$ stands for the length of α and α_j for the j-th letter of α. A *factor* of a word α is a contiguous subword of α. The factor containment relation is a quasi-order, but not a well-quasi-order, since it contains infinite antichains, for instance, $\{101, 1001, 10001, \ldots\}$. A language is *factorial* if it is closed under taking factors. It is well-known (and not difficult to see) that a factorial language L can be uniquely characterized by a set of minimal forbidden words, also known as the *antidictionary* of L, i.e. the set of minimal (with respect to the factor containment relation) words that do not belong to L.

All graphs in this paper are undirected, without loops and multiple edges. A graph G is an *induced subgraph* of H if G can be obtained from H by vertex deletions. The induced subgraph relation is a quasi-order, but not a well-quasi-order, since the cycles C_k, $k \geq 3$, constitute an infinite antichain. A class X

© Springer Nature Switzerland AG 2019
C. Martín-Vide et al. (Eds.): LATA 2019, LNCS 11417, pp. 43–54, 2019.
https://doi.org/10.1007/978-3-030-13435-8_3

of graphs, also known as a graph property, is *hereditary* if it is closed under taking induced subgraphs. Clearly, if X is hereditary, then it can be uniquely characterized by a set M of minimal forbidden induced subgraphs, in which case we say that graphs in X are M-free. The *speed* of X is the number of n-vertex labelled graphs in X, studied as a function of n.

2   Words, Graphs and Well-quasi-ordering

Well-quasi-ordering (WQO) is a highly desirable property and frequently discovered concept in mathematics and theoretical computer science [19,26]. A simple but powerful tool for proving well-quasi-orderability is the celebrated Higman's Lemma, which can be stated as follows. Let M be a set with a quasi-order $\leq$. We extend $\leq$ from M to M^* as follows: $a_1 \ldots a_m \leq b_1 \ldots b_n$ if and only if there is an order-preserving injection $f : \{a_1, \ldots, a_m\} \to \{b_1, \ldots, b_n\}$ with $a_i \leq f(a_i)$ for each $i = 1, \ldots, m$. Higman's Lemma states the following.

Lemma 1. ([21]) *If* $(M, \leq)$ *is a WQO, then* $(M^*, \leq)$ *is a WQO.*

Kruskal [25] extended this result to the set of finite trees partially ordered under homeomorphic embedding. In other words, Kruskal's tree theorem restricted to paths becomes Higman's lemma. Later, Robertson and Seymour [31] generalized Kruskal's tree theorem to the set of all graphs partially ordered under the minor relation. However, the induced subgraph relation is not a well-quasi-order. Other examples of important relations that are not well-quasi-orders are pattern containment relation on permutations [35], embeddability relation on tournaments [16], minor ordering of matroids [22], factor containment relation on words [27]. On the other hand, each of these relations may become a well-quasi-order under some additional restrictions. Below we present some examples and discuss a number of open problems related to this topic.

2.1   An Introductory Example

A word can be interpreted as a graph in various ways. Consider, for instance, a binary word $\alpha = \alpha_1 \ldots \alpha_n$, and let us associate with this word a graph G_α with vertices $v_1, \ldots, v_{n+1}$ such that for each $i = 2, \ldots, n+1$, vertex v_i is adjacent to the vertices $v_1, \ldots, v_{i-1}$ if $\alpha_i = 1$ and v_i is not adjacent to the vertices $v_1, \ldots, v_{i-1}$ if $\alpha_i = 0$. In other words, G_α can be constructed from a single vertex recursively applying one of the following two operations: adding a dominating vertex (i.e. a vertex adjacent to every other vertex in the graph) or adding an isolated vertex. The graphs that can be constructed by means of these two operations are known in the literature as *threshold graphs*.

Obviously, not every graph is threshold. For instance, none of the following graphs is threshold, since none of them contains a dominating or isolated vertex: the path on 4 vertices P_4, the cycle on 4 vertices C_4 and the complement of C_4, denoted $\overline{C}_4$. Moreover, no graph containing P_4, C_4 or $\overline{C}_4$ as an induced subgraph is threshold, i.e. threshold graphs are $(P_4, C_4, \overline{C}_4)$-free. The inverse inclusion also is true, which leads to the following conclusion: a graph is threshold if and only if it is $(P_4, C_4, \overline{C}_4)$-free.

Let us note that the original definition of threshold graphs differs from both characterizations presented in the two previous paragraphs. The notion of threshold graphs was introduced in [17] and was inspired by the notion of *threshold Boolean function*. Since its original introduction, the notion of threshold graphs gave rise to a vast literature on the topic, including the book [28].

The relationship between binary words and threshold graphs described earlier is a bijection and it provides an easy way for counting *unlabelled* threshold graphs (observe that counting unlabelled graphs is generally a more difficult task than counting labelled graphs). This relationship also shows, with the help of Higman's Lemma, that the class of threshold graphs is well-quasi-ordered under the induced subgraph relation. The same conclusion can be derived from two other seemingly unrelated results, which we discuss in the next section.

2.2   Geometric Grid Classes of Permutations and Letter Graphs

Any collection of n points on the plane, with no two on a common vertical or horizontal line, uniquely defines a permutation π of n elements. This can be done, for instance, by labeling the points from 1 to n from bottom to top and then recording the labels reading left to right (see Fig. 1 for an illustration). By deleting any point, we obtain a permutation π' of $n-1$ elements, in which case we say that π contains π' as a *pattern*.

The notion of a pattern defines a partial order on the set of permutations known as the *pattern containment* relation. A *pattern class of permutations*, or simply a *permutation class*, is any set of permutations which is downward closed under the pattern containment relation.

The pattern containment relation is not a well-quasi-order, since it contains infinite antichains [35]. However, under certain restrictions, this relation may become a well-quasi-order. To give an example, let us introduce the notion of monotone grid classes of permutations.

Let M be an $s \times t$ matrix with entries in $\{0, \pm 1\}$. An M-*gridding* of a permutation π represented by a collection of points on the plane is a partition of the plane into $s \times t$ cells by means of vertical and horizontal lines so that the cell in column i and row j of the partition is empty if $M(i, j) = 0$, contains the elements of π in an increasing order if $M(i, j) = +1$, and contains the elements of π in a decreasing order if $M(i, j) = -1$. Figure 1 represents two M-griddings of the permutation 351624 with $M = \begin{pmatrix} +1 & -1 \\ 0 & +1 \end{pmatrix}$ and $M = \begin{pmatrix} +1 & -1 \\ -1 & +1 \end{pmatrix}$. The grid class of M consists of all permutations which admit an M-gridding and is known as a *monotone grid class*.

The restriction to monotone grid classes is a strong restriction, but it is not strong enough to guarantee well-quasi-ordering. To describe more restrictions, we define the *cell graph* of M as follows: the vertices of this graph are the non-zero entries of M, in which two vertices are adjacent if and only if the corresponding entries share a row or a column and there are no non-zero entries between them in this row or column.

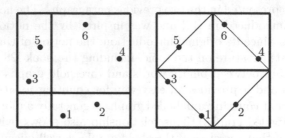

Fig. 1. Two griddings of the permutation 351624.

Theorem 1. ([38]) *For a $0/\pm1$ matrix M, the grid class of M is well-quasi-ordered under the pattern containment relation if and only if the cell graph of M is a forest, i.e. a graph without cycles.*

Cycles in cell graphs can give rise to infinite antichains of permutations. However, if we require the elements of each cell in a monotone grid class to belong to a diagonal of the respective cell (see the gridding on the left of Fig. 1), then infinite antichains "magically" disappear and the class becomes well-quasi-ordered. This is known as *geometric gridding*, a notion introduced in [2]. The authors of [2] characterized geometric grid classes of permutations in various ways, of which we quote the following two results.

Theorem 2. ([2]) *Every geometrically griddable class of permutations is well-quasi-ordered and is in bijection with a regular language.*

Now we move from words to graphs and define the notion of letter graphs introduced in [29]. Let Σ be a finite alphabet and let $\mathcal{P} \subseteq \Sigma^2$ be a set of ordered pairs of symbols from Σ, known as a decoder. With each word $w = w_1w_2 \cdots w_n$ over Σ we associate a graph $G(\mathcal{P}, w)$, called the *letter graph* of w, by defining the vertex set of this graph to be $\{1, 2, \ldots, n\}$ with i being adjacent to $j > i$ if and only if the ordered pair (w_i, w_j) belongs to $\mathcal{P}$.

It is not difficult to see that every graph G is a letter graph in a sufficiently large alphabet with an appropriate decoder $\mathcal{P}$. The minimum ℓ such that G is a letter graph in an alphabet of ℓ letters is the *lettericity* of G. A graph is a k-letter graph if its lettericity is at most k.

With the help of Higman's Lemma it is not difficult to conclude that for each fixed value of k, the set of all k-letter graphs is well-quasi-ordered by the induced subgraph relation, which was formally proved in [29].

The notion of letter graphs was introduced 11 years earlier than the notion of geometric grid classes of permutations, and nothing in the definitions of these two notions suggests any connection between them. However, there is intriguing relationship between these notions revealed recently in [3]. To describe this relationship, we define the permutation graph G_π of a permutation π on the set $\{1, 2, \ldots, n\}$ to be the graph with vertex set $\{1, 2, \ldots, n\}$ in which two vertices i and j are adjacent if and only if $(i - j)(\pi(i) - \pi(j)) < 0$.

Theorem 3. ([3]) *Let X be a class of permutations and $\mathcal{G}_X$ the corresponding class of permutation graphs. If X is a geometrically griddable class, then $\mathcal{G}_X$ is a class of k-letter graphs for a finite value of k.*

This theorem suggests the idea that geometrically griddable classes of permutations and letter graphs are two languages describing the same concept in the universe of permutations and permutation graphs, respectively. However, the inverse of Theorem 3 remains an open problem, which we state below as a conjecture.

Conjecture 1. Let X be a class of permutations and $\mathcal{G}_X$ the corresponding class of permutation graphs. If $\mathcal{G}_X$ is a class of k-letter graphs for a finite value of k, then X is a geometrically griddable class.

To support this conjecture, we return to the notion of threshold graphs introduced in Sect. 2.1 and observe that every threshold graph is a 2-letter graph. Indeed, consider the alphabet $\Sigma = \{0, 1\}$ and the decoder $\mathcal{P} = \{(1, 1), (1, 0)\}$, and let $w = w_1 w_2 \cdots w_n$ be any binary word. In the graph $G(\mathcal{P}, w)$, if $w_i = 1$, then i is adjacent to every vertex $j > i$ (since both pairs $(1, 1)$ and $(1, 0)$ belong to $\mathcal{P}$), and if $w_i = 0$ then i is not adjacent to any vertex $j > i$ (since neither $(0, 0)$ nor $(0, 1)$ belong to $\mathcal{P}$). Therefore, $G(\mathcal{P}, w)$ is a threshold graph. Notice that this representation is similar but not identical to the correspondence between graphs and words described in Sect. 2.1.

On the other hand, the *geometric* grid class of $M = \begin{pmatrix} -1 & +1 \\ +1 & -1 \end{pmatrix}$, also known as the X-class, consists of permutations that avoid the following four permutations as patterns: 2143, 3412, 2413, 3142 (see e.g. [18]). The permutation graphs of the first two of these permutations are, respectively, $\overline{C}_4$ and C_4, while the last two permutations both represent a P_4. Since a graph is threshold if and only if it is $(P_4, C_4, \overline{C}_4)$-free, we conclude that the permutations graphs corresponding to the X-class are precisely the threshold graphs.

2.3 Deciding WQO

Deciding whether a permutation class is well-quasi-ordered is a difficult question. Decidability of this question for classes defined by finitely many forbidden permutations was stated as an open problem in [15]. Similar questions have been studied for the induced subgraph relation on graphs [24], the embeddability relation on tournaments [16], the minor ordering of matroids [22]. However, the decidability of this problem has been shown only for one or two forbidden elements (graphs, permutations, tournaments, matroids).

A breakthrough result in this area was recently obtained in [9], where decidability was proved for factorial languages. The solution is based on the analysis of the structure of an automaton describing the input language. The authors of [9] also discuss an alternative approach, which suggests a possible way to approach the same problem for graphs and permutations. This approach is based on the notion of a *periodic* infinite antichain. Speaking informally, an infinite antichain of words is called periodic of period p if each element of this set becomes a factor of some infinite periodic word of period p after dropping some prefix and suffix. For instance, the set $\{101, 1001, 10001, \ldots\}$ is a periodic infinite antichain of period 1. The following theorem was proved in [9].

Theorem 4. ([9]) *Let $D = \{\alpha^1, \alpha^2, \ldots, \alpha^k\}$ be a finite set of pairwise incomparable words and X be the factorial language with the antidictionary D. Then X is well-quasi-ordered by the factor containment relation if and only if it contains no periodic infinite antichains of period at most $|\alpha^1| + |\alpha^2| + \ldots + |\alpha^k| + 1$.*

To apply the idea of periodic infinite antichains to graphs, we modify the notion of letter graphs by distinguishing between consecutive and nonconsecutive vertices corresponding to a word $w = w_1 w_2 \cdots w_n$. For nonconsecutive vertices $i < j$ the definition remains the same: i and j are adjacent if and only if $(w_i, w_j) \in \mathcal{P}$. For consecutive vertices, we change the definition to the opposite: i and $i+1$ are adjacent if and only if $(w_i, w_{i+1}) \notin \mathcal{P}$. Let us denote the graph obtained in this way from a word w by $G^*(\mathcal{P}, w)$. For instance, if a is a letter of Σ and $(a, a) \notin \mathcal{P}$, then the word $aaaaa$ defines a path on 5 vertices. With some restrictions, the induced subgraph relation on graphs defined in this way corresponds to the factor containment relation on words, i.e. $G^*(\mathcal{P}, w)$ is an induced subgraph of $G^*(\mathcal{P}, w')$ if and only if w is a factor of w'.

The graph $G^*(\mathcal{P}, w)$ constructed from a periodic word w is called a *periodic graph*. The period of w is called the period of $G^*(\mathcal{P}, w)$. To construct periodic antichains, we break the periodicity on both ends of the graph (word) by inserting an appropriate prefix and suffix. The following conjecture was proposed in [9] and was inspired by Theorem 4.

Conjecture 2. There is a function $f : \mathbb{N} \to \mathbb{N}$ such that the class X of graphs defined by a finite collection F of forbidden induced subgraphs is well-quasi-ordered by the induced subgraph relation if and only if X contains no periodic infinite antichains of period at most $f(t(F))$, where $t(F)$ stands for the total number of vertices of graphs in F.

To support this conjecture, let us mention the following decidability problem, which was recently solved in [8]: given a finite collection F of graphs, decide whether the speed of the class of F-free graphs is above or below the Bell number. The solution is based on a characterization of minimal classes with speeds above the Bell number by means of almost periodic words.

Definition 1. *A word w is* almost periodic *if for any factor f of w there is a constant k_f such that any factor of w of size at least k_f contains f as a factor.*

A jump to the Bell number for hereditary graph properties was identified in [12]. This paper distinguishes classes of graphs of two types: the classes where a certain graph parameter, called in [8] the *distinguishing number*, is finite and the classes where this parameter is infinite. For the case where the distinguishing number is infinite, the paper [12] provides a complete description of minimal classes above the Bell number, of which there are precisely 13. In the case where this parameter is finite, the family of minimal classes is infinite and all of them have been characterized in [8] via the notion of almost periodic words as follows.

Let A be a finite alphabet and $\mathcal{P}$ a symmetric decoder, i.e. a decoder containing with each pair (a_i, a_j) the pair (a_j, a_i). Also, let w be a word over A and $G^*(\mathcal{P}, w)$ the letter graph of w distinguishing between consecutive and non-consecutive letters, as defined earlier. Finally, let $X^*(\mathcal{P}, w)$ be the class of graphs containing all induced subgraphs of $G^*(\mathcal{P}, w)$.

Theorem 5. ([8]) *Let X be a hereditary class of graphs with a finite distinguishing number. Then X is a minimal class of speed above the Bell number if and only if there exists an infinite almost periodic word w over a finite alphabet and a symmetric decoder $\mathcal{P}$ such that $X = X^*(\mathcal{P}, w)$.*

In [8], it was shown that for hereditary classes defined by a *finite* collection F of minimal forbidden induced subgraphs, the word "almost" can be omitted from this theorem. Moreover, the period of w in this case is bounded by a function of $t(F)$, where $t(F)$, as before, is the total number of vertices of graphs in F. This leads to a procedure deciding the Bell number for hereditary graph properties, as was shown in [8].

Interestingly, the same procedure decides well-quasi-ordering by induced subgraphs for classes with a finite distinguishing number, as was recently shown in [10]. In other words, this result verifies Conjecture 2 for classes with a finite distinguishing number.

We conclude this section by observing that Theorem 5 brings us back from graphs to words, and it also makes a bridge to the next section, where the speed of a hereditary graph property is an important issue.

3 Representing Graphs by Words

Representing graphs by words in a finite alphabet, or graph coding, is important in computer science for representing graphs in computer memory [20,23,37]. Without loss of generality we will assume that our alphabet is binary.

For a class X of graphs, we denote by X_n the set of graphs in X with the vertex set $\{1, 2, \ldots, n\}$. *Coding* of graphs in the class X is a family of bijective mappings $\Phi = \{\phi_n : n = 1, 2, 3, \ldots\}$, where $\phi_n : X_n \to \{0, 1\}^*$. A coding Φ is called *asymptotically optimal* if[1]

$$\lim_{n \to \infty} \frac{\max\limits_{G \in X_n} |\phi_n(G)|}{\log |X_n|} = 1.$$

---

[1] All logarithms are of base 2.

Every labelled graph G with n vertices can be represented by a binary word of length $\binom{n}{2}$, one bit per each pair of vertices, with 1 standing for an edge and 0 for an non-edge. Such a word can be obtained by reading the elements of the adjacency matrix above the main diagonal. The word obtained by reading these elements row by row, starting with the first row, is called the *canonical code* of G and is denoted $\phi_n^c(G)$.

If no a priori information about the graph is available, then $\binom{n}{2}$ is the minimum number of bits needed to represent the graph. However, if we know that our graph possesses some special properties, then this knowledge may lead to a shorter representation. For instance,

- if we know that our graph is bipartite, then we do not need to describe the adjacency of vertices that belong to the same part in its bipartition. Therefore, we need at most $n^2/4$ bits to describe the graph, the worst case being a bipartite graph with $n/2$ vertices in each of its parts.
- if we know that our graph is not an arbitrary bipartite graph but *chordal* bipartite, then we can further shorten the code and describe any graph in this class with at most $O(n \log^2 n)$ bits [36].
- a further restriction to trees (a proper subclass of chordal bipartite graphs) enables us to further shorten the code to $(n-2) \log n$ bits, which is the length of binary representation of the Prüfer code for trees [30].

How much can the canonical representation be shortened for graphs with a property X? For hereditary properties this question can be answered through the notion of entropy.

3.1 Entropy of Hereditary Properties

In order to represent graphs in a class X, we need at least $|X_n|$ different binary words. Therefore, in the worst case the length of a binary code of an n-vertex graph in X cannot be shorter than $\log |X_n|$. Thus, the ratio $\log |X_n|/\binom{n}{2}$ can be viewed as the coefficient of compressibility for representing n-vertex graphs in X. Its limit value, for $n \to \infty$, was called by Alekseev in [4] the *entropy* of X. Moreover, in the same paper Alekseev showed that for every hereditary property X the entropy necessarily exists and in [5] he proved that its value takes the following form:

$$\lim_{n \to \infty} \frac{\log |X_n|}{\binom{n}{2}} = 1 - \frac{1}{k(X)}, \tag{1}$$

where $k(X)$ is a natural number, called the *index* of X. To define this notion let us denote by $\mathcal{E}_{i,j}$ the class of graphs whose vertices can be partitioned into at most i independent sets and j cliques. In particular, $\mathcal{E}_{2,0}$ is the class of bipartite graphs and $\mathcal{E}_{1,1}$ is the class of split graphs. Then $k(X)$ is the largest k such that X contains $\mathcal{E}_{i,j}$ with $i + j = k$. Independently, this result was obtained by Bollobás and Thomason [13,14] and is known nowadays as the Alekseev-Bollobás-Thomason Theorem (see e.g. [7]).

3.2   Coding of Graphs in Classes of High Speed

In [4], Alekseev proposed a universal algorithm which gives an asymptotically optimal coding for graphs in every hereditary class X of index $k > 1$, i.e. of non-zero entropy. Below we present an adapted version of this algorithm.

Let $n > 1$ be a natural number and let p be a prime number between $\lfloor n/\sqrt{\log n} + 1 \rfloor$ and $2\lfloor n/\sqrt{\log n} \rfloor$. Such a number always exists by the Bertrand-Chebyshev theorem (see e.g. [1]). Define $k = \lfloor n/p \rfloor$. Then

$$p \le 2n/\sqrt{\log n}, \quad k \le \sqrt{\log n}, \quad n - kp < p. \tag{2}$$

Let G be an arbitrary graph with n vertices. Denote by D_n the set of all pairs of vertices of G. We split D_n into two disjoint subsets R_1 and R_2 as follows: R_1 consists of the pairs (a, b) such that $a \le kp$, $b \le kp$ and $\lfloor (a-1)/p \rfloor \ne \lfloor (b-1)/p \rfloor$, and R_2 consists of all the remaining pairs. Let us denote by $\mu^{(1)}$ the binary word consisting of the elements of the canonical code corresponding to the pairs of R_2. This word will be included in the code of G we construct.

Now let us take care of the pairs in R_1. For all $x, y \in \{0, 1, \dots, p-1\}$, we define

$$Q_{x,y} = \{pi + 1 + res_p(xi + y) \ i = 0, 1, \dots, k-1\},$$

where $res_p(z)$ is the remainder on dividing z by p. Let us show that every pair of R_1 appears in exactly one set $Q_{x,y}$. Indeed, if $(a, b) \in Q_{x,y}$ $(a < b)$, then

$$xi_1 + y \equiv a \ (\mod p), \quad xi_2 + y \equiv b \ (\mod p),$$

where $i_1 = \lfloor (a-1)/p \rfloor$, $i_2 = \lfloor (b-1)/p \rfloor$. Since $i_1 \ne i_2$ (by definition of R_1), there exists a unique solution of the following system

$$\begin{aligned} x(i_1 - i_2) &\equiv a - b \ (\mod p) \\ y(i_1 - i_2) &\equiv ai_2 - bi_1 \ (\mod p). \end{aligned} \tag{3}$$

Therefore, by coding the graphs $G_{x,y}$ induced by $Q_{x,y}$ and combining their codes with the word $\mu^{(1)}$ (that describes the pairs in R_2) we obtain a complete description of G.

To describe the graphs $G_{x,y}$ induced by $Q_{x,y}$ we first relabel their vertices according to

$$z \to \lfloor (z-1)/p \rfloor + 1.$$

In this way, we obtain p^2 graphs $G'_{x,y}$, each on the vertex set $\{1, 2, \dots, k\}$. Some of these graphs may coincide. Let m $(m \le p^2)$ denote the number of pairwise different graphs in this set and $H_0, H_1, \dots, H_{m-1}$ an (arbitrarily) ordered list of m pairwise different graphs in this set. In other words, for each graph $G'_{x,y}$ there is a unique number i such that $G'_{x,y} = H_i$. We denote the binary representation of this number i by $\omega(x, y)$ and the length of this representation by ℓ, i.e. $\ell = \lceil \log m \rceil$. Also, denote

$$\mu^{(2)} = \phi_k^c(H_0)\phi_k^c(H_1)\dots\phi_k^c(H_{m-1}),$$

$$\mu^{(3)} = \omega(0,0)\omega(0,1)\ldots\omega(0,p-1)\omega(1,0)\ldots\omega(p-1,p-1).$$

The word $\mu^{(2)}$ describes all graphs H_i and the word $\mu^{(3)}$ indicates for each pair x, y the interval in the word $\mu^{(2)}$ containing the information about $G'_{x,y}$. Therefore, the words $\mu^{(2)}$ and $\mu^{(3)}$ completely describe all graphs $G_{x,y}$. In order to separate the word $\mu^{(2)}\mu^{(3)}$ into $\mu^{(2)}$ and $\mu^{(3)}$, it suffices to know the number ℓ, because $|\mu^{(2)}| = \ell p^2$ and the number p is uniquely defined by n. Since $m \leq 2^{\binom{k}{2}}$, the number ℓ can be described by at most

$$\lceil \log \ell \rceil = \lceil \log\lceil \log m \rceil \rceil \leq \left\lceil \log \binom{k}{2} \right\rceil \leq \lceil \log k^2 \rceil \leq \lceil \log \log n \rceil$$

binary bits. Let $\mu^{(0)}$ be the binary representation of the number ℓ of length $\lceil \log \log n \rceil$, and let

$$\phi_n^*(G) = \mu^{(0)}\mu^{(1)}\mu^{(2)}\mu^{(3)}, \qquad \Phi^* = \{\phi_n^* \ n = 2, 3, \ldots\}.$$

Theorem 6. [4] Φ^* *is an asymptotically optimal coding for any hereditary class* X *with* $k(X) > 1$.

3.3 Representing Graphs in Hereditary Classes of Low Speed

The universal algorithm presented in the previous section is, unfortunately, not optimal for classes X of index $k(X) = 1$, also known as *unitary classes*, since equation (1) does not provide the asymptotic behavior of $\log|X_n|$ in this case. This is unfortunate, because the family of unitary classes contains a variety of properties of theoretical or practical importance, such as line graphs, interval graphs, permutation graphs, threshold graphs, forests, planar graphs and, even more generally, all proper minor-closed graph classes, all classes of graphs of bounded vertex degree, of bounded tree- and clique-width, etc.

A systematic study of hereditary properties of low speed was initiated by Scheinerman and Zito in [32]. In particular, they distinguished the first four lower layers in the family of unitary classes: constant (classes X with $|X_n| = \Theta(1)$), polynomial ($|X_n| = n^{\Theta(1)}$), exponential ($|X_n| = 2^{\Theta(n)}$) and factorial ($|X_n| = n^{\Theta(n)}$). Independently, similar results have been obtained by Alekseev in [6]. Moreover, Alekseev described the set of minimal classes in all the four lower layers and the asymptotic structure of properties in the first three of them. A more detailed description of the polynomial and exponential layers was obtained by Balogh, Bollobás and Weinreich in [11]. However, the factorial layer remains largely unexplored and the asymptotic structure is known only for properties at the bottom of this layer, below the Bell numbers [11,12]. On the other hand, the factorial properties constitute the core of the unitary family, as all the interesting classes mentioned above (and many others) are factorial.

We conclude the paper with an important conjecture, which deals with representing graphs in the factorial classes.

Definition 2. *A representation of an n-vertex graph G is said to be* implicit *if it assigns to each vertex of G a binary code of length $O(\log n)$ so that the adjacency of two vertices is a function of their codes.*

This notion was introduced in [23], where the authors identified a variety of graph classes admitting an implicit representation. Clearly, not every class admits an implicit representation, since a bound on the total length of the code implies a bound on the number of graphs admitting such a representation. More precisely, only classes containing $2^{O(n \log n)}$ graphs with n vertices can admit an implicit representation. This restriction, however, is not sufficient to represent graphs implicitly. A simple counter-example can be found in [37]. This example deals with a non-hereditary graph property, which leaves the question of implicit representation for *hereditary* classes of speed $2^{O(n \log n)}$ open. These are precisely the classes with at most factorial speed of growth. It is known that every hereditary class with a sub-factorial speed admits an implicit representation [33]. For hereditary classes with factorial speeds the question of implicit representation is generally open and is known as the implicit graph representation conjecture (see e.g. [37]).

Conjecture 3. Any hereditary class of speed $n^{\Theta(n)}$ admits an implicit representation.

References

1. Aigner, M., Ziegler, G.M.: Proofs from THE BOOK, 4th edn. Springer, Berlin (2010). https://doi.org/10.1007/978-3-642-00856-6
2. Albert, M.H., Atkinson, M.D., Bouvel, M., Ruskuc, N., Vatter, V.: Geometric grid classes of permutations. Trans. Am. Math. Soc. **365**, 5859–5881 (2013)
3. Alecu, B., Lozin, V., Zamaraev, V., de Werra, D.: Letter graphs and geometric grid classes of permutations: characterization and recognition. In: Brankovic, L., Ryan, J., Smyth, W.F. (eds.) IWOCA 2017. LNCS, vol. 10765, pp. 195–205. Springer, Cham (2018). https://doi.org/10.1007/978-3-319-78825-8_16
4. Alekseev, V.E.: Coding of graphs and hereditary classes. Probl. Cybernet. **39**, 151–164 (1982). (in Russian)
5. Alekseev, V.E.: Range of values of entropy of hereditary classes of graphs. Diskret. Mat. **4**(2), 148–157 (1992). (in Russian; translation in Discrete Mathematics and Applications, 3 (1993), no. 2, 191–199)
6. Alekseev, V.E.: On lower layers of a lattice of hereditary classes of graphs. Diskretn. Anal. Issled. Oper. Ser. **4**(1), 3–12 (1997). (Russian)
7. Alon, N., Balogh, J., Bollobás, B., Morris, R.: The structure of almost all graphs in a hereditary property. J. Comb. Theory B **79**, 131–156 (2011)
8. Atminas, A., Collins, A., Foniok, J., Lozin, V.: Deciding the Bell number for hereditary graph properties. SIAM J. Discrete Math. **30**, 1015–1031 (2016)
9. Atminas, A., Lozin, V., Moshkov, M.: WQO is decidable for factorial languages. Inf. Comput. **256**, 321–333 (2017)
10. Atminas, A., Brignall, R.: Well-quasi-ordering and finite distinguishing number. https://arxiv.org/abs/1512.05993
11. Balogh, J., Bollobás, B., Weinreich, D.: The speed of hereditary properties of graphs. J. Comb. Theory Ser. B **79**(2), 131–156 (2000)
12. Balogh, J., Bollobás, B., Weinreich, D.: A jump to the Bell number for hereditary graph properties. J. Comb. Theory B **95**, 29–48 (2005)
13. Bollobás, B., Thomason, A.: Projections of bodies and hereditary properties of hypergraphs. Bull. London Math. Soc. **27**, 417–424 (1995)

14. Bollobás, B., Thomason, A.: Hereditary and monotone properties of graphs. In: Graham, R.L., Nesetril, J. (eds.) The Mathematics of Paul Erdős, II. AC, vol. 14, pp. 70–78. Springer, Berlin (1997). https://doi.org/10.1007/978-3-642-60406-5_7

15. Brignall, R., Ruškuc, N., Vatter, V.: Simple permutations: decidability and unavoidable substructures. Theor. Comput. Sci. **391**, 150–163 (2008)

16. Cherlin, G.L., Latka, B.J.: Minimal antichains in well-founded quasi-orders with an application to tournaments. J. Comb. Theory B **80**, 258–276 (2000)

17. Chvátal, V., Hammer, P.L.: Aggregation of inequalities in integer programming. Ann. Discrete Math. **1**, 145–162 (1977)

18. Elizalde, S.: The X-class and almost-increasing permutations. Ann. Comb. **15**, 51–68 (2011)

19. Finkel, A., Schnoebelen, P.: Well-structured transition systems everywhere!. Theor. Comput. Sci. **256**, 63–92 (2001)

20. Galperin, H., Wigderson, A.: Succinct representations of graphs. Inf. Control **56**, 183–198 (1983)

21. Higman, G.: Ordering by divisibility of abstract algebras. Proc. London Math. Soc. **2**, 326–336 (1952)

22. Hine, N., Oxley, J.: When excluding one matroid prevents infinite antichains. Adv. Appl. Math. **45**, 74–76 (2010)

23. Kannan, S., Naor, M., Rudich, S.: Implicit representation of graphs. SIAM J. Discrete Math. **5**, 596–603 (1992)

24. Korpelainen, N., Lozin, V.V.: Two forbidden induced subgraphs and well-quasi-ordering. Discrete Math. **311**, 1813–1822 (2011)

25. Kruskal, J.B.: Well-quasi-ordering, the tree theorem, and Vazsonyi's conjecture. Trans. Am. Math. Soc. **95**, 210–225 (1960)

26. Kruskal, J.B.: The theory of well-quasi-ordering: a frequently discovered concept. J. Comb. Theory A **13**, 297–305 (1972)

27. de Luca, A., Varricchio, S.: Well quasi-orders and regular languages. Acta Informatica **31**, 539–557 (1994)

28. Mahadev, N.V.R., Peled, U.N.: Threshold graphs and related topics. Annals of Discrete Mathematics, vol. 56. North-Holland, Amsterdam (1995)

29. Petkovšek, M.: Letter graphs and well-quasi-order by induced subgraphs. Discrete Math. **244**, 375–388 (2002)

30. Prüfer, H.: Neuer Beweis eines Satzes über Permutationen. Arch. Math. Phys. **27**, 742–744 (1918)

31. Robertson, N., Seymour, P.D.: Graph minors. XX. Wagner's conjecture. J. Comb. Theory Ser. B **92**, 325–357 (2004)

32. Scheinerman, E.R., Zito, J.: On the size of hereditary classes of graphs. J. Comb. Theory Ser. B **61**(1), 16–39 (1994)

33. Scheinerman, E.R.: Local representations using very short labels. Discrete Math. **203**, 287–290 (1999)

34. Sonderegger, M.: Applications of graph theory to an English rhyming corpus. Comput. Speech Lang. **25**, 655–678 (2011)

35. Spielman, D.A., Bóna, M.: An infinite antichain of permutations. Electron. J. Comb. **7**, 2 (2000)

36. Spinrad, J.P.: Nonredundant 1's in Γ-free matrices. SIAM J. Discrete Math. **8**, 251–257 (1995)

37. Spinrad, J.P.: Efficient Graph Representations. Fields Institute Monographs, vol. 19, xiii+342 pp. American Mathematical Society, Providence (2003)

38. Vatter, V., Waton, S.: On partial well-order for monotone grid classes of permutations. Order **28**, 193–199 (2011)

Automata

An Oracle Hierarchy
for Small One-Way Finite Automata

M. Anabtawi, S. Hassan, C. Kapoutsis$^{(\boxtimes)}$, and M. Zakzok

Carnegie Mellon University in Qatar, Doha, Qatar
{maleka,sabith,cak,mzakzok}@qatar.cmu.edu

Abstract. We introduce a *polynomial-size oracle hierarchy* for one-way finite automata. In it, a problem is in level k (resp., level 0) if itself or its complement is solved by a polynomial-size nondeterministic finite automaton with access to an oracle for a problem in level $k-1$ (resp., by a polynomial-size deterministic finite automaton with no oracle access). This is a generalization of the *polynomial-size alternating hierarchy* for one-way finite automata, as previously defined using polynomial-size alternating finite automata with a bounded number of alternations; and relies on an original definition of what it means for a nondeterministic finite automaton to access an oracle, which we carefully justify. We prove that our hierarchy is strict; that every problem in level k is solved by a deterministic finite automaton of $2k$-fold exponential size; and that level 1 already contains problems beyond the entire alternating hierarchy. We then identify five restrictions to our oracle-automaton, under which the oracle hierarchy is proved to coincide with the alternating one, thus providing an oracle-based characterization for it. We also show that, given all others, each of these restrictions is necessary for this characterization.

1   Introduction

In 2009, a plan was proposed to develop a *size-complexity theory for two-way finite automata* [3] (or *"minicomplexity theory"* [4]), by analogy to the standard time-complexity theory for Turing machines. An important part of that plan was to introduce and study a *"polynomial-size hierarchy"*, as direct analogue of the *polynomial-time hierarchy*, i.e., of the well-studied hierarchy

$$\mathsf{P} \subseteq (\Sigma_1\mathsf{P} \cap \Pi_1\mathsf{P}) \underset{\subsetneq}{\overset{\subsetneq}{\lessgtr}} \genfrac{}{}{0pt}{}{\Sigma_1\mathsf{P} = \mathsf{NP}}{\Pi_1\mathsf{P} = \mathsf{coNP}} \underset{\subsetneq}{\overset{\subsetneq}{\lessgtr}} (\Sigma_2\mathsf{P} \cap \Pi_2\mathsf{P}) \underset{\subsetneq}{\overset{\subsetneq}{\lessgtr}} \genfrac{}{}{0pt}{}{\Sigma_2\mathsf{P}}{\Pi_2\mathsf{P}} \cdots \subseteq \mathsf{PH} = \bigcup_{k \geq 1} \Sigma_k\mathsf{P},$$

where $\Sigma_k\mathsf{P}$ (resp., $\Pi_k\mathsf{P}$) is the class of problems decidable in polynomial time by alternating Turing machines (ATMs) which start in an existential (resp., universal) state and alternate fewer than k times between the existential and universal mode; and P, NP, coNP have their standard meanings. By direct transposition to minicomplexity, the following hierarchy was described

© Springer Nature Switzerland AG 2019
C. Martín-Vide et al. (Eds.): LATA 2019, LNCS 11417, pp. 57–69, 2019.
https://doi.org/10.1007/978-3-030-13435-8_4

$$2\mathsf{D} \subseteq (2\Sigma_1 \cap 2\Pi_1) \overset{\subsetneq}{\underset{\subsetneq}{}} \underset{2\Pi_1}{\overset{2\Sigma_1 = 2\mathsf{N}}{}} \overset{\subsetneq}{\underset{\subsetneq}{}} (2\Sigma_2 \cap 2\Pi_2) \overset{\subsetneq}{\underset{\subsetneq}{}} \underset{2\Pi_2}{\overset{2\Sigma_2}{}} \overset{\subsetneq}{\underset{\subsetneq}{}} \cdots \subseteq 2\mathsf{H} = \bigcup_{k\geq 1} 2\Sigma_k,$$

where $2\Sigma_k$ (resp., $2\Pi_k$) is the class of minicomplexity problems decidable by two-way alternating finite automata (2AFAs) with polynomially many states, an existential (resp., universal) start state, and fewer than k alternations; and, as usual, 2D and 2N are the respective classes for deterministic and nondeterministic automata (2DFAs and 2NFAs) with polynomially many states [9]. Soon later, Geffert proved that this hierarchy is strict above the first level [2].

When it was introduced [7,10], the polynomial-time hierarchy was defined not by ATMs, but by nondeterministic Turing machines with oracles (oracle-NTMs): each $\Sigma_k\mathsf{P}$ consisted of every problem which is solved by a polynomial-time NTM with an oracle for a problem in $\Sigma_{k-1}\mathsf{P}(\Sigma_k\mathsf{P} := \mathsf{NP}^{\Sigma_{k-1}\mathsf{P}})$, while $\Sigma_0\mathsf{P}$ was P; and each $\Pi_k\mathsf{P}$ consisted of the complements of problems in $\Sigma_k\mathsf{P}$ ($\Pi_k\mathsf{P} := \mathsf{co}\Sigma_k\mathsf{P}$). Only later [1], was it proved that this natural "polynomial-time *oracle* hierarchy" can also be viewed, as above, as the "polynomial-time *alternating* hierarchy" that we get from polynomial-time ATMs when we bound their number of alternations.

With this in mind, one observes that in minicomplexity, too, the "polynomial-size hierarchy" above is just the "polynomial-size *alternating* hierarchy" that we get from polynomial-size 2AFAs when we bound their number of alternations. One then wonders what the "polynomial-size *oracle* hierarchy" is: What is a natural definition of it? Does it, too, coincide with its alternating counterpart? [3]

Here, we study the special case where all automata involved are one-way. We introduce a "*one-way polynomial-size oracle hierarchy*" as an analogue of the polynomial-time oracle hierarchy, and compare it with the "*one-way polynomial-size alternating hierarchy*" of the classes $1\Sigma_k, 1\Pi_k, 1\mathsf{H}$, defined like $2\Sigma_k, 2\Pi_k, 2\mathsf{H}$ above but for one-way automata (1AFAs). We quickly find that, contrary to what is true for polynomial-time Turing machines, our oracle hierarchy is far stronger than its alternating counterpart: level 1 already contains problems outside the entire 1H (Theorem 2). So we ask: (i) What is really the power of this new hierarchy? and (ii) Are there restrictions under which it matches the alternating one?

For (i), we show that levels 2 and 3 contain even harder problems, beyond the power of polynomial-size alternating and Boolean finite automata (Theorem 2); that the hierarchy stands strict (Theorem 5); and that every problem in level k is solved by a 1DFA of $2k$-fold exponential size, and thus the full hierarchy lies within the power of 1DFAs of elementary size (Theorem 4). For (ii), we identify five restrictions to our underlying oracle-automaton model and prove that: under all five, our hierarchy coincides with the alternating one, thus providing an oracle-based characterization for it (Theorem 6); and that, in the presence of the other four, each of these restrictions is necessary for this characterization (Theorem 7).

Prior Work. Just as the polynomial-time oracle hierarchy is based on oracle-NTMs, our hierarchy rests on an analogous model of an "*oracle-1NFA*": a nondeterministic one-way finite automaton that can query an oracle. This can also be viewed as a "1NFA *reduction*" (just as an oracle-NTM can be viewed as a *nondeterministic Turing reduction*, of the problem solved by the machine to the problem solved by the oracle), and is then called *many-one*, if every computation branch makes exactly 1 query and then halts with the received answer as its own. Easily, many-one oracle-1NFAs are just *one-way nondeterministic finite transducers*.

Oracle-1NFAs have been studied before. In [8], their many-one restriction is used in defining oracle hierarchies over the context-free languages. In [11], the full-fledged variant is used for the same purpose, this time as a restriction of the stronger model of an oracle-NPDA (i.e., a one-way nondeterministic pushdown automaton with an oracle). Our work differs from these studies in three fundamental respects. First, we do not study computability on the involved automata ("*Can a problem be solved?*"), but complexity ("*Can a solvable problem be solved with few states?*"). Second, our hierarchy is not built on context-free problems but on regular problems that are solved by polynomial-size one-way deterministic finite automata. Finally, and most crucially, what the oracle in our model reads at the moment of query is not just the string x printed on the query tape, but the string uxv, where u and v the prefix and suffix of the input determined by the current head position. This makes our oracle-1NFAs far more powerful than the previous variants—and we do justify why this extra power is natural.

2   Preparation

If S is a finite set, then $|S|$ and $\overline{S}$ are its size and complement. If $h \geq 0$, then $[h] := \{0, 1, \ldots, h - 1\}$, and $[\![h]\!] := \{\alpha \mid \alpha \subseteq [h]\}$. If $k \geq 0$, then $\exp_k(h)$ is the k-fold exponential function in h: $\exp_0(h) = h$, $\exp_1(h) = 2^h$, $\exp_2(h) = 2^{2^h}$, etc.

If Σ is an alphabet, then Σ^* is all finite strings over it. If $z \in \Sigma^*$ is a string, then $|z|$, z_i, and z^R are its length, i-th symbol (if $1 \leq i \leq |z|$), and reverse $z_{|z|} \cdots z_2 z_1$. If $L \subseteq \Sigma^*$ is a set of strings, then $L^R := \{z^R \mid z \in L\}$ is its reverse.

Problems. A *(promise) problem* over Σ is a pair $\mathfrak{L} = (L, \tilde{L})$ of disjoint subsets of Σ^*. Its *positive* and *negative instances* are all $w \in L$ and all $w \in \tilde{L}$. A machine *solves* $\mathfrak{L}$ if it accepts every $w \in L$ but no $w \in \tilde{L}$. If $\tilde{L} = \overline{L}$, then $\mathfrak{L}$ is a *language*.

Let $h \geq 1$. Over the alphabet $[\![h]\!]$, the problem of *set equality* is: "Given two sets α, β, check that they are equal" [4]. Formally, this is the promise problem:

$$\mathrm{SEQ}_h := \big(\{\alpha\beta \mid \alpha = \beta\}, \{\alpha\beta \mid \alpha \neq \beta\}\big) = \{\alpha\beta \mid \alpha, \beta \subseteq [h] \mid \alpha = \beta\}.$$

Note that the input is promised to be only 2 symbols, each an entire subset of $[h]$. Also note our notation {*format* | *condition* | *test*}, where *format* and *condition* specify all instances, and *test* distinguishes the positive ones.

Over the alphabet $[\![h]\!] \cup \{\check{\alpha} \mid \alpha \subseteq [h]\}$, where each set α induces two symbols, α and $\check{\alpha}$ (a ticked variant), we define the following extensions of SEQ_h:

$$\exists\mathrm{SEQ}_h := \{\ \check{\alpha}\beta_1\beta_2 \cdots \beta_t \qquad \mid t \geq 0 \mid (\exists i)(\alpha = \beta_i)\ \}$$
$$\mathrm{OSEQ}_h := \{\ \check{\alpha}_L \beta_1 \beta_2 \cdots \beta_t \check{\alpha}_R \mid t \geq 0 \mid (\exists i)(\exists j)(i < j\ \&\ \beta_i = \alpha_R\ \&\ \beta_j = \alpha_L)\ \}.$$

We also need the variant of $\exists\mathrm{SEQ}_h$ where the sequence $\beta_1\beta_2 \cdots \beta_t$ is replaced by a single symbol, for the collection of all sets appearing in it; and, for each $k \geq 0$, the *equality* problem over the k-fold exponential alphabet:

$$\text{COMPACT } \exists\text{SEQ}_h := \{ \, \breve{\alpha}B \mid \alpha \subseteq [h] \, \& \, B \subseteq [\![h]\!] \mid \alpha \in B \, \}$$
$$\text{EQ}[k]_h := \{ \, ab \mid a, b \in [\exp_k(h)] \mid a = b \, \}.$$

The *complement* and *reverse* of $\mathfrak{L} = (L, \tilde{L})$ are the problems $\neg\mathfrak{L} := (\tilde{L}, L)$ and $\mathfrak{L}^{\text{R}} := (L^{\text{R}}, \tilde{L}^{\text{R}})$. The *conjunctive* and *disjunctive star* of $\mathfrak{L}$ are the problems:

$$\bigwedge\mathfrak{L} := \{ \, x_1 \# x_2 \# \cdots \# x_t \mid t \geq 0 \, \& \, (\forall i)(x_i \in L \cup \tilde{L}) \mid (\forall i)(x_i \in L) \, \}$$
$$\bigvee\mathfrak{L} := \{ \, x_1 \# x_2 \# \cdots \# x_t \mid t \geq 0 \, \& \, (\forall i)(x_i \in L \cup \tilde{L}) \mid (\exists i)(x_i \in L) \, \},$$

where # is a fresh delimiter. Using these, we build the sequence of problems:

$$\text{GEF}[2]_h := (\exists\text{SEQ}_h)^{\text{R}} \qquad \text{and} \qquad \text{for all } k \geq 3: \quad \text{GEF}[k]_h := \bigvee\neg\text{GEF}[k-1]_h \quad (1)$$

(mimicking Geffert's witnesses $E_{k,h}$ from [2]), and then let $\text{GEF}[\omega]_h := \text{GEF}[h]_h$.

Note that each problem above is really a *family of problems*: $\text{SEQ} = (\text{SEQ}_h)_{h \geq 1}$, $\text{EQ}[k] = (\text{EQ}[k]_h)_{h \geq 1}$, etc. To operate on a family $\mathcal{L} := (\mathfrak{L}_h)_{h \geq 1}$ means to operate on its individual components: $\neg\mathcal{L} := (\neg\mathfrak{L}_h)_{h \geq 1}$, $\mathcal{L}^{\text{R}} := (\mathfrak{L}_h^{\text{R}})_{h \geq 1}$, etc.

Machines. A *one-way alternating finite automaton* (1AFA) is any sextuple $A = (S, \Sigma, \delta, q_{\text{s}}, q_{\text{a}}, U)$, where S is a set of *states*, Σ is an *alphabet*, $q_{\text{s}}, q_{\text{a}} \in S$ are the *start* and *accept* states, $U \subseteq S$ and $S \setminus U$ are the *universal* and *existential* states, and $\delta \subseteq S \times (\Sigma \cup \{\vdash, \dashv\}) \times S \times \{\text{S}, \text{R}\}$ is a *transition relation*, with $\vdash, \dashv \notin \Sigma$ being the two endmarkers and S, R the head instructions "stay" and "move right".

An input $w \in \Sigma^*$ is presented to A as $\vdash w \dashv$. The computation starts in q_{s} on $\vdash$. In each step, a next state and head instruction are derived from δ and the current state and symbol. We never violate $\dashv$, except if the next state is q_{a}. So, every computation branch either *loops*, if it repeats a state on a cell; or *hangs*, if it reaches a state and symbol with no tuple in δ; or *accepts*, if it falls off $\dashv$.

Formally, a *configuration* is a pair $(q, i) \in S \times \{0, 1, \ldots, |w| + 2\}$, representing A being in q on cell i (hosting $\vdash$, w_i, $\dashv$, or nothing, if i is respectively 0, from 1 to $|w|$, $|w| + 1$, or $|w| + 2$). The resulting *computation* is a tree of configurations, with root $(q_{\text{s}}, 0)$ and each node's children determined by δ and w. A node (q, i) is *accepting* if $(q, i) = (q_{\text{a}}, |w| + 2)$; or $q \in U$ and all children are accepting; or $q \notin U$ and at least one child is accepting. If the root is accepting, then A accepts w.

If $U = \emptyset$, then A is *nondeterministic* (1NFA). If moreover δ has at most one tuple $(q, a, ., .)$ for all q, a, then A is *deterministic* (1DFA).

Our 1AFAs are essentially the two-way alternating finite automata of [5] minus the ability to reverse. They differ substantially from the 1AFAs of [1], where each state-symbol pair maps to a Boolean function. We call those automata *Boolean* (1BFAs). Clearly, 1AFAs restrict 1BFAs to the two functions AND and OR.

Complexity. For $k \geq 1$, $1\Sigma_k$ is all (families of) problems solvable by (families of) polynomially-large 1AFAs with existential start and fewer than k alternations [4]:

$$1\Sigma_k := \left\{ (\mathfrak{L}_h)_{h \geq 1} \; \middle| \; \begin{array}{l} \text{there exist 1AFAs } (M_h)_{h \geq 1} \text{ and polynomial } s \\ \text{such that every } M_h \text{ solves } \mathfrak{L}_h \text{ with } s(h) \text{ states, an} \\ \text{existential start state, and fewer than } k \text{ alternations} \end{array} \right\};$$

and similarly for $1\Pi_k$ (resp., 1A), except that the start is universal (resp., the alternations are unbounded). If the M_h are 1DFAs, 1NFAs, or 1BFAs, then we get 1D, $1N = 1\Sigma_1$ [9], and 1B. Easily: $\neg SEQ \in 1\Sigma_1$; $\exists SEQ^R \in 1\Sigma_2$.

The following inclusions and equalities in the *one-way polynomial-size alternating hierarchy* are easy to verify (the strictness of the inclusions is from [2,9]):

$$1D \subsetneq (1\Sigma_1 \cap 1\Pi_1) \begin{smallmatrix} \subsetneq \\ \subset \end{smallmatrix} \begin{matrix} 1\Sigma_1 = 1N \\ 1\Pi_1 = co1N \end{matrix} \subseteq (1\Sigma_2 \cap 1\Pi_2) \begin{smallmatrix} \subsetneq \\ \subset \end{smallmatrix} \begin{matrix} 1\Sigma_2 = co1\Pi_2 \\ 1\Pi_2 = co1\Sigma_2 \end{matrix} \subseteq \cdots \subseteq 1H,$$

where $1H := \bigcup_{k \geq 1} 1\Sigma_k$. Note the standard notation $coC := \{\neg L \mid L \in C\}$ for the complements of problems in C. Similarly, $reC := \{L^R \mid L \in C\}$. Also, the classes 2^C, $\exp_k(C)$, and e^C are defined like C, except now the automata are larger than in C by an exponential, k-fold exponential, or elementary function.[1]

The next facts are easy or known ([6, Corollary 1]), except for $GEF[\omega] \in 1A \setminus 1H$ and for $(COMPACT \; \exists SEQ)^R \notin 1A$ (which are of independent interest).

Theorem 1. $1H \subsetneq 1A \subsetneq 1B = re2^{1D} \subsetneq 2^{1N} \subsetneq 2^{2^{1D}}$ *and the inclusions' strictness is witnessed respectively by* $GEF[\omega]$, $(COMPACT \; \exists SEQ)^R$, $\neg EQ[2]$, *and* $EQ[2]$.

3   Oracle-1NFAs

Intuitively, an oracle-1NFA is a 1NFA which can pause its computation to ask an oracle whether an instance of a problem $\mathfrak{X}$ is positive; receive an answer YES/NO; and resume accordingly. For example, such a device with $O(1)$ states and an oracle for SEQ_h should be able to implement the following nondeterministic algorithm $\mathcal{A}$ for $OSEQ_h$: "Scan the input $\breve{\alpha}_L \beta_1 \cdots \beta_t \breve{\alpha}_R$ and guess a β_i; ask whether $\beta_i = \alpha_R$; if NO, reject (in this branch); otherwise, resume scanning and guess a β_j; ask whether $\beta_j = \alpha_L$; if NO, reject (in this branch); otherwise, accept."

In prior work [8,11], an oracle-1NFA N is a 1NFA augmented by a write-only, one-way query tape. At every step of its usual computation on the input, N may also append a string to the query tape. Also at every step, N may enter a query state q where, in a single step: (i) the oracle (for a language X) reads the string x on the query tape and answers $a := YES/NO$ based on whether $x \in X$ or not; (ii) the query tape is reset; and (iii) N resumes its computation from q based on a. In the end, N^X *solves* a language L if it accepts exactly all $w \in L$.

This model is a natural restriction of an oracle-NTM to a single, read-only, one-way tape, and serves well the purposes of [8,11]. But does it serve well *our* purposes? For example, can it implement our earlier algorithm $\mathcal{A}$? Unfortunately, the answer is no. The natural approach starts by scanning the input $\breve{\alpha}_L \beta_1 \cdots \beta_t \breve{\alpha}_R$, guessing a β_i, and printing it, as a first step towards preparing the first query to the oracle (is $\beta_i = \alpha_R$?). To complete this query, however, we must then print α_R, to form the SEQ_h instance $\beta_i \alpha_R$; for this, we must know α_R; hence, we must resume scanning until α_R. But, by then, we have consumed the input, and we cannot implement the rest of $\mathcal{A}$ (i.e., guess a β_j and ask whether $\beta_j = \alpha_L$).

---

[1] A function is *elementary* if it is the composition of finitely many arithmetic operations, exponentials, logarithms, constants, and solutions of algebraic equations.

We conclude that our oracle-1NFAs must be stronger than in [8,11]. We choose to enhance them as follows: Whenever the machine enters a query state, the oracle reads not just the current string x on the query tape, but the string uxv, where u is the prefix of the input w up to (and not including) the current cell i of the input head, and v the suffix of w from the i-th cell onwards (so that $w = uv$).

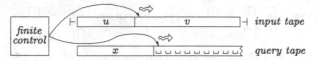

With this modification, $\mathcal{A}$ is now easy to implement: "Scan the input, *guess* a β_i, print $\triangleright$, and enter a query state; at this point, $uxv = \breve{\alpha}_\mathrm{L}\beta_1 \cdots \triangleright \beta_i \cdots \beta_t\breve{\alpha}_\mathrm{R}$ and the query is whether the set after $\triangleright$ equals the right ticked set ($\beta_i = \alpha_\mathrm{R}$). If NO, reject. Otherwise, resume scanning, *guess* a β_j, print $\triangleleft$, and enter a query state; now $uxv = \breve{\alpha}_\mathrm{L}\beta_1 \cdots \triangleleft\beta_j \cdots \beta_t\breve{\alpha}_\mathrm{R}$ and the query is whether the set after $\triangleleft$ equals the left ticked set ($\beta_j = \alpha_\mathrm{L}$). If NO, then reject; else accept."

Hence, the enhanced model supports our initial intuition. Note, however, that we had to modify the oracle. Instead of SEQ_h, it now answers for the "marked" variant MSEQ: "Given a string $\breve{\alpha}_\mathrm{L}\beta_1 \cdots *\beta_i \cdots \beta_t\breve{\alpha}_\mathrm{R}$ where $* \in \{\triangleright, \triangleleft\}$, check that $\beta_i = \alpha_\mathrm{R}$, if $* = \triangleright$; or that $\beta_i = \alpha_\mathrm{L}$, if $* = \triangleleft$." Still, this only modifies the formatting of the instance (the sets to compare are not alone, but marked in a list) and not the essence of the test. Such variations should be tolerated.

Definition 1. An oracle-1NFA is a tuple $N = (S, Q, \Sigma, \Gamma, \delta, q_\mathrm{s}, q_\mathrm{a})$, where S, Q are disjoint sets of *control* and *query states*; $q_\mathrm{s}, q_\mathrm{a} \in S$ are the *start* and *accept states*; Σ, Γ are the *input* and *query alphabets*; and δ is the *transition relation*:

$$\delta \subseteq (S \cup Q) \times (\Sigma \cup \{\vdash, \dashv\} \cup \{\mathrm{Y}, \mathrm{N}\}) \times (S \cup Q) \times \{\mathrm{S}, \mathrm{R}\} \times \Gamma^*$$

with $\vdash, \dashv \notin \Sigma$ being the two endmarkers; $\mathrm{Y}, \mathrm{N} \notin \Sigma \cup \{\vdash, \dashv\}$ being the two oracle answers; and S, R being the two head instructions.

For any $w \in \Sigma^*$ and $\mathfrak{X} = (X, \tilde{X})$ over Γ, the *computation of $N^{\mathfrak{X}}$ on w* starts in q_s with the input tape containing $\vdash w \dashv$; the input head reading $\vdash$; every query tape cell containing the blank symbol $\sqcup \notin \Gamma$; and the query head reading the leftmost $\sqcup$. In every step, N examines the current state p and *action symbol* a:

- If $p \in S$, then $a \in \Sigma \cup \{\vdash, \dashv\}$ is the symbol read by the input head.
- If $p \in Q$, then $a = \mathrm{Y}$ or N, based on whether $uxv \in X$ or $uxv \in \tilde{X}$, where $x \in \Gamma^*$ is the string on the query tape; and $u = w_1 \cdots w_{i-1}$ and $v = w_i \cdots w_{|w|}$ are the partition of w by the current head position $1 \le i \le |w| + 1$.

For these p, a and every $(p, a, q, d, y) \in \delta$, the machine: switches to state q; has its input head stay, if $d = \mathrm{S}$, or advance, if $d = \mathrm{R}$; and appends y to the query tape. If $p \in Q$ then, before y is appended, the query tape is reset to blank.

If more than one $(p, a, ., ., .) \in \delta$ exist, then N nondeterministically follows all of them in different branches; if no such $(p, a, ., ., .) \in \delta$ exists, then the current branch *hangs*. If the input head ever falls off $\dashv$ (into q_a), then the current branch *accepts*; if the input head stays forever on a cell, then the current branch *loops*. We say that $N^{\mathfrak{X}}$ *accepts* w if at least one of the branches accepts.

We say N *respects* $\mathfrak{X}$ *on problem* $\mathfrak{L} = (L, \tilde{L})$ if every query uxv asked in every computation of $N^{\mathfrak{X}}$ on every $w \in L \cup \tilde{L}$ is an instance of $\mathfrak{X}$. If so and $N^{\mathfrak{X}}$ accepts w for all $w \in L$ but no $w \in \tilde{L}$, then $N^{\mathfrak{X}}$ *solves* $\mathfrak{L}$. Finally, N is *deterministic* (oracle-1DFA) if δ has at most 1 tuple $(p, a, ., ., .)$, for all p, a. ○

4   The One-Way Polynomial-Size Oracle Hierarchy

We now introduce oracle-based minicomplexity classes, and then the hierarchy.

Definition 2. Let $\mathcal{X} = (\mathfrak{X}_h)_{h \geq 1}$ be a family of problems. The class of families of problems which are "*solvable by small* 1NFAS *with access to* $\mathcal{X}$" is:

$$1N^{\mathcal{X}} := \left\{ (\mathfrak{L}_h)_{h \geq 1} \,\middle|\, \begin{array}{l} \text{there exist oracle-1NFAs } (N_h)_{h \geq 1} \text{ and polynomials } s, q \\ \text{such that every } N_h \text{ has } s(h) \text{ states, and } N_h^{\mathfrak{X}_{q(h)}} \text{ solves } \mathfrak{L}_h \end{array} \right\}.$$

If $\mathcal{X}$ is arbitrary in a class C of problem families, we write $1N^C := \bigcup_{\mathcal{X} \in C} 1N^{\mathcal{X}}$. If the N_h are actually deterministic, then we write $1D^{\mathcal{X}}$ and $1D^C$.   ○

E.g., the earlier implementation of $\mathcal{A}$ shows that OSEQ is in $1N^{\text{MSEQ}}$; and thus also in $1N^{1\Pi_1}$, since MSEQ $\in 1\Pi_1$. Now, our hierarchy is built as follows:

Definition 3. The *one-way polynomial-size oracle hierarchy* is the collection of classes: $1\hat{\Delta}_0 = 1\hat{\Sigma}_0 = 1\hat{\Pi}_0 := 1D$ and, for all $k \geq 0$:

$$1\hat{\Delta}_{k+1} := 1D^{1\hat{\Sigma}_k}, \qquad 1\hat{\Sigma}_{k+1} := 1N^{1\hat{\Sigma}_k}, \qquad 1\hat{\Pi}_{k+1} := \text{co}1\hat{\Sigma}_{k+1}.$$

We also let $1\hat{H} := \bigcup_{k \geq 0} 1\hat{\Sigma}_k$.   ○

Easily, $1D \subseteq 1\hat{\Delta}_k \subseteq 1\hat{\Sigma}_k, 1\hat{\Pi}_k \subseteq 1\hat{\Delta}_{k+1} \subseteq 1\hat{H}$ for all k; therefore, this is indeed a hierarchy. So, e.g., our earlier algorithm $\mathcal{A}$ implies that OSEQ $\in 1\hat{\Sigma}_2$.

This hierarchy is quite strong: its first, second, and third levels already contain problems which are respectively outside 1H, 1A, and 1B (cf. Theorem 1.)

Theorem 2. GEF[ω] $\in 1\hat{\Delta}_1$ *and* (COMPACT $\exists$SEQ)$^R \in 1\hat{\Delta}_2$ *and* EQ[2] $\in 1\hat{\Delta}_3$.

At the same time, the hierarchy is not arbitrarily strong, as we now show.

Lemma 3. *If* $\mathfrak{L}$ *is solved by* $N^{\mathfrak{X}}$, *where* N *is an* s-*state oracle-1NFA and* $\mathfrak{X}$ *a problem solved by an* r-*state* 1DFA, *then* $\mathfrak{L}$ *is solved by an* 1AFA *with* $O(sr^3)$ *states.*

Proof. Let $N = (S, Q, \Sigma, \Gamma, ., q_s, .)$. Let $D = (R, \Gamma, ., p_s, ., \emptyset)$ be a 1DFA for $\mathfrak{X}$ with $|R| = r$. We build a 1AFA A for $\mathfrak{L}$. On every input w, A simulates $N^{\mathfrak{X}}$ on w.

To see how A works, fix w and follow one branch of the computation of $N^{\mathfrak{X}}$ on w. Using existential states, A can easily simulate N along that branch up to the time t when N asks the first query. At that point, the input head partitions w into a prefix u and suffix v, and the query tape contains a string x. To continue the simulation, A needs to know the oracle's answer a, i.e., whether uxv is in X or $\tilde{X}$ (since N respects $\mathfrak{X}$ on $\mathfrak{L}$, there is no third case), i.e., whether D accepts uxv. But knowing this at time t is impossible, as A has not seen v yet.

To overcome this problem, A guesses and stores a value $\tilde{a} \in \{Y, N\}$ for a and then branches universally into: (1) a deterministic branch $D(\tilde{a})$, which uses D to verify that $\tilde{a}$ is correct; and (2) an alternating tree $N(\tilde{a})$, which resumes the simulation of $N^{\mathfrak{X}}$ on w assuming $\tilde{a}$ as the oracle's answer. This way, A accepts iff the actual answer a (i.e., the value $\tilde{a}$ for which (1) accepts) leads the rest of the simulation to accept. Let us see how each of (1) and (2) is implemented.

(1) To verify $\tilde{a}$, the branch $D(\tilde{a})$ determines whether D accepts uxv. For this, it deterministically simulates D from $p' := D(ux)$ (i.e., D's state after reading $\vdash ux$) on v (which, at time t, is the rest of the input). On $\dashv$, $D(\tilde{a})$ accepts iff $\tilde{a} = Y$ and the simulation accepted; or $\tilde{a} = N$ and the simulation rejected.

Of course, this requires knowing p' at time t. For this, during its computation on u, A performs two more simulations, parallel to its simulation of N on u:

(i) A simulation D of D on $\vdash u$ (from p_s); this way, at time t, A knows $p := D(u)$.

(ii) A piecemeal simulation $\tilde{D}$ of D from $p = D(u)$ on x, which is advanced every time the simulation of N on u produces the next piece of x; this way, at time t, A knows the required state p', since $p_s \dashv \vdash u \mapsto p \dashv x \mapsto p'$.

Note, however, that N starts producing x strictly before it finishes reading u; so, $p = D(u)$ is needed by $\tilde{D}$ before it is produced by D. To overcome this problem, A implements $\tilde{D}$ as follows: First, it guesses and stores a value $\tilde{p}$ for p; then simulates D on x from $\tilde{p}$, eventually producing a corresponding value $\tilde{p}'$ for p' at time t; then, it compares the stored $\tilde{p}$ with the actual p produced by D. If $\tilde{p} \neq p$, then the guess was wrong, and this branch rejects; otherwise, the guess was correct, and thus $\tilde{p}'$ is safe to use as $p' = D(ux)$ at the start of $D(\tilde{a})$.

(2) The branch $N(\tilde{a})$ works recursively. Using existential states, A simulates N along a branch of its computation after the first query has been answered by $\tilde{a}$ and up to the time when N asks the next query. At that point, it again guesses a value in $\{Y, N\}$ for the answer and branches universally into a deterministic branch which verifies the correctness of the guessed answer; and an alternating tree which resumes the computation assuming that answer. And so on.

Formally, A uses states of two different forms: $(\tilde{a}, \tilde{p}; p, \tilde{p}', q) \in \{Y, N\} \times R^3 \times S$ and $(\tilde{a}, p) \in \{Y, N\} \times R$. In every state $(\tilde{a}, \tilde{p}; p, \tilde{p}', q)$:

p is the current state of the simulation D of D from p_s on w. By advancing D every time an input symbol is read, A guarantees that p is always $D(u)$, where u the current prefix of w that has been read.

$\tilde{p}'$ is the current state of the simulation $\tilde{D}$ of D from $\tilde{p}$ (the second component) on the string x which is produced for the query tape by the simulation of N between times t_{i-1} (when the $(i-1)$-st query is asked; or the computation starts, if $i = 1$) and t_i (when the i-th query is asked). By advancing $\tilde{D}$ every time an infix of x is produced, A guarantees that, at time t_i, $\tilde{p}'$ is the state produced by D when run from $\tilde{p}$ on x. So, if $p = \tilde{p}$ also holds at t_i, then $\tilde{p}'$ is the state $D(ux)$ of D after reading the prefix $\vdash ux$ of the i-th query.

q is the current state of the simulation $N(\tilde{a}_1, \ldots, \tilde{a}_{i-1})$ of one branch of N on w and on the guessed values $\tilde{a}_1, \ldots, \tilde{a}_{i-1}$ for the answers to all queries so far.

Right after the $(i-1)$-st query is answered (or the computation starts, if $i = 1$), A is at a state $(. , . ; p, . , q)$, where p, q the current states of simulations D and N (if $i = 1$, this is the start state $(Y, p_s; p_s, p_s, q_s)$).

Then, A guesses and stores two values $\tilde{a}_* \in \{Y, N\}$ and $\tilde{p}_* \in R$, for the answer to the i-th query and D's state at the time t_i of that query. Then, using states $(\tilde{a}_*, \tilde{p}_*; p, \tilde{p}', q)$, it continues D (in p) and N (in q) on the incoming input symbols; and starts $\tilde{D}$ (in $\tilde{p}'$, from $\tilde{p}' = \tilde{p}_*$) on the query symbols produced by N.

At time t_i, the input head partitions w into a prefix u and suffix v, and the query symbols produced by N since t_{i-1} form a string x. Then A checks whether $\tilde{p}_* = p$, i.e., the guess for $D(u)$ was correct. If not, then A rejects (in this branch). Otherwise, the actual state $D(ux)$ is $\tilde{p}'$ and thus, to check whether $\tilde{a}_*$ is correct, A can simulate D from $\tilde{p}'$ on the suffix v. So, A universally splits into:

- A simulation $D(\tilde{a}_*)$ which, using states $(\tilde{a}_*, p)$ (starting with $p = \tilde{p}'$), simulates D from $\tilde{p}'$ on v and accepts iff $\tilde{a}_*$ matches D's final decision.
- A continuation of the simulations D and N from state $(\tilde{a}_*, . ; p, . , q)$, which first uses $\tilde{a}_*$ as oracle answer to advance N into a state q' while keeping D in p; then resumes from $(. , . ; p, . , q')$ towards the $(i+1)$-st query (by guessing new $\tilde{a}_*, \tilde{p}^*$, restarting $\tilde{D}$, etc.). If no such query occurs, then $D, \tilde{D}$ are ignored and the branch accepts iff $N(\tilde{a}_1, \ldots, \tilde{a}_{i-1}, \tilde{a}_*)$ falls off $\dashv$ into the accept state.

Overall, from the two forms of states, we see A has size $2r^3 s + 2r = O(sr^3)$. $\square$

Now, by Lemma 3 and $1A \subseteq 2^{1N}$ (Theorem 1), we gradually get the following upper bounds. In the end, 1DFAs of elementary size contain the entire hierarchy.

Theorem 4. (i) $1\hat{\Sigma}_1, 1\hat{\Pi}_1 \subseteq 1A$.   (iii) *For* $k \geq 0$: $1\hat{\Sigma}_k, 1\hat{\Pi}_k \subseteq \exp_{2k}(1D)$.
(ii) *For* $k \geq 1$: $1\hat{\Sigma}_k \subseteq \exp_{2k-1}(1N)$.   (iv) $1\hat{H} \subseteq e^{1D}$.

Finally, we prove that our hierarchy is strict, in the sense that successive levels are distinct. We arrive at this gradually, in the next theorem. All parts are straightforward, except (ii), which generalizes the last part of Theorem 2.

Theorem 5. *The following statements are true:*
(i) *For* $k \geq 0$: $1\hat{\Sigma}_k = 1\hat{\Sigma}_{k+1} \implies 1\hat{\Sigma}_k = 1\hat{H}$.   (iii) *For* $k \geq 1$: $1\hat{\Sigma}_k \subsetneq 1\hat{\Delta}_{4k-1}$.
(ii) *For* $k \geq 1$: $\text{EQ}[k] \in 1\hat{\Delta}_{2k-1}$.   (iv) *For* $k \geq 0$: $1\hat{\Sigma}_k \subsetneq 1\hat{\Sigma}_{k+1}$.

Proof (sketch). The proof of (i) is standard; and (iv) follows from (i) and (iii).

In (iii), the witness is $\text{EQ}[2k]$, which is in $1\hat{\Delta}_{4k-1}$, by (ii); but not in $1\hat{\Sigma}_k$, by Theorem 4 (ii) and because $\text{EQ}[2k] \notin \exp_{2k-1}(1N)$ (by a generalization of $\text{SEQ} \notin 1N$).

For (ii), we first introduce two auxiliary problems, for each $k \geq 0$:

- $\mathcal{Z}_k = (\mathfrak{Z}_{k,h})_{h \geq 1}$: "Given a string of the form $w \triangleright ab$, where $a, b \in [\exp_k(h)]$ and w contains no $\triangleright$, check that $a = b$."

– $\mathcal{Y}_k = (\mathfrak{Y}_{k,h})_{h \geq 1}$: "Given a string of the form $w\#u\#u'\#v\#v'$ where u, u', v, v' are strings over $[\exp_k(h)]$ and w contains no #, check that some symbol in u appears in v'; or some symbol in u' appears in v."

We then prove that (a) $\mathcal{Y}_k \in 1\mathrm{N}^{\mathcal{Z}_k}$ for $k \geq 0$; and (b) $\mathrm{EQ}[k] \in 1\mathrm{D}^{\mathcal{Y}_{k-1}}$, for $k \geq 1$. Then, (ii) follows from (a) and (b) with a straightforward induction on k.   □

5   A Characterization for the Alternating Hierarchy

We now identify a set of restrictions to our oracle-1NFAs from Definition 1, under which the polynomial-size oracle hierarchy coincides with the alternating one.

Definition 4. An oracle-1NFA is called:

1. *many-one*: if it makes at most 1 query, and then accepts iff the answer is YES;
2. *synchronous*: if it prints on the oracle tape only right before asking a query;
3. *laconic*: if, at every query, the query tape contains at most 1 symbol;
4. *omitting*: if, at every query with input partition u, v and query tape string x, the query string omits the prefix u, and is thus only xv;
5. *query-deterministic*: if, at every query, for every possible prior state-symbol-move triple, there is at most 1 string that can be printed on the query tape.

If N satisfies all of these restrictions, then it is a *weak* oracle-1NFA.   ∘

Definition 5. The classes $(1\mathrm{N})^{\mathcal{X}}$ and $(1\mathrm{N})^{C}$ are defined exactly as the respective classes in Definition 2, except that now all oracle-1NFAs are weak.   ∘

Definition 6. The *one-way polynomial-size weak-oracle hierarchy* consists of the classes: $1\check{\Sigma}_0 = 1\check{\Pi}_0 := 1\mathrm{D}$ and, for all $k \geq 0$:

$$1\check{\Sigma}_{k+1} := (1\mathrm{N})^{\mathrm{co}1\check{\Sigma}_k} \quad \text{and} \quad 1\check{\Pi}_{k+1} := \mathrm{co}1\check{\Sigma}_{k+1}.$$

We also let $1\check{\mathrm{H}} := \bigcup_{k \geq 0} 1\check{\Sigma}_k$.   ∘

We now prove the promised characterization of the alternating hierarchy.

Theorem 6. *For all $k \geq 1$: $1\check{\Sigma}_k = 1\Sigma_k$ and $1\check{\Pi}_k = 1\Pi_k$.*

Proof. Since $1\check{\Pi}_k = \mathrm{co}1\check{\Sigma}_k$ (by definition) and $1\Pi_k = \mathrm{co}1\Sigma_k$, it is enough to prove $1\check{\Sigma}_k = 1\Sigma_k$. We use induction on k. (Always let $\mathcal{L} = (\mathfrak{L}_h)_{h \geq 1}$ and $\mathcal{X} = (\mathfrak{X}_h)_{h \geq 1}$.)

   Base case $(k = 1)$: By definition, $1\check{\Sigma}_1 = (1\mathrm{N})^{\mathrm{co}1\mathrm{D}} = (1\mathrm{N})^{1\mathrm{D}}$ and $1\Sigma_1 = 1\mathrm{N}$. Obviously $(1\mathrm{N})^{1\mathrm{D}} \supseteq 1\mathrm{N}$; so, we just prove $(1\mathrm{N})^{1\mathrm{D}} \subseteq 1\mathrm{N}$. Pick any $\mathcal{L} \in (1\mathrm{N})^{1\mathrm{D}}$. Then $\mathcal{L} \in (1\mathrm{N})^{\mathcal{X}}$ for some $\mathcal{X} \in 1\mathrm{D}$. So, every $\mathfrak{L}_h$ is solved by an $s(h)$-state weak oracle-1NFA N with access to an oracle for $\mathfrak{X}_{q(h)}$, which is solved by an $r(q(h))$-state 1DFA D, for polynomial s, q, r. We build a 1NFA N' which simulates $N^{\mathfrak{X}_{q(h)}}$.

   On input w, N' starts simulating some branch of N on w. If N ever enters a query state q, then this is the only query in the branch and the branch will halt immediately, accepting iff the answer is YES (since N is many-one), namely iff

D accepts bv, where b the symbol on the transition into q (since N is synchronous and laconic) and v the current remaining suffix of w (since N is omitting). So, to find out this branch's decision, N' just simulates D on bv. Equivalently, N' just simulates D from $q_b := D(b)$ on the rest of the input. This is clearly doable using a copy of the states of D. Overall, N' is of size $O(s(h) + r(q(h))) = \mathrm{poly}(h)$.

Inductive step $(k \geq 1)$. Assume $1\check{\Sigma}_k = 1\Sigma_k$. We prove that $1\check{\Sigma}_{k+1} = 1\Sigma_{k+1}$.

$[\subseteq]$ Let $\mathcal{L} \in 1\check{\Sigma}_{k+1}$. By Definition 6, the inductive hypothesis, and the known fact $\mathrm{col}\Sigma_k = 1\Pi_k$, this means $\mathcal{L} \in (1\mathrm{N})^{1\Pi_k}$, i.e., $\mathcal{L} \in (1\mathrm{N})^{\mathcal{X}}$ for some $\mathcal{X} \in 1\Pi_k$. So, $\mathcal{L}_h$ is solved by an $s(h)$-state weak oracle-1NFA N with access to an oracle for $\mathcal{X}_{q(h)}$, which is solved by an $r(q(h))$-state 1AFA U with universal start and fewer than k alternations, for polynomial s, q, r. We build a 1AFA A which simulates $N^{\mathcal{X}_{q(h)}}$.

On input w, A existentially simulates some branch of N on w until (if ever) N is ready to make a transition $\tau = (p, c, q, d, b)$ into a query state q. As in the base case (and since 1–4 in Definition 4 hold), A can then find out the branch's decision by running a simulation U of U on bv, where u, v is the current partition of w. Note that c is the first symbol of v, if $d = \mathrm{S}$; or the last symbol of u, if $d = \mathrm{R}$.

An important difference is that now there is not a unique q_b to serve as starting state for simulating U on the rest of the input. Instead, A must run U on $\vdash b$ more carefully. For as long as U is within $\vdash b$, A keeps its input head on the last cell before executing τ, i.e., the cell containing c; consumes no input symbols at all; and stores in its state the string $\vdash b$ itself and the current state and position $i \in \{0, 1\}$ of U within $\vdash b$. Only when U exits from $\vdash b$ into v, does A start reading input symbols again, to continue U on the rest of the input.

But how does A store $\vdash b$? The naive approach, of storing b, works only if the query alphabet has $\mathrm{poly}(h)$ symbols. But this is not guaranteed. So, A works differently: it stores the states p, q and direction d of the transition $\tau = (p, c, q, d, b)$; then, from p, q, d (stored) and c (current input symbol), it finds b as the last component of the unique (since N is query-deterministic) transition $(p, c, q, d, \cdot)$.

Overall, A starts existentially, performs fewer than $k + 1$ alternations (as U has fewer than k), and uses $O(s(h) + s(h)^2 \cdot r(q(h)) + r(q(h))) = \mathrm{poly}(h)$ states.

$[\supseteq]$ Let $\mathcal{L} \in 1\Sigma_{k+1}$. Then $\mathcal{L}_h$ is solved by a 1AFA $A_h = A = (S, \Sigma, ., ., q_s, ., U)$ with fewer than $k + 1$ alternations, $q_s \notin U$, $|S| = \mathrm{poly}(h)$. By [2, Theorem 2.3], we may assume that S is partitioned into sets $S_{k+1}, S_k, \ldots, S_1$ such that $q_s \in S_{k+1}$; $U = S_k \cup S_{k-2} \cup \cdots$; and every alternation switches from a state in S_i to a state in S_j for some $i > j$. We build a weak oracle-1NFA N and a problem $\mathcal{X}_h$ such that $N^{\mathcal{X}_h}$ simulates A.

Intuitively, N simulates a branch of A for as long as A stays in S_{k+1}. If it ever enters a state $q \notin S_{k+1}$, then N prints symbol q and enters query state q (N is synchronous; laconic; and query-deterministic, as every $(., ., q, ., y)$ uniquely has $y = q$), to ask (now the query string is qv, where v the current remaining suffix; so N is omitting) if the computation of A from q on v is accepting, and accepts iff the answer is YES (N is many-one). This correctly simulates A.

To implement this, we use $N = (S_{k+1}, (S \setminus S_{k+1}), \Sigma, (S \setminus S_{k+1}), ., q_s, .)$ with oracle $\mathcal{X}_h := $ "Given $qv \in (S \setminus S_{k+1})\Sigma^*$, check that the computation of A_h from q

on v is accepting." Easily, $\mathfrak{X}_h$ is solved by a 1AFA A'_h which, on input qv, simply simulates A_h from q on v. This A'_h can be made to start universally, makes fewer than k alternations, and has at most $|S| = \mathrm{poly}(h)$ states. So, $\mathcal{X} \in 1\Pi_k$. By $1\Pi_k = \mathrm{co}1\Sigma_k$ and the inductive hypothesis, this means $\mathcal{X} \in \mathrm{co}1\check{\Sigma}_k$. Overall, $\mathfrak{L}_h$ is solved by the weak oracle-1NFA N with $|S_{k+1}| = \mathrm{poly}(h)$ states and an oracle for $\mathcal{X}_h$. So, $\mathcal{L} \in (1\mathrm{N})^{\mathcal{X}}$. Since $\mathcal{X} \in \mathrm{co}1\check{\Sigma}_k$, we have $\mathcal{L} \in (1\mathrm{N})^{\mathrm{co}1\check{\Sigma}_k} = 1\check{\Sigma}_{k+1}$.   □

Finally, we prove that all five restrictions in Definition 4 are necessary for Theorem 6: lifting any of them results in oracle-1NFAs which are strictly stronger than weak. We say an oracle-1NFA is *i-weak* if it satisfies Definition 4 except for restriction i.

Theorem 7. *For all $i \in \{1, \ldots, 5\}$, there exists $j \in \{1, 2\}$ such that small i-weak oracle-1NFAs accessing $1\Pi_j$ are strictly stronger than weak ones accessing $1\Pi_j$.*

Proof (sketch). Small weak oracle-1NFAs with access to some $1\Pi_j$ cannot solve (COMPACT ∃SEQ)$^\mathrm{R}$—or else the problem would be in 1H, and thus in 1A, contradicting Theorem 1. However, the problem is solved by small 3-weak oracle-1NFAs with access to $1\Pi_2$; and by small 5-weak oracle-1NFAs with access to $1\Pi_1$.

Small weak oracle-1NFAs with access to some $1\Pi_j$ cannot solve ∃SEQ—or else the problem would be in 1H, and so in $\mathrm{re}2^{1\mathrm{D}}$ (Theorem 1), so $\exists\mathrm{SEQ}^\mathrm{R} \in 2^{1\mathrm{D}}$, which is false (by a pigeonhole argument, $\exists\mathrm{SEQ}_h^\mathrm{R}$ needs $\geq 2^{2^h}$ states on a 1DFA). But 2-weak and 4-weak small oracle-1NFAs with access to $1\Pi_1$ can solve the problem.

Small weak oracle-1NFAs with access to $1\Pi_1$ cannot solve SEQ $\wedge$ $(\bigwedge \neg\mathrm{SEQ})$ = "Given $x\#y$, where x an instance of SEQ and y an instance of $\bigwedge\neg\mathrm{SEQ}$, check that both instances are positive."—or else the problem is in $1\Sigma_2$, which we can prove is false (by an involved proof of independent interest). However, the problem is solved by small 1-weak oracle-1NFAs with the same access.   □

6   Conclusion

We introduced a new oracle-1NFA model, studied the polynomial-size hierarchy induced by it, and used it to characterize the one-way polynomial-size alternating hierarchy as a special case. A natural next step is to repeat this study for two-way automata. Before that, the question $1\hat{\Sigma}_k \overset{?}{=} 1\hat{\Pi}_k$ is still open, for all $k \geq 1$. Another direction is to investigate whether our oracle hierarchy can host analogues of known theorems about the polynomial-time hierarchy (e.g., $\mathsf{BPP} \subseteq \Sigma_2\mathsf{P}$).

References

1. Chandra, A.K., Kozen, D.C., Stockmeyer, L.J.: Alternation. J. ACM **28**(1), 114–133 (1981)
2. Geffert, V.: An alternating hierarchy for finite automata. Theoret. Comput. Sci. **445**, 1–24 (2012)
3. Kapoutsis, C.A.: Size complexity of two-way finite automata. In: Diekert, V., Nowotka, D. (eds.) DLT 2009. LNCS, vol. 5583, pp. 47–66. Springer, Heidelberg (2009). https://doi.org/10.1007/978-3-642-02737-6_4

4. Kapoutsis, C.: Minicomplexity. J. Automata Lang. Comb. **17**(2–4), 205–224 (2012)
5. Ladner, R.E., Lipton, R.J., Stockmeyer, L.J.: Alternating pushdown and stack automata. SIAM J. Comput. **13**(1), 135–155 (1984)
6. Leiss, E.: Succinct representation of regular languages by Boolean automata. Theoret. Comput. Sci. **13**, 323–330 (1981)
7. Meyer, A.R., Stockmeyer, L.J.: The equivalence problem for regular expressions with squaring requires exponential space. In: Proceedings of the Symposium on Switching and Automata Theory, pp. 125–129 (1972)
8. Reinhardt, K.: Hierarchies over the context-free languages. In: Dassow, J., Kelemen, J. (eds.) IMYCS 1990. LNCS, vol. 464, pp. 214–224. Springer, Heidelberg (1990). https://doi.org/10.1007/3-540-53414-8_44
9. Sakoda, W.J., Sipser, M.: Nondeterminism and the size of two-way finite automata. In: Proceedings of STOC, pp. 275–286 (1978)
10. Stockmeyer, L.J.: The polynomial-time hierarchy. Theoret. Comput. Sci. **3**, 1–22 (1977)
11. Yamakami, T.: Oracle pushdown automata, nondeterministic reducibilities, and the hierarchy over the family of context-free languages. In: Geffert, V., Preneel, B., Rovan, B., Štuller, J., Tjoa, A.M. (eds.) SOFSEM 2014. LNCS, vol. 8327, pp. 514–525. Springer, Cham (2014). https://doi.org/10.1007/978-3-319-04298-5_45

Orbits of Abelian Automaton Groups

Tim Becker[1][(✉)] and Klaus Sutner[2]

[1] Department of Computer Sciences, University of Wisconsin-Madison,
1210 W. Dayton St., Madison, WI 53706, USA
tbecker@cs.wisc.edu
[2] Computer Science Department, Carnegie Mellon University, 5000 Forbes Avenue,
Pittsburgh, PA 15213, USA
sutner@cs.cmu.edu

Abstract. Automaton groups are a class of self-similar groups generated by invertible finite-state transducers [11]. Extending the results of Nekrashevych and Sidki [12], we describe a useful embedding of abelian automaton groups into a corresponding algebraic number field, and give a polynomial time algorithm to compute this embedding. We apply this technique to study iteration of transductions in abelian automaton groups. Specifically, properties of this number field lead to a polynomial-time algorithm for deciding when the orbits of a transduction are a rational relation. These algorithms were implemented in the SageMath computer algebra system and are available online [2].

Keywords: Automaton groups · Embedding · Number field · Orbits · Rational relation · Classification

1 Introduction

An *invertible binary transducer* $\mathcal{A}$ is a Mealy automaton over the binary alphabet where each state has an invertible output function. The transductions of $\mathcal{A}$ are therefore length-preserving invertible functions on binary strings. These transductions (along with their inverses) naturally generate a group under composition, denoted $\mathcal{G}(\mathcal{A})$. Such groups, over a general alphabet, are called automaton groups or self-similar groups; these groups have been studied in great detail, see [6,11] for extensive studies.

Automaton groups have many interesting properties and are capable of surprising complexity. A number of well-known groups can be generated by fairly simple transducers, indicating that transducers may be a useful semantic interpretation for many groups. Bartholdi's recent book review in the Bulletin of the AMS about the relationship between syntactic and semantic approaches to algebra gives some examples where transducers play such a role [1]. For instance, after Grigorchuk famously solved the long-open problem of finding a group of intermediate growth, it was realized that his group can be generated by the 5-state invertible binary transducer. In fact, even 3-state invertible binary transducers

© Springer Nature Switzerland AG 2019
C. Martín-Vide et al. (Eds.): LATA 2019, LNCS 11417, pp. 70–81, 2019.
https://doi.org/10.1007/978-3-030-13435-8_5

generate groups which are exceedingly complicated, see [3] for a classification of all such automata.

Here, we will be primarily concerned with a simpler class of transducers: those which generate abelian groups. This situation has been previously studied in [13,17]. It is known that all abelian automaton groups are either boolean or free abelian [12], and in the free abelian case, one can show that the underlying automata have nice structural properties. We will summarize and build upon these results in Sect. 2.2. A running example in this paper will be the transducer CC_2^3, shown in Fig. 1. This transducer generates a group isomorphic to $\mathbb{Z}^2$ and is perhaps the simplest nontrivial transducer generating an abelian group [18].

We will make connections between abelian automaton groups and other areas of algebra that will provide useful insight into their structure and complexity. A result of Nekrashevych and Sidki shows that these groups admit embeddings into $\mathbb{Z}^m$ where transitions in the transducer correspond to affine maps [12]. In this paper, we describe a related embedding of abelian automaton groups into associated algebraic number fields and describe how these may be efficiently computed.

Properties of this embedding can be used to study computational problems arising in automaton groups. We demonstrate this by studying the complexity of iteration of transductions. Given a transduction $f \in \mathcal{G}(\mathcal{A})$, we write $f^* \subseteq \mathbf{2}^* \times \mathbf{2}^*$ for the transitive closure of f. Note that f^* is a length-preserving equivalence relation on $\mathbf{2}^*$. The complexity of this relation was studied in [18], where it was shown that a certain class of abelian transductions have rational orbit relations. We will refer to such transductions as *orbit-rational*. In [4], the authors study a related problem in a generalization of autmaton groups. In Proposition 6 they demonstrate a sequential automatic algebra with an orbit relation which is not context-free. Our results in this paper in a similar spirit. We give a precise characterization of orbit-rational abelian transducers and a corresponding decision procedure.

Throughout this paper, we will utilize the theory of free abelian groups, linear algebra, field theory, and some algebraic number theory. See [7] for the necessary material on the latter subjects, and [8,16] for background on algebraic number theory.

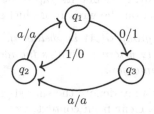

Fig. 1. The cycle-cum-chord transducer CC_2^3

2   Background

2.1   Automata and Automaton Groups

A *binary transducer* is a Mealy automaton of the form $\mathcal{A} = \langle Q, \mathbf{2}, \delta, \lambda \rangle$ where Q is a finite state set, $\delta : Q \times \mathbf{2} \to Q$ is the transition function, and $\lambda : Q \times \mathbf{2} \to \mathbf{2}$ is the output function. Such a machine is *invertible* if for each state $q \in Q$, the output function $\lambda(q, \cdot)$ is a permutation of $\mathbf{2}$. A state q is called a *toggle* state if $\lambda(q, \cdot)$ is the transposition and a *copy* state otherwise. We define the transduction of q, $\underline{q} : \mathbf{2}^* \to \mathbf{2}^*$ recursively as follows: $\underline{q}(\epsilon) = \epsilon$ and $\underline{q}(a \cdot w) = \lambda(q, a) \cdot \underline{\delta(q, a)}(w)$, where ϵ denotes the empty string, $\cdot$ denotes concatenation, and $a \in \mathbf{2}$. Note that invertibility of transductions follows from invertibility of the transition functions. We let a boldface character, e.g. $\mathbf{a}$ denote an element of $\mathbf{2}$, and $\overline{\mathbf{a}}$ denote the flipped character. The inverse machine $\mathcal{A}^{-1}$ is computed by simply flipping the edge labels of $\mathcal{A}$: if $p \xrightarrow{\mathbf{a}/\mathbf{b}} q$ in $\mathcal{A}$ then $p^{-1} \xrightarrow{\mathbf{b}/\mathbf{a}} q^{-1}$ in $\mathcal{A}^{-1}$.

Invertible transducers define a subclass of automaton groups. The group $\mathcal{G}(\mathcal{A})$ is formed by taking all transductions and their inverses under composition. As described in [18] the group $\mathcal{G}(\mathcal{A})$ can be seen as a subgroup of the automorphism group of the infinite binary tree, denoted $\mathbf{Aut}(\mathbf{2}^*)$. Clearly any automorphism $f \in \mathbf{Aut}(\mathbf{2}^*)$ can be written in the form $f = (f_0, f_1)\pi$ where $\pi \in S_2$. Here π describes the action of f on the root, and f_0 and f_1 are the automorphisms induced by f on the two subtrees. We call $(f_0, f_1)\pi$ the *wreath representation* of f; this name is derived from the fact that $\mathbf{Aut}(\mathbf{2}^*) \cong \mathbf{Aut}(\mathbf{2}^*) \wr S_2$, where $\wr$ denotes the wreath product. Let $\sigma \in S_2$ denote the transposition. A transduction f is called *odd* if $f = (f_0, f_1)\sigma$ and *even* otherwise. In the even case, we'll write $f = (f_0, f_1)$. Here f_0 and f_1 are called the *residuals* of f, a concept first introduced by Raney [14]. We call the maps $f \mapsto f_\mathbf{a}$ for $\mathbf{a} \in \mathbf{2}$ the *residuation maps*. Residuals can be extended to arbitrary length words by $f_\epsilon = f$ and $f_{\mathbf{w} \cdot \mathbf{a}} = (f_\mathbf{w})_\mathbf{a}$, where $w \in \mathbf{2}^*$ and $a \in \mathbf{2}$. The complete group automaton for $\mathcal{A}$, denoted $\mathfrak{C}(\mathcal{A})$, has as its state set $\mathcal{G}(\mathcal{A})$ with transitions of the form $f \xrightarrow{\mathbf{a}/\mathbf{b}} f_\mathbf{a}$, where $\mathbf{b} = f\mathbf{a}$.

2.2   Abelian Automata

For any automorphism $f \in \mathcal{G}(\mathcal{A})$, define its gap to be $\gamma_f = (f_0)(f_1)^{-1}$, so that $f_0 = \gamma_f f_1$. An easy induction on the wreath product shows the following [13]:

Lemma 1. *An automaton group $\mathcal{G}(\mathcal{A})$ is abelian if, and only if, all even elements of G have gap value I, where I denotes the identity automorphism, and all odd elements have the same gap.*

Thus, for abelian groups, we may denote the shared gap value by $\gamma_\mathcal{A}$, and when the underlying automaton is clear from context, we will simply denote the gap value by γ. It follows that every odd f satisfies $f = (\gamma f_1, f_1)\sigma$ and every even f satisfies $f = (f_1, f_1)$.

If $\mathcal{G}(\mathcal{A})$ is abelian, we will call $\mathcal{A}$ an *abelian automaton*. It should be noted that Lemma 1 gives an easy decision procedure to determine if a given machine $\mathcal{A}$

is abelian. Let $\mathcal{B}$ be the minimization of the product machine $\mathcal{A} \times \mathcal{A}^{-1}$, which can be computed using a partition-refinement algorithm, where the initial partition is induced by even and odd states. Then $\mathcal{A}$ is abelian if and only if the gap of each even state is collapsed to the identity state in $\mathcal{B}$ and if the gap of each odd state is collapsed to the same state in $\mathcal{B}$.

3   Affine Residuation Parametrization

In this section we will discuss embeddings of abelian automaton groups where residuation corresponds to an affine map.

3.1   Residuation Pairs

When $\mathcal{G}(\mathcal{A}) \cong \mathbb{Z}^m$, elements of the group may be represented as integer vectors in $\mathbb{Z}^m$. This section will use this interpretation, and explore the linear-algebraic properties of the residuation maps. Let $H \leq \mathcal{G}(\mathcal{A})$ be the subgroup of even automorphisms. It's clear that H is a subgroup of index 2 and that the residuation maps restricted to H are homomorphisms into $\mathcal{G}(\mathcal{A})$. Maps of this form are known as $1/2$-endomorphisms and were studied by Nekrashevych and Sidki in [12]. The authors proved that when $\mathcal{G}(\mathcal{A})$ is free abelian, the residuation maps take the form of an affine map. The following theorem summarizes their results to this end.

Theorem 2. *If $\mathcal{G}(\mathcal{A}) \cong \mathbb{Z}^m$, then there exists an isomorphism $\phi : \mathcal{G}(\mathcal{A}) \to \mathbb{Z}^m$, an $m \times m$ rational matrix A, and a rational vector r which satisfy*

$$\phi(f_{\mathbf{a}}) = \begin{cases} A \cdot \phi(f) & \text{if } f \text{ is even}, \\ A \cdot \phi(f) + (-1)^a r & \text{if } f \text{ is odd}. \end{cases} \tag{1}$$

Also, the matrix A satisfies several interesting properties:

- *A is contracting, i.e., its spectral radius is less than 1.*
- *The characteristic polynomial $\chi(z)$ of A is irreducible over $\mathbb{Q}$, and has the form $\chi(z) = z^m + \frac{1}{2}g(z)$, where $g(z) \in \mathbb{Z}[z]$ is of degree at most $m - 1$.*

We'll call the pair A, r a *residuation pair* for $\mathcal{A}$. Then $\mathcal{G}(\mathcal{A})$ (and its residuation relations) is completely determined by the image of one state under ϕ and a residuation pair.

Example 3. CC_2^3 admits the following residuation pair:

$$A = \begin{pmatrix} -1 & 1 \\ -1/2 & 0 \end{pmatrix}, \qquad r = \begin{pmatrix} -1 \\ -3/2 \end{pmatrix}, \qquad \phi(s_1) = \begin{pmatrix} 1 \\ 0 \end{pmatrix}$$

This parameterization is a useful tool for performing computations in $\mathcal{G}(\mathcal{A})$. Transduction composition becomes vector addition and residuation becomes an affine map over $\mathbb{Z}^m$. However, the residuation pair is not unique. In fact, the matrix A may not be unique even up to $GL(m, \mathbb{Z})$ similarity. A theorem of Latimer and MacDuffee implies that the $GL(m, \mathbb{Z})$ similarity classes of matrices with characteristic polynomial $\chi_A(z)$ are in one-to-one correspondence with the ideal classes of $\mathbb{Z}[z]/(\chi_A(z))$ [9]. Utilizing computer algebra, we can find an example with multiple similarity classes.

Example 4. The residuation matrices of the automaton CC_8^{15} have 2 $GL(m, \mathbb{Z})$ similarity classes.

Furthermore, it is unclear at this point how one may compute a residuation pair for a general abelian automaton.

3.2 Number Field Embedding

We introduce a parametrization which addresses the above concerns, i.e. it will be unique for $\mathcal{A}$, and we will give a method to compute it efficiently. We will show that $\mathcal{G}(\mathcal{A})$ can be embedded as an additive subgroup of an algebraic number field $F(\mathcal{A})$. At this point, it is not clear that $F(\mathcal{A})$ is unique, but this will indeed be the case, as shown in Theorem 9. In this section, we will use some basic results from algebraic number theory, see [8,16] for the requisite background.

Suppose $\mathcal{A}$ has states $q_1, \ldots, q_n$. For each state q_i, we introduce an unknown x_i, and let $R = \mathbb{Q}[z, x_1, \ldots, x_n]$. For each transition $q_i \xrightarrow{\text{a/b}} q_j$ in $\mathcal{A}$, we define the polynomial $p_{i,j} \in R$ as $p_{i,j} = zx_i - x_j + a - b$, where we interpret a and b as integers. Let $\mathcal{I}$ be the ideal of R generated by the set of all such polynomials, and let $\mathcal{S}$ be the system of equations defined by $\mathcal{I}$, i.e. by setting each $p_{i,j} = 0$.

Lemma 5. *The polynomial system $\mathcal{S}$ has a solution.*

Proof. Let A, r be a residuation pair of $\mathcal{A}$ and let $\chi(z)$ be the characteristic polynomial of A. Define $F = \mathbb{Q}(\alpha)$, where α is any root of $\chi(z)$, and let $\phi : \mathcal{G}(\mathcal{A}) \to \mathbb{Z}^m$ be the isomorphism from Theorem 2. We will construct a map $\psi : \mathbb{Z}^m \to F$ such that applying $\psi \circ \phi$ to the states of $\mathcal{A}$ yields a solution to $\mathcal{S}$.

Since $\chi(z)$ is irreducible, it's clear that $\mathcal{B} = \{r, Ar, \ldots, A^{m-1}r\}$ is a basis for $\mathbb{Q}^m$. Define $\psi : \mathbb{Q}^m \to F$ on $\mathcal{B}$ as $\psi(A^k r) = \alpha^k$. Then we have an injective homomorphism $\Psi : \mathcal{G}(\mathcal{A}) \to F$, where $\Psi = \psi \circ \phi$. Now applying ψ to the terms in Eq. (1) gives

$$\Psi(f_{\mathbf{a}}) = \begin{cases} \alpha\Psi(f) & \text{if } f \text{ is even,} \\ \alpha\Psi(f) + (-1)^a & \text{if } f \text{ is odd.} \end{cases} \tag{2}$$

It follows that $\alpha, \Psi(q_1), \ldots, \Psi(q_n)$ is a solution to $\mathcal{S}$. $\square$

We now analyze the structure of a general solution to $\mathcal{S}$, and show that up to conjugates, the above solution is unique. For an example of such a solution, look forward to Example 11. For the following results, let $\alpha, \beta_1, \ldots, \beta_n \in \mathbb{C}$ be solutions for $z, x_1, \ldots, x_n$ respectively. Define the map Ψ on the generators of $\mathcal{G}(\mathcal{A})$ as $\Psi(q_i) = \beta_i$.

Lemma 6. *For each $f \in \mathcal{G}(\mathcal{A})$, Eq. (2) holds.*

Proof. The definition of the generators of $\mathcal{I}$ ensures that it holds for the generators of $\mathcal{G}(\mathcal{A})$. This can be extended to arbitrary products by induction on the length of the product. Let $f \in \mathcal{G}(\mathcal{A})$, and write $f = s \cdot g$, where s is a generator and $g \neq I$. By induction we have that both s and g obey Eq. (2). Consider the possible parities of s and g; e.g. if s is odd and g is even, then note $\alpha \Psi(s) = \Psi(s_{\mathbf{a}}) - (-1)^a$, and so

$$\alpha \Psi(f) + (-1)^a = \alpha \Psi(s) + \alpha \Psi(g) + (-1)^a = \Psi(s_{\mathbf{a}}) + \Psi(g_{\overline{\mathbf{a}}}) = \Psi(f_{\mathbf{a}}).$$

The other cases follow similarly. □

Lemma 7. *α is unique up to conjugates, i.e. it is a root of $\chi(z)$, the characteristic polynomial of a residuation matrix of $\mathcal{A}$.*

Proof. Let γ be the gap value for $\mathcal{A}$ discussed in Sect. 3.1. From Lemma 6, we see $\Psi(\gamma) = 2$. Let m be the rank of $\mathcal{G}(\mathcal{A})$, and let ζ_k be a non-identity length-k residual of γ, so that for $k = 1, \ldots, m$, $\Psi(\zeta_i)$ is a polynomial in α of degree k. Then $\gamma, \zeta_1, \ldots, \zeta_m$ are linearly dependent in $\mathcal{G}(\mathcal{A})$, and hence are also under Ψ. This shows that α is a root of a degree m polynomial, which is the same degree as the irreducible $\chi(z)$ from Lemma 5, implying that α satisfies $\chi(\alpha) = 0$. □

Lemma 8. *Let $\mathcal{L}$ be the integral span of $\beta_1, \ldots, \beta_n$, and let $\Psi : \mathcal{G}(\mathcal{A}) \to \mathcal{L}$ be the homomorphism defined on the generators as $q_i \mapsto \beta_i$. Then Ψ is an isomorphism.*

Proof. Suppose for the sake of contradiction that there is a non-identity $f \in \mathcal{G}(\mathcal{A})$ such that $\Psi(f) = 0$. If f is even, then some finite residual $f_{\mathbf{w}}$ must be odd (because f is non-identity), and $\Psi(f_{\mathbf{w}}) = \alpha^{|\mathbf{w}|} \Psi(f) = 0$. Thus without loss of generality, we may assume f is odd. It follows from Lemma 6 that $\Psi(f_0) = 1$, and thus $1 \in \mathcal{L}$.

Then, by induction, we can show that $\alpha^k \in \mathcal{L}$ for all $k \in \mathbb{N}$. The base case of $k = 0$ follows from the above, and for the inductive case let us assume $\alpha^k = \sum_{i=1}^n c_i \beta_i$. Let $\partial_0 \beta_i$ denote the 0-residual of β_i. Then, if q_i is even, we have $\alpha q_i = \partial_0 q_i$ and if q_i is odd, we have $\alpha q_i = \partial_0 q_i - 1$. It follows that

$$\alpha^{k+1} = \alpha \sum_{i=1}^n c_i \beta_i = \sum_{i=1}^n c_i \partial_0 \beta_i - \sum_{i=1}^n c_i,$$

Because $1 \in \mathcal{L}$, we conclude that the constant term $\sum_{i=1}^n c_i \in \mathcal{L}$, implying that $\alpha^{k+1} \in \mathcal{L}$. Thus, since $|\alpha| < 1$, there are arbitrarily small nonzero elements in $\mathcal{L}$. But $\mathcal{L}$ is a discrete subgroup of a number field, and hence has a smallest nonzero element [16], so we have a contradiction. □

We summarize these results in the following theorem:

Theorem 9. *There exists a unique (up to conjugates) algebraic number α such that $\mathcal{G}(\mathcal{A})$ embeds into the field $F(\mathcal{A}) = \mathbb{Q}(\alpha)$, such that if $\Psi : \mathcal{G}(\mathcal{A}) \to F(\mathcal{A})$ is the embedding, then for all $f \in \mathcal{G}(\mathcal{A})$,*

$$\Psi(f_{\mathbf{a}}) = \begin{cases} \alpha \Psi(f) & \text{if } f \text{ is even,} \\ \alpha \Psi(f) + (-1)^a & \text{if } f \text{ is odd.} \end{cases}$$

The existence of a unique solution addresses one of the issues mentioned with the residuation matrix. What remains is to show the number field embedding is efficiently computable.

Theorem 10. *$F(\mathcal{A})$ and the embedding $\Psi : \mathcal{G}(\mathcal{A}) \to F(\mathcal{A})$ from Theorem 9 can be computed in time $O(n^6)$.*

Proof. We seek to compute $\chi(z)$ along with the unique solution to $\mathcal{S}$ as elements of $F(\mathcal{A})$. By Theorem 9, computing a triangular decomposition of $\mathcal{I}$, with respect to the lexicographic monomial ordering on $x_1 < \cdots < x_n < z$, would yield $\chi(z)$ as the first element [10]. The values for x_i may then be computed by solving the linear system in $F(\mathcal{A})$. The work required to compute a triangular decomposition is dominated by the calculation of a Gröbner basis for $\mathcal{I}$ [10]. In general, Gröbner basis calculation is known to be EXPSPACE-complete. However, the bilinear structure of the equations allow for better upper bounds on the complexity. Thus the F_5 algorithm from [5] computes a Gröbner basis for $\mathcal{I}$ in time $O(n^6)$. □

Example 11. The polynomial ideal for CC_2^3 is

$$\mathcal{I} = (zx_1 + 1 - x_3, zx_1 - 1 - x_2, zx_3 - x_2, zx_2 - x_1).$$

A triangular decomposition gives the minimal polynomial $\chi(z) = z^2 + z + 1/2$. Letting α denote a root of $\chi(z)$, we have

$$\Psi(q_1) = \frac{1}{5}(-6\alpha - 2), \qquad \Psi(q_2) = \frac{1}{5}(4\alpha - 2), \qquad \Psi(q_3) = \frac{1}{5}(4\alpha + 8).$$

4   Orbit Rationality

4.1   Background

We briefly return to the case of a general (possibly nonabelian) automaton group. For $f \in \mathcal{G}(\mathcal{A})$ and $\mathbf{x} \in 2^*$, we define the *orbit* of $\mathbf{x}$ under f, denoted $f^*(\mathbf{x})$, as the set of iterates of f applied to $\mathbf{x}$, $\{f^t\mathbf{x} \mid t \in \mathbb{Z}\}$. Following this, we define the *orbit language*,

$$\mathbf{orb}(f) = \{\mathbf{x}{:}\mathbf{y} \mid \exists t \in \mathbb{Z} \text{ such that } f^t\mathbf{x} = \mathbf{y}\},$$

where the *convolution* $\mathbf{x}{:}\mathbf{y}$ of two words $\mathbf{x}, \mathbf{y} \in 2^k$ is defined by

$$\mathbf{x}{:}\mathbf{y} = \begin{array}{|c|c|c|c|} \hline x_1 & x_2 & \dots & x_k \\ \hline y_1 & y_2 & \dots & y_k \\ \hline \end{array} \in (2 \times 2)^k.$$

We concern ourselves with the following question: Given $\mathbf{x}, \mathbf{y} \in 2^*$, is $\mathbf{x}{:}\mathbf{y} \in \mathbf{orb}(f)$? We'll call automorphisms *orbit-rational* if their orbit language is regular (and hence their orbit relation is rational). Consider the *orbit with translation language* as defined in [18]:

$$\mathbf{R}(f, g) = \{\mathbf{x}{:}\mathbf{y} \mid \exists t \in \mathbb{Z} \text{ such that } gf^t\mathbf{x} = \mathbf{y}\}.$$

It was shown that $\mathbf{R}$ is closed under quotients. If $f, g \in \mathcal{G}(\mathcal{A})$ and $\mathbf{b} = g\mathbf{a}$, then

$$(\mathbf{a{:}b})^{-1}\mathbf{R}(f, g) = \begin{cases} \mathbf{R}(f_{\mathbf{a}}, g_{\mathbf{a}}) & \text{if } f \text{ is even,} \\ \mathbf{R}(f_{\mathbf{a}} f_{\overline{\mathbf{a}}}, g_{\mathbf{a}}) & \text{if } f \text{ is odd.} \end{cases}$$

$$(\mathbf{a{:}\overline{b}})^{-1}\mathbf{R}(f, g) = \begin{cases} \emptyset & \text{if } f \text{ is even,} \\ \mathbf{R}(f_{\mathbf{a}} f_{\overline{\mathbf{a}}}, f_{\mathbf{a}} g_{\overline{\mathbf{a}}}) & \text{if } f \text{ is odd.} \end{cases}$$

Consider the infinite transition system $M_{\mathcal{A}}$ over $\mathbf{2} \times \mathbf{2}$ and with transitions

$$\mathbf{R}(f, g) \xrightarrow{\mathbf{a{:}b}} (\mathbf{a{:}b})^{-1}\mathbf{R}(f, g).$$

For any $f \in \mathcal{G}(\mathcal{A})$, $\mathbf{R}(f, I)$ is the orbit language for f. By Brzozowski's theorem, f is orbit-rational if and only if the subautomaton of $M_{\mathcal{A}}$ reachable from (f, I) is finite. Because $\mathcal{G}(\mathcal{A})$ is contracting (see e.g. [11]), this occurs if and only if finitely many first arguments to $\mathbf{R}$ appear in the closure of $\mathbf{R}(f, I)$ under residuation. The first arguments of the quotients depend only on the input bit $\mathbf{a}$, which leads us to consider the maps

$$\varphi_{\mathbf{a}}(f) = \begin{cases} f_{\mathbf{a}} & \text{if } f \text{ is even,} \\ f_{\mathbf{a}} f_{\overline{\mathbf{a}}} & \text{if } f \text{ is odd.} \end{cases}$$

Hence to determine if f is orbit-rational, it suffices to determine the cardinality of the set resulting from iterating φ_0, φ_1 starting at f.

4.2   The Abelian Case

Throughout this section we will assume $\mathcal{A}$ is abelian. In this case, we have $\varphi_a = \varphi_{\overline{a}}$, so we will drop the subscript and simply refer to φ. If f is odd, then $f = (\gamma f_{\mathbf{1}}, f_{\mathbf{1}})\sigma$, where γ is the gap value of $\mathcal{A}$. Then, $\varphi(f) = \gamma f_{\mathbf{1}}^2$, and

$$\varphi(f) = \begin{cases} f_{\mathbf{0}} & \text{if } f \text{ is even,} \\ \gamma f_{\mathbf{1}}^2 & \text{if } f \text{ is odd.} \end{cases} \tag{3}$$

We seek to understand the behavior of iterating φ on an automorphism, and in particular, determine when $\varphi^*(f) = \{\varphi^t(f) \mid t \in \mathbb{N}\}$ is finite. To accomplish this, will return to the wreath representation for automorphisms and relate φ to an extension of parity for automorphisms in $\mathcal{G}(\mathcal{A})$.

Definition 12. *The even rank of an automorphism $f \in \mathcal{G}(\mathcal{A})$, denoted $|f|$, is defined as the minimum integer k such that $\varphi^k(f)$ is odd. If there is no such integer, then $|f| = \infty$.*

When the context is clear, we will abbreviate "even rank" as "rank". It is clear that when f is even, $\varphi(f) = f_{\mathbf{0}} = f_{\mathbf{1}}$, so the rank equivalently measures the distance from f to its first odd residual. If f has infinite rank, then for every $w \in \mathbf{2}^*$, the residual $f_{\mathbf{w}}$ is even. Thus $f\mathbf{x} = \mathbf{x}$ for all $\mathbf{x} \in \mathbf{2}^*$, implying that the only automorphism with infinite rank is the identity. We will now prove the primary connection between rank and φ: that rank equality is preserved under φ.

Lemma 13. *If $f, g \in \mathcal{G}(\mathcal{A})$ with $|f| = |g|$, then $|\varphi(f)| = |\varphi(g)|$.*

Proof. The case when $|f| > 0$ is clear, but if $|f| = 0$, then we may write f in wreath representation as $f = (\gamma h, h)\sigma$, where γ is the gap value discussed in Sect. 3.1, and it follows that $\varphi(f) = \gamma h^2$. If we had $|\gamma| < |h^2|$, it would follow that $|\varphi(f)| = |\gamma|$, so it suffices to show this inequality. Indeed, since h^2 is even and $h^2 = (\gamma h_1^2, \gamma h_1^2)$, we have $|h^2| \geq 1 + \min(|\gamma|, |h_1^2|)$. This inequality would hold for any square h^2; in particular, it also holds for h_1^2. It follows that the min takes value $|\gamma|$, so $|h^2| \geq 1 + |\gamma|$. Thus, for any odd f, $|\varphi(f)| = |\gamma|$, which completes the proof. $\qquad\square$

Corollary 14. *If f is an odd automorphism and $t = |\varphi^*(f)|$ is finite, then $\varphi^t(f) = f$.*

Proof. Because $|\varphi^*(f)|$ is finite, the sequence $\{\varphi^n(f) | n \geq 0\}$ is eventually periodic. Lemma 13 shows iterating φ on f produces a cyclic sequence of ranks of the form $0, |\gamma|, |\gamma| - 1, \ldots, 0, \ldots$. We note that φ is invertible when restricted to the automorphisms of rank at most $|\gamma|$. Indeed, for any automorphism g, if $|g| < |\gamma|$, then the unique inverse is $\varphi^{-1}(g) = (g, g)$. If instead $|g| = |\gamma|$, there is a unique odd h such that $g = \varphi(h) = \gamma h_1^2$. It follows that the first repeated automorphism in $\varphi^*(f)$ is f itself, so $\varphi^t(f) = f$. $\qquad\square$

The preceding results can be interpreted in the corresponding number field. Recall the map $\Psi : \mathcal{G}(\mathcal{A}) \to F(\mathcal{A})$ satisfying the properties described in Sect. 3.2. Let $\chi(z)$ be the unique characteristic polynomial for $\mathcal{A}$, and let α be a root of χ such that $F(\mathcal{A}) = \mathbb{Q}(\alpha)$. Let $\mathcal{L} = \Psi(\mathcal{G}(\mathcal{A}))$ be the image of the group elements in $F(\mathcal{A})$. Then $\Gamma = \Psi\varphi\Psi^{-1}$ is the orbit residuation map in $\mathcal{L}$, so $|\varphi^*(f)| = |\Gamma^*(\Psi(f))|$, and it follows from Eq. (3) that for any $\beta \in \mathcal{L}$,

$$\Gamma(\beta) = \begin{cases} \alpha\beta & \text{if } \Psi^{-1}(\beta) \text{ is even,} \\ 2\alpha\beta & \text{if } \Psi^{-1}(\beta) \text{ is odd.} \end{cases}$$

Lemma 15. *If $f \in \mathcal{G}(\mathcal{A})$, $f \neq I$, and $\varphi^*(f)$ is finite, then $(2\alpha^k)^n = 1$ for some $k, n \in \mathbb{N}$. Furthermore, $\varphi^*(g)$ is finite for any $g \in \mathcal{G}(\mathcal{A})$.*

Proof. Suppose $\varphi^*(f)$ is finite. Because any non-identity f has finite rank, if we let $f' = \varphi^{|f|}(f)$, then f' is odd and $\varphi^*(f')$ is finite.

By Corollary 14, we may write $\varphi^t(f') = f'$. Let h be the first odd automorphism after f' in the sequence $\{\varphi^n(f') \mid n \geq 0\}$, say $\varphi^k(f') = h$. So in $F(\mathcal{A})$, $\Gamma^k\Psi(f') = 2\alpha^k\Psi(h)$. Then by Lemma 13, the sequence of parities starting from f' and h are identical, meaning that any odd state reachable by f' must be of the form $\varphi^{kn}(f')$. Thus taking $n = \frac{t}{k}$ shows $(2\alpha^k)^n \Psi(f') = \Psi(f')$. Since $f' \neq I$, it follows that $\Psi(f') \neq 0$, and so $(2\alpha^k)^n = 1$ in $F(\mathcal{A})$. Now if $g \in \mathcal{G}(\mathcal{A})$ with $g \neq I$, then $g' = \varphi^{|g|}$ is odd, and

$$\Gamma^{kn}(\Psi(g)) = (2\alpha^k)^n \Psi(g) = \Psi(g),$$

so $\varphi^{kn}(g) = g$, and hence $\varphi^*(g)$ is finite. $\qquad\square$

Lemma 16. *Some power of α is rational if and only if for some $k, n \in \mathbb{N}$, $(2\alpha^k)^n = 1$. In this case, α has magnitude $2^{-\frac{1}{m}}$, where m is the rank of the free abelian group $\mathcal{G}(\mathcal{A})$.*

Proof. First assume that $(2\alpha^k)^n = 1$ for some integers k and n. Then $\alpha^{kn} = 2^{-n}$. Conversely let ℓ be smallest such that $\alpha^\ell = r$ is rational. Then α is a root of $p(z) = z^\ell - r$. Let $\chi(z)$ be the irreducible characteristic polynomial of $\mathcal{A}$. Since χ is the minimal polynomial of λ_0, then $\chi(z) \mid p(z)$. Thus all roots of χ have equal magnitude, and since the constant term of $\chi(z)$ is $\pm\frac{1}{2}$, this magnitude is $|\alpha| = \pm 2^{-\frac{1}{m}}$, where m is the rank of $\mathcal{G}(\mathcal{A})$. Since $\lambda^\ell = r$ has rational norm, m divides ℓ. Setting $k = m$ and $n = \frac{2\ell}{m}$ guarantees that $(2\alpha^k)^n = 1$. $\square$

Our main result follows from the preceding lemmas:

Theorem 17. *Let $\chi(z)$ be the unique characteristic polynomial for $\mathcal{A}$, and let α be a root of χ such that $F(\mathcal{A}) = \mathbb{Q}(\alpha)$. Then for any $f \in \mathcal{G}(\mathcal{A})$, f is orbit-rational if and only if some power of α is rational.*

Example 18. CC_2^3 is orbit rational. Recall from Example 11 that $F(CC_2^3) = \mathbb{Q}[z]/(\chi(z))$ for $\chi(z) = z^2 + z + 1/2$. If α is a root of $\chi(z)$, then $\alpha^4 - -1/4$.

4.3   Decision Procedure

We aim to turn Theorem 17 into a decision procedure for orbit rationality.

Lemma 19. *Some power of α is rational if and only if $\alpha^{4\ell}$ is rational, where $\ell = \frac{m}{2}$ if m is even and $\ell = m$ if m is odd.*

Proof. By Lemma 16, all roots of $\chi(z)$ have norm $2^{-\frac{1}{m}}$ and therefore lie on the complex disk of radius $2^{-\frac{1}{m}}$. We will follow a technique of Robinson in [15] to show $\chi(z)$ is of the form $P(z^\ell)$, where P has degree at most 2. We write

$$\chi(z) = a_m x^m + a_{m-1} x^{m-1} + \cdots + a_1 x + a_0,$$

where $a_m = 1$ and $a_0 = \pm\frac{1}{2}$. Now if β is any root of $\chi(z)$, then the conjugate $\bar{\beta} = 2^{-2m}\beta^{-1}$ is also a root of $\chi(z)$. Consider the polynomial $p(z) = z^m \chi\left(\frac{2^{-2m}}{z}\right)$. Then, $p(z)$ has the same roots and same degree as $\chi(z)$, so $\chi(z)$ is a constant multiple of $p(z)$. Computing the leading coefficient shows $a_0 \chi(z) = p(z)$, and equating the remaining coefficients gives for all $k \leq m$, $a_0 a_{m-k} = a_k 2^{-\frac{2k}{m}}$. Thus $2^{-\frac{2k}{m}}$ is rational when $a_k \neq 0$. Let ℓ be the smallest integer such that $2^{-\frac{2k}{m}}$ is rational, i.e. $\ell = m$ if m is odd or $\ell = \frac{m}{2}$ if m is even. Then a_k is nonzero only if $\ell \mid k$, so there exists a degree $\frac{m}{\ell}$ polynomial $P(z)$ such that $\chi(z) = P\left(z^\ell\right)$. That is, the roots of $\chi(z)$ are of the form $\sqrt[\ell]{\beta}$ for β a root of P. Note that $P(z)$ is monic and irreducible, has constant term $\pm\frac{1}{2}$, and all of its roots have norm $2^{-\frac{\ell}{m}}$. This process reduces $\chi(z)$ to a degree 1 or 2 polynomial, depending on the parity of m. If m is odd, then the only possible polynomials are $P(z) = z \pm \frac{1}{2}$, both of which have a single rational root. Thus the only interesting case is if

m is even, where we claim there are only 4 possibilities for $P(z)$. The appendix of [13] lists the 6 polynomials over $\mathbb{Q}$ of degree 2 which are monic, irreducible, contracting, and have constant term $\pm\frac{1}{2}$:

$$P_1(z) = z^2 - \frac{1}{2}, \qquad P_2(z) = z^2 + \frac{1}{2}, \qquad P_3(z) = z^2 - z + \frac{1}{2},$$

$$P_4(z) = z^2 + z + \frac{1}{2}, \qquad P_5(z) = z^2 - \frac{1}{2}z + \frac{1}{2}, \qquad P_6(z) = z^2 + \frac{1}{2}z + \frac{1}{2}.$$

We claim that, in the orbit-rational case, $P(z)$ cannot be $P_5(z)$ or $P_6(z)$. The polynomial $P_5(z)P_6(z)$ has roots $\beta = \pm\frac{i}{4}(\sqrt{7} \pm i)$, which live in the degree 2 extension $\mathbb{Q}(\sqrt{-7})$. If one of these roots satisfied $\beta^k = r$ for some integer k and rational number r, then $\mathbb{Q}(\sqrt{-7})$ would contain an kth root of unity. Recall that a kth root of unity has degree $\varphi(k)$ over $\mathbb{Q}$, where φ denotes Euler's totient function. Thus we would have $\varphi(k) = 2$, so $k = 3$ or $k = 4$. It's straightforward to check that β^k is not rational for any of the above roots β where $k = 3, 4$. Thus, $P_5(z)$ and $P_6(z)$ are not possible. One can also verify any root β of $P_1(z)$, $P_2(z)$, $P_3(z)$, or $P_4(z)$ satisfies $\beta^4 = \pm\frac{1}{4}$. Since the roots of $\chi(z)$ satisfy $\lambda^\ell = \beta$ for a root β of $P(z)$, it follows that $\lambda^{4\ell}$ is rational. $\qquad\square$

Theorem 20. *Given an abelian binary invertible transducer $\mathcal{A}$, we can decide if $\mathcal{G}(\mathcal{A})$ is orbit-rational in polynomial time.*

Proof. By Theorem 10, we can compute the number field of $\mathcal{A}$ and find $\chi(z)$ in time $O(n^6)$. Let ℓ be as in Lemma 19. Using standard number field arithmetic techniques, we compute $z^{4\ell}$ in the field $\mathbb{Q}[z]/(\chi(z))$ and check if it is rational. $\quad\square$

5   Discussion and Open Problems

We extended the results of Nekrashevych and Sidki in [12]. This yielded an embedding of $\mathcal{G}(\mathcal{A})$ into a number field where residuation in $\mathcal{A}$ became an affine map in $F(\mathcal{A})$. This removed redundancies present in the residuation pairs, giving each automorphism a unique element in a number field. Additionally, we have demonstrated that this embedding is computable in polynomial time.

Phrasing computational problems about $\mathcal{A}$ in terms of $F(\mathcal{A})$ may yield efficient solutions. We demonstrated this with the question of deciding orbit-rationality, where the problem reduces to a simple computation in $F(\mathcal{A})$. We expect other computational problems in $\mathcal{A}$ can exploit the algebraic structure of $F(\mathcal{A})$ in a similar way to yield efficient solutions. It is not clear how these results may be generalized to non-abelian automaton groups, and this is the largest open question we raise. At this time we are not aware of any nonabelian orbit-rational automaton groups.

Acknowlegements. The authors would like to thank Eric Bach for his helpful feedback on a draft of this paper. We also thank Evan Bergeron and Chris Grossack for many helpful conversations on the results presented.

References

1. Bartholdi, L.: Book review: combinatorial algebra: syntax and semantics. Bull. AMS **54**(4), 681–686 (2017)
2. Becker, T.: Embeddings and orbits of abelian automaton groups (2018). https://github.com/tim-becker/thesis-code
3. Bondarenko, I., et al.: Classification of groups generated by 3-state automata over a 2-letter alphabet. Algebra Discrete Math. **1**, April 2008
4. Brough, M., Khoussainov, B., Nelson, P.: Sequential automatic algebras. In: Beckmann, A., Dimitracopoulos, C., Löwe, B. (eds.) CiE 2008. LNCS, vol. 5028, pp. 84–93. Springer, Heidelberg (2008). https://doi.org/10.1007/978-3-540-69407-6_9
5. Faugère, J.C., El Din, M.S., Spaenlehauer, P.J.: Gröbner bases of bihomogeneous ideals generated by polynomials of bidegree (1, 1): algorithms and complexity. J. Symbolic Comput. **46**(4), 406–437 (2011)
6. Grigorchuk, R.I., Nekrashevich, V.V., Sushchanskii, V.I.: Automata, dynamical systems and groups. Proc. Steklov Inst. Math. **231**, 128–203 (2000)
7. Hungerford, T.W.: Algebra. Springer, New York (1974). https://doi.org/10.1007/978-1-4612-6101-8
8. Ireland, K., Rosen, M., Rosen, M.: A Classical Introduction to Modern Number Theory. Graduate Texts in Mathematics. Springer, New York (1990). https://doi.org/10.1007/978-1-4757-2103-4
9. Latimer, C.G., MacDuffee, C.C.: A correspondence between classes of ideals and classes of matrices. Ann. Math. **34**(2), 313–316 (1933). http://www.jstor.org/stable/1968204
10. Lazard, D.: Solving zero-dimensional algebraic systems. J. Symbolic Comput. **13**(2), 117–131 (1992). https://doi.org/10.1016/S0747-7171(08)80086-7. http://www.sciencedirect.com/science/article/pii/S0747717108800867
11. Nekrashevych, V.: Self-similar groups. Mathematical Surveys and Monographs. American Mathematical Society, Providence (2014)
12. Nekrashevych, V., Sidki, S.: Automorphisms of the binary tree: state-closed subgroups and dynamics of 1/2-endomorphisms. London Math. Soc. Lect. Note Ser. **311**, 375–404 (2004)
13. Okano, T.: Invertible binary transducers and automorphisms of the binary tree. Master's thesis, Carnegie Mellon University (2015)
14. Raney, G.N.: Sequential functions. J. ACM (JACM) **5**(2), 177–180 (1958)
15. Robinson, R.M.: Conjugate algebraic integers on a circle. Mathematische Zeitschrift **110**(1), 41–51 (1969). https://doi.org/10.1007/BF01114639
16. Stein, W.: Algebraic Number Theory, a Computational Approach. Harvard, Massachusetts (2012)
17. Sutner, K.: Abelian invertible automata. In: Adamatzky, A. (ed.) Reversibility and Universality. ECC, vol. 30, pp. 37–59. Springer, Cham (2018). https://doi.org/10.1007/978-3-319-73216-9_3
18. Sutner, K., Lewi, K.: Iterating inverse binary transducers. J. Automata, Lang. Comb. **17**(2–4), 293–313 (2012)

Bounded Automata Groups are co-ET0L

Alex Bishop(iD) and Murray Elder$^{(\boxtimes)}$(iD)

School of Mathematical and Physical Sciences, University of Technology Sydney,
Ultimo, NSW 2007, Australia
{alexander.bishop,murray.elder}@uts.edu.au

Abstract. Holt and Röver proved that finitely generated bounded
automata groups have indexed co-word problem. Here we sharpen this
result to show they are in fact co-ET0L.

Keywords: Formal language theory · ET0L language ·
Check-stack pushdown automaton · Bounded automata group ·
Co-word problem

1 Introduction

A recurrent theme in group theory is to understand and classify group-theoretic
problems in terms of their formal language complexity [1, 9–11, 20]. Many authors
have considered the groups whose non-trivial elements, i.e. *co-word problem*,
can be described as a context-free language [3, 14, 16, 18]. Holt and Röver went
beyond context-free to show that a large class known as *bounded automata groups*
have an indexed co-word problem [15]. This class includes important examples
such as Grigorchuk's group of intermediate growth, the Gupta-Sidki groups,
and many more [12, 13, 21, 24]. For the specific case of the Grigorchuk group,
Ciobanu *et al.* [5] showed that the co-word problem was in fact ET0L. ET0L is
a class of languages coming from L-systems which lies strictly between context-
free and indexed [19, 22, 23]. Ciobanu *et al.* rely on the grammar description
of ET0L for their result. Here we are able to show that all finitely generated
bounded automata groups have ET0L co-word problem by instead making use
of an equivalent machine description: check-stack pushdown (cspd) automata.

ET0L languages, in particular their deterministic versions, have recently
come to prominence in describing solution sets to equations in groups and
monoids [4, 7, 8]. The present paper builds on the recent resurgence of interest,
and demonstrates the usefulness of a previously overlooked machine description.

For a group G with finite generating set X, we denote by $\mathrm{coW}(G, X)$ the set
of all words in the language $(X \cup X^{-1})^\star$ that represent non-trivial elements in G.
We call $\mathrm{coW}(G, X)$ the co-word problem for G (with respect to X). Given a class
$\mathcal{C}$ of formal languages that is closed under inverse homomorphism, if $\mathrm{coW}(G, X)$

Research supported by Australian Research Council grant DP160100486 and an Australian Government Research Training Program PhD Scholarship.

© Springer Nature Switzerland AG 2019
C. Martín-Vide et al. (Eds.): LATA 2019, LNCS 11417, pp. 82–94, 2019.
https://doi.org/10.1007/978-3-030-13435-8_6

is in $\mathcal{C}$ then so is $\mathrm{coW}(G, Y)$ for any finite generating set Y of G [14]. Thus, we say that a group is *co-$\mathcal{C}$* if it has a co-word problem in the class $\mathcal{C}$. Note that ET0L is a full AFL [6] and so is closed under inverse homomorphism.

2   ET0L Languages and CSPD Automata

An *alphabet* is a finite set. Let Σ and V be two alphabets which we will call the *terminals* and *non-terminals*, respectively. We will use lower case letters to represent terminals in Σ and upper case letters for non-terminals in V. By Σ^*, we will denote the set of words over Σ with $\varepsilon \in \Sigma^*$ denoting the empty word.

A *table*, τ, is a finite set of context-free replacement rules where each non-terminal, $X \in V$, has at least one replacement in τ. For example, with $\Sigma = \{a, b\}$ and $V = \{S, A, B\}$, the following are tables.

$$\alpha : \begin{cases} S \to SS \mid S \mid AB \\ A \to A \\ B \to B \end{cases} \qquad \beta : \begin{cases} S \to S \\ A \to aA \\ B \to bB \end{cases} \qquad \gamma : \begin{cases} S \to S \\ A \to \varepsilon \\ B \to \varepsilon \end{cases} \qquad (1)$$

We apply a table, τ, to the word $w \in (\Sigma \cup V)^*$ to obtain a word w', written $w \to^\tau w'$, by performing a replacement in τ to each non-terminal in w. If a table includes more than one rule for some non-terminal, we nondeterministically apply any such rule to each occurrence. For example, with $w = SSSS$ and α as in (1), we can apply α to w to obtain $w' = SABSSAB$. Given a sequence of tables $\tau_1, \tau_2, \ldots, \tau_k$, we will write $w \to^{\tau_1 \tau_2 \cdots \tau_k} w'$ if there is a sequence of words $w = w_1, w_2, \ldots, w_{k+1} = w'$ such that $w_j \to^{\tau_j} w_{j+1}$ for each j.

Definition 1 (Asveld [2]). *An* ET0L *grammar is a 5-tuple* $G = (\Sigma, V, T, \mathcal{R}, S)$, *where*

1. Σ *and* V *are the alphabets of* terminals *and* non-terminals, *respectively;*
2. $T = \{\tau_1, \tau_2, \ldots, \tau_k\}$ *is a finite set of* tables;
3. $\mathcal{R} \subseteq T^*$ *is a regular language called the* rational control; *and*
4. $S \in V$ *is the* start symbol.

The ET0L *language produced by the grammar* G, *denoted* $L(G)$, *is*

$$L(G) := \{w \in \Sigma^* : S \to^v w \ \text{for some} \ v \in \mathcal{R}\}.$$

For example, with α, β and γ as in (1), the language produced by the above grammar with rational control $\mathcal{R} = \alpha^* \beta^* \gamma$ is $\{(a^n b^n)^m : n, m \in \mathbb{N}\}$.

2.1   CSPD Automata

A *cspd automaton*, introduced in [17], is a nondeterministic finite-state automaton with a one-way input tape, and access to both a *check-stack* (with stack alphabet Δ) and a *pushdown stack* (with stack alphabet Γ), where access to these two stacks is tied in a very particular way. The execution of a cspd machine can be separated into two stages.

In the first stage the machine is allowed to push to its check-stack but *not* its pushdown, and further, the machine will not be allowed to read from its input tape. Thus, the set of all possible check-stacks that can be constructed in this stage forms a regular language which we will denote as $\mathcal{R}$.

In the second stage, the machine can no longer alter its check-stack, but is allowed to access its pushdown and input tape. We restrict the machine's access to its stacks so that it can only move along its check-stack by pushing and popping items to and from its pushdown. In particular, every time the machine pushes a value onto the pushdown it will move up the check-stack, and every time it pops a value off of the pushdown it will move down the check-stack; see Fig. 1 for an example of this behaviour.

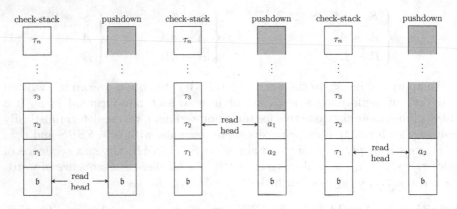

Fig. 1. An example of a cspd machine pushing $w = a_1 a_2$, where $a_1, a_2 \in \Delta$, onto its pushdown stack, then popping a_1

We define a cspd machine formally as follows.

Definition 2. *A cspd machine is a 9-tuple* $\mathcal{M} = (Q, \Sigma, \Gamma, \Delta, \flat, \mathcal{R}, \theta, q_0, F)$, *where*

1. *Q is the set of states;*
2. *Σ is the input alphabet;*
3. *Γ is the alphabet for the pushdown;*
4. *Δ is the alphabet for the check-stack;*
5. *$\flat \notin \Delta \cup \Gamma$ is the bottom of stack symbol;*
6. *$\mathcal{R} \subseteq \Delta^*$ is a regular language of allowed check-stack contents;*
7. *θ is a finite subset of*

$$(Q \times (\Sigma \cup \{\varepsilon\}) \times ((\Delta \times \Gamma) \cup \{(\varepsilon, \varepsilon), (\flat, \flat)\})) \times (Q \times (\Gamma \cup \{\flat\})^*),$$

and is called the transition relation (see below for allowable elements of θ);

8. $q_0 \in Q$ is the start state; and
9. $F \subseteq Q$ is the set of accepting states.

In its initial configuration, the machine will be in state q_0, the check-stack will contain $\flat w$ for some nondeterministic choice of $w \in \mathcal{R}$, the pushdown will contain only the letter $\flat$, the read-head for the input tape will be at its first letter, and the read-head for the machine's stacks will be pointing to the $\flat$ on both stacks. From here, the machine will follow transitions as specified by θ, each such transition having one of the following three forms, where $a \in \Sigma \cup \{\varepsilon\}$, $p, q \in Q$ and $w \in \Gamma^*$.

1. $((p, a, (\flat, \flat)), (q, w\flat)) \in \theta$ meaning that if the machine is in state p, sees $\flat$ on both stacks and is able to consume a from its input; then it can follow this transition to consume a, push w onto the pushdown and move to state q.
2. $((p, a, (d, g)), (q, w)) \in \theta$ where $(d, g) \in \Delta \times \Gamma$, meaning that if the machine is in state p, sees d on its check-stack, g on its pushdown, and is able to consume a from its input; then it can follow this transition to consume a, pop g, then push w and move to state q.
3. $((p, a, (\varepsilon, \varepsilon)), (q, w)) \in \theta$ meaning that if the machine is in state p and can consume a from its input; then it can follow this transition to consume a, push w and move to state q.

In the previous three cases, $a = \varepsilon$ corresponds to a transition in which the machine does not consume a letter from input. We use the convention that, if $w = w_1 w_2 \cdots w_k$ with each $w_j \in \Gamma$, then the machine will first push the w_k, followed by the w_{k-1} and so forth. The machine accepts if it has consumed all its input and is in an accepting state $q \in F$.

· In [17] van Leeuwen proved that the class of languages that are recognisable by cspd automata is precisely the class of ET0L languages.

3   Bounded Automata Groups

For $d \geqslant 2$, let $\mathcal{T}_d$ denote the d-regular rooted tree, that is, the infinite rooted tree where each vertex has exactly d children. We identify the vertices of $\mathcal{T}_d$ with words in Σ^* where $\Sigma = \{a_1, a_2, \dots, a_d\}$. In particular, we will identify the root with the empty word $\varepsilon \in \Sigma^*$ and, for each vertex $v \in V(\mathcal{T}_d)$, we will identify the k-th child of v with the word va_k, see Fig. 2.

Recall that an automorphism of a graph is a bijective mapping of the vertex set that preserves adjacencies. Thus, an automorphism of $\mathcal{T}_d$ preserves the root and "levels" of the tree. The set of all automorphisms of $\mathcal{T}_d$ is a group, which we denote by $\mathrm{Aut}(\mathcal{T}_d)$. We denote the permutation group of Σ as $\mathrm{Sym}(\Sigma)$. An important observation is that $\mathrm{Aut}(\mathcal{T}_d)$ can be seen as the wreath product $\mathrm{Aut}(\mathcal{T}_d) \wr \mathrm{Sym}(\Sigma)$, since any automorphism $\alpha \in \mathrm{Aut}(\mathcal{T}_d)$ can be written uniquely as $\alpha = (\alpha_1, \alpha_2, \dots, \alpha_d) \cdot \sigma$ where $\alpha_i \in \mathrm{Aut}(\mathcal{T}_d)$ is an automorphism of the sub-tree with root a_i, and $\sigma \in \mathrm{Sym}(\Sigma)$ is a permutation of the first level. Let $\alpha \in \mathrm{Aut}(\mathcal{T}_d)$ where $\alpha = (\alpha_1, \alpha_2, \dots, \alpha_d) \cdot \sigma \in \mathrm{Aut}(\mathcal{T}_d) \wr \mathrm{Sym}(\Sigma)$. For any $b = a_i \in \Sigma$, the restriction of α to b, denoted $\alpha|_b := \alpha_i$, is the action of α on the sub-tree

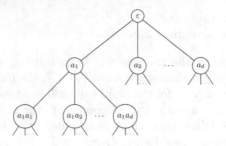

Fig. 2. A labelling of the vertices of $\mathcal{T}_d$ with the root labelled ε

with root b. Given any vertex $w = w_1w_2\cdots w_k \in \Sigma^*$ of $\mathcal{T}_d$, we can define the *restriction of α to w* recursively as

$$\alpha|_w = \left(\alpha|_{w_1w_2\cdots w_{k-1}}\right)\Big|_{w_k}$$

and thus describe the action of α on the sub-tree with root w.

A *Σ-automaton*, (Γ, v), is a finite directed graph with a distinguished vertex v, called the *initial state*, and a $(\Sigma \times \Sigma)$-labelling of its edges, such that each vertex has exactly $|\Sigma|$ outgoing edges: with one outgoing edge with a label of the form $(a, \cdot)$ and one with a label of the form $(\cdot, a)$ for each $a \in \Sigma$.

Given some Σ-automaton (Γ, v), where $\Sigma = \{a_1, \ldots, a_d\}$, we can define an automorphism $\alpha_{(\Gamma, v)} \in \text{Aut}(\mathcal{T}_d)$ as follows. For any given vertex $b_1b_2\cdots b_k \in \Sigma^* = V(\mathcal{T}_d)$, there exists a unique path in Γ starting from the initial vertex, v, of the form $(b_1, b_1')\,(b_2, b_2')\,\cdots\,(b_k, b_k')$, thus we will now define $\alpha_{(\Gamma, v)}$ such that $\alpha_{(\Gamma, v)}(b_1b_2\cdots b_k) = b_1'b_2'\cdots b_k'$. Notice that it follows from the definition of a Σ-automaton that $\alpha_{(\Gamma, v)}$ is a tree automorphism as required.

An *automaton automorphism*, α, of the tree $\mathcal{T}_d$ is an automorphism for which there exists a Σ-automaton, (Γ, v), such that $\alpha = \alpha_{(\Gamma, v)}$. The set of all automaton automorphisms of the tree $\mathcal{T}_d$ form a group which we will denote as $\mathcal{A}(\mathcal{T}_d)$. A subgroup of $\mathcal{A}(\mathcal{T}_d)$ is called an *automata group*.

An automorphism $\alpha \in \text{Aut}(\mathcal{T}_d)$ will be called *bounded* (originally defined in [24]) if there exists a constant $N \in \mathbb{N}$ such that for each $k \in \mathbb{N}$, there are no more than N vertices $v \in \Sigma^*$ with $|v| = k$ (i.e. at level k) such that $\alpha|_v \neq 1$. Thus, the action of such a bounded automorphism will, on each level, be trivial on all but (up to) N sub-trees. The set of all such automorphisms form a group which we will denote as $\mathcal{B}(\mathcal{T}_d)$. The group of all *bounded automaton automorphisms* is defined as the intersection $\mathcal{A}(\mathcal{T}_d) \cap \mathcal{B}(\mathcal{T}_d)$, which we will denote as $\mathcal{D}(\mathcal{T}_d)$. A subgroup of $\mathcal{D}(\mathcal{T}_d)$ is called a *bounded automata group*.

A *finitary automorphism* of $\mathcal{T}_d$ is an automorphism ϕ such that there exists a constant $k \in \mathbb{N}$ for which $\phi|_v = 1$ for each $v \in \Sigma^*$ with $|v| = k$. Thus, a finitary automorphism is one that is trivial after some k levels of the tree. Given a finitary automorphism ϕ, the smallest k for which this definition holds will be called its *depth* and will be denoted as $\text{depth}(\phi)$. We will denote the group formed by all finitary automorphisms of $\mathcal{T}_d$ as $\text{Fin}(\mathcal{T}_d)$. See Fig. 3 for examples of the actions of finitary automorphisms on their associated trees (where any unspecified sub-tree is fixed by the action).

Fig. 3. Examples of finitary automorphisms $a, b \in \mathrm{Fin}(\mathcal{T}_2)$

Let $\delta \in \mathcal{A}(\mathcal{T}_d) \setminus \mathrm{Fin}(\mathcal{T}_d)$. We call δ a *directed automaton automorphism* if

$$\delta = (\phi_1, \phi_2, \ldots, \phi_{k-1}, \delta', \phi_{k+1}, \ldots, \phi_d) \cdot \sigma \in \mathrm{Aut}(\mathcal{T}_d) \wr \mathrm{Sym}(\Sigma) \qquad (2)$$

where each ϕ_j is finitary and δ' is also directed automaton (that is, not finitary and can also be written in this form). We call $\mathrm{dir}(\delta) = b = a_k \in \Sigma$, where $\delta' = \delta|_b$ is directed automaton, the *direction* of δ; and we define the *spine* of δ, denoted $\mathrm{spine}(\delta) \in \Sigma^\omega$, recursively such that $\mathrm{spine}(\delta) = \mathrm{dir}(\delta)\,\mathrm{spine}(\delta')$. We will denote the set of all directed automaton automorphisms as $\mathrm{Dir}(\mathcal{T}_d)$. See Fig. 4 for examples of directed automaton automorphisms (in which a and b are the finitary automorphisms in Fig. 3).

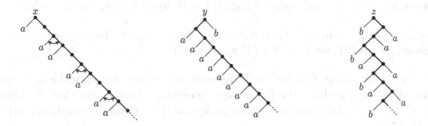

Fig. 4. Examples of directed automata automorphisms $x, y, z \in \mathrm{Dir}(\mathcal{T}_2)$

The following lemma is essential to prove our main theorem.

Lemma 3. *The spine,* $\mathrm{spine}(\delta) \in \Sigma^\omega$, *of a directed automaton automorphism,* $\delta \in \mathrm{Dir}(\mathcal{T}_d)$, *is eventually periodic, that is, there exists some* $\iota = \iota_1 \iota_2 \cdots \iota_s \in \Sigma^\star$, *called the* initial section, *and* $\pi = \pi_1 \pi_2 \cdots \pi_t \in \Sigma^\star$, *called the* periodic section, *such that* $\mathrm{spine}(\delta) = \iota\,\pi^\omega$; *and*

$$\delta|_{\iota\,\pi^k\,\pi_1\pi_2\cdots\pi_j} = \delta|_{\iota\,\pi_1\pi_2\cdots\pi_j} \qquad (3)$$

for each $k, j \in \mathbb{N}$ *with* $0 \leqslant j < t$.

Proof. Let (Γ, v) be a Σ-automaton such that $\delta = \alpha_{(\Gamma,v)}$. By the definition of Σ-automata, for any given vertex $w = w_1 w_2 \cdots w_k \in \Sigma^\star$ of T_d there exists a vertex $v_w \in \mathrm{V}(\Gamma)$ such that $\delta|_w = \alpha_{(\Gamma,v_w)}$. In particular, such a vertex v_w can be obtained by following the path with edges labelled $(w_1, w_1')(w_2, w_2') \cdots (w_k, w_k')$. Then, since there are only finitely many vertices in Γ, the set of all restrictions of δ is finite, that is,

$$\left| \left\{ \delta|_w = \alpha_{(\Gamma,v_w)} \,:\, w \in \Sigma^\star \right\} \right| < \infty. \tag{4}$$

Let $b = b_1 b_2 b_3 \cdots = \mathrm{spine}(\delta) \in \Sigma^\omega$ denote the spine of δ. Then, there exists some $n, m \in \mathbb{N}$ with $n < m$ such that

$$\delta|_{b_1 b_2 \cdots b_n} = \delta|_{b_1 b_2 \cdots b_n \cdots b_m} \tag{5}$$

as otherwise there would be infinitely many distinct restrictions of the form $\delta|_{b_1 b_2 \cdots b_k}$ thus contradicting (4). By the definition spine, it follows that

$$\mathrm{spine}\left(\delta|_{b_1 b_2 \cdots b_n}\right) = (b_{n+1} b_{n+2} \cdots b_m)\, \mathrm{spine}\left(\delta|_{b_1 b_2 \cdots b_n \cdots b_m}\right).$$

and hence, by (5),

$$\mathrm{spine}\left(\delta|_{b_1 b_2 \cdots b_n}\right) = (b_{n+1} b_{n+2} \cdots b_m)^\omega.$$

Thus,

$$\mathrm{spine}(\delta) = (b_1 b_2 \cdots b_n)\, \mathrm{spine}\left(\delta|_{b_1 b_2 \cdots b_n}\right) = (b_1 b_2 \cdots b_n)\, (b_{n+1} b_{n+2} \cdots b_m)^\omega.$$

Then, by taking $\iota = b_1 b_2 \cdots b_n$ and $\pi = b_{n+1} b_{n+2} \cdots b_m$, we have $\mathrm{spine}(\delta) = \iota \pi^\omega$. Moreover, from (5), we have Eq. (3) as required. $\qquad\square$

Notice that each finitary and directed automata automorphism is also bounded, in fact, we have the following proposition which shows that the generators of any given bounded automata group can be written as words in $\mathrm{Fin}(T_d)$ and $\mathrm{Dir}(T_d)$.

Proposition 4 (Proposition 16 in [24]). *The group $\mathcal{D}(T_d)$ of bounded automata automorphisms is generated by $\mathrm{Fin}(T_d)$ together with $\mathrm{Dir}(T_d)$.*

4   Main Theorem

Theorem 5. *Every finitely generated bounded automata group is co-ET0L.*

The idea of the proof is straightforward: we construct a cspd machine that chooses a vertex $v \in \mathrm{V}(T_d)$, writing its label on the check-stack and a copy on its pushdown; as it reads letters from input, it uses the pushdown to keep track of where the chosen vertex is moved; and finally it checks that the pushdown and check-stack differ. The full details are as follows.

Proof. Let $G \subseteq \mathcal{D}(\mathcal{T}_d)$ be a bounded automata group with finite symmetric generating set X. By Proposition 4, we can define a map

$$\varphi : X \rightarrow (\text{Fin}(\mathcal{T}_d) \cup \text{Dir}(\mathcal{T}_d))^\star$$

so that $x =_{\mathcal{D}(\mathcal{T}_d)} \varphi(x)$ for each $x \in X$. Let

$$Y = \left\{ \alpha \in \text{Fin}(\mathcal{T}_d) \cup \text{Dir}(\mathcal{T}_d) \ : \ \alpha \text{ or } \alpha^{-1} \text{ is a factor of } \varphi(x) \text{ for some } x \in X \right\}$$

which is finite and symmetric. Consider the group $H \subseteq \mathcal{D}(\mathcal{T}_d)$ generated by Y. Since ET0L is closed under inverse word homomorphism, it suffices to prove that $\text{coW}(H, Y)$ is ET0L, as $\text{coW}(G, X)$ is its inverse image under the mapping $X^\star \rightarrow Y^\star$ induced by φ. We construct a cspd machine $\mathcal{M}$ that recognises $\text{coW}(H, Y)$, thus proving that G is co-ET0L.

Let $\alpha = \alpha_1 \alpha_2 \cdots \alpha_n \in Y^\star$ denote an input word given to $\mathcal{M}$. The execution of the cspd will be separated into four stages; (1) choosing a vertex $v \in \Sigma^\star$ of $\mathcal{T}_d$ which witnesses the non-triviality of α (and placing it on the stacks); (2a) reading a finitary automorphism from the input tape; (2b) reading a directed automaton automorphism from the input tape; and (3) checking that the action of α on v that it has computed is non-trivial.

After Stage 1, $\mathcal{M}$ will be in state q_{comp}. From here, $\mathcal{M}$ nondeterministically decides to either read from its input tape, performing either Stage 2a or 2b and returning to state q_{comp}; or to finish reading from input by performing Stage 3.

We set both the check-stack and pushdown alphabets to be $\Delta = \Gamma = \Sigma \sqcup \{t\}$.

Stage 1: Choosing a Witness $v = v_1 v_2 \cdots v_m \in \Sigma^\star$.

If α is non-trivial, then there must exist a vertex $v \in \Sigma^\star$ such that $\alpha \cdot v \neq v$. Thus, we nondeterministically choose such a witness from $\mathcal{R} = \Sigma^\star t$ and store it on the check-stack, where the letter t represents the top of the check-stack.

From the start state, q_0, $\mathcal{M}$ will copy the contents of the check-stack onto the pushdown, then enter the state $q_{\text{comp}} \in Q$. Formally, this will be achieved by adding the transitions (for each $a \in \Sigma$):

$$((q_0, \varepsilon, (b, b)), (q_0, tb)), \ ((q_0, \varepsilon, (a, t)), (q_0, ta)), \ ((q_0, \varepsilon, (t, t)), (q_{\text{comp}}, t)).$$

This stage concludes with $\mathcal{M}$ in state q_{comp}, and the read-head pointing to (t, t). Note that whenever the machine is in state q_{comp} and $\alpha_1 \alpha_2 \cdots \alpha_k$ has been read from input, then the contents of pushdown will represent the permuted vertex $(\alpha_1 \alpha_2 \cdots \alpha_k) \cdot v$. Thus, the two stacks are initially the same as no input has been read and thus no group action has been simulated. In Stages 2a and 2b, only the height of the check-stack is impotant, that is, the exact contents of the check-stack will become relevant in Stage 3.

Stage 2a: Reading a Finitary Automorphism $\phi \in Y \cap \text{Fin}(\mathcal{T}_d)$.

By definition, there exists some $k_\phi = \text{depth}(\phi) \in \mathbb{N}$ such that $\phi|_u = 1$ for each $u \in \Sigma^\star$ for which $|u| \geqslant k_\phi$. Thus, given a vertex $v = v_1 v_2 \cdots v_m \in \Sigma^\star$, we have

$$\phi(v) = \phi(v_1 v_2 \cdots v_{k_\phi}) \, v_{(k_\phi+1)} \cdots v_m.$$

Given that $\mathcal{M}$ is in state q_{comp} with $tv_1v_2\cdots v_m\flat$ on its pushdown, we will read ϕ from input, move to state $q_{\phi,\varepsilon}$ and pop the t; we will then pop the next k_ϕ (or fewer if $m < k_\phi$) letters off the pushdown, and as we are popping these letters we visit the sequence of states q_{ϕ,v_1}, q_{ϕ,v_1v_2}, $\ldots$, $q_{\phi,v_1v_2\cdots v_{k_\phi}}$. From the final state in this sequence, we then push $t\phi(v_1\cdots v_{k_\phi})$ onto the pushdown, and return to the state q_{comp}.

Formally, for letters $a, b \in \Sigma$, $\phi \in Y \cap \text{Fin}(\mathcal{T}_d)$, and vertices $u, w \in \Sigma^\star$ where $|u| < k_\phi$ and $|w| = k_\phi$, we have the transitions

$$((q_{\text{comp}}, \phi, (t, t)), (q_{\phi,\varepsilon}, \varepsilon)), \quad ((q_{\phi,u}, \varepsilon, (a, b)), (q_{\phi,ub}, \varepsilon)),$$

$$((q_{\phi,w}, \varepsilon, (\varepsilon, \varepsilon)), (q_{\text{comp}}, t\phi(w)))$$

for the case where $m > k_\phi$, and

$$((q_{\phi,u}, \varepsilon, (\flat, \flat)), (q_{\text{comp}}, t\phi(u)\flat))$$

for the case where $m \leqslant k_\phi$. Notice that we have finitely many states and transitions since Y, Σ and each k_ϕ is finite.

Stage 2b: Reading a Directed Automorphism $\delta \in Y \cap \text{Dir}(\mathcal{T}_d)$.

By Lemma 3, there exists some $\iota = \iota_1\iota_2\cdots\iota_s \in \Sigma^\star$ and $\pi = \pi_1\pi_2\cdots\pi_t \in \Sigma^\star$ such that $\text{spine}(\delta) = \iota\pi^\omega$ and

$$\delta(\iota\pi^\omega) = I_1 I_2 \cdots I_s \, (\Pi_1\Pi_2\cdots\Pi_t)^\omega$$

where

$$I_i = \delta|_{\iota_1\iota_2\cdots\iota_{i-1}}(\iota_i) \qquad \text{and} \qquad \Pi_j = \delta|_{\iota\,\pi_1\pi_2\cdots\pi_{j-1}}(\pi_j).$$

Given some vertex $v = v_1v_2\cdots v_m \in \Sigma^\star$, let $\ell \in \mathbb{N}$ be largest such that $p = v_1v_2\cdots v_\ell$ is a prefix of the sequence $\iota\pi^\omega = \text{spine}(\delta)$. Then by definition of directed automorphism, $\delta' = \delta|_p$ is directed and $\phi = \delta|_a$, where $a = v_\ell$, is finitary. Then, either $p = \iota_1\iota_2\cdots\iota_\ell$ and

$$\delta(u) = (I_1 I_2 \cdots I_\ell) \, \delta'(a) \, \phi(v_{\ell+2}v_{\ell+3}\cdots v_m),$$

or $p = \iota\pi^k\pi_1\pi_2\cdots\pi_j$, with $\ell = |\iota| + k\cdot|\pi| + j$, and

$$\delta(u) = (I_1 I_2 \cdots I_s)\,(\Pi_1\Pi_2\cdots\Pi_t)^k\,(\Pi_1\Pi_2\cdots\Pi_j)\,\delta'(a)\,\phi(v_{\ell+2}v_{\ell+3}\cdots v_m).$$

Hence, from state q_{comp} with $tv_1v_2\cdots v_m\flat$ on its pushdown, $\mathcal{M}$ reads δ from input, moves to state $q_{\delta,\iota,0}$ and pops the t; it then pops pa off the pushdown, using states to remember the letter a and the part of the prefix to which the final letter of p belongs (i.e. ι_i or π_j). From here, $\mathcal{M}$ performs the finitary automorphism ϕ on the remainder of the pushdown (using the same construction as Stage 2a), then, in a sequence of transitions, pushes $t\delta(p)\delta'(a)$ and returns to state q_{comp}. The key idea here is that, using only the knowledge of the letter a, the part of ι or π to which the final letter of p belongs, and the height of the check-stack, that $\mathcal{M}$ is able to recover $\delta(p)\delta'(a)$.

We now give the details of the states and transitions involved in this stage of the construction.

We have states $q_{\delta,\iota,i}$ and $q_{\delta,\pi,j}$ with $0 \leqslant i \leqslant |\iota|$, $1 \leqslant j \leqslant |\pi|$; where $q_{\delta,\iota,i}$ represents that the word $\iota_1\iota_2\cdots\iota_i$ has been popped off the pushdown, and $q_{\delta,\pi,j}$ represents that a word $\iota\pi^k\pi_1\pi_2\cdots\pi_j$ for some $k \in \mathbb{N}$ has been popped of the pushdown. Thus, we begin with the transition

$$((q_{\mathrm{comp}}, \delta, (\mathsf{t}, \mathsf{t})), (q_{\delta,\iota,0}, \varepsilon)),$$

then for each $i, j \in \mathbb{N}$, $a \in \Sigma$ with $0 \leqslant i < |\iota|$ and $1 \leqslant j < |\pi|$, we have transitions

$$((q_{\delta,\iota,i}, \varepsilon, (a, \iota_{i+1})), (q_{\delta,\iota,(i+1)}, \varepsilon)), \quad ((q_{\delta,\iota,|\iota|}, \varepsilon, (a, \pi_1)), (q_{\delta,\pi,1}, \varepsilon)),$$
$$((q_{\delta,\pi,j}, \varepsilon, (a, \pi_{j+1})), (q_{\delta,\pi,(j+1)}, \varepsilon)), \quad ((q_{\delta,\pi,|\pi|}, \varepsilon, (a, \pi_1)), (q_{\delta,\pi,1}, \varepsilon))$$

to consume the prefix p.

After this, $\mathcal{M}$ will either be at the bottom of its stacks, or its read-head will see a letter on the pushdown that is not the next letter in the spine of δ. Thus, for each $i, j \in \mathbb{N}$ with $0 \leqslant i \leqslant |\iota|$ and $1 \leqslant j \leqslant |\pi|$ we have states $q_{\delta,\iota,i,a}$ and $q_{\delta,\pi,j,a}$; and for each $b \in \Sigma$ we have transitions

$$((q_{\delta,\iota,i}, \varepsilon, (b, a)), (q_{\delta,\iota,i,a}, \varepsilon))$$

where $a \neq \iota_{i+1}$ when $i < |\iota|$ and $a \neq \pi_1$ otherwise, and

$$((q_{\delta,\pi,j}, \varepsilon, (b, a)), (q_{\delta,\pi,j,a}, \varepsilon))$$

where $a \neq \pi_{j+1}$ when $j < |\pi|$ and $a \neq \pi_1$ otherwise.

Hence, after these transitions, $\mathcal{M}$ has consumed pa from its pushdown and will either be at the bottom of its stacks in some state $q_{\delta,\iota,i}$ or $q_{\delta,\pi,j}$; or will be in some state $q_{\delta,\iota,i,a}$ or $q_{\delta,\pi,j,a}$. Note here that, if $\mathcal{M}$ is in the state $q_{\delta,\iota,i,a}$ or $q_{\delta,\pi,j,a}$, then from Lemma 3 we know $\delta' = \delta|_p$ is equivalent to $\delta|_{\iota_1\iota_2\cdots\iota_i}$ or $\delta|_{\iota\pi_1\pi_2\cdots\pi_j}$, respectively; and further, we know the finitary automorphism $\phi = \delta|_{pa} = \delta'|_a$.

Thus, for each state $q_{\delta,\iota,i,a}$ and $q_{\delta,\pi,a}$ we will follow a similar construction to Stage 2a, to perform the finitary automorphism ϕ to the remaining letters on the pushdown, then push $\delta'(a)$ and return to the state $r_{\delta,\iota,i}$ or $r_{\delta,\pi,j}$, respectively. For the case where $\mathcal{M}$ is at the bottom of its stacks we have transitions

$$((q_{\delta,\iota,i}, \varepsilon, (\mathsf{b}, \mathsf{b})), (r_{\delta,\iota,i}, \mathsf{b})), \quad ((q_{\delta,\pi,i}, \varepsilon, (\mathsf{b}, \mathsf{b})), (r_{\delta,\pi,i}, \mathsf{b}))$$

with $0 \leqslant i \leqslant |\iota|$, $1 \leqslant j \leqslant |\pi|$.

Thus, after following these transitions, $\mathcal{M}$ is in some state $r_{\delta,\iota,i}$ or $r_{\delta,\pi,j}$ and all that remains is for $\mathcal{M}$ to push $\delta(p)$ with $p = \iota_1\iota_2\cdots\iota_i$ or $p = \iota\pi^k\pi_1\pi_2\cdots\pi_k$, respectively, onto its pushdown. Thus, for each $i, j \in \mathbb{N}$ with $0 \leqslant i \leqslant |\iota|$ and $1 \leqslant j \leqslant |\pi|$, we have transitions

$$((r_{\delta,\pi,i}, \varepsilon, (\varepsilon, \varepsilon)), (q_{\mathrm{comp}}, \mathsf{t}I_1I_2\cdots I_i)), \quad ((r_{\delta,\pi,j}, \varepsilon, (\varepsilon, \varepsilon)), (r_{\delta,\pi}, \Pi_1\Pi_2\cdots\Pi_j))$$

where from the state $r_{\delta,\pi}$, through a sequence of transitions, $\mathcal{M}$ will push the remaining $I\Pi^k$ onto the pushdown. In particular, we have transitions

$$((r_{\delta,\pi}, \varepsilon, (\varepsilon, \varepsilon)), (r_{\delta,\pi}, \Pi)), \quad ((r_{\delta,\pi}, \varepsilon, (\varepsilon, \varepsilon)), (q_{\mathrm{comp}}, \mathsf{t}I)),$$

so that $\mathcal{M}$ can nondeterministically push some number of Π's followed by $\mathfrak{t}I$ before it finishes this stage of the computation. We can assume that the machine pushes the correct number of Π's onto its pushdown as otherwise it will not see $\mathfrak{t}$ on its check-stack while in state q_{comp} and thus would not be able to continue with its computation, as every subsequent stage (2a, 2b, 3) of the computation begins with the read-head pointing to $\mathfrak{t}$ on both stacks.

Once again it is clear that this stage of the construction requires only finitely many states and transitions.

Stage 3: Checking that the Action is Non-trivial.

At the beginning of this stage, the contents of the check-stack represent the chosen witness, v, and the contents of the pushdown represent the action of the input word, α, on the witness, i.e., $\alpha \cdot v$.

In this stage $\mathcal{M}$ checks if the contents of its check-stack and pushdown differ. Formally, we have states q_{accept} and q_{check}, with q_{accept} accepting; for each $a \in \Sigma$, we have transitions

$$((q_{comp}, \varepsilon, (\mathfrak{t}, \mathfrak{t})), (q_{check}, \varepsilon)), \quad ((q_{check}, \varepsilon, (a, a)), (q_{check}, \varepsilon))$$

to pop identical entries of the pushdown; and for each $(a, b) \in \Sigma \times \Sigma$ with $a \neq b$ we have a transition

$$((q_{check}, \varepsilon, (a, b)), (q_{accept}, \varepsilon))$$

to accept if the stacks differ by a letter.

Observe that if the two stacks are identical, then there is no path to the accepting state, q_{accept}, and thus $\mathcal{M}$ will reject. Notice also that by definition of cspd automata, if $\mathcal{M}$ moves into q_{check} before all input has been read, then $\mathcal{M}$ will not accept, i.e., an accepting state is only effective if all input is consumed.

Soundness and Completeness.

If α is non-trivial, then there is a vertex $v \in \Sigma^*$ such that $\alpha \cdot v \neq v$, which $\mathcal{M}$ can nondeterministically choose to write on its check-stack and thus accept α. If α is trivial, then $\alpha \cdot v = v$ for each vertex $v \in \Sigma^*$, and there is no choice of checking stack for which $\mathcal{M}$ will accept, so $\mathcal{M}$ will reject.

Thus, $\mathcal{M}$ accepts a word if and only if it is in $\mathrm{coW}(H, Y)$. Hence, the co-word problem $\mathrm{coW}(H, Y)$ is ET0L, completing our proof. $\square$

Acknowledgments. The authors wish to thank Claas Röver, Michal Ferov and Laura Ciobanu for helpful comments.

References

1. Anīsīmov, A.V.: The group languages. Kibernetika (Kiev) **7**(4), 18–24 (1971)
2. Asveld, P.R.J.: Controlled iteration grammars and full hyper-AFL's. Inf. Control **34**(3), 248–269 (1977)
3. Bleak, C., Matucci, F., Neunhöffer, M.: Embeddings into Thompson's group V and $\mathrm{co}\mathcal{CF}$ groups. J. Lond. Math. Soc. (2) **94**(2), 583–597 (2016). https://doi.org/10.1112/jlms/jdw04

4. Ciobanu, L., Diekert, V., Elder, M.: Solution sets for equations over free groups are EDT0L languages. Int. J. Algebra Comput. **26**(5), 843–886 (2016). https://doi.org/10.1142/S0218196716500363
5. Ciobanu, L., Elder, M., Ferov, M.: Applications of L systems to group theory. Int. J. Algebra Comput. **28**(2), 309–329 (2018). https://doi.org/10.1142/S0218196718500145
6. Culik II, K.: On some families of languages related to developmental systems. Int. J. Comput. Math. **4**, 31–42 (1974). https://doi.org/10.1080/00207167408803079
7. Diekert, V., Elder, M.: Solutions of twisted word equations, EDT0L languages, and context-free groups. In: 44th International Colloquium on Automata, Languages, and Programming, LIPIcs. Leibniz International Proceedings in Informatics, vol. 80, Article No. 96, 14. Schloss Dagstuhl. Leibniz-Zent. Inform., Wadern (2017)
8. Diekert, V., Jeż, A., Kufleitner, M.: Solutions of word equations over partially commutative structures. In: 43rd International Colloquium on Automata, Languages, and Programming, LIPIcs. Leibniz International Proceedings in Informatics, vol. 55, Article No. 127, 14. Schloss Dagstuhl. Leibniz-Zent. Inform., Wadern (2016)
9. Elder, M., Kambites, M., Ostheimer, G.: On groups and counter automata. Int. J. Algebra Comput. **18**(8), 1345–1364 (2008). https://doi.org/10.1142/S0218196708004901
10. Epstein, D.B.A., Cannon, J.W., Holt, D.F., Levy, S.V.F., Paterson, M.S., Thurston, W.P.: Word Processing in Groups. Jones and Bartlett Publishers, Boston (1992)
11. Gilman, R.H.: Formal languages and infinite groups. In: Geometric and Computational Perspectives on Infinite Groups, Minneapolis, MN and New Brunswick, NJ, 1994. DIMACS: Series in Discrete Mathematics and Theoretical Computer Science, vol. 25, pp. 27–51. American Mathematical Society, Providence (1996)
12. Grigorchuk, R.: On Burnside's problem on periodic groups. Funktsional. Anal. i Prilozhen. **14**(1), 53–54 (1980)
13. Gupta, N., Sidki, S.: On the Burnside problem for periodic groups. Mathematische Zeitschrift **182**(3), 385–388 (1983). https://doi.org/10.1007/BF01179757
14. Holt, D.F., Rees, S., Röver, C.E., Thomas, R.M.: Groups with context-free co-word problem. J. Lond. Math. Soc. (2) **71**(3), 643–657 (2005). https://doi.org/10.1112/S002461070500654X
15. Holt, D.F., Röver, C.E.: Groups with indexed co-word problem. Int. J. Algebra Comput. **16**(5), 985–1014 (2006). https://doi.org/10.1142/S0218196706003359
16. König, D., Lohrey, M., Zetzsche, G.: Knapsack and subset sum problems in nilpotent, polycyclic, and co-context-free groups. In: Algebra and Computer Science, Contemporary Mathematics, vol. 677, pp. 129–144. American Mathematical Society, Providence (2016)
17. van Leeuwen, J.: Variations of a new machine model. In: 17th Annual Symposium on Foundations of Computer Science, Houston, Texas 1976, pp. 228–235. IEEE Computer Society, Long Beach, October 1976. https://doi.org/10.1109/SFCS.1976.35
18. Lehnert, J., Schweitzer, P.: The co-word problem for the Higman-Thompson group is context-free. Bull. Lond. Math. Soc. **39**(2), 235–241 (2007). https://doi.org/10.1112/blms/bdl043
19. Lindenmayer, A.: Mathematical models for cellular interactions in development I. Filaments with one-sided inputs. J. Theoret. Biol. **18**(3), 280–99 (1968). https://doi.org/10.1016/0022-5193(68)90079-9

20. Muller, D.E., Schupp, P.E.: The theory of ends, pushdown automata, and second-order logic. Theoret. Comput. Sci. **37**(1), 51–75 (1985). https://doi.org/10.1016/0304-3975(85)90087-8

21. Nekrashevych, V.: Self-similar Groups, Mathematical Surveys and Monographs, vol. 117. American Mathematical Society, Providence (2005). https://doi.org/10.1090/surv/117

22. Rozenberg, G., Salomaa, A. (eds.): Handbook of Formal Languages. Springer, Berlin (1997). https://doi.org/10.1007/978-3-642-59126-6

23. Rozenberg, G.: Extension of tabled 0L-systems and languages. Int. J. Comput. Inf. Sci. **2**, 311–336 (1973)

24. Sidki, S.: Automorphisms of one-rooted trees: growth, circuit structure, and acyclicity. J. Math. Sci. (New York) **100**(1), 1925–1943 (2000). https://doi.org/10.1007/BF02677504. Algebra, 12

Decidability of Sensitivity and Equicontinuity for Linear Higher-Order Cellular Automata

Alberto Dennunzio[1], Enrico Formenti[2], Luca Manzoni[1(✉)], Luciano Margara[3], and Antonio E. Porreca[1,4]

[1] Dipartimento di Informatica, Sistemistica e Comunicazione,
Università degli Studi di Milano-Bicocca, Viale Sarca 336/14, 20126 Milan, Italy
{dennunzio,luca.manzoni,porreca}@disco.unimib.it
[2] Universite Côte d'Azur, CNRS, I3S, Nice Cedex, France
enrico.formenti@unice.fr
[3] Department of Computer Science and Engineering, University of Bologna,
Cesena Campus, Via Sacchi 3, Cesena, Italy
luciano.margara@unibo.it
[4] Aix Marseille Université, Université de Toulon, CNRS, LIS, Marseille, France
antonio.porreca@lis-lab.fr

Abstract. We study the dynamical behavior of linear higher-order cellular automata (HOCA) over $\mathbb{Z}_m$. In standard cellular automata the global state of the system at time t only depends on the state at time $t-1$, while in HOCA it is a function of the states at time $t-1, \ldots, t-n$, where $n \geq 1$ is the memory size. In particular, we provide easy-to-check necessary and sufficient conditions for a linear HOCA over $\mathbb{Z}_m$ of memory size n to be sensitive to the initial conditions or equicontinuous. Our characterizations of sensitivity and equicontinuity extend the ones shown in [23] for linear cellular automata (LCA) over $\mathbb{Z}_m^n$ in the case $n = 1$. We also prove that linear HOCA over $\mathbb{Z}_m$ of memory size n are indistinguishable from a subclass of LCA over $\mathbb{Z}_m^n$. This enables to decide injectivity and surjectivity for linear HOCA over $\mathbb{Z}_m$ of memory size n by means of the decidable characterizations of injectivity and surjectivity provided in [2] and [20] for LCA over $\mathbb{Z}_m^n$.

1 Introduction

Cellular automata (CA) are well-known formal models of natural computing which have been successfully applied in a wide number of fields to simulate complex phenomena involving local, uniform, and synchronous processing (for recent results and an up-to date bibliography on CA, see [1,6,7,16,25], while for other models of natural computing see for instance [9,12,17]). More formally, a CA is made of an infinite set of identical finite automata arranged over a regular cell grid (usually $\mathbb{Z}^d$ in dimension d) and all taking a state from a finite set S called the *set of states*. In this paper, we consider one-dimensional CA. A *configuration* is a snapshot of all states of the automata, i.e., a function $c : \mathbb{Z} \to S$.

© Springer Nature Switzerland AG 2019
C. Martín-Vide et al. (Eds.): LATA 2019, LNCS 11417, pp. 95–107, 2019.
https://doi.org/10.1007/978-3-030-13435-8_7

A *local rule* updates the state of each automaton on the basis of its current state and the ones of a finite set of neighboring automata. All automata are updated synchronously. In the one-dimensional settings, a CA is a structure $\langle S, r, f \rangle$ where $r \in \mathbb{N}$ is the *radius* and $f : S^{2r+1} \to S$ is the local rule which updates, for each $i \in \mathbb{Z}$, the state of the automaton in the position i of the grid $\mathbb{Z}$ on the basis of states of the automata in the positions $i - r, \ldots, i + r$. A configuration is an element of $S^{\mathbb{Z}}$ and describes the (global) state of the CA. The feature of synchronous updating induces the following *global rule* $F : S^{\mathbb{Z}} \to S^{\mathbb{Z}}$ defined as

$$\forall c \in S^{\mathbb{Z}}, \forall i \in \mathbb{Z}, \qquad F(c)_i = f(c_{i-r}, \ldots c_{i+r}) \ .$$

As such, the global map F describes the change from any configuration c at any time $t \in \mathbb{N}$ to the configuration $F(c)$ at $t + 1$ and summarises the main features of the CA model, namely, the fact that it is defined through a local rule which is applied uniformly and synchronously to all cells.

Because of a possible inadequacy, in some contexts, of every single one of the three defining features, variants of the original CA model started appearing, each one relaxing one among these three features. Asynchronous CA relax synchrony (see [8,10,11,18,26] for instance), non-uniform CA relax uniformity [13–15], while hormonal CA (for instance) relax locality [4]. However, from the mathematical point of view all those systems, as well as the original model, fall in the same class, namely, the class of autonomous discrete dynamical systems (DDS) and one could also precise *memoryless* systems.

In [27], Toffoli introduced *higher-order CA* (HOCA), i.e., variants of CA in which the updating of the state of a cell also depends on the past states of the cell itself and its neighbours. In particular, he showed that any arbitrary reversible *linear* HOCA can be embedded in a reversible *linear CA* (LCA), where linear means that the local rule is linear. Essentially, the trick consisted in memorizing past states and recover them later on. Some years later, Le Bruyn and Van Den Bergh explained and generalized the Toffoli construction and proved that any linear HOCA having the ring $S = \mathbb{Z}_m$ as alphabet and memory size n can be simulated by a linear CA over the alphabet $\mathbb{Z}_m^n$ (see the precise definition in Sect. 2) [2]. In this way, as we will see, a practical way to decide injectivity (which is equivalent to reversibility in this setting) and surjectivity of HOCA can be easily derived by the characterization of the these properties for the corresponding LCA simulating them. Indeed, in [2] and [20], characterizations of injectivity and surjectivity of a LCA over $\mathbb{Z}_m^n$ are provided in terms of properties of the determinant of the matrix associated with it, where the determinant turns out to be another LCA (over $\mathbb{Z}_m$). Since the properties of LCA over $\mathbb{Z}_m$ (i.e., LCA over $\mathbb{Z}_m^n$ with $n = 1$) have been extensively studied and related decidable characterizations have been obtained [3,5,24], one derives the algorithms to decide injectivity and surjectivity for LCA over $\mathbb{Z}_m^n$ and, then, as we will see, also for HOCA over $\mathbb{Z}_m$ of memory size n, by means of the associated matrix. The purpose of the present paper is to study, in the context of linear HOCA, sensitivity to the initial conditions and equicontinuity, where the former is the well-known basic component and essence of the chaotic behavior of a DDS, while the latter represents a strong form of stability. To do that, we put in evidence that any linear HOCA over $\mathbb{Z}_m$ of memory size n is not only simulated by but also topologically conjugated to a LCA over $\mathbb{Z}_m^n$ defined by a matrix having a specific form. Thus, in order to decide injectivity and surjectivity for linear HOCA over $\mathbb{Z}_m$ of memory size n, one can use the decidable characterization provided in [2] and [20] for deciding the same properties for LCA over $\mathbb{Z}_m^n$ by means of that specific matrix. As main result, we prove that sensitivity to the initial condition and equicontinuity are decidable properties for linear HOCA over $\mathbb{Z}_m$ of memory size n (Theorem 2). In particular we provide a decidable characterization of those properties, in terms of the matrix associated with a linear HOCA. Remark that if $n = 1$, starting from our characterizations one recovers exactly the well known characterizations of sensitivity and equicontinuity for LCA over $\mathbb{Z}_m$.

2 Higher-Order CA and Linear CA

We begin by reviewing some general notions and introducing notations we will use throughout the paper.

A *discrete dynamical system* (DDS) is a pair $(\mathcal{X}, \mathcal{F})$ where $\mathcal{X}$ is a space equipped with a metric, i.e., a metric space, and $\mathcal{F}$ is a transformation on $\mathcal{X}$ which is continuous with respect to that metric. The *dynamical evolution* of a DDS $(\mathcal{X}, \mathcal{F})$ starting from the initial state $x^{(0)} \in \mathcal{X}$ is the sequence $\{x^{(t)}\}_{t \in \mathbb{N}} \subseteq \mathcal{X}$ where $x^{(t)} = \mathcal{F}^t(x^{(0)})$ for any $t \in \mathbb{N}$. When $\mathcal{X} = S^{\mathbb{Z}}$ for some set finite S, $\mathcal{X}$ is usually equipped with the metric d defined as follows $\forall c, c' \in S^{\mathbb{Z}}, d(c, c') = \frac{1}{2^n}$, where $n = \min\{i \geq 0 : c_i \neq c'_i \text{ or } c_{-i} \neq c'_{-i}\}$. Recall that $S^{\mathbb{Z}}$ is a Cantor space.

Any CA $\langle S, r, f \rangle$ defines the DDS $(S^{\mathbb{Z}}, F)$, where F is the CA global rule (which is continuous). From now on, for the sake of simplicity, we will sometimes identify a CA with its global rule F or with the DDS $(S^{\mathbb{Z}}, F)$.

Recall that two DDS $(\mathcal{X}, \mathcal{F})$ and $(\mathcal{X}', \mathcal{F}')$ are *topologically conjugated* if there exists a homeomorphism $\phi : \mathcal{X} \mapsto \mathcal{X}'$ such that $\mathcal{F}' \circ \phi = \phi \circ \mathcal{F}$, while the *product* of $(\mathcal{X}, \mathcal{F})$ and $(\mathcal{X}', \mathcal{F}')$ is the DDS $(\mathcal{X} \times \mathcal{X}', \mathcal{F} \times \mathcal{F}')$ where $\mathcal{F} \times \mathcal{F}'$ is defined as $\forall (x, x') \in \mathcal{X} \times \mathcal{X}', (\mathcal{F} \times \mathcal{F}')(x, x') = (\mathcal{F}(x), \mathcal{F}'(x'))$.

Notation 1. *For all $i, j \in \mathbb{Z}$ with $i \leq j$, we write $[i, j] = \{i, i + 1, \ldots, j\}$ to denote the interval of integers between i and j. For any $n \in \mathbb{N}$ and any set Z the set of all $n \times n$ matrices with coefficients in Z and the set of Laurent polynomials with coefficients in Z will be noted by $Mat(n, Z)$ and $Z[X, X^{-1}]$, respectively. In the sequel, bold symbols are used to denote vectors, matrices, and configurations over a set of states which is a vectorial space. Moreover, m will be an integer bigger than 1 and $\mathbb{Z}_m = \{0, 1, \ldots, m-1\}$ the ring with the usual sum and product modulo m. For any $x \in \mathbb{Z}^n$ (resp., any matrix $M(X) \in Mat(n, \mathbb{Z}[X, X^{-1}])$), we will denote by $[x]_m \in \mathbb{Z}_m^n$ (resp., $[M(X)]_m$), the vector (resp., the matrix) in which each component x^i of x (resp., every coefficient of each element of $M(X)$) is taken modulo m. Finally, for any matrix $M(X) \in \mathbb{Z}_m[X, X^{-1}]$ and any $t \in \mathbb{N}$, the t-th power of $M(X)$ will be noted more simply by $M^t(X)$ instead of $(M(X))^t$.*

Definition 1 (Higher-Order Cellular Automata). A *Higher-Order Cellular Automata (HOCA)* is a structure $\mathcal{H} = \langle k, S, r, h \rangle$ where $k \in \mathbb{N}$ with $k \geq 1$ is the *memory* size, S is the *alphabet*, $r \in \mathbb{N}$ is the *radius*, and $h \colon S^{(2r+1)k} \to S$ is the *local rule*. Any HOCA $\mathcal{H}$ induces the *global rule* $H \colon \left(S^{\mathbb{Z}}\right)^k \to \left(S^{\mathbb{Z}}\right)^k$ associating any vector $e = (e^1, \dots, e^k) \in \left(S^{\mathbb{Z}}\right)^k$ of k configurations of $S^{\mathbb{Z}}$ with the vector $H(e) \in \left(S^{\mathbb{Z}}\right)^k$ such that $H(e)^j = e^{j+1}$ for each $j \neq k$ and $\forall i \in \mathbb{Z}, H(e)_i^k = h\begin{pmatrix} e_{[i-r,i+r]}^1 \\ \vdots \\ e_{[i-r,i+r]}^k \end{pmatrix}$. In this way, $\mathcal{H}$ defines the DDS $\left(\left(S^{\mathbb{Z}}\right)^k, H\right)$. As with CA, we identify a HOCA with its global rule or the DDS defined by it.

Remark 1. It is easy to check that for any HOCA $\mathcal{H} = \langle k, S, r, h \rangle$ there exists a CA $\langle S^k, r, f \rangle$ which is topologically conjugated to $\mathcal{H}$.

The study of the dynamical behaviour of HOCA is still at its early stages; a few results are known for the class of *linear HOCA*, namely, those HOCA defined by a local rule f which is *linear*, i.e., S is $\mathbb{Z}_m$ and there exist coefficients $a_i^j \in \mathbb{Z}_m$ ($j = 1, \dots, k$ and $i = -r, \dots, r$) such that for any element

$$x = \begin{pmatrix} x_{-r}^1 & \cdots & x_r^1 \\ \vdots & \vdots & \vdots \\ x_{-r}^k & \cdots & x_r^k \end{pmatrix} \in \mathbb{Z}_m^{(2r+1)k}, \quad f(x) = \left[\sum_{j=1}^{k} \sum_{i=-r}^{r} a_i^j x_i^j \right]_m .$$

Clearly, linear HOCA are additive, i.e., $\forall c, d \in \left(\mathbb{Z}_m^{\mathbb{Z}}\right)^k, H(c) + H(d)$, where, with an abuse of notation, $+$ denotes the extension of the sum over $\mathbb{Z}_m$ to both $\mathbb{Z}_m^{\mathbb{Z}}$ and $\left(\mathbb{Z}_m^{\mathbb{Z}}\right)^k$.

In [2], a much more convenient representation is introduced for the case of linear HOCA (in dimension $d = 1$) by means of the following notion.

Definition 2 (Linear Cellular Automata). A *Linear Cellular Automaton* (LCA) is a CA $\mathcal{L} = \langle \mathbb{Z}_m^n, r, f \rangle$ where the local rule $f \colon (\mathbb{Z}_m^n)^{2r+1} \to \mathbb{Z}_m^n$ is defined by $2r + 1$ matrices $M_{-r}, \dots, M_0, \dots, M_r \in Mat\,(n, \mathbb{Z}_m)$ as follows: $f(x_{-r}, \dots, x_0, \dots, x_r) = \left[\sum_{i=-r}^{r} M_i \cdot x_i \right]_m$ for any $(x_{-r}, \dots, x_0, \dots, x_r) \in (\mathbb{Z}_m^n)^{2r+1}$.

Remark 2. LCA have been strongly investigated in the case $n = 1$ and all the dynamical properties have been characterized in terms of the 1×1 matrices (i.e., coefficients) defining the local rule, in any dimension too [3, 24].

We recall that any linear HOCA $\mathcal{H}$ can be simulated by a suitable LCA, as shown in [2]. Precisely, given a linear HOCA $\mathcal{H} = \langle k, \mathbb{Z}_m, r, h \rangle$, where h is defined by the coefficients $a_i^j \in \mathbb{Z}_m$, the LCA simulating $\mathcal{H}$ is $\mathcal{L} = \langle \mathbb{Z}_m^k, r, f \rangle$ with f defined by following matrices

$$M_0 = \begin{bmatrix} 0 & 1 & 0 & \cdots & 0 & 0 \\ 0 & 0 & 1 & \ddots & 0 & 0 \\ 0 & 0 & 0 & \ddots & 0 & 0 \\ \vdots & \vdots & \vdots & \ddots & \ddots & \vdots \\ 0 & 0 & 0 & \cdots & 0 & 1 \\ a_0^1 & a_0^2 & a_0^3 & \cdots & a_0^{k-1} & a_0^k \end{bmatrix} , \text{and} \quad M_i = \begin{bmatrix} 0 & 0 & 0 & \cdots & 0 & 0 \\ 0 & 0 & 0 & \cdots & 0 & 0 \\ 0 & 0 & 0 & \cdots & 0 & 0 \\ \vdots & \vdots & \vdots & \ddots & \vdots & \vdots \\ 0 & 0 & 0 & \cdots & 0 & 0 \\ a_i^1 & a_i^2 & a_i^3 & \cdots & a_i^{k-1} & a_i^k \end{bmatrix} , \quad (1)$$

for each $i \in [-r, r]$ with $i \neq 0$.

Remark 3. We want to put in evidence that a stronger result actually holds (easy proof, important remark): any linear HOCA $\mathcal{H}$ is topologically conjugated to the LCA $\mathcal{L}$ defined by the matrices in (1). Clearly, the converse also holds: for any LCA defined by the matrices in (1) there exists a linear HOCA which is topologically conjugated to it. In other words, up to a homeomorphism the whole class of linear HOCA is identical to the subclass of LCA defined by the matrices above introduced. In the sequel, we will call $\mathcal{L}$ the *matrix presentation* of $\mathcal{H}$.

We are now going to show a stronger and useful new fact, namely, that the class of linear HOCA is nothing but the subclass of LCA represented by a formal power series which is a matrix in Frobenius normal form. Before proceeding, let us recall the *formal power series* (fps) which have been successfully used to study the dynamical behaviour of LCA in the case $n = 1$ [19,24]. The idea of fps is that configurations and global rules are represented by suitable polynomials and the application of the global rule turns into multiplications of polynomials. In the more general case of LCA over $\mathbb{Z}_m^n$, a configuration $c \in (\mathbb{Z}_m^n)^{\mathbb{Z}}$ can be associated with the fps

$$P_c(X) = \sum_{i \in \mathbb{Z}} c_i X^i = \begin{bmatrix} c^1(X) \\ \vdots \\ c^n(X) \end{bmatrix} = \begin{bmatrix} \sum_{i \in \mathbb{Z}} c_i^1 X^i \\ \vdots \\ \sum_{i \in \mathbb{Z}} c_i^n X^i \end{bmatrix}.$$

Then, if F is the global rule of a LCA defined by $M_{-r}, \ldots, M_0, \ldots, M_r$, one finds $P_{F(c)}(X) = [M(X)P_c(X)]_m$ where

$$M(X) = \left[\sum_{i=-r}^{r} M_i X^{-i} \right]_m$$

is the finite fps associated with the LCA F. In this way, for any integer $t > 0$ the fps associated with F^t is $M(X)^t$, and then $P_{F^t(c)}(X) = [M(X)^t P_c(X)]_m$. Throughout this paper, $M(X)^t$ will refer to $[M(X)^t]_m$.

A matrix $M(X) \in Mat\left(n, Z\left[X, X^{-1}\right]\right)$ is in *Frobenius normal form* if

$$M(X) = \begin{bmatrix} 0 & 1 & 0 & \cdots & 0 & 0 \\ 0 & 0 & 1 & \ddots & 0 & 0 \\ 0 & 0 & 0 & \ddots & 0 & 0 \\ \vdots & \vdots & \vdots & \ddots & \ddots & \vdots \\ 0 & 0 & 0 & \cdots & 0 & 1 \\ m_0(X) & m_1(X) & m_2(X) & \cdots & m_{n-2}(X) & m_{n-1}(X) \end{bmatrix} \tag{2}$$

where each $m_i(X) \in Z\left[X, X^{-1}\right]$. From now on, $m(X)$ will always make reference to the n-th row of a matrix $M(X) \in Mat\left(n, Z\left[X, X^{-1}\right]\right)$ in Frobenius normal form.

Definition 3 (Frobenius LCA). A LCA F over the alphabet $\mathbb{Z}_m^n$ is said to be a *Frobenius LCA* if the fps $M(X) \in Mat\left(n, \mathbb{Z}_m\left[X, X^{-1}\right]\right)$ associated with F is in Frobenius normal form.

It is immediate to see that a LCA is a Frobenius one iff it is defined by the matrices in (1), i.e., iff it is topologically conjugated to a linear HOCA. This fact together with Remark 3 and Definition 3, allow us to state the following

Proposition 1. *Up to a homeomorphism, the class of linear HOCA over $\mathbb{Z}_m$ of memory size n is nothing but the class of Frobenius LCA over $\mathbb{Z}_m^n$.*

Remark 4. Actually, in literature a matrix is in Frobenius normal form if either it or its transpose has a form as in (2). Since any matrix in Frobenius normal form is conjugated to its transpose, any Frobenius LCA F is topologically conjugated to a LCA G such that the fps associated with G is the transpose of the fps associated with G. In other words, up to a homeomorphism, such LCA G, linear HOCA, and Frobenius LCA form the same class.

From now on, we will focus on Frobenius LCA, i.e., matrix presentations of linear HOCA. Indeed, they allow convenient algebraic manipulations that are very useful to study formal properties of linear HOCA. For example, in [2] and [20], the authors proved decidable characterization for injectivity and surjectivity for LCA in terms of the matrix $M(X)$ associated to them. We want to stress that, by Remark 3 and Definition 3, one can use these characterizations for deciding injectivity and surjectivity of linear HOCA. In this paper we are going to adopt a similar attitude, i.e., we are going to characterise the dynamical behaviour of linear HOCA by the properties of the matrices in their matrix presentation.

3   Dynamical Properties

In this paper we are particularly interested to the so-called *sensitivity to the initial conditions* and *equicontinuity*. As dynamical properties, they represent the main features of instable and stable DDS, respectively. The former is the well-known basic component and essence of the chaotic behavior of DDS, while the latter is a strong form of stability.

Let $(\mathcal{X}, \mathcal{F})$ be a DDS. The DDS $(\mathcal{X}, \mathcal{F})$ is *sensitive to the initial conditions* (or simply *sensitive*) if there exists $\varepsilon > 0$ such that for any $x \in \mathcal{X}$ and any $\delta > 0$ there is an element $y \in \mathcal{X}$ such that $d(y, x) < \delta$ and $d(\mathcal{F}^n(y), \mathcal{F}^n(x)) > \varepsilon$ for some $n \in \mathbb{N}$. Recall that, by Knudsen's Lemma [21], $(\mathcal{X}, \mathcal{F})$ is sensitive iff $(\mathcal{Y}, \mathcal{F})$ is sensitive where $\mathcal{Y}$ is any dense subset of $\mathcal{X}$ which is $\mathcal{F}$-invariant, i.e., $\mathcal{F}(\mathcal{Y}) \subseteq \mathcal{Y}$.

In the sequel, we will see that in the context of LCA an alternative way to study sensitivity is via equicontinuity points. An element $x \in \mathcal{X}$ is an *equicontinuity point* for $(\mathcal{X}, \mathcal{F})$ if $\forall \varepsilon > 0$ there exists $\delta > 0$ such that for all $y \in \mathcal{X}$, $d(x, y) < \delta$ implies that $d(\mathcal{F}^n(y), \mathcal{F}^n(x)) < \varepsilon$ for all $n \in \mathbb{N}$. The system $(\mathcal{X}, \mathcal{F})$ is said to be *equicontinuous* if $\forall \varepsilon > 0$ there exists $\delta > 0$ such that for all $x, y \in \mathcal{X}$, $d(x, y) < \delta$ implies that $\forall n \in \mathbb{N}$, $d(\mathcal{F}^n(x), \mathcal{F}^n(y)) < \varepsilon$. Recall that any CA $(S^{\mathbb{Z}}, F)$ is equicontinuous if and only if there exist two integers $q \in \mathbb{N}$ and $p > 0$ such that $F^q = F^{q+p}$ [22]. Moreover, for the subclass of LCA defined by $n = 1$ the following result holds:

Theorem 1 ([24]). *Let $(\mathbb{Z}_m^{\mathbb{Z}}, F)$ be a LCA where the local rule $f \colon (\mathbb{Z}_m)^{2r+1} \to \mathbb{Z}_m$ is defined by $2r + 1$ coefficients $m_{-r}, \ldots, m_0, \ldots, m_r \in \mathbb{Z}_m$. Denote by $\mathcal{P}$ the set of prime factors of m. The following statements are equivalent: (i) F is sensitive to the initial conditions; (ii) F is not equicontinuous; (iii) there exists a prime number $p \in \mathcal{P}$ which does not divide $\gcd(m_{-r}, \ldots, m_{-1}, m_1, \ldots, m_r)$.*

The dichotomy between sensitivity and equicontinuity still holds for general LCA.

Proposition 2. *Let $\mathcal{L} = \langle \mathbb{Z}_m^n, r, f \rangle$ be a LCA where the local rule $f \colon (\mathbb{Z}_m^n)^{2r+1} \to \mathbb{Z}_m^n$ is defined by $2r + 1$ matrices $\mathbf{M}_{-r}, \ldots, \mathbf{M}_0, \ldots, \mathbf{M}_r \in \mathrm{Mat}(n, \mathbb{Z}_m)$. The following statements are equivalent: (i) F is sensitive to the initial conditions; (ii) F is not equicontinuous; (iii) $|\{\mathbf{M}(X)^i, i \geq 1\}| = \infty$.*

Proof. It is clear that conditions *(ii)* and *(iii)* are equivalent. The equivalence between *(i)* and *(ii)* is a consequence of linearity of F and Knudsen's Lemma applied on the subset of the finite configurations, i.e., those having a state different from the null vector only in a finite number of cells. □

An immediate consequence of Proposition 2 is that any characterization of sensitivity to the initial conditions in terms of the matrices defining LCA over $\mathbb{Z}_m^n$ would also provide a characterization of equicontinuity. In the sequel, we are going to show that such a characterization actually exists. First of all, we recall a result that helped in the study of dynamical properties in the case $n = 1$ and we now state it in a more general form for LCA over $\mathbb{Z}_m^n$ (immediate generalisation of the result in [3,5]).

Let $((\mathbb{Z}_m^n)^{\mathbb{Z}}, F)$ be a LCA and let q be any factor of m. We will denote by $[F]_q$ the map $[F]_q : (\mathbb{Z}_q^n)^{\mathbb{Z}} \to (\mathbb{Z}_q^n)^{\mathbb{Z}}$ defined as $[F]_q(c) = [F(c)]_q$, for any $c \in (\mathbb{Z}_q^n)^{\mathbb{Z}}$.

Lemma 1 ([3,5]). *Consider any LCA $((\mathbb{Z}_m^n)^{\mathbb{Z}}, F)$ with $m = pq$ and $\gcd(p,q) = 1$. It holds that the given LCA is topologically conjugated to $\left((\mathbb{Z}_p^n)^{\mathbb{Z}} \times (\mathbb{Z}_q^n)^{\mathbb{Z}}, [F]_p \times [F]_q\right)$.*

As a consequence of Lemma 1, if $m = p_1^{k_1} \cdots p_l^{k_l}$ is the prime factor decomposition of m, any LCA over $\mathbb{Z}_m^n$ is topologically conjugated to the product of LCAs over $\mathbb{Z}_{p_i^{k_i}}^n$. Since sensitivity is preserved under topological conjugacy for DDS over a compact space and the product of two DDS is sensitive if and only if at least one of them is sensitive, we will study sensitivity for Frobenius LCA over $\mathbb{Z}_{p^k}^n$. We will show a decidable characterization of sensitivity to the initial conditions for Frobenius LCA over $\mathbb{Z}_{p^k}^n$ (Lemma 8). Such a decidable characterization together with the previous remarks about the decomposition of m, the topological conjugacy involving any LCA over $\mathbb{Z}_m^n$ and the product of LCAs over $\mathbb{Z}_{p_i^{k_i}}^n$, and how sensitivity behaves with respect to a topological conjugacy and the product of DDS, immediately lead to state the main result of the paper.

Theorem 2. *Sensitivity and Equicontinuity are decidable for Frobenius LCA over $\mathbb{Z}_m^n$, or, equivalently, for linear HOCA over $\mathbb{Z}_m$ of memory size n.*

4   Sensitivity of Frobenius LCA over $\mathbb{Z}_{p^k}^n$

In order to study sensitivity of Frobenius LCA over $\mathbb{Z}_{p^k}^n$, we introduce two concepts about Laurent polynomials.

Definition 4 (deg^+ and deg^-). Given any polynomial $\mathrm{p}(X) \in \mathbb{Z}_{p^k}[X, X^{-1}]$, the *positive* (resp., *negative*) *degree of* $\mathrm{p}(X)$, denoted by $deg^+[\mathrm{p}(X)]$ (resp., $deg^-[\mathrm{p}(X)]$) is the maximum (resp., minimum) degree among those of the monomials having both positive (resp., negative) degree and coefficient which is not multiple of p. If there is no monomial satisfying both the required conditions, then $deg^+[\mathrm{p}(X)] = 0$ (resp., $deg^-[\mathrm{p}(X)] = 0$).

Definition 5 (Sensitive Polynomial). A polynomial $\mathrm{p}(X) \in \mathbb{Z}_{p^k}[X, X^{-1}]$ is *sensitive* if either $deg^+[\mathrm{p}(X)] > 0$ or $deg^-[\mathrm{p}(X)] < 0$. As a consequence, a Laurent polynomial $\mathrm{p}(X)$ is not sensitive iff $deg^+[\mathrm{p}(X)] = deg^-[\mathrm{p}(X)] = 0$.

Trivially, it is decidable to decide whether a Laurent polynomial is sensitive.

Remark 5. Consider a matrix $M(X) \in Mat\left(n, \mathbb{Z}_{p^k}[X, X^{-1}]\right)$ in Frobenius normal form. The characteristic polynomial of $M(X)$ is then $\mathscr{P}(y) = (-1)^n(-\mathrm{m}_0(X) - \mathrm{m}_1(X)y - \cdots - \mathrm{m}_{n-1}(X)y^{n-1} + y^n)$. By the Cayley-Hamilton Theorem, one obtains

$$M^n(X) = \mathrm{m}_{n-1}(X)M(X)^{n-1} + \cdots + \mathrm{m}_1(X)M(X)^1 + \mathrm{m}_0(X)I . \qquad (3)$$

We now introduce two further matrices that will allow us to access the information hidden inside $M(X)$.

Definition 6 ($U(X)$, $L(X)$, d^+, and d^-). For any matrix $M(X) \in Mat\left(n, \mathbb{Z}_{p^k}\left[X, X^{-1}\right]\right)$ in Frobenius normal form the matrices $U(X), L(X) \in Mat\left(n, \mathbb{Z}_{p^k}\left[X, X^{-1}\right]\right)$ associated with $M(X)$ are the matrices in Frobenius normal form where each component $u_i(X)$ and $l_i(X)$ (with $i = 0, \ldots, n-1$) of the n-th row $u(X)$ and $l(X)$ of $U(X)$ and $L(X)$, respectively, is defined as follows:

$$u_i(X) = \begin{cases} \text{monomial of degree } deg^+[m_i(X)] \text{ inside } m_i(X) & \text{if } d_i^+ = d^+ \\ 0 & \text{otherwise} \end{cases}$$

$$l_i(X) = \begin{cases} \text{monomial of degree } deg^-[m_i(X)] \text{ inside } m_i(X) & \text{if } d_i^- = d^- \\ 0 & \text{otherwise} \end{cases},$$

where $d_i^+ = \frac{deg^+[m_i(X)]}{n-i}$, $d_i^- = \frac{deg^-[m_i(X)]}{n-i}$, $d^+ = \max\{d_i^+\}$, and $d^- = \min\{d_i^-\}$.

Definition 7 ($\widehat{M}(X)$ and $\overline{M}(X)$). For any Laurent polynomial $p(X) \in \mathbb{Z}_{p^k}\left[X, X^{-1}\right]$, $\widehat{p}(X)$ and $\overline{p}(X)$ are defined as the Laurent polynomial obtained from $p(X)$ by removing all the monomials having coefficients that are multiple of p and $\overline{p}(X) = p(X) - \widehat{p}(X)$, respectively. These definitions extend component-wise to vectors. For any matrix $M(X) \in Mat\left(n, \mathbb{Z}_{p^k}\left[X, X^{-1}\right]\right)$ in Frobenius normal form, $\widehat{M}(X)$ and $\overline{M}(X)$ are defined as the matrix obtained from $M(X)$ by replacing its n-th row $m(X)$ with $\widehat{m}(X)$ and $\overline{M}(X) = M(X) - \widehat{M}(X)$, respectively.

Definition 8 (Graph G_M). Let $M(X) \in Mat\left(n, \mathbb{Z}_{p^k}\left[X, X^{-1}\right]\right)$ be any matrix in Frobenius normal form. The graph $G_M = \langle V_M, E_M \rangle$ associated with $M(X)$ is such that $V_M = \{1, \ldots, n\}$ and $E_M = \{(h, k) \in V_M^2 \mid M(X)_k^h \neq 0\}$. Moreover, each edge $(h, k) \in E_M$ is labelled with $M(X)_k^h$.

Clearly, for any matrix $M(X) \in Mat\left(n, \mathbb{Z}_{p^k}\left[X, X^{-1}\right]\right)$ in Frobenius normal form, any natural $t > 0$, and any pair (h, k) of entries, the element $M^t(X)_k^h$ is the sum of the weights of all paths of length t starting from h and ending to k, where the weight of a path is the product of the labels of its edges.

Lemma 2. *Let $p > 1$ be a prime number and $a, b \geq 0$, $k > 0$ be integers such that $1 \leq a < p^k$ and $\gcd(a, p) = 1$. Then, $[a + pb]_{p^k} \neq 0$.*

Lemma 3. *Let $p > 1$ be a prime number and h, k be two positive integers. Let $l_1, \ldots, l_h$ and $\alpha_1, \ldots, \alpha_h$ be positive integers such that $l_1 < l_2 < \cdots < l_h$ and for each $i = 1, \ldots, h$ both $1 \leq \alpha_i < p^k$ and $\gcd(\alpha_i, p) = 1$ hold. Consider the sequence $b : \mathbb{Z} \to \mathbb{Z}_{p^k}$ defined for any $l \in \mathbb{Z}$ as $b_l = [\alpha_1 b_{l-l_1} + \cdots + \alpha_h b_{l-l_h}]_{p^k}$ if $l > 0$, $b_0 = 1$, and $b_l = 0$, if $l < 0$. Then, it holds that $[b_l]_p \neq 0$ for infinitely many $l \in \mathbb{N}$.*

For any matrix $M(X) \in Mat\left(n, \mathbb{Z}_{p^k}\left[X, X^{-1}\right]\right)$ in Frobenius normal form, we are now going to study the behavior of $U^t(X)$ and $L^t(X)$, and, in particular, of their elements $U^t(X)^n_n$ and $L^t(X)^n_n$. These will turn out to be crucial in order to establish the sensitivity of the LCA defined by $M(X)$.

Notation 2. *For a sake of simplicity, for any matrix $M(X) \in Mat$ $\left(n\mathbb{Z}_{p^k}\left[X, X^{-1}\right]\right)$ in Frobenius normal form, from now on we will denote by $\mathrm{u}^{(t)}(X)$ and $\mathbb{l}^{(t)}(X)$ the elements $(U^t(X))^n_n$ and $L^t(X)^n_n$, respectively.*

Lemma 4. *Let $M(X) \in Mat\left(n, \mathbb{Z}_{p^k}\left[X, X^{-1}\right]\right)$ be a matrix such that $M(X) = \widehat{N}(X)$ for some $N(X) \in Mat\left(n, \mathbb{Z}_{p^k}\left[X, X^{-1}\right]\right)$ in Frobenius normal form. For any natural $t > 0$, $\mathrm{u}^{(t)}(X)$ (resp., $\mathbb{l}^{(t)}(X)$) is either null or a monomial of degree td^+ (resp., td^-).*

Proof. We show that the statement is true for $\mathrm{u}^{(t)}(X)$ (the proof concerning $\mathbb{l}^{(t)}(X)$ is identical by replacing d^+, $U(X)$ and related elements with d^-, $L(X)$ and related elements). For each $i \in V_U$, let γ_i be the simple cycle of G_U from n to n and passing through the edge (n, i). Clearly, γ_i is the path $n \to i \to i+1 \ldots \to n-1 \to n$ (with γ_n the self-loop $n \to n$) of length $n - i + 1$ and its weight is the monomial $\mathrm{u}_{i-1}(X)$ of degree $(n - i + 1)d^+$. We know that $\mathrm{u}^{(t)}(X)$ is the sum of the weights of all cycles of length t starting from n and ending to n in G_U if at least one of such cycles exists, 0, otherwise. In the former case, each of these cycles can be decomposed in a certain number $s \geq 1$ of simple cycles $\gamma^1_{j_1}, \ldots, \gamma^s_{j_s}$ of lengths giving sum t, i.e., such that $\sum_{i=1}^{s}(n - j_i + 1) = t$. Therefore, $(U^t(X))^n_n$ is a monomial of degree $\sum_{i=1}^{s}(n - j_i + 1)d^+ = td^+$. $\square$

Lemma 5. *Let $M(X) \in Mat\left(n, \mathbb{Z}_{p^k}\left[X, X^{-1}\right]\right)$ be any matrix in Frobenius normal form. For every integer $t \geq 1$ both the following recurrences hold*

$$\mathrm{u}^{(t)}(X) = \mathrm{u}_{n-1}(X)\mathrm{u}^{(t-1)}(X) + \cdots + \mathrm{u}_1(X)\mathrm{u}^{(t-n+1)}(X) + \mathrm{u}_0(X)\mathrm{u}^{(t-n)}(X) \quad (4)$$

$$\mathbb{l}^{(t)}(X) = \mathbb{l}_{n-1}(X)\mathbb{l}^{(t-1)}(X) + \cdots + \mathbb{l}_1(X)\mathbb{l}^{(t-n+1)}(X) + \mathbb{l}_0(X)\mathbb{l}^{(t-n)}(X) \quad (5)$$

with initial conditions $\mathrm{u}^{(0)}(X) = \mathbb{l}^{(0)}(X) = 1$, and $\mathrm{u}^{(l)}(X) = \mathbb{l}^{(l)}(X) = 0$ for $l < 0$.

Proof. We show the recurrence involving $\mathrm{u}^{(t)}(X)$ (the proof for $\mathbb{l}^{(t)}(X)$ is identical by replacing $U(X)$ and its elements with $L(X)$ and its elements). Since $U(X)$ is in Frobenius normal form too, by (3), Recurrence (4) holds for every $t \geq n$. It is clear that $\mathrm{u}^{(0)}(X) = 1$. Furthermore, by the structure of the graph G_U and the meaning of $U(X)^n_n$, Equation (4) is true under the initial conditions for each $t = 1, \ldots, n-1$. $\square$

Lemma 6. *Let $M(X) \in Mat\left(n, \mathbb{Z}_{p^k}\left[X, X^{-1}\right]\right)$ be a matrix such that $M(X) = \widehat{N}(X)$ for some matrix $N(X) \in Mat\left(n, \mathbb{Z}_{p^k}\left[X, X^{-1}\right]\right)$ in Frobenius normal form. Let $v(t)$ (resp., $\lambda(t)$) be the coefficient of $\mathrm{u}^{(t)}(X)$ (resp., $\mathbb{l}^{(t)}(X)$). It holds that $\gcd[v(t), p] = 1$ (resp., $\gcd[\lambda(t), p] = 1$), for infinitely many $t \in \mathbb{N}$.*
In particular, if the value d^+ (resp., d^-) associated with $M(X)$ is non null, then

for infinitely many $t \in \mathbb{N}$ *both* $\left[u^{(t)}(X)\right]_{p^k} \neq 0$ *and* $deg(\left[u^{(t)}(X)\right]_{p^k}) \neq 0$ *(resp.,* $\left[\mathbb{1}^{(t)}(X)\right]_{p^k} \neq 0$ *and* $deg(\left[\mathbb{1}^{(t)}(X)\right]_{p^k}) \neq 0)$ *hold. In other terms, if* $d^+ > 0$ *(resp.,* $d^- < 0)$ *then* $|\{u^{(t)}(X), t \geq 1\}| = \infty$ *(resp.,* $|\{\mathbb{1}^t(X), t \geq 1\}| = \infty)$.

Proof. We show the statements concerning $v(t)$, $\boldsymbol{U}(X)$, $u^{(t)}(X)$, and d^+. Replace X by 1 in the matrix $\boldsymbol{U}(X)$. Now, the coefficient $v(t)$ is just the element of position (n, n) in the t-th power of the obtained matrix $\boldsymbol{U}(1)$. Over $\boldsymbol{U}(1)$, the thesis of Lemma 5 is still valid replacing $u^{(t)}(X)$ by $v(t)$. Thus, for every $t \in \mathbb{N}$, $v(t) = u_{n-1}(1)v(t-1) + \cdots + u_1(1)v(t-n+1) + u_0(1)v(t-n)$ with initial conditions $v(0) = 1$ and $v(l) = 0$, for $l < 0$, where each $u_i(1)$ is the coefficient of the monomial $u_i(X)$ inside $\boldsymbol{U}(X)$. Thus, it follows that $[v(t)]_{p^k} = [u_{n-1}(1)v(t-1) + \cdots + u_1(1)v(t-n+1) + u_0(1)v(t-n)]_{p^k}$. By Lemma 3 we obtain that $\gcd[v(t), p] = 1$ (and so $[v(t)]_{p^k} \neq 0$, too) for infinitely many $t \in \mathbb{N}$. In particular, if the value d^+ associated with $\boldsymbol{M}(X)$ is non null, then, by the structure of G_U and Lemma 4, both $\left[u^{(t)}(X)\right]_{p^k} \neq 0$ and $deg(\left[u^{(t)}(X)\right]_{p^k}) \neq 0$ hold for infinitely many $t \in \mathbb{N}$, too. Therefore, $|\{u^{(t)}(X), t \geq 1\}| = \infty$. The same proof runs for the statements involving $\lambda(t)$, $\boldsymbol{L}(X)$, $u^{(t)}(X)$, and d^- provided that these replace $v(t)$, $\boldsymbol{U}(X)$, $u^{(t)}(X)$, and d^+, respectively.   □

Lemma 7. *Let* $\boldsymbol{M}(X) \in Mat\left(n, \mathbb{Z}_{p^k}\left[X, X^{-1}\right]\right)$ *be a matrix in Frobenius normal form. If either* $|\{u^{(t)}(X), t \geq 1\}| = \infty$ *or* $|\{\mathbb{1}^{(t)}(X), t \geq 1\}| = \infty$ *then* $|\{\widehat{\boldsymbol{M}}^t(X)_n^n, t \geq 1\}| = \infty$.

Proof. Assume that $|\{u^{(t)}(X), t \geq 1\}| = \infty$. Since G_U is a subgraph of $G_{\widehat{\boldsymbol{M}}}$ (with different labels), for each integer t from Lemma 6 applied to $\widehat{\boldsymbol{M}}(X)$, the cycles of length t in $G_{\widehat{\boldsymbol{M}}}$ with weight containing a monomial of degree td^+ are exactly the cycles of length t in G_U. Therefore, it follows that $|\{\widehat{\boldsymbol{M}}^t(X)_n^n, t \geq 1\}| = \infty$. The same argument on G_L and involving d^- allows to prove the thesis if $|\{\mathbb{1}^{(t)}(X), t \geq 1\}| = \infty$.

We are now able to present and prove the main result of this section. It shows a decidable characterization of sensitivity for Frobenius LCA over $\mathbb{Z}_{p^k}^n$.

Lemma 8. *Let* $\left((\mathbb{Z}_{p^k}^n)^{\mathbb{Z}}, F\right)$ *be any Frobenius LCA over* $\mathbb{Z}_{p^k}^n$ *and let* $(m_0(X),$ $\ldots, m_{n-1})$ *be the n-th row of the matrix* $\boldsymbol{M}(X) \in Mat\left(n, \mathbb{Z}_{p^k}\left[X, X^{-1}\right]\right)$ *in Frobenius normal form associated with F. Then, F is sensitive to the initial conditions if and only if $m_i(X)$ is sensitive for some $i \in [0, n-1]$.*

Proof. Let us prove the two implications separately.
Assume that all $m_i(X)$ are not sensitive. Then, $\widehat{\boldsymbol{M}}(X) \in Mat\left(n, \mathbb{Z}_{p^k}\right)$, i.e., it does not contain the formal variable X, and $\boldsymbol{M}(X) = \widehat{\boldsymbol{M}}(X) + p\boldsymbol{M}'(X)$, for some $\boldsymbol{M}'(X) \in Mat\left(n, \mathbb{Z}_{p^k}\left[X, X^{-1}\right]\right)$ in Frobenius normal form. Therefore, for any integer $t > 0$, $\boldsymbol{M}^t(X)$ is the sum of terms, each of them consisting of a product in which p^j appears as factor, for some natural j depending on t and on the specific term which p^j belongs to. Since every element of $\boldsymbol{M}^t(X)$ is taken modulo p^k,

for any natural $t > 0$ it holds that in each term of such a sum p^j appears with $j \in [0, k-1]$ (we stress that j may depend on t and on the specific term of the sum, but it is always bounded by k). Therefore, $|\{M^i(X) : i > 0\}| < \infty$ and so, by Proposition 2, F is not sensitive to the initial conditions.

Conversely, suppose that $m_i(X)$ is sensitive for some $i \in [0, n-1]$ and $d^+ > 0$ (the case $d^- < 0$ is identical). By Definition 7, for any natural $t > 0$ there exists a matrix $M'(X) \in Mat\left(n, \mathbb{Z}_{p^k}\left[X, X^{-1}\right]\right)$ such that $M^t(X) = \widehat{M}^t(X) + pM'(X)$. By a combination of Lemmata 6 and 7, we get $|\{\widehat{M}^t(X)_n^n, t \geq 1\}| = \infty$ and so, by Lemma 2, $|\{M^t(X)_n^n, t \geq 1\}| = \infty$ too. Therefore, it follows that $|\{M^t(X), t \geq 1\}| = \infty$ and, by Proposition 2, we conclude that F is sensitive to the initial conditions.   □

5   Conclusions

In this paper we have studied equicontinuity and sensitivity to the initial conditions for linear HOCA over $\mathbb{Z}_m$ of memory size n, providing decidable characterizations for these properties. We also proved that linear HOCA over $\mathbb{Z}_m$ of memory size n form a class that is indistinguishable from a subclass of LCA (namely, the subclass of Frobenius LCA) over $\mathbb{Z}_m^n$. This enables to decide injectivity and surjectivity for linear HOCA over $\mathbb{Z}_m$ of memory size n by means of the decidable characterizations of injectivity and surjectivity provided in [2] and [20] for LCA over $\mathbb{Z}_m^n$. A natural and pretty interesting research direction consists of investigating other chaotic properties for linear HOCA and all the mentioned dynamical properties, including sensitivity and equicontinuity, for the whole class of LCA over $\mathbb{Z}_m^n$.

References

1. Acerbi, L., Dennunzio, A., Formenti, E.: Shifting and lifting of cellular automata. In: Cooper, S.B., Löwe, B., Sorbi, A. (eds.) CiE 2007. LNCS, vol. 4497, pp. 1–10. Springer, Heidelberg (2007). https://doi.org/10.1007/978-3-540-73001-9_1
2. Bruyn, L.L., den Bergh, M.V.: Algebraic properties of linear cellular automata. Linear Algebra Appl. **157**, 217–234 (1991)
3. Cattaneo, G., Dennunzio, A., Margara, L.: Solution of some conjectures about topological properties of linear cellular automata. Theor. Comput. Sci. **325**(2), 249–271 (2004)
4. Cervelle, J., Lafitte, G.: On shift-invariant maximal filters and hormonal cellular automata. In: LICS: Logic in Computer Science, Reykjavik, Iceland, pp. 1–10, June 2017
5. d'Amico, M., Manzini, G., Margara, L.: On computing the entropy of cellular automata. Theor. Comput. Sci. **290**(3), 1629–1646 (2003)
6. Dennunzio, A.: From one-dimensional to two-dimensional cellular automata. Fundam. Informaticae **115**(1), 87–105 (2012)
7. Dennunzio, A., Di Lena, P., Formenti, E., Margara, L.: Periodic orbits and dynamical complexity in cellular automata. Fundam. Informaticae **126**(2–3), 183–199 (2013)

8. Dennunzio, A., Formenti, E., Manzoni, L.: Computing issues of asynchronous CA. Fundam. Informaticae **120**(2), 165–180 (2012)
9. Dennunzio, A., Formenti, E., Manzoni, L.: Reaction systems and extremal combinatorics properties. Theor. Comput. Sci. **598**, 138–149 (2015)
10. Dennunzio, A., Formenti, E., Manzoni, L., Mauri, G.: m-asynchronous cellular automata: from fairness to quasi-fairness. Nat. Comput. **12**(4), 561–572 (2013)
11. Dennunzio, A., Formenti, E., Manzoni, L., Mauri, G., Porreca, A.E.: Computational complexity of finite asynchronous cellular automata. Theor. Comput. Sci. **664**, 131–143 (2017)
12. Dennunzio, A., Formenti, E., Manzoni, L., Porreca, A.E.: Ancestors, descendants, and gardens of Eden in reaction systems. Theor. Comput. Sci. **608**, 16–26 (2015)
13. Dennunzio, A., Formenti, E., Provillard, J.: Non-uniform cellular automata: classes, dynamics, and decidability. Inf. Comput. **215**, 32–46 (2012)
14. Dennunzio, A., Formenti, E., Provillard, J.: Local rule distributions, language complexity and non-uniform cellular automata. Theor. Comput. Sci. **504**, 38–51 (2013)
15. Dennunzio, A., Formenti, E., Provillard, J.: Three research directions in non-uniform cellular automata. Theor. Comput. Sci. **559**, 73–90 (2014)
16. Dennunzio, A., Formenti, E., Weiss, M.: Multidimensional cellular automata: closing property, quasi-expansivity, and (un)decidability issues. Theor. Comput. Sci. **516**, 40–59 (2014)
17. Dennunzio, A., Guillon, P., Masson, B.: Stable dynamics of sand automata. In: Ausiello, G., Karhumäki, J., Mauri, G., Ong, L. (eds.) TCS 2008. IIFIP, vol. 273, pp. 157–169. Springer, Boston, MA (2008). https://doi.org/10.1007/978-0-387-09680-3_11
18. Ingerson, T., Buvel, R.: Structure in asynchronous cellular automata. Phys. D Nonlinear Phenomena **10**(1), 59–68 (1984)
19. Ito, M., Osato, N., Nasu, M.: Linear cellular automata over $\mathbb{Z}_m$. J. Comput. Syst. Sci. **27**, 125–140 (1983)
20. Kari, J.: Linear cellular automata with multiple state variables. In: Reichel, H., Tison, S. (eds.) STACS 2000. LNCS, vol. 1770, pp. 110–121. Springer, Heidelberg (2000). https://doi.org/10.1007/3-540-46541-3_9
21. Knudsen, C.: Chaos without nonperiodicity. Am. Math. Monthly **101**, 563–565 (1994)
22. Kůrka, P.: Languages, equicontinuity and attractors in cellular automata. Ergodic Theor. Dyn. Syst. **17**, 417–433 (1997)
23. Manzini, G., Margara, L.: Attractors of linear cellular automata. J. Comput. Syst. Sci. **58**(3), 597–610 (1999)
24. Manzini, G., Margara, L.: A complete and efficiently computable topological classification of D-dimensional linear cellular automata over Zm. Theor. Comput. Sci. **221**(1–2), 157–177 (1999)
25. Mariot, L., Leporati, A., Dennunzio, A., Formenti, E.: Computing the periods of preimages in surjective cellular automata. Nat. Comput. **16**(3), 367–381 (2017)
26. Schönfisch, B., de Roos, A.: Synchronous and asynchronous updating in cellular automata. Biosystems **51**(3), 123–143 (1999)
27. Toffoli, T.: Computation and construction universality. J. Comput. Syst. Sci. **15**, 213–231 (1977)

On Varieties of Ordered Automata

Ondřej Klíma[✉] and Libor Polák

Department of Mathematics and Statistics, Masaryk University,
Kotlářská 2, 611 37 Brno, Czech Republic
{klima,polak}@math.muni.cz

Abstract. The Eilenberg correspondence relates varieties of regular languages with pseudovarieties of finite monoids. Various modifications of this correspondence have been found with more general classes of regular languages on one hand and classes of more complex algebraic structures on the other hand. It is also possible to consider classes of automata instead of algebraic structures as a natural counterpart of classes of languages. Here we deal with the correspondence relating positive $\mathcal{C}$-varieties of languages to positive $\mathcal{C}$-varieties of ordered automata and we demonstrate various specific instances of this correspondence. These bring certain well-known results from a new perspective and also some new observations. Moreover, complexity aspects of the membership problem are discussed both in the particular examples and in a general setting.

Keywords: Algebraic language theory · Ordered automata

1 Introduction

Algebraic theory of regular languages is a well-established field in the theory of formal languages. The basic ambition of this theory is to obtain effective characterizations of various natural classes of regular languages. First examples of significant classes of languages, which were effectively characterized by properties of syntactic monoids, were the star-free languages by Schützenberger [17] and the piecewise testable languages by Simon [18]. A general framework for discovering relationships between properties of regular languages and properties of monoids was provided by Eilenberg [5], who established a one-to-one correspondence between the so-called *varieties* of regular languages and *pseudovarieties* of finite monoids. Here varieties of languages are classes closed for taking quotients, preimages under homomorphisms and Boolean operations. Thus a membership problem for a given variety of regular languages can be translated to a membership problem for the corresponding pseudovariety of finite monoids. An advantage of this approach is that pseudovarieties of monoids are exactly classes of finite monoids which have an equational description by pseudoidentities – see Reiterman [16]. For a thorough introduction to that theory we refer to surveys by Pin [13] and by Straubing and Weil [20].

The paper was supported by grant GA15-02862S of the Czech Science Foundation.

C. Martín-Vide et al. (Eds.): LATA 2019, LNCS 11417, pp. 108–120, 2019.
https://doi.org/10.1007/978-3-030-13435-8_8

Since not every natural class of languages is closed for taking all mentioned operations, various generalizations of the notion of varieties of languages have been studied. One possible generalization is the notion of *positive varieties* of languages introduced by Pin [12] – the classes need not be closed for taking complementation. Their equational characterization was given by Pin and Weil [15]. Another possibility is to weaken the closure property concerning preimages under homomorphisms – only homomorphisms from a certain fixed class $\mathcal{C}$ are used. In this way, one can consider $\mathcal{C}$-varieties of regular languages which were introduced by Straubing [19] and whose equational description was presented by Kunc [9]. These two generalizations could be combined as suggested by Pin and Straubing [14].

In our contribution we do not use syntactic structures at all. We consider classes of automata as another natural counterpart to classes of regular languages. In fact, we deal with classes of semiautomata, which are exactly automata without the specification of initial nor final states. Characterizing of classes of languages by properties of minimal automata is quite natural, since usually we assume that an input of a membership problem for a fixed class of languages is given exactly by the minimal deterministic automaton. For example, if we want to test whether an input language is piecewise testable, we do not need to compute its syntactic monoid which could be quite large (see Brzozowski and Li [2]). Instead of that, we check a condition which must be satisfied by its minimal automaton and which was also established in [18]. In [7], Simon's condition was reformulated and the so-called *confluent acyclic (semi)automata* were defined. In this setting, this characterization can be viewed as an instance of Eilenberg type theorem between varieties of languages and varieties of semiautomata.

Moreover, each minimal automaton is implicitly equipped with an order in which the final states form an upward closed subset. This leads to a notion of ordered automata. Then positive $\mathcal{C}$-varieties of ordered semiautomata can be defined as classes which are closed for taking certain natural closure operations. We recall here the general Eilenberg type theorem, namely Theorem 4 from Sect. 3, which states that positive $\mathcal{C}$-varieties of ordered semiautomata correspond to positive $\mathcal{C}$-varieties of languages.

Summarizing, there are three worlds:
(L) classes of regular languages,
(S) classes of finite monoids, sometimes enriched by an additional structure like the ordered monoids, monoids with distinguished generators, etc.,
(A) classes of semiautomata, sometimes ordered semiautomata, etc.

Most variants of Eilenberg correspondence relate (L) and (S), the relationship between (A) and (S) was studied by Chaubard et al. [3], and finally the transitions between (L) and (A) were initiated by Ésik and Ito [6]. In the first version of [8], we continue in the last approach, to establish Theorem 4. In fact, this result is a combination of Theorem 5.1 of [14] (only some hints to a possible proof are given there) and the main result of [3] relating worlds (S) and (A). In contrary, in [8], one can find a self-contained proof which does not go through the classes of monoids.

In the present contribution we concentrate on series of examples, which are various instances of Theorem 4. These bring certain well-known results from a new perspective and also some new observations. The complexity aspects of the membership problem are discussed both in the specific examples and also in a general setting. Moreover, a construction of the minimal ordered automaton of a given language is presented here.

Due to space limitations some proofs are omitted – the corresponding results are marked by the symbol ■. All missing proofs could be find in the second version of [8].

2   Ordered Automata

All automata which are considered in the paper are finite, deterministic and complete. Moreover, we use the term *semiautomaton* when the initial and final states are not explicitly given. Formally saying, a *deterministic finite automaton* (DFA) over the finite alphabet A is a five-tuple $\mathcal{A} = (Q, A, \cdot, i, F)$, where Q is a non-empty finite set of *states*, $\cdot : Q \times A \to Q$ is a complete *transition function*, $i \in Q$ is the *initial* state and $F \subseteq Q$ is the set of *final* states. The transition function can be extended to a mapping $\cdot : Q \times A^* \to Q$ by $q \cdot \lambda = q$, $q \cdot (ua) = (q \cdot u) \cdot a$, for every $q \in Q$, $u \in A^*$, $a \in A$ and the empty word λ. The automaton $\mathcal{A}$ *accepts* a word $u \in A^*$ if and only if $i \cdot u \in F$ and the language *recognized* by $\mathcal{A}$ is $\mathscr{L}(\mathcal{A}) = \{ u \in A^* \mid i \cdot u \in F \}$. More generally, for $q \in Q$, we denote $\mathscr{L}(\mathcal{A}, q) = \{ u \in A^* \mid q \cdot u \in F \}$.

We recall the construction of a minimal automaton of a regular language which was introduced by Brzozowski [1]. Since this automaton is uniquely determined we use also the adjective *canonical* for it. For a language $L \subseteq A^*$ and a pair of words $u, v \in A^*$, we denote by $u^{-1}Lv^{-1}$ the *quotient* of L by these words, i.e. the set $u^{-1}Lv^{-1} = \{ w \in A^* \mid uwv \in L \}$. In particular, a *left quotient* is defined as $u^{-1}L = \{ w \in A^* \mid uw \in L \}$. The *canonical deterministic automaton* of a regular language L is $\mathcal{D}_L = (D_L, A, \cdot, L, F_L)$, where $D_L = \{ u^{-1}L \mid u \in A^* \}$, $K \cdot a = a^{-1}K$, for each $K \in D_L$, $a \in A$, and $F_L = \{ K \in D_L \mid \lambda \in K \}$. A useful observation concerning the canonical automaton is that, for each state $K \in D_L$, we have $\mathscr{L}(\mathcal{D}_L, K) = K$. Since states of the canonical automaton are languages, they are ordered naturally by the set-theoretical inclusion. The action by each letter $a \in A$ is an isotone mapping: for each pair of states p, q such that $p \subseteq q$, we have $p \cdot a = a^{-1}p \subseteq a^{-1}q = q \cdot a$. Moreover, the set F_L of all final states is an upward closed subset with respect to $\subseteq$. These observations motivate the following definition.

Definition 1. *An* ordered automaton *over the alphabet A is a six-tuple $\mathcal{O} = (Q, A, \cdot, \leq, i, F)$, where (i) $(Q, A, \cdot, i, F)$ is DFA, (ii) $\leq$ is a partial order on the set Q, (iii) the action by every letter $a \in A$ is an isotone mapping from the ordered set $(Q, \leq)$ to itself and (iv) F is an upward closed subset of Q with respect to $\leq$.*

If an ordered automaton $\mathcal{O} = (Q, A, \cdot, \leq, i, F)$ is given, then we denote by $\overline{\mathcal{O}}$ the corresponding ordered semiautomaton $(Q, A, \cdot, \leq)$. In particular, for the canonical ordered automaton $\mathcal{D}_L = (D_L, A, \cdot, \subseteq, L, F)$ of the language L, we have $\overline{\mathcal{D}_L} = (D_L, A, \cdot, \subseteq)$.

We could recall, that the transition monoid of the minimal automaton of a regular language L is isomorphic to the syntactic monoid of L (see e.g. [13, Sect. 3]). Similarly, the ordered transition monoid of the minimal ordered automaton of L is isomorphic to the syntactic ordered monoid of L.

There is a natural question how the minimal ordered (semi)automaton can be computed from a given automaton.

Proposition 2. *There exists an algorithm which computes, for a given automaton $\mathcal{A} = (Q, A, \cdot, i, F)$, the minimal ordered automaton of the language $\mathscr{L}(\mathcal{A})$.*

Proof. Our construction is based on Hopcroft minimization algorithm for DFA's. We may assume that all states of $\mathcal{A}$ are reachable from the initial state i. Let $R = (Q \times F) \cup ((Q \setminus F) \times Q)$. Then we construct the relation $\overline{R}$ from R by removing unsuitable pairs of states step by step. At first, we put $R_1 = R$. Then for each integer k, if we find $(p, q) \in R_k$ and a letter $a \in A$ such that $(p \cdot a, q \cdot a) \notin R_k$, then we remove (p, q) from the current relation R_k, that is, we put $R_{k+1} = R_k \setminus \{(p, q)\}$. This construction stops after, say, m steps. So, $R_{m+1} = \overline{R}$ satisfies $(p, q) \in \overline{R} \implies (p \cdot a, q \cdot a) \in \overline{R}$, for every $p, q \in Q$ and $a \in A$. Now, we observe that, $(p, q) \in \overline{R}$ if and only if, for every $u \in A^*$, $(p \cdot u, q \cdot u) \in R$. Thus, the condition can be equivalently written as

$$(p, q) \in \overline{R} \text{ if and only if } (\ \forall u \in A^* \ : \ p \cdot u \in F \implies q \cdot u \in F \ ). \tag{1}$$

It follows that $\overline{R}$ is a transitive relation. So, $\overline{R}$ is a quasiorder on Q and we can consider the corresponding equivalence relation $\rho = \{ (p, q) \mid (p, q) \in \overline{R}, (q, p) \in \overline{R} \}$ on the set Q. Then the quotient set $Q/\rho = \{ [q]_\rho \mid q \in Q \}$ has a structure of the automaton: the rule $[q]_\rho \cdot_\rho a = [q \cdot a]_\rho$, for each $q \in Q$ and $a \in A$, defines correctly actions by letters using (1). Furthermore, the relation $\leq$ on Q/ρ defined by the rule $[p]_\rho \leq [q]_\rho$ iff $(p, q) \in \overline{R}$, is a partial order on Q/ρ compatible with actions by letters. So, $\mathcal{A}_\rho = (Q/\rho, A, \cdot_\rho, \leq, [i]_\rho, F_\rho)$, where $F_\rho = \{ [f]_\rho \mid f \in F \}$, is an ordered automaton recognizing $\mathscr{L}(\mathcal{A})$. Moreover, if there are two states $[p]_\rho, [q]_\rho \in Q/\rho$ such that $\mathscr{L}(\mathcal{A}_\rho, p) = \mathscr{L}(\mathcal{A}_\rho, q)$, then $(p, q) \in \rho$. Thus, the ordered automaton $\mathcal{A}_\rho$ is isomorphic to the minimal ordered automaton of the language $\mathscr{L}(\mathcal{A})$. □

Note also that the classical power-set construction makes from a nondeterministic automaton an ordered deterministic automaton which is ordered by the set-theoretical inclusion. Thus, for the purpose of a construction of the minimal ordered automaton, one may also use Brzozowski's minimization algorithm using power-set construction for the reverse of the given language.

The last technical notion concerning ordered semiautomata is related to a homomorphism of free monoids $f : B^* \to A^*$. For a language $L \subseteq A^*$, we denote the *preimage* under the homomorphism f by $f^{-1}(L) = \{ v \in B^* \mid f(v) \in L \}$.

This language can be recognized by a semiautomaton given in the following construction.

For a homomorphism $f : B^* \to A^*$ and an ordered semiautomaton $\mathcal{O} = (Q, A, \cdot, \leq)$, we denote by $\mathcal{O}^f$ the semiautomaton $(Q, B, \cdot^f, \leq)$ where $q \cdot^f b = q \cdot f(b)$ for every $q \in Q$ and $b \in B$. We call $\mathcal{O}^f$ a f-renaming of $\mathcal{O}$. Furthermore, we say that $(P, B, \circ, \preceq)$ is an f-subsemiautomaton of $(Q, A, \cdot, \leq)$ if it is a subsemiautomaton of $\mathcal{O}^f$.

3   Positive $\mathcal{C}$-Varieties of Ordered Semiautomata

For the purpose of this paper, following Straubing [19], the *category of homomorphisms* $\mathcal{C}$ is a category where objects are all free monoids over non-empty finite alphabets and morphisms are certain monoid homomorphisms among them. This "categorical" definition means that $\mathcal{C}$ satisfies the following properties: For each finite alphabet A, the identity mapping id_A belongs to $\mathcal{C}(A^*, A^*)$; If $f \in \mathcal{C}(B^*, A^*)$ and $g \in \mathcal{C}(C^*, B^*)$, then their composition gf belongs to $\mathcal{C}(C^*, A^*)$. As first examples, we mention the category of *literal* homomorphisms $\mathcal{C}_l$, where $f \in \mathcal{C}_l(B^*, A^*)$ if and only if $f(B) \subseteq A$; and the category of surjective homomorphisms $\mathcal{C}_s$, where $f \in \mathcal{C}_s(B^*, A^*)$ if and only if $A \subseteq f(B)$.

Now, for a category of homomorphisms, a *positive $\mathcal{C}$-variety of languages* $\mathcal{V}$ associates to every non-empty finite alphabet A a class $\mathcal{V}(A)$ of regular languages over A in such a way that (i) $\mathcal{V}(A)$ is closed for taking finite unions and intersections; (ii) $\mathcal{V}(A)$ is closed for taking quotients, (iii) $\mathcal{V}$ is closed for taking preimages under homomorphisms of $\mathcal{C}$, i.e. $f \in \mathcal{C}(B^*, A^*)$, $L \in \mathcal{V}(A)$ imply $f^{-1}(L) \in \mathcal{V}(B)$. We talk about $\mathcal{C}$-*variety of languages* if every $\mathcal{V}(A)$ is also closed for taking complements.

If $\mathcal{C}$ consists of all homomorphisms we get exactly the notion of the *positive varieties of languages*. When adding "each $\mathcal{V}(A)$ is closed for taking complements", we get exactly the notion of the *variety of languages*.

Definition 3. *Let $\mathcal{C}$ be a category of homomorphisms. A positive $\mathcal{C}$-variety of ordered semiautomata* $\mathbb{V}$ *associates to every non-empty finite alphabet A a class* $\mathbb{V}(A)$ *of ordered semiautomata over A in such a way that (i)* $\mathbb{V}(A) \neq \emptyset$ *is closed for taking disjoint unions and direct products of non-empty finite families, and homomorphic images and (ii)* $\mathbb{V}$ *is closed for taking f-subsemiautomata for all* $f \in \mathcal{C}(B^*, A^*)$.

For each positive $\mathcal{C}$-variety of ordered semiautomata $\mathbb{V}$, we denote by $\alpha(\mathbb{V})$ the class of regular languages given by the following formula

$$(\alpha(\mathbb{V}))(A) = \{\, L \subseteq A^* \mid \overline{\mathcal{D}_L} \in \mathbb{V}(A) \,\}.$$

For each positive $\mathcal{C}$-variety of regular languages $\mathcal{V}$, we denote by $\beta(\mathcal{V})$ the positive $\mathcal{C}$-variety of ordered semiautomata generated by all ordered semiautomata $\overline{\mathcal{D}_L}$, where $L \in \mathcal{V}(A)$ for some alphabet A. Now we are ready to state the Eilenberg type correspondence for positive $\mathcal{C}$-varieties of ordered semiautomata.

Theorem 4 ([14, Theorem 5.1] **together with** [3]). *Let $\mathcal{C}$ be a category of homomorphisms. The mappings α and β are mutually inverse isomorphisms between the lattice of all positive $\mathcal{C}$-varieties of ordered semiautomata and the lattice of all positive $\mathcal{C}$-varieties of regular languages.*

For a detailed proof, we refer to [8, Sect. 6]. If we consider the positive $\mathcal{C}$-variety $\mathcal{V}$ of regular languages which is closed for taking complements, then the corresponding positive $\mathcal{C}$-variety $\beta(\mathcal{V}) = \mathbb{V}$ of ordered semiautomata is closed for taking dual ordered semiautomata. Since the ordered semiautomaton $(Q, A, \cdot, =)$ is isomorphic to a subsemiautomaton of the product of the ordered semiautomata $(Q, A, \cdot, \leq)$ and $(Q, A, \cdot, \geq)$, the $\mathcal{C}$-variety $\beta(\mathcal{V}) = \mathbb{V}$ is fully described by the classes of semiautomata $\{ (Q, A, \cdot) \mid (Q, A, \cdot, \leq) \in \mathbb{V}(A) \}$.

Then we can define a $\mathcal{C}$-*variety of semiautomata* in the same manner as in Definition 3. From Theorem 4, it follows that there exists one to one correspondence between $\mathcal{C}$-varieties of regular languages and $\mathcal{C}$-varieties of semiautomata. The details can be found in [8, Sect. 7].

4   Examples

In this section we present several instances of Eilenberg type correspondence. Some of them are just reformulations of examples already mentioned in existing literature. In particular, the first three subsections correspond to pseudovarieties of aperiodic, $\mathcal{R}$-trivial and $\mathcal{J}$-trivial monoids, respectively. Also Subsect. 4.4 has a natural counterpart in pseudovarieties of ordered monoids satisfying the inequality $1 \leq x$. In all these cases, $\mathcal{C}$ is the category of all homomorphisms denoted by $\mathcal{C}_{all}$. Nevertheless, we believe that these correspondences viewed from the perspective of varieties of (ordered) semiautomata are of some interest. Another four subsections works with different categories $\mathcal{C}$ and Subsects. 4.6 and 4.7 bring new examples of (positive) $\mathcal{C}$-varieties of (ordered) automata.

4.1   Counter-Free Automata

The star free languages were characterized by Schützenberger [17] as the languages having aperiodic syntactic monoids. Here we recall the subsequent characterization of McNaughton and Papert [10] by counter-free automata. We say that a semiautomaton $(Q, A, \cdot)$ is *counter-free* if, for each $u \in A^*$, $q \in Q$ and $n \in \mathbb{N}$ such that $q \cdot u^n = q$, we have $q \cdot u = q$.

Proposition 5. *All counter-free semiautomata form a variety of semiautomata.*

Proof. It is easy to see that disjoint unions, subsemiautomata, products and f-renamings of counter-free semiautomata are again counter-free.

Let $\varphi : (Q, A, \cdot) \to (P, A, \circ)$ be a surjective homomorphism of semiautomata and let $(Q, A, \cdot)$ be counter-free. We prove that also $(P, A, \circ)$ is a counter-free semiautomaton. Take $p \in P, u \in A^*$ and $n \in \mathbb{N}$ such that $p \circ u^n = p$. Let $q \in Q$ be an arbitrary state such that $\varphi(q) = p$. Then, for each $j \in \mathbb{N}$, we have $\varphi(q \cdot u^{jn}) = p \circ u^{jn} = p$. Since the set $\{q, q \cdot u^n, q \cdot u^{2n}, \ldots\}$ is finite, there exist $k, \ell \in \mathbb{N}$ such that $q \cdot u^{kn} = q \cdot u^{(k+\ell)n}$. If we take $r = q \cdot u^{kn}$ then $r \cdot u^{\ell n} = r$. Since $(Q, A, \cdot)$ is counter-free, we get $r \cdot u = r$. Consequently, $p \circ u = \varphi(r) \circ u = \varphi(r \cdot u) = \varphi(r) = p$. $\square$

The promised link between languages and automata follows.

Proposition 6 (McNaughton, Papert [10]). *Star free languages are exactly the languages recognized by counter-free semiautomata.*

Note that this characterization is effective, although testing whether a regular language given by a DFA is aperiodic is even PSPACE-complete problem by Cho and Huynh [4].

4.2   Acyclic Automata

The *content* $c(u)$ of a word $u \in A^*$ is the set of all letters occurring in u. We say that a semiautomaton $(Q, A, \cdot)$ is *acyclic* if, for every $u \in A^+$ and $q \in Q$, we have that $q \cdot u = q$ implies $\forall a \in c(u) : q \cdot a = q$.

One can prove the following result in a similar way as Proposition 5.

Proposition 7. *All acyclic semiautomata form a variety of semiautomata.* $\square$

According to Pin [11, Chapt. 4, Sect. 3], a semiautomaton $(Q, A, \cdot)$ is called *extensive* if there exists a linear order $\preceq$ on Q such that $(\forall q \in Q, a \in A)\ q \preceq q \cdot a$. Note that such an order need not to be compatible with actions of letters. One can easily show that a semiautomaton is acyclic if and only if it is extensive. We prefer to use the term acyclic, since we consider extensive actions by letters (compatible with ordering of a semiautomaton) later in the paper. Anyway, testing whether a given semiautomaton is acyclic can be decided using the breadth-first search algorithm.

Proposition 8 (Pin [11]). *The languages over the alphabet A accepted by acyclic semiautomata are exactly disjoint unions of the languages of the form*

$$A_0^* a_1 A_1^* a_2 A_2^* \ldots A_{n-1}^* a_n A_n^* \text{ where } a_i \notin A_{i-1} \subseteq A \text{ for } i = 1, \ldots, n.$$

4.3   Acyclic Confluent Automata

In our paper [7] concerning piecewise testable languages, we introduced a certain condition on automata being motivated by the terminology from the theory of rewriting systems. We say that a semiautomaton $(Q, A, \cdot)$ is *confluent*, if for each state $q \in Q$ and every pair of words $u, v \in A^*$, there is a word $w \in A^*$ such that $c(w) \subseteq c(uv)$ and $(q \cdot u) \cdot w = (q \cdot v) \cdot w$. In paper [7] this definition was studied only in the context of acyclic (semi)automata, in which case several equivalent conditions were described. One of them can be rephrased in the following way.

Lemma 9. *Let $\mathcal{Q} = (Q, A, \cdot)$ be an acyclic semiautomaton. Then $\mathcal{Q}$ is confluent if and only if, for each $q \in Q$, $u, v \in A^*$, we have $q \cdot u \cdot (uv)^{|Q|} = q \cdot v \cdot (uv)^{|Q|}$.* ∎

Using the condition from Lemma 9, one can prove that the class of all acyclic confluent semiautomata is a variety of semiautomata similarly as in Proposition 5. Finally, the main result from [7] can be formulated in the following way. It is mentioned in [7] that the defining condition is testable in a polynomial time.

Proposition 10 (Klíma and Polák [7]). *The variety of all acyclic confluent semiautomata corresponds to the variety of all piecewise testable languages.*

4.4 Ordered Automata with Extensive Actions

We say that an ordered semiautomaton $(Q, A, \cdot, \leq)$ has *extensive actions* if, for every $q \in Q$, $a \in A$, we have $q \leq q \cdot a$. Clearly, the defining condition is testable in a polynomial time. The transition ordered monoids of such ordered semiautomata are characterized by the inequality $1 \leq x$. It is known [13, Proposition 8.4] that the last inequality characterizes the positive variety of all finite unions of languages of the form

$$A^* a_1 A^* a_2 A^* \ldots A^* a_\ell A^*, \quad \text{where} \quad a_1, \ldots, a_\ell \in A, \ \ell \geq 0.$$

Therefore we call them *positive piecewise testable languages*. In this way one can obtain the following statement.

Proposition 11. *The class of all ordered semiautomata with extensive actions is a positive variety of ordered semiautomata and corresponds to the positive variety of all positive piecewise testable languages.* ∎

Note that a usual characterization of the class of positive piecewise testable languages is given by a forbidden pattern for DFA (see e.g. [20, p. 531]). This pattern consists of two words $v, w \in A^*$ and two states p and $q = p \cdot v$ such that $p \cdot w \in F$ and $q \cdot w \notin F$. In view of (1) from Sect. 2, the presence of the pattern is equivalent to the existence of two states $[p]_\rho \not\leq [q]_\rho$, such that $[p]_\rho \cdot_\rho v = [q]_\rho$ in the minimal automaton of the language. The membership for the class of positive piecewise testable languages is decidable in polynomial time – see [13, Corollary 8.5] or [20, Theorem 2.20].

4.5 Autonomous Automata

We recall examples from the paper [6]. We call a semiautomaton $(Q, A, \cdot)$ *autonomous* if for each state $q \in Q$ and every pair of letters $a, b \in A$, we have $q \cdot a = q \cdot b$. For a positive integer d, let $\mathbb{V}_d$ be the class of all autonomous semiautomata being disjoint unions of cycles whose lengths divide d. Clearly, the defining conditions are testable in a linear time.

Proposition 12 (Ésik and Ito [6]). *(i) All autonomous semiautomata form a $\mathcal{C}_l$-variety of semiautomata and the corresponding $\mathcal{C}_l$-variety of languages consists of regular languages L such that, for all $u, v \in A^*$, if $u \in L$, $|u| = |v|$ then $v \in L$.*

(ii) The class $\mathbb{V}_d$ forms a $\mathcal{C}_l$-variety of semiautomata and the corresponding $\mathcal{C}_l$-variety of languages consists of all unions of $(A^d)^* A^i$, $i \in \{0, \ldots, d - 1\}$.

4.6   Synchronizing and Weakly Confluent Automata

Synchronizing automata are intensively studied in the literature. A semiautomaton $(Q, A, \cdot)$ is *synchronizing* if there is a word $w \in A^*$ such that the set $Q \cdot w$ is a one-element set. We use an equivalent condition, namely, for each pair of states $p, q \in Q$, there exists a word $w \in A^*$ such that $p \cdot w = q \cdot w$ (see e.g. Volkov [21, Proposition 1]). In this paper we consider the classes of semiautomata which are closed for taking disjoint unions. So, we need to study disjoint unions of synchronizing semiautomata. Those automata can be equivalently characterized by the following weaker version of confluence. We say that a semiautomaton $(Q, A, \cdot)$ is *weakly confluent* if, for each state $q \in Q$ and every pair of words $u, v \in A^*$, there is a word $w \in A^*$ such that $(q \cdot u) \cdot w = (q \cdot v) \cdot w$.

Proposition 13. *A semiautomaton is weakly confluent if and only if it is a disjoint union of synchronizing semiautomata.* ∎

Since the synchronization property can be tested in the polynomial time (see [21]), Proposition 13 implies that the weak confluence of a semiautomaton can be tested in the polynomial time, as well.

Proposition 14. *All weakly confluent semiautomata form a $\mathcal{C}_s$-variety of semiautomata.* ∎

4.7   Automata for Languages Closed Under Inserting Segments

We know that a language $L \subseteq A^*$ is positive piecewise testable if, for every pair of words $u, w \in A^*$ such that $uw \in L$ and for a letter $a \in A$, we have $uaw \in L$. So, we can add an arbitrary letter into each word from the language (at an arbitrary position) and the resulting word stays in the language. Now we consider an analogue, where we put into the word a word of a given fixed length.

For each positive integer n, we consider the following property of a given regular language $L \subseteq A^*$:

$$\forall u, v, w \in A^* \ : \ uw \in L, |v| = n \text{ implies } uvw \in L.$$

We say that L is closed under n-*insertions* whenever L satisfies this property. We show that the class of all regular languages closed under n-insertions form a positive $\mathcal{C}$-variety of languages by describing the corresponding positive $\mathcal{C}$-variety of ordered semiautomata.

At first, we need to describe an appropriate category of homomorphisms. Let $\mathcal{C}_{lm}$ be the category consisting of so-called *length-multiplying* (see [19]) homomorphisms: $f \in \mathcal{C}_{lm}(B^*, A^*)$ if there exists a positive integer k such that $|f(b)| = k$

for every $b \in B$. Let n be a positive integer and $\mathcal{Q} = (Q, A, \cdot, \leq)$ be an ordered semiautomaton. We say that $\mathcal{Q}$ has n-*extensive* actions if, for every $q \in Q$ and $u \in A^*$ such that $|u| = n$, we have $q \leq q \cdot u$.

Note that ordered semiautomata from Subsect. 4.4 are ordered semiautomata which have 1-extensive actions. Of course, these ordered semiautomata have n-extensive actions for every n. More generally, if n divides m and an ordered semiautomaton $\mathcal{Q}$ has n-extensive actions, then $\mathcal{Q}$ has m-extensive actions.

Proposition 15. *Let n be a positive integer. The class of all ordered semiautomata which have n-extensive actions forms a positive $\mathcal{C}_{lm}$-variety of ordered semiautomata. The corresponding positive $\mathcal{C}_{lm}$-variety of languages consists of all regular languages closed under n-insertions.* ∎

For a fixed n, it is decidable in polynomial time whether a given ordered semiautomaton has n-extensive actions, because the relation $q \leq q \cdot u$ has to be checked only for polynomially many words u.

4.8 Automata for Finite and Prefix-Testable Languages

Finite languages do not form a variety, because their complements, the so-called *cofinite languages*, are not finite. Moreover, the class of all finite languages is not closed for taking preimages under all homomorphisms. However, one can restrict the category of homomorphisms to the so-called *non-erasing* ones: we say that a homomorphism $f : B^* \to A^*$ is *non-erasing* if $f^{-1}(\lambda) = \{\lambda\}$. The category of all non-erasing homomorphisms is denoted by $\mathcal{C}_{ne}$. Note that $\mathcal{C}_{ne}$-varieties of languages correspond to +-varieties of languages (see [19]).

We use certain technical terminology for states of a given semiautomaton $(Q, A, \cdot)$: we say that a state $q \in Q$ *has a cycle*, if there is a word $u \in A^+$ such that $q \cdot u = q$ and we say that the state q is *absorbing* if for each letter $a \in A$ we have $q \cdot a = q$. We call a semiautomaton $(Q, A, \cdot)$ *strongly acyclic*, if each state which has a cycle is absorbing. It is evident that every strongly acyclic semiautomaton is acyclic.

Proposition 16. *(i) All strongly acyclic semiautomata form a $\mathcal{C}_{ne}$-variety.*
(ii) All strongly acyclic confluent semiautomata form a $\mathcal{C}_{ne}$-variety. ∎

Proposition 17. *The $\mathcal{C}_{ne}$-variety of all finite and all cofinite languages corresponds to the $\mathcal{C}_{ne}$-variety of all strongly acyclic confluent semiautomata.* ∎

Naturally, one can try to describe the corresponding $\mathcal{C}_{ne}$-variety of languages for the $\mathcal{C}_{ne}$-variety of strongly acyclic semiautomata. Following Pin [13, Sect. 5.3], we call $L \subseteq A^*$ a *prefix-testable* language if L is a finite union of a finite language and languages of the form uA^*, with $u \in A^*$. One can prove the following statement in a similar way as Proposition 17. Note that one can find also a characterization via syntactic semigroups in [13, Sect. 5.3].

Proposition 18. *The $\mathcal{C}_{ne}$-variety of all prefix-testable languages corresponds to the $\mathcal{C}_{ne}$-variety of all strongly acyclic semiautomata.* □

The characterization from Proposition 17 can be modified for the positive C_{ne}-variety consisting of all finite languages (see [8] for details). Note that, all considered conditions on semiautomata discussed in this subsection can be checked in polynomial time.

5 Membership Problem for C-Varieties of Semiautomata

In the previous section, the membership problem for (positive) C-varieties of semiautomata was always solved by an ad hoc argument. Here we discuss whether it is possible to give a general result in this direction. For that purpose, recall that ω-identity is a pair of ω-terms, which are constructed from variables by (repeated) successive application of concatenation and the unary operation $u \mapsto u^\omega$. In a particular monoid, the interpretation of this unary operation assigns to each element s its uniquely determined power which is idempotent.

In the case of C_{all} consisting of all homomorphisms, we mention Theorem 2.19 from [20] which states the following result: if the corresponding pseudovariety of monoids is defined by a finite set of ω-identities then the membership problem of the corresponding variety of languages is decidable by a polynomial space algorithm in the size of the input automaton. Thus, Theorem 2.19 slightly extends the case when the pseudovariety of monoids is defined by a finite set of identities. The algorithm checks the defining ω-identities in the syntactic monoid M_L of a language L and uses the basic fact that M_L is the transition monoid of the minimal automaton of L. This extension is possible, because the unary operation $(\ )^\omega$ can be effectively computed from the input automaton.

We should mention that checking a fixed identity in an input semiautomaton can be done in a better way. Such a (NL) algorithm (a folklore algorithm in the theory, see [8] for details) guesses a pair of finite sequences of states for two sides of a given identity $u = v$ which are visited during reading the word u (and v respectively) letter by letter. These sequences have the same first states and distinct last states. Then the algorithm checks whether for each variable, there is a transition of the automaton given by a word, which transforms all states in the sequence in the right way, when every occurrence of the variable is considered. If, for every used variable, there is such a word, we obtained a counterexample disproving the identity $u = v$.

Whichever algorithm is used, we can immediately get the generalization to the case of positive varieties of languages, because checking inequalities can be done in the same manner as checking identities. However, we want to use the mentioned algorithms to obtain a corresponding result for positive C-varieties of ordered semiautomata for the categories used in this paper. For such a result we need the following formal definition. An ω-inequality $u \leq v$ *holds in an ordered semiautomaton* $\mathcal{O} = (Q, A, \cdot, \leq)$ *with respect to a category* C if, for every $f \in C(X^*, A^*)$ with X being the set of variables occurring in uv, and for every $p \in Q$, we have $p \cdot f(u) \leq p \cdot f(v)$. Here $f(u)$ is equal to $f(u')$, where u' is a word obtained from u if all occurrences of ω are replaced by an exponent n satisfying the equality $s^\omega = s^n$ in the transition monoid of $\mathcal{O}$ for its arbitrary element s.

Theorem 19. *Let $\mathcal{O} = (Q, A, \cdot, \leq)$ be an ordered semiautomaton, let $u \leq v$ be an ω-inequality and $\mathcal{C}$ be one of the categories $\mathcal{C}_{ne}$, $\mathcal{C}_l$, $\mathcal{C}_s$ and $\mathcal{C}_{lm}$. The problem whether $u \leq v$ holds in $\mathcal{O}$ with respect to $\mathcal{C}$ is decidable.* ∎

The cases of $\mathcal{C}_{ne}$, $\mathcal{C}_l$ and $\mathcal{C}_s$ are modifications of the algorithms for $\mathcal{C}_{all}$. As those, they are decidable by a polynomial space algorithms or NL algorithms in the case when u and v are products of variables. For the case of $\mathcal{C}_{lm}$ we are not able to bound the complexity.

References

1. Brzozowski, J.: Canonical regular expressions and minimal state graphs for definite events. In: Mathematical Theory of Automata, vol. 12, pp. 529–561. Research institute, Brooklyn (1962)
2. Brzozowski, J., Li, B.: Syntactic complexity of $\mathcal{R}$- and $\mathcal{J}$-trivial regular languages. In: Jurgensen, H., Reis, R. (eds.) DCFS 2013. LNCS, vol. 8031, pp. 160–171. Springer, Heidelberg (2013). https://doi.org/10.1007/978-3-642-39310-5_16
3. Chaubard, L., Pin, J.É., Straubing, H.: Actions, wreath products of C-varieties and concatenation product. Theor. Comput. Sci. **356**(1–2), 73–89 (2006)
4. Cho, S., Huynh, D.T.: Finite automaton aperiodicity is PSPACE-complete. Theor. Comput. Sci. **88**(1), 99–116 (1991)
5. Eilenberg, S.: Automata, Languages and Machines, vol. B. Academic Press, Cambridge (1976)
6. Ésik, Z., Ito, M.: Temporal logic with cyclic counting and the degree of aperiodicity of finite automata. Acta Cybern. **16**(1), 1–28 (2003)
7. Klíma, O., Polák, L.: Alternative automata characterization of piecewise testable languages. In: Béal, M.-P., Carton, O. (eds.) DLT 2013. LNCS, vol. 7907, pp. 289–300. Springer, Heidelberg (2013). https://doi.org/10.1007/978-3-642-38771-5_26
8. Klíma, O., Polák, L.: On varieties of ordered automata. CoRR (2017(v1), 2019(v2)). http://arxiv.org/abs/1712.08455
9. Kunc, M.: Equational description of pseudovarieties of homomorphisms. RAIRO Theor. Inf. Appl. **37**(3), 243–254 (2003)
10. McNaughton, R., Papert, S.: Counter-Free Automata. MIT Press, Cambridge (1971)
11. Pin, J.É.: Varieties of Formal Languages. Plenum Publishing Co., New York (1986)
12. Pin, J.É.: A variety theorem without complementation. Russ. Math. **39**, 80–90 (1995)
13. Pin, J.-E.: Syntactic semigroups. In: Rozenberg, G., Salomaa, A. (eds.) Handbook of Formal Languages, pp. 679–746. Springer, Heidelberg (1997). https://doi.org/10.1007/978-3-642-59136-5_10
14. Pin, J.É., Straubing, H.: Some results on C-varieties. RAIRO Theor. Inf. Appl. **39**(1), 239–262 (2005)
15. Pin, J.É., Weil, P.: A Reiterman theorem for pseudovarieties of finite first-order structures. Algebra Univers. **35**(4), 577–595 (1996)
16. Reiterman, J.: The Birkhoff theorem for finite algebras. Algebra Univers. **14**, 1–10 (1982)
17. Schützenberger, M.P.: On finite monoids having only trivial subgroups. Inf. Control **8**(2), 190–194 (1965)

18. Simon, I.: Piecewise testable events. In: Brakhage, H. (ed.) GI-Fachtagung 1975. LNCS, vol. 33, pp. 214–222. Springer, Heidelberg (1975). https://doi.org/10.1007/3-540-07407-4_23

19. Straubing, H.: On logical descriptions of regular languages. In: Rajsbaum, S. (ed.) LATIN 2002. LNCS, vol. 2286, pp. 528–538. Springer, Heidelberg (2002). https://doi.org/10.1007/3-540-45995-2_46

20. Straubing, H., Weil, P.: Varieties. CoRR (2015). http://arxiv.org/abs/1502.03951

21. Volkov, M.V.: Synchronizing automata and the Černý conjecture. In: Martín-Vide, C., Otto, F., Fernau, H. (eds.) LATA 2008. LNCS, vol. 5196, pp. 11–27. Springer, Heidelberg (2008). https://doi.org/10.1007/978-3-540-88282-4_4

Automata over Infinite Sequences of Reals

Klaus Meer[⊠] and Ameen Naif

Brandenburg University of Technology Cottbus-Senftenberg,
Platz der Deutschen Einheit 1, 03046 Cottbus, Germany
{meer,naif}@b-tu.de

Abstract. Gandhi, Khoussainov, and Liu introduced and studied a generalized model of finite automata able to work over algebraic structures, in particular the real numbers. The present paper continues the study of (a variant) of this model dealing with computations on infinite strings of reals. Our results support the view that this is a suitable model of finite automata over the real numbers. We define Büchi and Muller versions of the model and show, among other things, several closure properties of the languages accepted, a real number analogue of McNaughton's theorem, and give a metafinite logic characterizing the infinite languages acceptable by non-deterministic Büchi automata over $\mathbb{R}$.

Keywords: Automata and logic · Real number computations

1 Introduction

Given the success of finite automata in the realm of computation over finite alphabets there have been several attempts to generalize the concept of finite automata to deal with more general structures or data types. One particular such approach has been introduced by Gandhi, Khoussainov, and Liu [2] and further extended in [6]. It provides both a finite automata model over algebraic structures and is able to deal with infinite alphabets as underlying universes. In the words of the authors of those papers their work fits into several lines of research and there are different motivations to study such generalizations. In [2] the authors discuss different previous approaches to design finite automata over infinite alphabets and their role in program verification and database theory. One goal is to look for a generalized framework that is able to homogenize at least some of these approaches. As the authors remark, many classical automata models like pushdown automata, Petri nets, visibly pushdown automata can be simulated by the new model. Another major motivation results from work on algebraic models of computation over structures like the real and complex numbers. Here, the authors suggested their model as a finite automata variant of the Blum-Shub-Smale BSS model [1]. They then asked to analyze such automata over structures like real or algebraically closed fields.

In [10,11] we developed a theory of finite automata for the BSS model over $\mathbb{R}$ and $\mathbb{C}$ being based on the approach of Gandhi, Khoussainov, and Liu - henceforth

C. Martín-Vide et al. (Eds.): LATA 2019, LNCS 11417, pp. 121–133, 2019.
https://doi.org/10.1007/978-3-030-13435-8_9

GKL for short. The main outcomes of those papers are as follows: If the original model of GKL is considered in such an algebraic framework the model turns out to be too powerful due to its abilities to perform - though in a restricted way - computations. Many of the classical questions about finite automata become undecidable, many structural properties of acceptable languages over the reals are lost, see [10]. Therefore, in [11] the original GKL model is restricted; the automata are equipped with a period after which all computed results are 'forgotten' by the automaton. The results in [11] show that this causes the model to become a reasonable real number variant of a finite automata model.

In the present paper we continue research into the direction of studying the GKL approach over the structure of real numbers. More precisely, we extend the periodic GKL automata introduced in [11] to a model working over infinite sequences of real numbers. We define real number versions of Büchi and Muller automata and show, among other things, several structural properties of the languages accepted, an analogue of McNaughton's theorem, and give a metafinite logic characterizing the infinite languages acceptable by non-deterministic real number Büchi automata. Here, 'metafinite' refers to a framework extending finite model theory to deal as well with structures defined over infinite universes [3].

We suppose the reader to be familiar with the classical theory of finite automata over infinite words, see for example [5,7,13]. Due to space restriction all missing proofs are postponed to the full version.

2   The Automaton Model

We recall the definition of a periodic GKL automaton over $\mathbb{R}$ from [11] working over finite strings of real numbers. The extension to computations on infinite strings then is immediate.

Definition 1. *(Periodic GKL automata over $\mathbb{R}$).* *(a) A deterministic periodic $\mathbb{R}$-automaton $\mathcal{A}$, also called periodic GKL $\mathbb{R}$-automaton, consists of the following objects:*

- *a finite state space Q and an initial state $q_0 \in Q$,*
- *a set $F \subseteq Q$ of final (accepting) states,*
- *a set of $\ell \in \mathbb{N}$ registers which contain fixed given constants $c_1, \ldots, c_\ell \in \mathbb{R}$,*
- *a set of $k \in \mathbb{N}$ registers which can store real numbers denoted by $v_1, \ldots, v_k$,*
- *a counter containing a number $t \in \{0, 1, \ldots, T-1\}$; we call $T \in \mathbb{N}$ the period of the automaton,*
- *a transition function $\delta : Q \times \{0,1\}^{k+\ell} \to Q \times \{\circ_1, \ldots, \circ_k\}$.*

A configuration of $\mathcal{A}$ is a tuple from $Q \times \mathbb{R}^k \times \{0, \ldots, T-1\}$ specifying the current data during a computation.

(b) The automaton processes elements of $\mathbb{R}^* := \bigsqcup_{n \geq 1} \mathbb{R}^n$, i.e., sequences of finite length with real numbers as components. For an input $(x_1, \ldots, x_n) \in \mathbb{R}^n$ it works as follows. The computation starts in state q_0 with initial assignment $0 \in \mathbb{R}$ for the values $v_1, \ldots, v_k \in \mathbb{R}$ and value 0 for the counter. $\mathcal{A}$ reads the

input components step by step. Suppose a value x is read in state $q \in Q$ with counter value t. The next state together with an update of the values v_i and t are computed as follows:

A performs $k + \ell$ comparisons $x\Delta_1 v_1?, x\Delta_2 v_2?, \ldots, x\Delta_k v_k?, x\Delta_{k+1} c_1?, \ldots,$ $x\Delta_{k+\ell} c_\ell?$, where $\Delta_i \in \{\geq, \leq, =\}$. This gives a vector $b \in \{0,1\}^{k+\ell}$, where a component 0 indicates that the comparison that was tested is violated whereas 1 codes that it is valid. Depending on state q and b the automaton moves to a state $q' \in Q$ (which could again be q) and performs on each of the k registers an operation. If $t = T - 1$ the counter and all register values v_i are reset to 0 and a new period starts. In all other cases the counter is increased by 1 and the values of all v_i are updated applying one of the operations in the structure: $v_i \leftarrow x \circ_i v_i$. Here, $\circ_i \in \{+, -, \times, null, pr_1, pr_2\}, 1 \leq i \leq k$, where pr_1, pr_2 denote the projections onto the respective component and null sets the register to 0. Which operation is chosen depends on q and b only. When the final component of an input is read A performs the tests for this component and moves to its last state without any further computation. It accepts the input if this final state belongs to F, otherwise A rejects.

(c) Non-deterministic periodic $\mathbb{R}$-automata are defined similarly with the only difference that δ becomes a relation in the following sense: If in state q the tests result in $b \in \{0,1\}^{k+\ell}$ the automaton can non-deterministically choose a next state and as update operations one among finitely many tuples $(q', \circ_1, \ldots, \circ_k) \in Q \times \{+, -, \times, null, pr_1, pr_2\}^k$.[1]

As usual, a non-deterministic automaton accepts an input if there is at least one accepting computation.

(d) The language of finite sequences accepted by A is denoted by $L(A) \subseteq \mathbb{R}^*$.

Note that by using T copies of each state it is easy to make the transitions also depending on the counter value. Now consider an automaton working on inputs from $\mathbb{R}^\omega := \prod_{i=1}^{\infty} \mathbb{R}$. For such periodic Büchi $\mathbb{R}$-automata it remains to define an accepting condition. Given the periodicity there are several options.

Definition 2. *A (non-)deterministic periodic Büchi $\mathbb{R}$-automaton (N)DBA$_\mathbb{R}$ A with period T is a (non-)deterministic periodic GKL $\mathbb{R}$-automaton with period T working on $w \in \mathbb{R}^\omega$ under one of the following acceptance conditions:*

Acc_1 : *A accepts $w \in \mathbb{R}^\omega$ iff there is a computation of A and a state $q \in F$ (the final state set) which is visited infinitely often during the computation;*

Acc_2 : *similar, but A is required to pass infinitely often through a $q \in F$ at a full period.*

Lemma 3. *NDBA$_\mathbb{R}$'s and DBA$_\mathbb{R}$'s with acceptance conditions Acc_1 accept the same languages as with Acc_2.*

---

[1] Note that if the counter has value $T-1$ also non-deterministic automata have to reset all register values to 0; only the next state can then be chosen non-deterministically.

Lemma 4. *The class of languages accepted by a DBA$_\mathbb{R}$ is strictly included in the class of languages accepted by an NDBA$_\mathbb{R}$. The former is not closed under complementation.*

We first establish some closure and decomposition properties of languages accepted by NDBA$_\mathbb{R}$.

Proposition 5. *Let $\mathcal{A}, \mathcal{B}$ be NDBA$_\mathbb{R}$ with periods T_A, T_B and accepted languages $L(\mathcal{A}), L(\mathcal{B})$, respectively. Then both $L(\mathcal{A}) \cup L(\mathcal{B})$ and $L(\mathcal{A}) \cap L(\mathcal{B})$ are accepted by NDBA$_\mathbb{R}$'s with period $T := T_A \cdot T_B$.*

Next, we want to derive a characterization for Büchi languages over $\mathbb{R}$ using concatenations and the Kleene-star of languages accepted by periodic GKL $\mathbb{R}$-automata. However, in order to do so we have to be a bit careful how to define concatenation and the Kleene star operation. The reason is that when concatenating two sequences accepted by two automata, the first might finish its computation with a counter value different from 0. It is then not clear how to choose a period of an automaton for the concatenation of the respective languages. The same problem occurs with the Kleene star. Therefore we define a restricted form of the two operations depending on a given period. When dealing with Büchi languages in $\mathbb{R}^\omega$ this is appropriate since one can split each computation on an input $w \in \mathbb{R}^\omega$ into blocks of length T or a multiple of T, no matter how T looks like. This motivates the following definition.

Definition 6. *For $A \subseteq \mathbb{R}^*, T \in \mathbb{N}$ define $A^{(T)} := \{x \in A \mid |x| = kT$ for some $k \in \mathbb{N}\}$. Moreover, $(A^{(T)})^*$ and $(A^{(T)})^\omega$ then are defined as usual concatenating finitely or countably infinitely many sequences from $A^{(T)}$, respectively.*

Lemma 7. *Let $\mathcal{A}$ be a non-deterministic periodic GKL automaton and $\mathcal{B}$ be an NDBA$_\mathbb{R}$ both having the same period $T \in \mathbb{N}$. Then the languages $L(\mathcal{A})^T.L(\mathcal{B})$ and $\left(L(\mathcal{A})^T\right)^\omega$ are accepted by an NDBA$_\mathbb{R}$.*

Given the previous lemma a decomposition theorem for languages accepted by Büchi $\mathbb{R}$-automata follows. Before stating it we recall from [11] the definition of certain semi-algebraic sets attached to the computation of periodic automata.[2]

Definition 8. *Let $\mathcal{A}$ be an NDBA$_\mathbb{R}$ with period T. For two states p, q of $\mathcal{A}$ define the set $S(p, q)$ as those $x \in \mathbb{R}^T$ for which there is a computation in $\mathcal{A}$ on x which moves from p in T steps to q assuming a new period is started in p.*

It is not hard to see that the sets $S(p, q)$ actually are semi-algebraic [11].

Theorem 9. *(a) Let $L \subseteq \mathbb{R}^\omega$ be accepted by an NDBA$_\mathbb{R}$ $\mathcal{A}$ with period T using Acc_2. Let q_0 denote the start state and F_A the final state set. Then*

$$L = \bigcup_{p \in F_A} S(q_0, p)^{(T)} . \left(S(p, p)^{(T)}\right)^\omega.$$

---

[2] A semi-algebraic set in $\mathbb{R}^n$ is a set that can be defined as a Boolean combination (using finite unions, intersections, and complements) of solution sets of polynomial equalities and inequalities.

Moreover, the sets involved in the above description of L are accepted by a periodic GKL $\mathbb{R}$-automaton with period T.

(b) Vice versa, let $L \subseteq \mathbb{R}^\omega$ be such that there exists a representation of form

$$L = \bigcup_{i=1}^{s} (U_i)^{(T_i)} \cdot \left(V_i^{(T_i)} \right)^\omega,$$

where $s, T_i \in \mathbb{N}$ and U_i, V_i are accepted by a periodic GKL $\mathbb{R}$-automaton with period $T_i, 1 \leq i \leq s$. Then L is accepted by an NDBA$_\mathbb{R}$ with period $T := \prod_{i=1}^{s} T_i$.

The next structural result will deal with the closure of languages accepted by NDBA$_\mathbb{R}$ under complementation. Also in the real number framework this can be done along different approaches. We shall extend Büchi's original construction as well as using Muller $\mathbb{R}$-automata.

Definition 10. *Let $\mathcal{A}$ be an NDBA$_\mathbb{R}$ with period T.*

(i) For states p, q and a sequence $u \in \mathbb{R}^*$ of length kT for some $k \in \mathbb{N}$, define $p \xrightarrow{u} q$ iff $\mathcal{A}$ can move from p to q when starting a new period in p and reading u.
Define $p \xrightarrow{u}_f q$ iff in addition there is such a computation from p to q reading u along which $\mathcal{A}$ passes through $F_\mathcal{A}$.
(ii) Two sequences $u, v, \in \mathbb{R}^*$ with lengths a multiple of T are called $\mathcal{A}$-equivalent, in terms $u \sim_\mathcal{A} v$, iff for all pairs (p, q) of states in $Q_\mathcal{A}$ it holds

$$p \xrightarrow{u}_f q \Leftrightarrow p \xrightarrow{v}_f q$$

Obviously, $\sim_\mathcal{A}$ defines an equivalence relation of finite index on $\bigcup_{k \in \mathbb{N}} \mathbb{R}^{kT} \subset \mathbb{R}^*$.

Lemma 11. *Let $\mathcal{A}$ be an NDBA$_\mathbb{R}$ with period T.*

(a) Each equivalence class of $\sim_\mathcal{A}$ is accepted by a periodic GKL $\mathbb{R}$-automaton with period T.
(b) For two equivalence classes $U, V \subseteq \bigcup_{k \in \mathbb{N}} \mathbb{R}^{kT}$ either it holds $UV^\omega \subseteq L(\mathcal{A})$ or $UV^\omega \subseteq \overline{L(\mathcal{A})}$.
(c) For every $w \in \mathbb{R}^\omega$ there are equivalence classes U, V of $\sim_\mathcal{A}$ such that $w \in UV^\omega$.

Theorem 12. *The set of languages accepted by NDBA$_\mathbb{R}$'s is closed under complementation.*

3   Periodic Muller $\mathbb{R}$-automata

Another classical way to prove closure of Büchi acceptable languages under complementation is by changing the acceptance condition. The resulting non-deterministic Muller automata have been shown to be equivalent to non-deterministic Büchi automata on the one hand side and to deterministic Muller automata on the other, thus yielding the complementation result. In the present section we show that a periodic version of Muller $\mathbb{R}$-automata can be defined and used to prove once again Theorem 12, this time by extending the well known concept of Safra trees.

Definition 13. *A (non-)deterministic Muller $\mathbb{R}$-automaton (N)DMA$_{\mathbb{R}}$ $\mathcal{A}$ with period T is defined like a (N)DBA$_{\mathbb{R}}$, but with a family $\mathcal{F} = \{F_1, \ldots, F_s\}$ of finite state sets $F_i \subseteq Q$. Again, two acceptance conditions will be considered:*

Acc_1 : *$\mathcal{A}$ accepts $w \in \mathbb{R}^\omega$ iff there is an $F_i \in \mathcal{F}$ and a computation of $\mathcal{A}$ on w which passes through all $q \in F_i$ infinitely often, but only finitely often through all other states $q \notin F_i$.*
Acc_2 : *The same as Acc_1, but precisely the states in F_i occur infinitely often along the computation when the counter value is 0.*

Lemma 14. *Non-deterministic Muller $\mathbb{R}$-automata and non-deterministic Büchi $\mathbb{R}$-automata accept the same languages in $\mathbb{R}^\omega$. This holds for both acceptance conditions. Consequently, NDMA$_{\mathbb{R}}$'s accept the same languages under both acceptance conditions.*

Lemma 15. *The class of languages accepted by DMA$_{\mathbb{R}}$'s is closed under complementation and intersection.*

The main result in this section is a real number analogue of McNaughton's theorem [9]. It says that each non-deterministic Büchi $\mathbb{R}$-automaton has an equivalent deterministic Muller $\mathbb{R}$-automaton. Thus, in contrast to NDBA$_{\mathbb{R}}$'s, deterministic and non-deterministic Muller $\mathbb{R}$-automata have the same computational power by Lemma 14. The proof for periodic Muller $\mathbb{R}$-automata extends Safra's construction for the finite alphabet setting [12]. We suppose the reader to be familiar with this proof, see [5,13]. An important ingredient of Safra's proof is the powerset construction to determinize non-deterministic finite automata. The corresponding construction for periodic GKL $\mathbb{R}$-automata has been given in [11] and will be used here as well.

Theorem 16. *Let $\mathcal{A}$ be an NDBA$_{\mathbb{R}}$ with period T. Then there is a DMA$_{\mathbb{R}}$ $\mathcal{B}$ with period T and $L(\mathcal{A}) = L(\mathcal{B})$.*

Proof. Though the proof proceeds along Safra's original one, the extension to $\mathbb{R}$-automata has some complications. This mainly holds because the determinization of non-deterministic periodic GKL $\mathbb{R}$-automata is more involved than that for classical non-deterministic finite automata. The reason for this is the fact that in the powerset construction the evolvement of register assignments has to be protocolled as well. Without recalling the construction from [11] completely the following aspects are important. Given a non-deterministic periodic automaton $\mathcal{A}$ with period T there is an integer N only depending on $\mathcal{A}$ and an equivalent deterministic periodic automaton $\mathcal{A}'$ with period T such that $\mathcal{A}'$ uses N registers. For each input $w \in \mathbb{R}^*$ $\mathcal{A}'$ protocols every possible computation of $\mathcal{A}$ on w in the following sense: For each prefix $w_1 \ldots w_i$ of w $\mathcal{A}'$ codes in its configuration states reachable by $\mathcal{A}$ when reading the prefix together with the corresponding register assignments. In order not to repeat completely Safra's construction we give a brief description only and then point out the decisive additional aspects needed here. In Safra's proof the deterministic Muller automaton $\mathcal{M}$ has as its states so called Safra trees. These are trees with at most $n := |Q_\mathcal{A}|$ many nodes, also called macrostates. Each such macrostate has an integer name chosen from the set $\{1, \ldots, 2n\}$ and a label that is a subset of $Q_\mathcal{A}$. A step of $\mathcal{M}$ transforms one Safra tree to another as follows: For each macrostate of the current state (i.e., Safra tree) it basically performs the powerset construction. However, there is a clever way how to dynamically enlarge and reduce the number of macrostates in a Safra tree. In order to guarantee that accepting runs again and again hit final state sets there is a procedure of introducing sons of a macrostate; a son gets as name a currently non-used one from $\{1, \ldots, 2n\}$ (note that at each step at most n macrostates are contained in a Safra tree) and as label all final states of $\mathcal{A}$ contained in the label of the father macrostate. In the next step the powerset construction is also performed on the son macrostate. The construction is done in such a way that a state is at most contained in one son of a macrostate; if the union of the labels of all sons of a macrostate equals the label of the father node, then all nodes below the father are deleted and the father gets for the duration of one step a flash mark $\sharp$. It can be shown that an accepting run of $\mathcal{A}$ is equivalent to the existence of a run of $\mathcal{M}$ in which there is a persistent macrostate, i.e., a macrostate with an integer name j that from a certain point in time on is never deleted, and flashes infinitely often.

When simulating an NDBA$_\mathbb{R}$ by a DMA$_\mathbb{R}$ we want to use the same idea. However, some aspects become more involved. For example, in the classical powerset construction, if a state occurs in several brother nodes of a macrostate only the one in the oldest such macrostate survives. In the real number framework this is not possible since the same state might occur in several brother nodes with different corresponding register assignments, and it is necessary to keep track of all of them. A way to solve this problem is based on Lemma 14. Since DMA$_\mathbb{R}$'s with both acceptance conditions accept the same languages we can enforce to apply main parts of Safra's construction only after full periods of the computation. This allows to restrict to a discrete situation even for $\mathbb{R}$-automata. More precisely, we define Safra $\mathbb{R}$-trees similarly to discrete ones but with additional information about the realizable register assignments included. Thus, a macrostate again has a name in $\{1, \ldots, 2n\}$ and a label. The latter now is a collection of pairs of reachable states of $\mathcal{A}$ together with corresponding register assignments. If we project those pairs onto their first component we therefore get a subset of $Q_\mathcal{A}$ of the states reachable by $\mathcal{A}$. Note, however, that the same state can occur several times as result of the powerset construction in [11]. Given the above mentioned fact that a fixed number of N registers suffice to store this information, such a label is well defined. The maximal number of macrostates in a Safra $\mathbb{R}$-tree just as in the discrete setting is the number n of states of $\mathcal{A}$. For a Safra $\mathbb{R}$-tree representing the current state of a computation of $\mathcal{A}$ by coding in its macrostates all reachable states and corresponding register assignments, the

Muller $\mathbb{R}$-automaton performs on all macrostates the new powerset construction. As long as the counter value is different from 0 this is repeated without creating new or merging current macrostates or deleting states in labels. The decisive point is the behaviour at full periods. It is here where the original Safra construction is applied. Note that after full periods all register assignments are 0; thus each macrostate of a Safra $\mathbb{R}$-tree after a full period basically has a subset of Q_A as its label. This is the justification in [11] that the powerset construction does not explode. Now when a period is finished we adjust the $\mathbb{R}$-tree by appending to each macrostate a son with label all the final states of Q_A that occur in the label of the given state. The register values in the corresponding label are set to 0. The powerset construction for $\mathbb{R}$-automata is performed for an entire period on all macrostates of the Safra $\mathbb{R}$-tree. When the period is finished, the original merging and deletion steps are performed. This includes two actions:

- if a state from Q_A occurs in the labels of several macrostates of the current tree, it is deleted from all but the oldest one and
- if the union of the states in the labels of all sons of a macrostate equals the states in the label of this macrostate, then all nodes below that macrostate are deleted. In this situation the macrostate flashes for one time step.

A new round starts with adding new sons and repeating the above. The Muller $\mathbb{R}$-automaton accepts with condition Acc_2. Its final state sets are all families of Safra $\mathbb{R}$-trees of macrostates occurring after full time periods such that at least one node name $k \in \{1, \ldots, 2n\}$ occurs as name of a macrostate in all trees of the family, and at least in one tree this node flashes.   □

4   A Logic for Periodic $\mathbb{R}$-automata

In this section we show that also in our setting there is a strong relation between the languages accepted by periodic $\mathbb{R}$-automata - both for sequences in $\mathbb{R}^*$ and infinite sequences in $\mathbb{R}^\omega$ - and certain logics defined to deal with real numbers as input data. The latter will be done using the framework of metafinite model theory. Metafinite model theory was introduced in [3] and studied for the BSS model in [4]. It provides a way to describe real number computations and complexity in a logical framework by separating its discrete from real number aspects.

Below, a blending of a monadic second order logic on the discrete part and a certain restricted logic on the real number part will be important. A metafinite MSO logic in the BSS model over $\mathbb{R}$ was introduced and studied in [8]. Here, due to periodicity and the limited computational abilities of periodic automata a much weaker fragment of such a real number MSO logic will be considered.

The main result of this section will be a kind of analogue of Büchi's theorem for periodic GKL and periodic Büchi $\mathbb{R}$-automata. We suppose the reader to be familiar with 'classical' MSO logic and Büchi's theorem. For this and proofs of corresponding classical theorems along Büchi's work see once more any of the texts [5,7,13].

4.1   From Periodic $\mathbb{R}$-automata to Formulas

We first develop a logical description of the computation of a periodic $\mathbb{R}$-automaton. It uses a mixture of a classical discrete MSO logic and a metafinite part for dealing with real number data. This outlines a way how to define in the following subsection in a more precise way a metafinite $MSO_{\mathbb{R}}$ logic which precisely reflects the abilities of periodic $\mathbb{R}$-automata. For case of simplicity we deal with automata for finite sequences in $\mathbb{R}^*$. The generalization to $\mathbb{R}^\omega$ is easy.

In Büchi's theorem, weak MSO logic over $\langle \mathbb{N}, <, succ, 0 \rangle$ is used to describe the computation of a finite automaton; 'weak' refers to monadic second order quantifiers only ranging over *finite* subsets of $\mathbb{N}$. Similarly, MSO logic characterizes the computation of Büchi automata on infinite words over a finite alphabet. In particular, assignments of monadic second order variables are used to code in which state the computation is at a specific step.

For describing the discrete part of periodic $\mathbb{R}$-automata we thus also use MSO logic. However, we additionally equip it with a unary relation symbol $P \subset \mathbb{N}$. Throughout, it is interpreted as a set $P_T := \{kT \mid k \in \mathbb{N}\}$ for a fixed $T \in \mathbb{N}$, describing the time steps after which an automaton with period T resets the register values to 0. For automata over $\mathbb{R}^*$ only a finite initial part of such a set P_T is considered. Thus we work with the discrete part $\langle \mathbb{N}, <, succ, P, 0 \rangle$. In addition, a metafinite real number part for modelling inputs $w \in \mathbb{R}^*$ or $w \in \mathbb{R}^\omega$, respectively, is added in form of a function symbol C interpreted for input w as a function $C_w : \mathbb{N} \to \mathbb{R}, C_w(i) := w_i$. The logic then has to reflect the (restricted) way the automaton is allowed to compute with reals represented by the $C_w(i), i \in \mathbb{N}$. The metafinite part also involves symbols for the automaton's constants.

Suppose $\mathcal{A}$ to be a (deterministic) periodic GKL $\mathbb{R}$-automaton with period T, k registers and real constants $c_1, \ldots, c_\ell$. Let $Q := \{q_1, \ldots, q_s\}$ denote its state set and $F \subseteq Q$ the accepting states, q_1 the start state. Each input $w \in \mathbb{R}^*$ is modelled as a function $C_w : \mathbb{N} \to \mathbb{R}$, where $C_w(i) := w_i$ for $1 \le i \le |w|$. $\mathcal{A}$'s computation on a w is expressed via a formula $\varphi_{\mathcal{A}}$ over vocabulary $\langle \mathbb{N}, <, succ, P, C, c_1, \ldots, c_\ell \rangle$. Here, P is always interpreted as a $P_T \subset \mathbb{N}$ as explained before. The interpretation of the constants is always assumed to include 0 and 1.

On $\langle \mathbb{N}, <, succ, P \rangle$ formula $\varphi_{\mathcal{A}}$ is in MSO logic, i.e., built from $P(i), X(i)$ for first order variables i and second order variables X, quantifications $\exists i, \forall i, \exists X, \forall X$ and Boolean combinations thereof. The use of quantification of first order variables i when applied to $C(i)$ is limited to $\forall i \in P : C(i)$ and $\exists i \in P : C(i)$. Beside this, the metafinite parts of $\varphi_{\mathcal{A}}$ reflect the way $\mathcal{A}$ can compute. If below we deal with inputs $w \in \mathbb{R}^*$ all discrete second order variables X, P_T are finite subsets of $\mathbb{N}$ and the corresponding logic is called weak $MSO_{\mathbb{R}}$ logic. For $w \in \mathbb{R}^\omega$ monadic second order variables range over arbitrary subsets of $\mathbb{N}$.

We now give a high-level description of $\varphi_{\mathcal{A}}$. In Subsect. 4.2 we then precisely define the logical framework characterizing periodic GKL $\mathbb{R}$-automata.

Let $\varphi_{\mathcal{A}}(C) \equiv \exists X_1, \ldots, X_s \varphi_1(X_1, \ldots, X_s) \wedge \varphi_2(X_1, \ldots, X_s, C)$, where $X_1, \ldots, X_s$ are monadic variables ranging over $\mathbb{N}$, and being finite in case we deal with inputs from $\mathbb{R}^*$. Here, φ_1 is a (weak) MSO formula on $\langle \mathbb{N}, <, succ, 0 \rangle$ expressing that $X_1, \ldots, X_s$ is a partition of the set $\{1, \ldots, |w|\}$ for inputs $w \in \mathbb{R}^*$ or of $\mathbb{N}$ for inputs from $\mathbb{R}^\omega$, respectively. The X_j's interpretation is: $X_j(i) = 1$ iff $\mathcal{A}$ is in state j after it read input $w(i)$; for $i = 0$ it holds $X_1(0) = 1$.

If $\mathcal{A}$ works over $\mathbb{R}^*$ the formula has to express the length $|w|$ of an input w; this easily can be defined and is left to the reader. Also the accepting condition $\bigvee_{j \in F} X_j(|w|)$ has to be added in φ_1. If $\mathcal{A}$ is a periodic Büchi $\mathbb{R}$-automaton instead

we add $\left(\exists X_{inf} : X_{inf} \subseteq P \wedge X_{inf} \text{ is infinite} \wedge \bigvee_{j \in F} X_{inf} \subseteq X_j \right)$. This eas-

ily can be formalized in MSO logic over $\langle \mathbb{N}, <, succ, P, 0 \rangle$ and expresses that at least one final state occurs infinitely often after a full period has been finished.

φ_2 expresses a correct computation of $\mathcal{A}$ on w. In order to do so, for all $i \leq |w|$ if $X_j(i)$, then $\mathcal{A}$'s state j' after having processed w_{i+1} must describe a permitted transition of $\mathcal{A}$. This involves $\mathcal{A}$'s computation on the real number data. In order to start from register assignments 0 we describe the corresponding conditions in blocks of length T of a full period. More precisely, for $i \in P$ a formula $\tilde{\varphi}_2(X_1, \ldots, X_s, C, i)$ consists of a finite collection of formulas containing for $C(i), C(succ(i)), C(succ^2(i)), \ldots, C(succ^{T-1}(i))$ MSO$_\mathbb{R}$-terms that represent the register entries generated by a computation of T steps, starting in the correct state (expressed via $X_j(i)$ for a unique j) with register values 0. The MSO$_\mathbb{R}$-terms describe current register contents and are built from 0 along the computations that $\mathcal{A}$ performs, each time combining an already computed term with the next input component. Then, $\tilde{\varphi}_2$ expresses that tests performed by $\mathcal{A}$, the subsequent computations and the state visited next correspond to the partition $X_1, \ldots, X_s$ of $\mathbb{N}$ or $\{1, \ldots, |w|\}$, respectively. Finally, the formula $\varphi_2(X_1, \ldots, X_s, C)$ is obtained as $\forall i \, P \, \tilde{\varphi}_2(X_1, \ldots, X_s, C, i)$. This shows

Proposition 17. *For each periodic GKL $\mathbb{R}$-automaton $\mathcal{A}$ with period $T \in \mathbb{N}$ and the corresponding formula $\varphi_\mathcal{A}$ as constructed above, we have $w \in L(\mathcal{A}) \Leftrightarrow C_w \models \varphi(C)$, when P is interpreted as P_T. This holds mutatis mutandis for periodic Büchi $\mathbb{R}$-automata, inputs $w \in \mathbb{R}^\omega$ and the MSO$_\mathbb{R}$-formula $\varphi_\mathcal{A}$.* $\square$

The above description intends to give an idea of the fragments that can be used in our logic. For the direction from formulas to automata we have of course to be more precise. This holds in particular with respect to the way the logic deals with terms and formulas involving reals. One important aspect that shows up above is the limited way in which terms $C(i)$ enter into formulas. Here, we only allow quantifiers of the form $\forall i \in P, \exists i \in P$, where P has to be interpreted as a periodic subset of $\mathbb{N}$, and the quantified formula only depends on i in that it contains terms $C(i), C(succ(i)), \ldots, C(succ^{T-1}(i))$.

4.2   From MSO$_\mathbb{R}$ Logic to Periodic Automata

We now introduce more precisely a logic which exactly captures the computations of periodic GKL $\mathbb{R}$-automata. Since the latter works with inputs from $\mathbb{R}^*$

and $\mathbb{R}^\omega$, respectively, we always consider the universe $\mathbb{N}$ and a function symbol C as part of the vocabulary for the logic. For input w it is interpreted as (initial fragment) $C_w : \mathbb{N} \to \mathbb{R}$ with $C_w(i) := w_i$. Formulas $\varphi_\mathcal{A}$ attached to an automaton $\mathcal{A}$ have C as (only) free variable; the goal is to construct $\varphi_\mathcal{A}$ in such a way that $C_w \models \varphi_\mathcal{A}$ iff $w \in L(\mathcal{A})$. The logic to be defined is an example of a metafinite logic as introduced in [3] and studied for real number computations in [4]. In [8] a monadic second order logic for BSS computations was defined and analysed in view of the design of efficient algorithms for structures of bounded treewidth. The logic introduced below also has MSO constructs combining finite structures and the real numbers; it is, however, (much) weaker than the MSO logic from [8].

As said above we consider usual MSO logic on $\langle \mathbb{N}, <, succ, P, 0 \rangle$, where the unary symbol P is always be interpreted as a set $P_T = \{kT \mid k \in \mathbb{N}_0\}$ for a fixed $T \in \mathbb{N}$. For the metafinite part a function symbol $C : \mathbb{N} \to \mathbb{R}$ and constants $c_1, \ldots, c_\ell$ are added. We assume 0 and 1 to be among the constants.

We first define real number terms. Each such term has an index $i \in \mathbb{N}$. Terms depend on the interpretation of P as particular relation P_T as explained above.

Definition 18. *Real number terms are defined as follows:*

(i) $0 \in \mathbb{R}$ is a real number term of index 0.

(ii) Let t_i be a real number term of index i; then $[1 - P(succ(i))] \cdot [C(succ(i)) \circ t_i]$ is a term of index $i+1$. Here, $\circ$ is any of the allowed binary operations. Note that 0 is a term for each index and the only term for indexes j where $j \in P$.

Definition 19. *Monadic Second Order logic $MSO_\mathbb{R}$ is defined as follows:*

1. *All MSO formulas over the discrete part $\langle \mathbb{N}, <, succ, P, 0 \rangle$ are $MSO_\mathbb{R}$-formulas.*
2. *Purely real formulas prf on the metafinite part are built as follows:*
   2.1 For all $i \in \mathbb{N}$ comparisons $C(i) \Delta c_j$ for $\Delta \in \{\geq, =, \leq\}$ and c_j one of the constants are atomic prf.
   2.2 Let $i \in \mathbb{N}, t_i$ be a term of index i, then $C(succ(i)) \Delta t_i$ is a prf.
   2.3 For any $i \in \mathbb{N}$ a finite Boolean conjunction of atomic prf as in 2.2 with indexes $i, succ(i), succ^2(i), \ldots, succ^m(i)$ for some constant m is a prf. We denote such a formula by $\varphi_\mathbb{R}(i)$ and say that i is bound by a prf.
3. *Any finite combination of $\wedge, \vee, \neg$ of purely real formulas $\varphi_\mathbb{R}(i)$ is an $MSO_\mathbb{R}$-formula; the indexes that are bound by one of the building prf remain bound.*
4. *If $\varphi_\mathbb{R}(i, X)$ is an $MSO_\mathbb{R}$-formula with i bound by a prf and X a discrete free variable (first or second order), then the following are $MSO_\mathbb{R}$-formulas:*
   $\forall X : \varphi_\mathbb{R}(i, X), \exists X : \varphi_\mathbb{R}(i, X), \forall i \in P : \varphi_\mathbb{R}(i)$ and $\exists i \in P : \varphi_\mathbb{R}(i)$.
   In the first two cases if X is a first order variable it is not allowed to be bound by a prf and i remains bound.
5. *All Boolean combinations by $\wedge, \vee, \neg$ of $MSO_\mathbb{R}$-formulas are $MSO_\mathbb{R}$-formulas. Indexes i bound by a prf remain bound under these operations.*

It is important to stress that above first order variables i that occur in real number terms and formulas can only be quantified via $\forall i \in P$ and $\exists i \in P$.

Obviously, this is necessary if the period of an automaton should be modelled somehow. Thus, a formula like $\forall i \in \mathbb{N} \, C(i){\cdot}C(succ(i)) > 0$ is forbidden. Similarly, for different i, j that are bound by a prf there is no mix in the formula of real number terms relating i or one of its successors with j or one of its successors. Both such constructions would require a related automaton to remember input components without respecting the reset operation required after each period. From the discussion in Subsect. 4.1 part (a) of the following theorem should be obvious now. The proof of harder part (b) has to be postponed to the full version.

Theorem 20. *(a) Let $\mathcal{A}$ be a periodic Büchi $\mathbb{R}$-automaton with period $T \in \mathbb{N}$ working over $\mathbb{R}^\omega$ and using constants $c_1, \ldots, c_\ell$. Then there exists a $\mathrm{MSO}_\mathbb{R}$-formula $\varphi_\mathcal{A}(C, P)$ over $\langle \mathbb{N}, <, succ, P, C, c_1, \ldots, c_\ell \rangle$ such that $w \in L(\mathcal{A})$ if and only if $(C_w, P_T) \models \varphi_\mathcal{A}(C, P)$ with $P_T = \{kT \mid k \in \mathbb{N}_0\}$.*

(b) Let $\varphi(C, P)$ be a $\mathrm{MSO}_\mathbb{R}$-formula involving real number constants $c_1 \ldots, c_\ell$. Fix the interpretation of P to be some P_T, so C is the only free variable. Then there is a periodic Büchi $\mathbb{R}$-automaton $\mathcal{A}_\varphi$ using the above constants such that $w \in L(\mathcal{A})$ iff $(C_w, P_T) \models \varphi(C, P)$.

The statement holds similarly if computations on finite sequences in $\mathbb{R}^*$ and weak $\mathrm{MSO}_\mathbb{R}$ logic are considered. □

Acknowledgement. We thank all reviewers for their very thorough reading and many useful comments.

References

1. Blum, L., Cucker, F., Shub, M., Smale, S.: Complexity and Real Computation. Springer, New York (1998). https://doi.org/10.1007/978-1-4612-0701-6
2. Gandhi, A., Khoussainov, B., Liu, J.: Finite automata over structures. In: Agrawal, M., Cooper, S.B., Li, A. (eds.) TAMC 2012. LNCS, vol. 7287, pp. 373–384. Springer, Heidelberg (2012). https://doi.org/10.1007/978-3-642-29952-0_37
3. Grädel, E., Gurevich, Y.: Metafinite model theory. Inf. Comput. **140**(1), 26–81 (1998)
4. Grädel, E., Meer, K.: Descriptive complexity theory over the real numbers. In: Leighton, F.T., Borodin, A. (eds.) Proceedings of the 27th STOC, pp. 315–324 (1995)
5. Hofmann, M., Lange, M.: Automatentheorie und Logik. Springer, Berlin (2011). https://doi.org/10.1007/978-3-642-18090-3
6. Khoussainov, B., Liu, J.: Decision problems for finite automata over infinite algebraic structures. In: Han, Y.-S., Salomaa, K. (eds.) CIAA 2016. LNCS, vol. 9705, pp. 3–11. Springer, Cham (2016). https://doi.org/10.1007/978-3-319-40946-7_1
7. Khoussainov, B., Nerode, A.: Automata Theory and Its Applications. Birkhäuser, Springer, Berlin (2001)
8. Makowsky, J.A., Meer, K.: Polynomials of bounded tree-width. In: Foundations of Computational Mathematics, Proceedings of the Smalefest, pp. 211–250. World Scientific (2002)
9. McNaughton, R.: Testing and generating infinite sequences by a finite automaton. Inf. Control **9**(5), 521–530 (1966)

10. Meer, K., Naif, A.: Generalized finite automata over real and complex numbers. Theor. Comput. Sci. **591**, 85–98 (2015)
11. Meer, K., Naif, A.: Periodic generalized automata over the reals. In: Dediu, A.-H., Janoušek, J., Martín-Vide, C., Truthe, B. (eds.) LATA 2016. LNCS, vol. 9618, pp. 168–180. Springer, Cham (2016). https://doi.org/10.1007/978-3-319-30000-9_13
12. Safra, S.: On the complexity of omega-automata. In: 29th Symposium on FOCS, New York, pp. 319–327 (1988)
13. Thomas, W.: Languages, automata, and logic. In: Rozenberg, G., Salomaa, A. (eds.) Handbook of Formal Languages, pp. 389–455. Springer, Heidelberg (1997). https://doi.org/10.1007/978-3-642-59126-6_7

Nonuniform Families of Polynomial-Size Quantum Finite Automata and Quantum Logarithmic-Space Computation with Polynomial-Size Advice

Tomoyuki Yamakami[✉]

Faculty of Engineering, University of Fukui, 3-9-1 Bunkyo, Fukui 910-8507, Japan
TomoyukiYamakami@gmail.com

Abstract. The state complexity of a finite(-state) automaton intuitively measures the size of the description of the automaton. Sakoda and Sipser [STOC 1972, pp. 275–286] were concerned with nonuniform families of finite automata and they discussed the behaviors of nonuniform complexity classes defined by families of such finite automata having polynomial-size state complexity. In a similar fashion, we introduce nonuniform state complexity classes using families of quantum finite automata. Our primarily concern is one-way quantum finite automata empowered by garbage tapes. We show inclusion and separation relationships among nonuniform state complexity classes of various one-way finite automata, including deterministic, nondeterministic, probabilistic, and quantum finite automata of polynomial size. For two-way quantum finite automata equipped with garbage tapes, we discover a close relationship between the nonuniform state complexity of such a polynomial-size quantum finite automata family and the parameterized complexity class induced by quantum logarithmic-space computation assisted by polynomial-size advice.

Keywords: Quantum finite automata · State complexity ·
Quantum Turing machine · Bounded-error probability ·
Quantum advice

1 Prelude: Quick Overview

This exposition reports a collection of preliminary results obtained by the currently on-going study on the state complexity of nonuniform families of quantum finite automata, which is briefly referred to as the *nonuniform state complexity*.

1.1 Nonuniform State Complexity of Finite Automata Families

Each finite(-state) automaton is completely described by a set of transitions of its inner states. The number of inner states is thus crucial to measure the descriptional size of the automaton and it works as a complexity measure, known as

© Springer Nature Switzerland AG 2019
C. Martín-Vide et al. (Eds.): LATA 2019, LNCS 11417, pp. 134–145, 2019.
https://doi.org/10.1007/978-3-030-13435-8_10

the *state complexity* of the automaton. This complexity measure thus naturally regards as an indicator for the computational power of the automaton. Instead of taking a single automaton, in this exposition, we consider a "family" of finite automata in a way similar to a family of Boolean circuits. Such a family of finite automata may be generated in a uniform way by a certain deterministic algorithm. Unlike Boolean circuits, nevertheless, inputs of automata are *not limited* to certain fixed lengths. For brevity, the term "uniform sate complexity" refers to the state complexity of such a uniform family of finite automata. Opposed to the uniform state complexity, here we intend to study its "nonuniform" counterpart under the name of *nonuniform state complexity*. This nonuniform complexity measure turns out to be closely related to a nonuniform model of Turing-machine computations.

Nonuniform state complexity has played various roles in automata theory. An early discussion that attempted to relate certain state complexity issues to the collapses of known space-bounded complexity classes dates back to late 1970s. Sakoda and Sipser [9], following Berman and Lingas [2], argued on the state complexity of transforming one family of 2-way nondeterministic finite automata (or 2nfa's, for short) into another family of 2-way deterministic finite automata (or 2dfa's). From their works, we have come to know that the state complexity of a family of automata is related to the work-tape space of a Turing machine. In this line of study, after a long recess, Kapoutsis [6] and Kapoutsis and Pighizzini [7] lately revitalized a discussion on the relationships between logarithmic-space (or log-space, for short) complexity classes and state complexity classes in connection to the L = NL question (in fact, the NL ⊆ L/poly question, where L/poly is the nonuniform analogue of L).

Taking a complexity-theoretic approach, Kapoutsis [4,5] earlier discussed relationships among the nonuniform state complexity classes 1D, 1N, 2D, and 2N of families of "promise" decision problems, each of which is solved by a nonuniform family of deterministic and nondeterministic finite automata of polynomially many inner states (see Sect. 2 for their definitions). Along the same line of study, Yamakami [14] recently gave a characterization of the polynomial-time sub-linear-space "parameterized" complexity class, known as PsubLIN, and an NL-complete problem 3DSTCON parameterized by the number of vertices of an input graph (which is generally referred to as a *size parameter*) in terms of the state complexities of restricted 2nfa's and narrow 2-way alternating finite automata.

An important discovery of [14] is the fact that a nonuniform family of promise decision problems is more closely related to parameterized decision problems than "standard" decision problems (whose complexities are measured by the binary encoding size of inputs). A decision problem (or a language) L over an alphabet Σ and a reasonable size parameter m from Σ^* to $\mathbb{N}$ (the set of all natural numbers) form a *parameterized decision problem* (L, m) [13]. We can naturally translate such a parameterized decision problem (L, m) into a uniform family $\{(L_n, \overline{L}_n)\}_{n \in \mathbb{N}}$ of promise decision problems and also translate $\{(L_n, \overline{L}_n)\}_{n \in \mathbb{N}}$ back into another parameterized decision problem (K, m), which is "almost" the same as (L, m). See Sect. 2.2 for more details. These translations between parameterized decision problems and families of promise decision problems play an essential role in this exposition. For notational readability, we use the special prefix "para-" and write, for example, para-NL to denote the parameterized analogue of NL, that is, the collection of parameterized NL languages.

After the study of state complexity classes was initiated in [9], a further elaboration has been long anticipated; however, there has been little research on how to expand the scope of these classes. Our purpose of this exposition is to enrich the world of nonuniform state complexity classes toward a whole new direction.

1.2   An Extension to Quantum Finite Automata

We intend to expand the scope of nonuniform state complexity theory to an emerging field of *quantum finite automata*. The behaviors of quantum finite automata, viewed as a natural extension of probabilistic finite automata, are governed by *quantum physics*. Moore and Crutchfield as well as Kondacs and Watrous modeled the quantumization of finite automata in two quite different ways. Lately, these definitions have been considered insufficient for implementation and advantages over classical finite automata and, for this reason, numerous generalizations have been proposed (see, e.g., a survey [1] for references). Here, we intend to use two distinct models: *measure-many 1-way*[1] *quantum finite automata with garbage tapes* (or 1qfa's, for short) and *measure-many 2-way quantum finite automata with garbage tapes* (or 2qfa's), where garbage tapes are write-only tapes used to discard unwanted information, which is considered to be released into a surrounding environment. For an early use of tape tracks to discard the unnecessary information, see [12]. The above models are simple to describe with no additional use of mixed states, ancilla qubits, superoperators, classical inner states, etc. and they are also as powerful as the generalized model cited in [1].

1.3   Overview of Main Contributions

In analogy to 1D and 2D, we introduce their probabilistic and quantum variants as follows. We write 1Q for the collection of families $\{(L_n, \overline{L}_n)\}_{n \in \mathbb{N}}$, each $(L_n, \overline{L}_n)$ of which is solved by a certain 2qfa M_n with unbounded-error probability using polynomially many inner states. If we relax the unbounded-error requirement to the bounded-error requirement (i.e., error probability is bounded from above by a certain constant in $[0, 1/2)$), we write 1BQ in place of 1Q. Similarly to Boolean circuits, we often limit the length of input strings fed to given finite automata. Furthermore, if we replace quantum finite automata by probabilistic finite automata, then we obtain 1BP and 1P from 1BQ and 1Q, respectively. By

---

[1] We use this term "1-way" *in a strict sense* that a tape head always moves to the right and is not allowed to stay still on the same cell. This term is called "real time" in certain literature.

allowing 1dfa's to have $2^{p(n)}$ states for a certain polynomial p, we obtain 2^{1D} from 1D. Using the 2-way models, we naturally obtain 2D and 2N from 1D and 1N, respectively. The nonuniform state complexity class 2BQ is introduced in a way similar to 1BQ but using bounded-error 2qfa's instead of bounded-error 1qfa's. When nondeterministic quantum computation is used, we obtain 1NQ. There are a few known separations: $1D \subsetneq 1N \subsetneq 2^{1D}$, $1N \neq$ co-1N [4,5], and $1D \subsetneq 2D \subseteq 2N \subsetneq 2^{1D}$ [4]. We also obtain $2BP \subseteq 2^{1D}$ from [3, Theorem 6.1] and $2BP \not\subseteq 2N$ from [3, Theorem 6.2].

The first part of our main result is summarized as follows.

Theorem 1. 1. $1D \subsetneq 1BP \subsetneq 1BQ \subsetneq 2^{1D} \cap 1Q$.
2. $1BQ = $ co-1BQ *and* $1P = $ co-1P.
3. $1D \subsetneq 1N \subsetneq 1Q = 1P$ *and* $1N \subsetneq 1NQ$.
4. $1D \subsetneq 2^{1D}$, $1BP \subsetneq 2^{1BP}$, $1BQ \subsetneq 2^{1BQ}$, *and* $1BQ \subseteq 2^{1D}$.
5. $2D \subsetneq 2BP \subseteq 2BQ$ *and* $2P \subseteq 2^{1P}$.

To introduce the nonuniformity notion into a model of quantum Turing machine (or QTM, for short), we equip QTMs with the Karp-Lipton type *advice* as supplemental external information to empower those QTMs (see, e.g., [8]).

When the input size $|x|$ of each string x in $L_n \cup \overline{L}_n$ is limited to at most $p(n)$ for a certain fixed polynomial p, we write 2N/poly and 2BQ/poly instead of 2N and 2BQ, respectively. We show the following close connections between advised complexity classes and nonuniform state complexity classes.

Corollary 2. 1. $2N/\text{poly} \subseteq 2BQ$ *iff* $NL \subseteq BQL/\text{poly}$.
2. $2BQ/\text{poly} \subseteq 2BP$ *iff* $BQL \subseteq BPL/\text{poly}$.

Corollary 2 is a consequence of a more general theorem (Theorem 5), which follows from the exact characterizations (Proposition 6) of parameterized complexity classes in terms of nonuniform state complexity classes, and vice versa. This proposition helps us translate nonuniform state complexity classes, such as 2D/poly, 2N/poly, 2BP/poly, and 2BQ/poly into their corresponding advised parameterized complexity classes, para-L/poly, para-NL/poly, para-BPL/poly, and para-BQL/poly, where the last class para-BQL/poly, for example, denotes the collection of parameterized decision problems (L, m) solvable by bounded-error QTMs using work tapes of space logarithmic in $|x|m(x)$ with (deterministic) advice of size polynomial in $|x|m(x)$ (see Sect. 2.2 for their definitions).

Nishimura and Yamakami [8] introduced the notion of *quantum advice* to enhance the ability of QTMs. Quantum advice manifests a quantumization of *randomized advice* (see, e.g., [11]). To emphasize the use of quantum advice, we write BQL/Qpoly in accordance with [8]. As discussed in [12], the rewriting of an advice tape provides extra power for quantum finite automata. We thus allow a QTM to "erase" advice symbols before terminating to make quantum interference to take place. In parallel to the change of deterministic advice to quantum advice, we also modify our basic model of 2qfa's as follows. Firstly, we express a (quantum) transition function as the form of a matrix or a table, which can be easily encoded into a string over a certain alphabet. For readability, we use the term "transition table" to address this encoded string. This encoding further makes it possible to consider a superposition of transition tables. Generally, we call by a *super quantum finite automaton* a quantum finite automaton obtained by substituting superpositions of transition tables for a quantum transition function. We further add a mechanism of "erasing" its transition tables without accessing the input before terminating. For convenience, we use the notion 2sBQ to express the nonuniform state complexity class obtained from 2BQ by substituting super 2qfa's for ordinary 2qfa's.

Theorem 3. 2sBQ/poly $\subseteq$ 2BQ *iff* BQL/poly = BQL/Qpoly.

Due to the page limit, in what follows, we are focused only on Corollary 2 and Theorem 3 and leave the rest to a complete version of this exposition. A further study on *relativization* (or *Turing reducibility*) was conducted in [15].

2   Preparations: Notions and Notation

Let $\mathbb{N}$ and $\mathbb{C}$ denote respectively the sets of all natural numbers (i.e., nonnegative integers) and of complex numbers. All *polynomials* in this exposition are assumed to have nonnegative integer coefficients. Assume that the *logarithms* are always to the base 2. Let Σ be any *alphabet*, which is a finite nonempty set. We use the notation λ to denote the *empty string* of length 0. A function $h : \mathbb{N} \to \Sigma^*$ is called *polynomially bounded* if there exists a polynomial p such that $|h(n)| \leq p(n)$ for all $n \in \mathbb{N}$.

2.1   Machine Models

Our finite automata are always equipped with read-only input tapes, which use two endmarkers ¢ (left endmarker) and $ (right endmarker), where a given input string is written initially in between the two endmarkers. In contrast, each Turing machine is equipped with a read-only input tape with the two endmarkers ¢ and $ as well as a rewritable work tape. Occasionally, we further equip a Turing machine with a read-only advice tape, which holds a given advice string, together with the two endmarkers. It is important to note that no machine modifies a given advice string during its computation (except for the quantum advice model in Sect. 4).

For clarity reason, we use the term "one way" only to refer to the condition of a given machine where its tape head always moves to the right without stopping (i.e., there is no λ-move). On the contrary, if we allow such "λ-moves," we instead use the term "1.5 way" to emphasize its difference from "one way" head moves.

An *advice function* is a function from $\mathbb{N}$ to Σ^* for a certain alphabet Σ. The advised nondeterministic complexity class NL/poly consists of languages, each L of which is recognized by a certain nondeterministic Turing machine (equipped with an advice tape) using a polynomially-bounded advice function.

We assume the reader's familiarity with the basics of quantum computation. Since Kondacs and Watrous's model of 1qfa's is strictly weaker in power than

1dfa's, there have been numerous generalizations proposed in the literature (see, e.g., a survey [1]). Here, we wish to empower their 1qfa's by simply equipping each of them with a write-only *garbage tape* in which a machine drops any symbol (called a garbage symbol) but never accesses any non-blank symbol written on the tape again. An early idea of 1qfa's discarding garbage information down to a portion of a read-once input tape was materialized in [12] and such a mechanism was shown to enhance the computational power of 1qfa's. The use of a garbage tape allows us to make 1qfa's simulate all 1dfa's. Each tape has the left endmarker $\xcent$, and input and advice tapes additionally have the right endmarker $. All tape cells are indexed by numbers in $\mathbb{N}$; in particular, $\xcent$ is always placed in cell 0.

Formally, a *1-way quantum finite automaton with a garbage tape* (where we hereafter use the term "1qfa" to indicate this particular model unless otherwise stated) M is a tuple $(Q, \Sigma, \{\xcent, \$\}, \Xi, \delta, q_0, Q_{acc}, Q_{rej})$, where Q is a finite set of inner states, Σ is an input alphabet, Ξ is a *garbage alphabet*, δ is a (quantum) transition function mapping to $\mathbb{C}$, q_0 ($\in Q$) is the initial inner state, and Q_{acc}, Q_{rej} are subsets of Q. All inner states in $Q_{acc} \cup Q_{rej}$ are called *halting states* and the rest of inner states are *non-halting states*. Let $\mathcal{H}_{halt}$ and $\mathcal{H}_{non}$ denote respectively the Hilbert spaces spanned by all halting states and by all non-halting states. The garbage tape can be considered as a surrounding environment that exists "externally," separated from the essential part of a computation. By observing the garbage tape at every step produces a mixed state of "internal" configurations of M and therefore, our model turns out to be as powerful as other generalized models of 1qfa's (see a survey [1]).

For our convenience, we use the notion of transition table [14, arXiv version], which is another way to describe δ. Let n be the number of inner states of M. Formally, letting $k = |Q||\Sigma \cup \{\xcent, \$\}|$ and $l = 3|Q||\Xi|$, a *transition table* T of M on input x is a $k \times l$ matrix, each row of which is indexed by (q, σ), each column is indexed by (q, d, ξ), and the $((q, \sigma), (p, d, \xi))$-entry is an approximation of a transition amplitude $\delta(q, \sigma \mid p, d, \xi)$.

In accordance with the aforementioned 1qfa's and 2qfa's, we equip quantum Turing machines with garbage tapes. We simply refer to *quantum Turing machines that are equipped with garbage tapes* as QTMs. Since we need to handle advice, we further supply an advice tape with an advice alphabet Θ. For convenience, we call a QTM with an advice tape by an *advised QTM*.

With the use of logarithmic work space, using one of the work tapes, we can implement an internal clock that helps quantum interference take place in a computation.

2.2 Parameterized Problems and Promise Problems

A *size parameter* is a function from Σ^* to $\mathbb{N}$ for a certain alphabet Σ. Typical examples include $m_{bin}(x) = |x|$ (binary size of input x) and $m_{ver}(G)$ indicates the number of vertices in a graph G. A *parameterized decision problem* over an alphabet Σ is a pair (L, m) with a language (or a decision problem) L over Σ and a size parameter $m : \Sigma^* \to \mathbb{N}$. Given a parameterized decision problem

(L, m), a family $\mathcal{L} = \{(L_n, \overline{L}_n)\}_{n \in \mathbb{N}}$ of promise decision problems is said to be *induced from* (L, m) if, for each index $n \in \mathbb{N}$, $L_n = L \cap \Sigma_n$ and $\overline{L}_n = \overline{L} \cap \Sigma_n$, where $\Sigma_n = \{x \in \Sigma^* \mid m(x) = n\}$.

A given size parameter $m : \Sigma^* \to \mathbb{N}$ is said to be *polynomially bounded* if there exists a polynomial p such that $m(x) \leq p(|x|)$ for all $x \in \sigma^*$. In contrast, m is *polynomially honest* if, for a certain fixed polynomial q, $|x| \leq p(m(x))$ holds for any $x \in \Sigma^*$. We use the notation PHSP to denote the set of all parameterized decision problems (L, m) such that m is polynomially-honest size parameters.

A *promise decision problem* is of the form (A, B) over an alphabet Σ satisfying both $A, B \subseteq \Sigma^*$ and $A \cap B = \emptyset$. As stated in Sect. 1.1, we deal with a "family" $\mathcal{L}$ of promise decision problems, having the form $\{(L_n, \overline{L}_n)\}_{n \in \mathbb{N}}$ over a certain fixed alphabet Σ. For such a family $\mathcal{L}$ of promise problems and a given family $\{M_n\}_{n\mathbb{N}}$ of certain specified machines that satisfy appropriate criteria for acceptance and rejection, we generally say that M_n *recognizes* (solves or computes) $(L_n, \overline{L}_n)$ if (1) for any $x \in L_n$, M_n accepts x and, (2) for all $x \in \overline{L}_n$, M_n rejects x. There is no requirement for the behavior of M_n on any string x outside of $L_n \cup \overline{L}_n$ and M_n possibly neither accepts nor rejects such an x.

On the contrary, let $\{(L_n, \overline{L}_n)\}_{n \in \mathbb{N}}$ be a family of promise decision problems over an alphabet Σ. We set $L_{all} = \bigcup_{n \in \mathbb{N}}(L_n \cup \overline{L}_n)$. Note that L_{all} is included in Σ^* but is not required to equal Σ^*. Let $\Sigma_\# = \Sigma \cup \{\#\}$. We define $K_n = \{1^n \# x \mid x \in L_n\}$ and $\overline{K}_n = \{1^n \# x \mid x \in \overline{L}_n\} \cup \{z \# x \mid z \in \Sigma^n - \{1^n\}, x \in \Sigma_\#^*\} \cup \{z \mid z \in \Sigma^n\}$ for each index $n \in \mathbb{N}$. Furthermore, we set $K = \bigcup_{n \in \mathbb{N}} K_n$ and $\overline{K} = \bigcup_{n \in \mathbb{N}} \overline{K}_n$. It follows that $K \cap \overline{K} = \emptyset$ and $K \cup \overline{K} = \Sigma_\#^*$. We define $m(w)$ as follows: $m(w) = n$ if $w = 1^n \# x$ for a certain x, and $m(w) = |w|$ otherwise. The pair (K, m) turns out to be a parameterized decision problem. We say that (K, m) is *induced from* $\{(L_n, \overline{L}_n)\}_{n \in \mathbb{N}}$.

As noted in Sect. 1.1, we use the prefix "para-" to describe parameterized complexity classes. We define para-BQL as the class of parameterized decision problems (L, m) solvable by bounded-error QTMs using $O(\log |x| m(x))$ space, where m is log-space computable *in unary* (see [13]). The probabilistic counterpart of para-BQL is denoted by para-BPL. Moreover, we write para-NL/poly to denote the parameterization of NL/poly, which is obtained by replacing languages L with parameterized decision problems (L, m).

2.3   Nonuniform State Complexity

Our purpose is to introduce nonuniform complexity classes defined by state complexities of quantum finite automata families. Related to these classes, we also consider classes based on probabilistic finite automata.

The state complexity generally refers to the number of inner states used in a given automaton. However, since we use a (uniform or nonuniform) family $\{M_n\}_{n \in \mathbb{N}}$ of finite automata, the state complexity of such a family becomes a function in n. More formally, the *state complexity* $sc(n)$ (or $sc(M_n)$) of a family $\{M_n\}_{n \in \mathbb{N}}$ of finite automata is a function defined by $sc(n) = |Q_n|$ for all indices $n \in \mathbb{N}$, where Q_n denotes a set of inner states of M_n [10]. In the rest of this paper, we use nonuniform families of finite automata.

The nonuniform state complexity class 1D is the collection of all nonuniform families $\{(L_n\overline{L}_n)\}_{n\in\mathbb{N}}$ over certain alphabets Σ satisfying the following: there are a polynomial p and a nonuniform family $\{M_n\}_{n\in\mathbb{N}}$ of 1dfa's such that, for each index $n \in \mathbb{N}$, (i) M_n has at most $p(n)$ states and (ii) M_n solves $(L_n, \overline{L}_n)$ on all inputs. In a similar way, we can define 1N using 1nfa's instead of 1dafa's. Moreover, the notation 2^{1D} indicates the collection of families $\{(L_n, \overline{L}_n)\}_{n\in\mathbb{N}}$ of promise decision problems, each of which is recognized by a certain 1dfa of at most $2^{p(n)}$ inner states for a certain polynomial p.

Formally, the notation 1BQ denotes the collection of nonuniform families $\{(L_n, \overline{L}_n)\}_{n\in\mathbb{N}}$ such that there exist a family $\{M_n\}_{n\in\mathbb{N}}$ of 1qfa's, a polynomial p, and a constant $\varepsilon \in [0, 1/2)$ satisfying the following: (1) for each $n \in \mathbb{N}$ and any x, if $x \in L_n$, then M_n accepts x with probability at least $1 - \varepsilon$; if $x \in \overline{L}_n$, then M_n rejects x with probability at least $1 - \varepsilon$, and (2) each M_n uses at most $p(n)$ inner states. When M_n satisfies Condition (1), we simply say that M_n *recognizes* $(L_n, \overline{L}_n)$ *with error probability at most* ε. In this case, M_n is also said to *make bounded-errors*. We obtain 1Q if we change Condition (1) into the following condition: (1') given any index $n \in \mathbb{N}$, for each $x \in L_n$, M_n accepts x with probability $>1/2$ and, for any $x \in \overline{L}_n$, M_n rejects x with probability $\geq 1/2$.

We define 1P in a similar way of defining 1D but using *one-way probabilistic finite automata with unbounded-error probability* (or 1pfa's, for short). By using the bounded-error criteria instead, we define 1BP (where "B" stands for "bounded error"). Similarly to 2^{1D}, we can define 2^{1BQ}, 2^{1Q}, 2^{1BP}, 2^{1P}, etc.

As shown below, quantum computation is different from deterministic one.

Proposition 4. $1D = 1.5D$, $1BQ \neq 1.5BQ$, *and* $1Q \neq 1.5Q$.

For two-way head moves, the length of accepting computation paths of 2dfa's and 2nfa's are always bounded linearly in input size. This fact shows that 2D and 2N are both included in 2^{1D}. We define 2BQ to be the collection of nonuniform families $\{(L_n, \overline{L}_n)\}_{n\in\mathbb{N}}$ such that there exist a family $\{M_n\}_{n\in\mathbb{N}}$ of 2qfa's and a polynomial p satisfying the following: (1) for each $n \in \mathbb{N}$, M_n makes bounded errors on all inputs in $\Sigma_n = L_n \cup \overline{L}_n$ and (2) each M_n uses at most $p(n)$ inner states. Let 2Q be defined similarly to 2BQ by using unbounded-error probability instead of bounded-error one.

In a similar fashion, we define 2BP to be the collection of nonuniform families $\{(L_n, \overline{L}_n)\}_{n\in\mathbb{N}}$, each $(L_n, \overline{L}_n)$ of which is recognized by 2-way bounded-error probabilistic finite automata (or 2bpfa, for short) of polynomially-many states with error probability at most $\varepsilon \in [0, 1/2)$. The unbounded-error analogue of 2BP is denoted by 2P.[2]

3   Advised QTMs and Quantum Finite Automata

Our goal in this section is to prove a general theorem, Theorem 5, from which Corollary 2 follows immediately. To achieve this goal, we first give a precise

---

[2] In [4], the polynomial-time 2BP was considered under the name of $2P_2$ and the polynomial-time 2P was studied under the name of 2P but it is restricted to so-called "regular" language families. Here, we do not demand such a condition.

characterization of parameterized decision problems solvable by polynomial-time logarithmic-space advised QTMs in terms of certain 2qfa's having polynomially many states.

Theorem 5. *Let* $\mathcal{A}, \mathcal{B} \in \{\mathrm{D}, \mathrm{N}, \mathrm{BP}, \mathrm{BQ}\}$. *It then follows that* $2\mathcal{A}/\mathrm{poly} \subseteq 2\mathcal{B}$ *iff* $\mathcal{A}\mathrm{L} \subseteq \mathcal{B}\mathrm{L}/\mathrm{poly}$, *where "DL" is understood as "L".*

3.1   The Roles of Advice and the Honesty Condition

We will prove a central claim, which establishes a close relation between nonuniform state complexity classes and parameterized complexity classes. Here, we state the claim in full generality.

Proposition 6. *Let* $(\mathcal{A}, \mathcal{B}) \in \{(2\mathrm{D}, \mathrm{L}), (2\mathrm{N}, \mathrm{NL}), (2\mathrm{BP}, \mathrm{BPL}), (2\mathrm{BQ}, \mathrm{BQL})\}$.

1. *For any parameterized decision problem* (L, m), *let* $\mathcal{L} = \{(L_n, \overline{L}_n)\}_{n \in \mathbb{N}}$ *be a family induced from* (L, m). *It then follows that* $(L, m) \in \text{para-}\mathcal{B}/\mathrm{poly} \cap \mathrm{PHSP}$ *iff* $\mathcal{L} \in \mathcal{A}/\mathrm{poly}$.
2. *Let* $\mathcal{L} = \{(L_n, \overline{L}_n)\}_{n \in \mathbb{N}}$ *be any family of promise decision problems and let* (K, m) *be a parameterized decision problem induced from* $\mathcal{L}$. *It then follows that* $\mathcal{L} \in \mathcal{A}/\mathrm{poly}$ *iff* $(K, m) \in \text{para-}\mathcal{B}/\mathrm{poly} \cap \mathrm{PHSP}$.

In this exposition, we will show the proposition only for the case of $\mathcal{A} = 2\mathrm{BQ}$ and $\mathcal{B} = \mathrm{BQL}$ since the other cases can be proven in a similar way. The proof of Proposition 6(1) is now split into two lemmas, Lemmas 7 and 8. Lemma 7 states that we can simulate an advised QTM by a certain nonuniform family of 2qfa's with appropriate state complexity.

Lemma 7. *Let* (L, m) *be a parameterized decision problem over an alphabet* Σ *and let* $\mathcal{L} = \{(L_n, \overline{L}_n)\}_{n \in \mathbb{N}}$ *be a family of promise decision problems induced from* (L, m). *Let* h *be an advice function and let* r *be a polynomial satisfying* $|h(n)| \leq r(n)$ *for all* $n \in \mathbb{N}$. *Assume that* m *is polynomially honest and that, with the help of* h, *an advised QTM* M *solves* L *with bounded-error probability. For any polynomial* p *and a function* ℓ, *there exists a family* $\{N_{n,l}\}_{n,l \in \mathbb{N}}$ *of 2qfa's with* $O(r(l) 2^{O(\ell(n,l))})$ *states such that, for any input* x *satisfying* $m(x) = n$, *M accepts* $(x, h(|x|))$ *in expected time* $p(m(x), |x|)$ *using space* $\ell(m(x), |x|)$ *with bounded-error probability iff* $N_{m(x),|x|}$ *accepts* x *in expected time* $O(p(m(x), |x|))$ *with bounded-error probability.*

Proof Sketch. Given M and h in the premise of the lemma, the desired 2qfa $N_{n,l}$ is designed to simulate M's computation using its inner states of the form (q, k, t, w, a), which indicates that M is in state q, scanning the kth cell of a work tape containing w, and the tth cell of an advice tape with $a = h(|x|)$.   □

The converse of Lemma 7 is shown in Lemma 8 by giving a simulation of a family of 2qfa's by advised QTMs. To make a quantum interference take place correctly, we need to avoid any time discrepancy caused by the different simulation speed, and thus we need to adjust the timing of reaching the same configurations. For this purpose, we need to implement an internal clock. This is possible because 2qfa's in question can use polynomially many inner states.

Lemma 8. *Let (L, m) be a parameterized decision problem over an alphabet Σ and let $\mathcal{L} = \{(L_n, \overline{L}_n)\}_{n \in \mathbb{N}}$ be induced from (L, m). Let r and p be functions and let $\{N_{n,l}\}_{n,l \in \mathbb{N}}$ be a family of $r(n, l)$-state 2qfa's such that $N_{m(x),|x|}$ solves $(L_n, \overline{L}_n)$ on all inputs x with bounded-error probability within $p(m(x), |x|)$ time. There exist an advised QTM M and an $O(r(m(x), |x|) \log r(m(x), |x|))$-bounded advice function h such that, for any $n \in \mathbb{N}$ and for any input x, $N_{m(x),|x|}$ accepts x within $p(m(x), |x|)$ time with bounded-error probability iff M accepts $(x, h(n))$ with bounded-error probability within $O(p(m(x), |x|))$ time using space $O(\log r(m(x), |x|))$.*

Proof Sketch. Taking $N_{n,l}$ in the premise of the lemma, we build the desired advised QTM M that simulates $N_{m(x),|x|}$ using an advice function h, which encodes a quantum circuit $C_{m(x),|x|}$ that approximately updates a configuration of $N_{m(x),|x|}$ at every step. Such $C_{n,l}$ can be made up of a universal set $\{HAD, CNOT, T\}$ of elementary quantum gates. □

We omit the proof of Proposition 6(2).

3.2 Proof of Theorem 5

We will give the proof of Theorem 5. For the intended proof, we need two supporting claims.

Lemma 9. *Let $(\mathcal{A}, \mathcal{B}) \in \{(\text{NL}, \text{L}), (\text{NL}, \text{BPL}), (\text{NL}, \text{BQL}), (\text{BQL}, \text{BPL})\}$. It then follows that $\mathcal{A}/\text{poly} \subseteq \mathcal{B}/\text{poly}$ iff $\mathcal{A} \subseteq \mathcal{B}/\text{poly}$.*

Lemma 10. *Let $(\mathcal{A}, \mathcal{B}) \in \{(\text{NL}, \text{L}), (\text{NL}, \text{BPL}), (\text{NL}, \text{BQL}), (\text{BQL}, \text{BPL})\}$. It then follows that para-$\mathcal{A} \cap \text{PHSP} \subseteq \text{para-}\mathcal{B}/\text{poly}$ iff $\mathcal{A} \subseteq \mathcal{B}/\text{poly}$.*

We are ready to give the proof of Theorem 5, which is now an easy consequence of Lemmas 9–10 and Proposition 6.

Proof of Theorem 5. Here, we show only the case of $\mathcal{A} = \text{N}$ and $\mathcal{B} = \text{BQ}$.

($\Leftarrow$) Assume that $\text{NL} \subseteq \text{BQL}/\text{poly}$. By Lemma 9, this is equivalent to $\text{NL}/\text{poly} \subseteq \text{BQL}/\text{poly}$. Lemma 10 then implies that para-$\text{NL}/\text{poly} \cap \text{PHSP} \subseteq \text{para-BQL}/\text{poly}$. Using this inclusion, we want to show that $2\text{N}/\text{poly} \subseteq 2\text{BQ}$.

Take any family $\mathcal{L} = \{(L_n, \overline{L}_n)\}_{n \in \mathbb{N}}$ in $2\text{N}/\text{poly}$. There are a polynomial s and a family $\{M_n\}_{n \in \mathbb{N}}$ of 2nfa's such that, for any index $n \in \mathbb{N}$, $|x| \leq s(n)$ for all $x \in \Sigma_n = L_n \cup \overline{L}_n$, and M_n solves $(L_n, \overline{L}_n)$. Consider (K, m), which is induced from $\{(L_n \overline{L}_n)\}_{n \in \mathbb{N}}$. By Proposition 6(2), (K, m) belongs to para-$\text{NL}/\text{poly} \cap \text{PHSP}$ since m is polynomially honest. By our assumption, (K, m) is also in para-BQL/poly. By Proposition 6(2), we conclude that $\mathcal{L} \in 2\text{BQ}/\text{poly} \subseteq 2\text{BQ}$.

($\Rightarrow$) Assume that $2\text{N}/\text{poly} \subseteq 2\text{BQ}$. It suffices to prove that para-$\text{NL}/\text{poly} \cap \text{PHSP} \subseteq \text{para-BQL}/\text{poly}$ because, once this is proven, Lemma 10 implies that $\text{NL}/\text{poly} \subseteq \text{BQL}/\text{poly}$ and Lemma 9 further concludes that $\text{NL} \subseteq \text{BQL}/\text{poly}$.

Let us take any parameterized decision problem (L, m) in para-$\text{NL}/\text{poly} \cap \text{PHSP}$. Let $\mathcal{L} = \{(L_n, \overline{L}_n)\}_{n \in \mathbb{N}}$ be induced from (L, m). Proposition 6(1) implies that $\mathcal{L} \in 2N/\text{poly}$. Our assumption then implies that $\mathcal{L} \in 2BQ/\text{poly}$. Proposition 6(1) then concludes that $(L, m) \in$ para-BQL/poly. □

From Theorem 5, Corollary 2 follows immediately. This theorem also leads to the main result of [6] (see also [7]), which relies on the property of a particular NL-complete problem, the directed graph s-t connectivity problem. Our argument instead uses the parameterized complexity classes para-L and para-NL as in Proposition 6.

Corollary 11. [6,7] $2N/\text{poly} \subseteq 2D$ *iff* $NL \subseteq L/\text{poly}$.

4   Quantum Advice and Quantum Transition Tables

For the proof of Theorem 3, let us consider quantum advice. It is shown in [8, Lemma 3.1] that a polynomial-time quantum Turing machine with quantum advice is translated into an equivalent polynomial-size quantum circuit family starting with additional quantum states. In the case of finite automata, we quantize transition tables and feed them to quantum finite automata. We abbreviate *2-way super quantum finite automata* as 2sqfa's.

Proposition 12. *1. Let (L, m) be any parameterized decision problem and let $\mathcal{L} = \{(L_n, \overline{L}_n)\}_{n \in \mathbb{N}}$ be induced from (L, m). It then follows that $(L, m) \in$ para-BQL/Qpoly $\cap$ PHSP iff $\mathcal{L} \in 2sBQ/\text{poly}$.*
2. Let $\mathcal{L} = \{(L_n, \overline{L}_n)\}_{n \in \mathbb{N}}$ be any family of promise decision problems and let (K, m) be induced from $\mathcal{L}$. It then follows that $\mathcal{L} \in 2sBQ/\text{poly}$ iff $(K, m) \in$ para-BQL/Qpoly $\cap$ PHSP.

Proof Sketch. We briefly state a key idea of how to prove (1). Assuming $(L, m) \in$ para-BQL/Qpoly $\cap$ PHSP, for any given advised QTM M and a quantum advice state $|\phi_n\rangle = \sum_s \alpha_s |s\rangle$ solving (L, m), we define M_s to run M on (x, s). We then convert it to a 2qfa $N_{m(x),s}$ that properly simulates M_s on input x. Take a polynomial p for which $m(x) \leq p(|x|)$ for all x. Let $T_{i,s}$ denote a transition table of $N_{i,s}$ and define $T'_s = T_{1,s} \# T_{2,s} \# \cdots \# T_{p(|x|)}$. Finally, we set $|\psi_{|x|}\rangle = \sum_s \alpha_s |T'_s\rangle$ and define the desired super 2qfa to operate according to $|\psi_{|x|}\rangle$.

For the converse, assume that $\mathcal{L} \in 2sBQ/\text{poly}$. Take $\{N_n\}_{n \in \mathbb{N}}$ and $\{|\psi_n\rangle\}_{n \in \mathbb{N}}$ that solve $\mathcal{L}$. We then prepare a quantum function f to indicate how to operate a single step of N_n. Finally, we define $h(x)$ to be $f(x) \otimes |\psi_{|x|}\rangle$ and design the desired advised QTM that simulates $N_{m(x)}$ on $(x, h(x))$.

Lemma 13. para-BQL/poly $\cap$ PHSP $=$ para-BQL/Qpoly $\cap$ PHSP *iff* BQL/poly $=$ BQL/Qpoly.

Theorem 3 follows from Propositions 6 and 12 and Lemma 13.

Proof of Theorem 3. ($\Rightarrow$) Assume that BQL/poly $=$ BQL/Qpoly. By Lemma 13, we obtain para-BQL/poly $\cap$ PHSP $=$ para-BQL/Qpoly $\cap$ PHSP. Let $\mathcal{L} = \{(L_n, \overline{L}_n)\}_{n \in \mathbb{N}}$ be any family in 2sBQ/poly and let (K, m) be induced from

$\mathcal{L}$. By Proposition 12(2), we obtain $(K, m) \in$ para-BQL/Qpoly. By our assumption, we obtain $(K, m) \in$ para-BQL/poly. By Proposition 6(2), it follows that $\mathcal{L} \in$ 2BQ/poly.

($\Leftarrow$) Assume that 2sBQ/poly $\subseteq$ 2BQ/poly. Owing to Lemma 13, we need to show that para-BQL/poly $\cap$ PHSP = para-BQL/Qpoly $\cap$ PHSP. Let (L, m) be any problem in para-BQL/Qpoly $\cap$ PHSP. Moreover, let $\mathcal{L} = \{(L_n, \overline{L}_n)\}_{n \in \mathbb{N}}$ be induced from (L, m). By Proposition 12(1), we obtain $\mathcal{L} \in$ 2sBQ/poly. Our assumption implies that $\mathcal{L} \in$ 2BQ/poly. Using Proposition 6(1), we conclude that $(L, m) \in$ para-BQL/poly $\cap$ PHSP. $\square$

References

1. Ambainis, A., Yakaryilmaz, A.S.: Automata and quantum computing. manuscript (2015). https://arxiv.org/abs/1507.01988
2. Berman, P., Lingas, A.: On complexity of regular languages in terms of finite automata. Technical Report 304, Institute of Computer Science, Polish Academy of Science, Warsaw (1977)
3. Dwork, C., Stockmeyer, L.: A time-complexity gap for two-way probabilistic finite state automata. SIAM J. Comput. **19**, 1011–1023 (1990)
4. Kapoutsis, C.A.: Size complexity of two-way finite automata. In: Diekert, V., Nowotka, D. (eds.) DLT 2009. LNCS, vol. 5583, pp. 47–66. Springer, Heidelberg (2009). https://doi.org/10.1007/978-3-642-02737-6_4
5. Kapoutsis, C.A.: Minicomplexity. J. Automat. Lang. Combin. **17**, 205–224 (2012)
6. Kapoutsis, C.A.: Two-way automata versus logarithmic space. Theory Comput. Syst. **55**, 421–447 (2014)
7. Kapoutsis, C.A., Pighizzini, G.: Two-way automata characterizations of L/poly versus NL. Theory Comput. Syst. **56**, 662–685 (2015)
8. Nishimura, H., Yamakami, T.: Polynomial time quantum computation with advice. Inf. Process. Lett. **90**, 195–204 (2004)
9. Sakoda, W.J., Sipser, M.: Nondeterminism and the size of two-way finite automata. In: Proceedings of the STOC 1978, pp. 275–286 (1978)
10. Villagra, M., Yamakami, T.: Quantum state complexity of formal languages. In: Shallit, J., Okhotin, A. (eds.) DCFS 2015. LNCS, vol. 9118, pp. 280–291. Springer, Cham (2015). https://doi.org/10.1007/978-3-319-19225-3_24
11. Yamakami, T.: The roles of advice to one-tape linear-time turing machines and finite automata. Int. J. Found. Comput. Sci. **21**, 941–962 (2010)
12. Yamakami, T.: One-way reversible and quantum finite automata with advice. Inf. Comput. **239**, 122–148 (2014)
13. Yamakami, T.: The 2CNF Boolean formula satsifiability problem and the linear space hypothesis. In: Proceedings of MFCS 2017. LIPIcs, vol. 83, pp. 62:1–62:14. Schloss Dagstuhl - Leibniz-Zentrum fuer Informatik (2017). A complete version is found at arXiv:1709.10453
14. Yamakami, T.: State complexity characterizations of parameterized degree-bounded graph connectivity, sub-linear space computation, and the linear space hypothesis. In: Konstantinidis, S., Pighizzini, G. (eds.) DCFS 2018. LNCS, vol. 10952, pp. 237–249. Springer, Cham (2018). https://doi.org/10.1007/978-3-319-94631-3_20. A complete and corrected version is found at arXiv:1811.06336
15. Yamakami, T.: Relativizations of nonuniform quantum finite automata families (2019, manuscript)

Equivalence Checking of Prefix-Free Transducers and Deterministic Two-Tape Automata

Vladimir A. Zakharov[✉]

Lomonosov Moscow State University, Leninskiye Gory, Moscow, Russia
zakh@cs.msu.su

Abstract. Although the equivalence problem for finite transducers is undecidable in the general case, it was shown that for some classes of transducers (bounded ambiguous, bounded valued, of bounded length degree) this problem has effective solutions which, however, require significant computational costs. In this paper we distinguish a new class of transducers (we call them prefix-free since their transitions are characterized by this property of languages) such that (1) the equivalence problem for transducers in this class is decidable in quadratic time, and (2) this class does not fall into the scope of previously known decidable cases. We also show that deterministic two-tape finite state automata (2-DFSAs) are convertible into prefix-free transducers. Due to this translation we obtain a simple procedure for checking equivalence of 2-DFSAs in polynomial time. We believe that the further development of this approach could bring us to an efficient equivalence checking algorithm for deterministic multi-tape automata with an arbitrary number of tapes.

Keywords: Transducer · Two-tape automaton ·
Equivalence checking · Prefix-free language · Language equation ·
Decision procedure

Finite transducers and two-tape automata stem from the same concept of Finite State Automaton. Both models compute the same class of rational relations on words but the computations are performed differently: transducers read words on the input tape and write on the output tape whereas two-tape automata read words alternately on two input tapes. Both models are easily convertible into each other, and, therefore, many authors do not distinguish them. However, when studying and applying transducers, it is customary to restrict ourselves to the consideration of real-time transducers which write on the output tape only in response to reading the next letter on the input tape. Real time transducers are widely used in text and speech processing, bioinformatics, verification and optimization of reactive system, etc. The application capabilities of two-tape automata turned out to be far more modest.

This research is supported by RFBR Grant 18-01-00854.

C. Martín-Vide et al. (Eds.): LATA 2019, LNCS 11417, pp. 146–158, 2019.
https://doi.org/10.1007/978-3-030-13435-8_11

Both models are hard for analysis: the inclusion, equivalence, and even totality problems are undecidable for real time transducers as well as for two-tape automata [5,7]. Meanwhile, many classes of transducers were discovered— sequential, functional, bounded ambiguous, bounded valued transducers—for which the equivalence problem was proved to be decidable. Decision algorithms are quite diverse, many of them require significant computational costs, but for some classes of transducers equivalence checking and minimization procedures are surprisingly simple and efficient [11]. Two-tape automata are not so much suitable for analysis: decidability of the equivalence problem was proved only for deterministic automata, and the only known algorithm [6] that allows solving this problem in polynomial time is rather sophisticated.

This unequal state of the art, when the diversity of decision procedures and techniques for the analysis of transducers and two-tape automata are concerned, gives rise to a natural question: is it possible to find a suitable relationship between two-tape automata and finite transducers as to allow one to analyze the behaviour of multi-tape machines with the help of decision procedures designed for real time transducers. Finite transducers are easily convertible into two-tape automata: just imagine that instead of writing the words to the output tape, a transducer reads them. But the reverse translation is less evident. Two-tape automata display much more freedom than real time transducers in handling their tapes. Therefore, even deterministic two-tape automata are able to recognize such binary relations on words that can not be computed by real time transducers with a bounded degree of nondeterminism (bounded ambiguous, bounded valued, etc.). As a result, all previously known classes of transducers, for which the decidability of the equivalence problem has been established, are useless when it comes to checking the equivalence of deterministic two-tape automata.

In this paper we show how to overcome these difficulties and to adjust equivalence checking techniques developed for finite transducers to verification of deterministic two-tape automata. To this end we propose to consider a new class of transducers. A distinctive feature of transducers from this class is that for any control state q and any input letter a the set of all output words w that can be written at the state q in response to a forms a prefix-free code. This new class of transducers does not fall into the scope of previously known decidable cases. We show that the equivalence of prefix-free transducers can be verified in quadratic time by checking the solvability of transduction equations. Another advantage of the new class of transducers is that every deterministic two-tape automaton can be quite simply converted into an equivalent prefix-free transducer. It should be noticed, however, that the corresponding prefix-free transducers, although being finite state machines, may have infinitely many transitions. Nevertheless, the above mentioned solvability checking techniques for transduction equations can be appropriately adapted for this case as well. Thus, we obtain an efficient procedure for checking equivalence of deterministic two-tape automata.

We begin with a brief exposition of the foundations of finite transducers and two-tape automata, and the main achievements in studying the equivalence problem for these models. Next we introduce prefix-free transducers and present the equivalence checking procedure for this class of transducers. Finally, we explain how to translate deterministic two-tape automata into finite transducers operating over prefix-free regular languages and how to modify the equivalence checking procedure to cope with prefix-free transducers of a more general type.

1   Preliminaries

A *word* over *alphabet* Σ is any finite sequence $w = a_1 a_2 \ldots a_k$ of letters in Σ. The empty word is denoted by ε. Given a pair of words u and v, we write uv for their *concatenation*. The set of all words over an alphabet Σ is denoted by Σ^*. A *language* over Σ is any subset of Σ^*. *Concatenation* of languages L_1 and L_2 is the language $L_1 L_2 = \{uv \ : \ u \in L_1, v \in L_2\}$. If $L_1 = \emptyset$ or $L_2 = \emptyset$ then $L_1 L_2 = \emptyset$. A *transduction* over alphabets Σ and Δ is any subset of $\Sigma^* \times \Delta^*$.

A *real time finite transducer* (briefly, transducer) over an input alphabet Σ and an output alphabet Δ is a quadruple $\pi = \langle Q, q_0, F, \longrightarrow \rangle$, where Q is a finite set of *states*, q_0 is an *initial state*, $F \subseteq Q$ is a subset of *final states*, and $\longrightarrow$ is a finite *transition relation* of the type $Q \times \Sigma \times \Delta^* \times Q$. Sometimes we will write $\pi(q_0)$ to emphasize that q_0 is the initial state of π. Quadruples (q, a, u, q') in $\longrightarrow$ are called *transitions* and depicted as $q \xrightarrow{a/u} q'$. A *run* of π on an input word $w = a_1 a_2 \ldots a_n$ is any finite sequence of transitions

$$q \xrightarrow{a_1/u_1} q_1 \xrightarrow{a_2/u_2} \cdots \xrightarrow{a_{n-1}/u_{n-1}} q_{n-1} \xrightarrow{a_n/u_n} q' \ . \tag{1}$$

The pair (w, u), where $u = u_1 u_2 \ldots u_n$, is called a *label* of a run (1). We will write $q \xrightarrow{w/u}_* q'$ to indicate that a transducer π has a run (1) labeled with (w, u) from a state q to a state q'. If $q' \in F$ then (1) is a *final* run. A *transduction relation realized by* a transducer π at its state q is the set of pairs $TR(\pi, q) = \{(w, u) : q \xrightarrow{w/u}_* q', q' \in F\}$. Transducers $\pi_1(q_1)$ and $\pi_2(q_2)$ are called *equivalent* ($\pi_1(q_1) \sim \pi_2(q_2)$ in symbols) iff $TR(\pi_1, q_1) = TR(\pi_2, q_2)$. *Equivalence problem* for transducers is that of checking, given a pair of transducers π_1 and π_2, whether $\pi_1 \sim \pi_2$ holds.

A transducer π is called

- *deterministic* if for every letter a and a state q it has at most one transition of the form $q \xrightarrow{a/u} q'$,
- *k-ambiguous* if for every input word w there is at most k final runs of π on w from the initial state q_0,
- *k-valued* if for every input word w the transduction relation $TR(\pi, q_0)$ contains at most k images of w,
- *of length-degree* k if for every input word w, the number of distinct lengths of the images u of w in $TR(\pi, q_0)$ is at most k.

The study of the equivalence problem for real time transducers began in the early 60s. First, it was shown that this problem is undecidable for non-deterministic transducers [7]. But the undecidability displays itself only when some input words may have arbitrary many images. The equivalence problem was shown to be decidable in polynomial time for deterministic transducers [3] and in polynomial space for single-valued transducers [2,14]. Later in [8] it was proved that this problem is decidable for bounded ambiguous transducers. Moreover, equivalence checking of unambiguous transducers can be performed in polynomial time. At the next stage bounded-valued transducers were studied. It was shown how to check in polynomial time whether the cardinality of the image of every word by a given transducer is bounded [18] and whether it is bounded by a given integer k [8]. The decidability of the equivalence problem for k-valued transducers was established in [4, 20]. In a series of papers [12,13,15] a construction to decompose k-valued transducers into a sum of functional and unambiguous ones was developed and used for checking bounded valuedness, k-valuedness and equivalence of k-valued transducers in exponential time. In [21] a straightforward techniques was developed to solve the same problems for transducers operating over semigroups. The largest among the known classes of transducers for which the equivalence problem is decidable in the class of transducers of bounded length-degree [19]. In [16] it was proved that the equivalence for bounded length-degree transducers is decidable in triple exponential time.

A *two-tape finite state automaton* (briefly, 2-FSA) over disjoint alphabets Σ and Δ is a 5-tuple $M = \langle S_1, S_2, s_0, F, \rightarrow \rangle$ such that S_1, S_2 is a partitioning of a finite set S of *states*, $s_0 \in S_1$ is an *initial state*, $F \subseteq S$ is a subset of *final states*, and $\rightarrow$ is a *transition relation* of the type $(S_1 \times \Sigma \times S) \cup (S_2 \times \Delta \times S)$. Transitions are depicted as $s \xrightarrow{x} s'$. A *run* of a 2-FSA M is any sequence of transitions

$$s \xrightarrow{z_1} s_1 \xrightarrow{z_2} \cdots \xrightarrow{z_{n-1}} s_{n-1} \xrightarrow{z_n} s' . \qquad (2)$$

A run (2) is *complete* if $s = s_0$ and $s' \in F$. We say that a 2-FSA M *accepts* a pair of words $(w, u) \in \Sigma^* \times \Delta^*$ if there is a complete run (2) of M such that w is the projection of the word $z_1 z_2 \ldots z_{n-1} z_n$ on the alphabet Σ and u is the projection of the same word $z_1 z_2 \ldots z_{n-1} z_n$ on the alphabet Δ. A *transduction relation recognized by* a 2-FSA M is the set $TR(M)$ of all pairs of words accepted by M. We say that 2-FSAs M' and M'' are *equivalent* if $TR(M') = TR(M'')$. A 2-FSA M is called *deterministic* (2-DFSA) if for every letter a and a state s it has at most one transition of the form $s \xrightarrow{a} s'$.

Undecidability of the equivalence problem for non-deterministic 2-FSAs was proved in [5]. But soon in [1] and [17] equivalence was shown to be decidable for 2-DFSAs. Later, in [6] it was discovered how to check equivalence of 2-DFSAs in polynomial time. And, finally, the decidability of equivalence problem for deterministic automata with arbitrary number of tapes was established in [10]. Strangely enough, since 1991 no new significant results on equivalence checking techniques for multi-tape automata have appeared.

2   Prefix-Free Transducers

A word u is a *prefix* of a word w if $w = uv$ holds for some word v. In this case w is called an *extension* of u and $v = u \setminus w$ a *left quotient* of u with w. We say that two words u_1 and u_2 are *compatible* if one of them is a prefix of the other. A language L is called *prefix-free* if all its words are pairwise incompatible. We say that two languages L' and L'' are *compatible* if every word in any of these languages is compatible with some word in the other language. Given a word u and a language L, we denote by $Pref(L)$ the set of all prefixes of the words in L, and by $u \setminus L$ a language $\{v : uv \in L\}$ which is a left quotient of u with L. Notice, that if $u \notin Pref(L)$ then $u \setminus L = \emptyset$.

Proposition 1. *Let L' and L'' be finite prefix-free compatible languages. Then there exist unique partitions $L' = \bigcup_{i=1}^{n} L'_i$ and $L'' = \bigcup_{i=1}^{n} L''_i$ of these languages such that for every $i, 1 \leq i \leq n$, one of the subsets L'_i or L''_i is a singleton $\{u\}$ and all words from the other are extensions of u.*

Such partitioning of a compatible pair of prefix-free languages L' and L'' will be called its *splitting*. The pairs of corresponding subsets L'_i and L''_i, $1 \leq i \leq n$, of a splitting will be called its *fractions*.

Given a transducer $\pi = \langle Q, q, F, \longrightarrow \rangle$ over languages Σ and Δ, a state $q \in Q$ and a letter $x \in \Sigma$, we denote by $Out_\pi(q, x)$ the set of all pairs (u, q') such that $q \xrightarrow{x/u} q'$ is a transition of π. A transducer π is called *prefix-free* if for every state q and an input letter x the language $L_\pi(q, x) = \{u : \exists p \ (u, p) \in Out_\pi(q, x)\}$ is prefix-free. Clearly, every deterministic transducer is prefix-free but not vice verse. Nevertheless, prefix-free transducers have a certain "deterministic" property: for every state q of a prefix-free transducer π and for every pair $(w, u) \in TR(\pi, q)$ there is the only run of π from the state q labeled with (u, w). It should be noticed also that some prefix-free transducers don't have such properties as bounded ambiguity, bounded valuedness, or bounded length-degree, i.e. the class of prefix-free transducers does not fall into the scope of previously known decidable cases of equivalence problem for real time finite transducers.

Example 2. Consider a prefix-free transducer $\pi = \langle \{q_0, q_1\}, q_0, \{q_1\}, \longrightarrow \rangle$, which has only 3 transitions $q_0 \xrightarrow{a/g} q_0$, $q_0 \xrightarrow{a/hh} q_0$, and $q_0 \xrightarrow{b/\varepsilon} q_1$. For every input $a^n b$ the lengths of outputs vary from n to $2n$. Hence, π is not a transducer of bounded length-degree.

Our equivalence checking technique for prefix-free transducers is based on manipulations with regular expressions. We associate with each transducer π a system of linear regular expression equations $\mathcal{E}(\pi)$ which specifies the behaviour of π. To check the equivalence $\pi' \sim \pi''$ we add to the set of equations $\mathcal{E}(\pi') \cup \mathcal{E}(\pi'')$ the equivalence requirement which is an equation of the form $X' = X''$ and then verify whether the resulting system of equations has a solution. To this end we use Gaussian elimination of variables. In general case this approach does not offer any advantages, but for prefix-free transducers splitting of compatible prefix-free languages (see Proposition 1) provides a suitable means for solving efficiently the systems of equations that specify the equivalence problem.

For the sake of clarity and to make the notation more simple we make some assumptions concerning transducers π' and π'' to be analyzed:

- the input alphabet $\Sigma = \{a_1, \ldots, a_k\}$ and the output alphabet Δ are disjoint; we will use symbols x, y, z to denote arbitrary letters from Σ and symbols u, v, w to denote words from Δ^*,
- $\pi' = \pi(q')$ and $\pi'' = \pi(q'')$, i.e. π' and π'' have the same transition relation but different initial states,
- the transducer π is *trim*, i.e. a final state is reachable from each state of π.

Regular expressions (regexes) are built of variables $X_1, X_2, \ldots$, constants $0, 1$, and letters from Σ and Δ by means of concatenation $\cdot$, and alternation $+$. They are interpreted on the semiring of transductions over Σ and Δ. The constants 0, 1, every letter x, and every word u are interpreted as transductions $\emptyset$, $\{(\varepsilon, \varepsilon)\}$, $\{(x, \varepsilon)\}$, and $\{(\varepsilon, u)\}$ respectively. Concatenation of transductions T_1 and T_2 is defined as expected: $T_1 T_2 = \{(h_1 h_2, u_1 u_2) : (h_1, u_1) \in T_1, (h_2, u_2) \in T_2\}$. It is important to mind that $x \cdot u = u \cdot x$ holds for every $x \in \Sigma$ and $u \in \Delta^*$.

We will focus on linear regexes of two types. A Δ-*regex* is any expression of the form $u_1 \cdot X_{i_1} + u_2 \cdot X_{i_2} + \cdots + u_m \cdot X_{i_m}$, where variables may have multiple occurrences. When a set of words $\{u_1, u_2, \ldots, u_m\}$ is prefix-free then such a Δ-regex will be also called prefix-free. A Σ-*regex* is any expression of the form $a_1 \cdot E_1 + a_2 \cdot E_2 + \cdots + a_k \cdot E_k$, where $E_i, 1 \le i \le k$, are Δ-regexes. When referring to such regexes as $E(X_1, \ldots, X_n)$ (for Δ-regex) or $G(X_1, \ldots, X_n)$ (for Σ-regex) we emphasize that $X_1, \ldots, X_n$ are the only variables involved in them.

Let $\pi = \langle Q, q_0, F, \longrightarrow \rangle$ be a finite transducer over Σ and Δ. With each state q of π we associate a variable X_q, and for every pair $q \in Q$ and $x \in \Sigma$ we build a Δ-regex $E_{q,x} = \displaystyle\sum_{(u,p)\in Out_\pi(q,x)} u \cdot X_p$. For every state q denote by c_q either the constant 1 if $q \in F$, or the constant 0 otherwise. Then the transducer π can be algebraically specified by the system of equations $\mathcal{E}_\pi$:

$$\left\{ X_q = \sum_{x \in \Sigma} x \cdot E_{q,x} + c_q : q \in Q \right\}.$$

Proposition 3. *For every finite transducer π the system of equation $\mathcal{E}_\pi$ has a unique solution $\{X_q = TR(\pi, q) : q \in Q\}$.*

The proof is based on Arden's Lemma adapted for transductions.

Corollary 4. *For every pair of states $p, q \in Q$ the equivalence $\pi(p) \sim \pi(q)$ holds iff the system of equations $\mathcal{E}_\pi \cup \{X_p = X_q\}$ has a solution.*

Thus, to verify the equivalence of transducers it suffices to learn how to check the solvability of certain systems of equations $\mathcal{E}$ which are the extensions of the systems $\mathcal{E}_\pi$ corresponding to transducers. The solvability of some extensions is quite obvious. We say that a system of linear equations

$$\mathcal{E} = \mathcal{E}_\pi(X_1, \ldots, X_n) \cup \{X'_j = E_j(X_1, \ldots, X_n) : 1 \le j \le m\},$$

is *reduced* if $\{X_1, \ldots, X_n\}$ and $\{X'_1, \ldots, X'_m\}$ are disjoint sets of variables and all right-hand sides E_j are Δ-regexes.

Proposition 5. *Every reduced system of equations $\mathcal{E}$ has the unique solution.*

Some other extensions of the systems $\mathcal{E}_\pi$ have no solutions.

Proposition 6. *If languages $L_1 = \{u_1, \ldots, u_\ell\}$ and $L_2 = \{v_1, \ldots, v_m\}$ are incompatible then a system of equations $\mathcal{E}_\pi(X_1, \ldots, X_n) \cup \{\sum_{i=1}^{\ell} u_i \cdot X_i = \sum_{j=1}^{m} v_j \cdot X_j\}$ has no solutions.*

Proposition 7. *If a set of words $\{u_1, \ldots, u_\ell\}$ is prefix-free and a system of equations $\mathcal{E}_\pi(X_1, \ldots, X_n) \cup \{X_1 = \sum_{i=1}^{\ell} u_i \cdot X_i\}$ has a solution then $\ell = 1$ and $u_1 = \varepsilon$.*

Below we present an iterative procedure which checks the solvability of the system of equations $\mathcal{E}_1 = \mathcal{E}_\pi \cup \{X_p = X_q\}$ for prefix-free transducer π by bringing this system to an equivalent reduced form. At the beginning of each iteration t the procedure gets at the input a system of equations of the form

$$\mathcal{E}_t = \mathcal{E}_{\pi_t} \cup \{X_i = E_i : 1 \le i \le N_t\},$$

where π_t is some prefix-free transducer (not necessarily π) and all Δ-regexes E_i are prefix-free. If a variable X occurs more than once in $\mathcal{E}_t$ then we call it *active*. At the t-th iteration the following equivalent transformations are applied to $\mathcal{E}_t$.

(1) *Removing of identities.* Equations of the form $X = X$ are removed from $\mathcal{E}_t$.
(2) *Checking the reducedness of $\mathcal{E}_t$.* If none of the variables $X_1, \ldots, X_{N_t}$ from left-hand sides of equations $X_i = E_i$ occurs elsewhere then the procedure terminates and announces the solvability of the system (due to Proposition 5).
(3) *Elimination of variables.* Consider an equation of the form $X_i = E_i$ in $\mathcal{E}_t$
   – if the variable X_i is involved in Δ-regex E_i then the procedure terminates and announces the unsolvability of the system (by Proposition 7);
   – otherwise in all other equations of the system $\mathcal{E}_t$ all the occurrences of X_i are replaced with the regex E_i.

After this step the number of active variables in $\mathcal{E}_t$ decreases. But the use of these substitutions has a side effect: non-standard equations of the form

(1) $E' = E''$, where E', E'' are non-variable Δ-regexes, and
(2) $E = G$, where E is a non-variable Δ-regex and G is a Σ-regex, may appear in $\mathcal{E}_t$. It may also happen that
(3) several equations of the form $X = G$ with the same variable X appear in $\mathcal{E}_t$.
(4) *Elimination of non-standard equations $E = G$.* At this step some equations which spoil the canonical form of the system are removed from $\mathcal{E}_t$. Then
   – for every equation of the form $E(X_1, \ldots, X_\ell) = G$ replace all the occurrences of variables $X_i, 1 \le i \le \ell$, in Δ-regex E with Σ-regexes G_i that correspond to these variables in the equations $X_i = G_i$ from the subsystem $\mathcal{E}_{\pi_t}$; then bring the resulting expression $E(G_1, \ldots, G_\ell)$ to the standard form of Σ-regex using commutativity law $u \cdot x = x \cdot u$ for letters from Σ and Δ;

– for every pair of equations $X = G'$ and $X = G''$ with the same left-hand side but different Σ-regexes G' and G'' replace one of these equations with the equation $G' = G''$.

After this step all equations of the form $E = G$ disappear and all equations of the form $X = G$ will have pairwise different left-hand side variables. But this is achieved by inserting to the system non-standard equations of the form 4) $G' = G''$ where G', G'' are Σ-regexes.

(5) *Elimination of nonstandard equations* $G' = G''$. The procedure removes from the system every equation of the form

$$\sum_{i=1}^{k} a_i \cdot E_i' = \sum_{i=1}^{k} a_i \cdot E_i''$$

and inserts instead of it k equations $E_i' = E_i'', 1 \le i \le k$. Thus, all equations of the form $G' = G''$ disappear from the system due to the introduction of new equations of the form $E' = E''$. After this step equations of this form are the only non-standard equations that remain in the system.

(6) *Elimination of nonstandard equations* $E' = E''$. The decisive importance here is that all Δ-regexes that occur in the equations of the system are prefix-free. This is due to the fact that the transducer π which gives rise to the initial system of equations $\mathcal{E}_1$ is prefix-free and that all transformations the equations of the system undergo preserve the prefix-free property of Δ-regexes that occur in the equations. For every equation

$$\sum_{i=1}^{\ell} u_i \cdot X_i' = \sum_{j=1}^{m} v_j \cdot X_j'' \tag{3}$$

check the compatiblity of the languages $L' = \{u_1, \ldots, u_\ell\}$ and $L'' = \{v_1, \ldots, v_m\}$.

– if the languages L' and L'' are incompatible then the procedure terminates and announces the unsolvability of the system (due to Proposition 6);
– otherwise the procedure makes a splitting $L' = \bigcup_{i=1}^{n} L_i'$ and $L'' = \bigcup_{i=1}^{n} L_i''$ of these languages, removes the Eq. (3) from $\mathcal{E}_t$, and inserts for every fraction $L_i' = \{u_{i_0}\}$ and $L_i'' = \{v_{i_1}, \ldots, v_{i_r}\}$ (or $L_i' = \{u_{i_1}, \ldots, u_{i_r}\}$ and $L_i'' = \{v_{i_0}\}$) an equation $X_{i_0}' = (u_{i_0} \setminus v_{i_1}) \cdot X_{i_1}'' + \cdots + (u_{i_0} \setminus v_{i_r}) \cdot X_{i_r}''$ (or $X_{i_0}'' = (v_{i_0} \setminus u_{i_1}) \cdot X_{i_1}' + \cdots + (v_{i_0} \setminus u_{i_r}) \cdot X_{i_r}'$ respectively).

After this step we obtain the system of equations $\mathcal{E}_{t+1}$ which is equivalent to $\mathcal{E}_t$ but has a smaller number of active variables than $\mathcal{E}_t$.

Proposition 8. *For every prefix-free transducer π and a pair of its states p, q the procedure above when being applied to the system of equations $\mathcal{E}_1 = \mathcal{E}_\pi \cup \{X_p = X_q\}$ terminates and correctly detects the solvability of $\mathcal{E}_1$.*

By choosing directed acyclic graphs to represent regexes and by using suitable reference data structures to represent the equations it is possible to perform all computations of one iteration of the described procedure in linear time without using additional memory. Thus, we arrive at

Theorem 9. *Equivalence problem for finite prefix-free transducers is decidable in quadratic time.*

3   Two-Tape Automata and Generalized Transducers

Let $M = \langle S_1, S_2, s_0, F, \rightarrow \rangle$ be a 2-DFSA over alphabets Σ and Δ. Without loss of generality we will assume that $F \subseteq S_1$. For every state $\widehat{s} \in S_1$ and a letter $x \in \Sigma$ define a set of pairs $Out_M(\widehat{s}, x) \subseteq \Delta^* \times S_1$ as follows. Consider the transition $\widehat{s} \xrightarrow{x} s$ of M. If $s \in S_1$ then $Out_M(\widehat{s}, x) = \{(\varepsilon, s)\}$. If $s \in S_2$ then $Out_M(\widehat{s}, x)$ is a set of all pairs $(z_1 z_2 \ldots z_{n-1} z_n, s')$ such that there exists a run (2) of M which passes only via states s_i of the set S_2 and ends at a state $s' \in S_1$.

Proposition 10. *For every 2-DFSA $M = \langle S_1, S_2, s_0, F, \rightarrow \rangle$ over alphabets Σ and Δ, a pair of states $\widehat{s}, s \in S_1$, and a letter $x \in \Sigma$ the set of words $L_M(\widehat{s}, s, x) = \{w : (w, s) \in Out_M(\widehat{s}, x)\}$ is a regular prefix-free language. Moreover, the union $L_M(\widehat{s}, x) = \bigcup_{s \in S_1} L_M(\widehat{s}, s, x)$ is also a regular prefix-free language.*

We associate with every 2-DFSA M a transducer $\pi_M = \langle S_1, s_0, F, \longrightarrow \rangle$ over the alphabets Σ and Δ such that the transition relation $\longrightarrow$ meets the requirement $s \xrightarrow{x/w} s' \Leftrightarrow (w, s') \in Out_M(s, x)$ for every quadruple $(s, x, w, s') \in S_1 \times \Sigma \times \Delta^* \times S_1$. Proposition below follows immediately from the definition of π_M.

Proposition 11. *The equality $TR(M) = TR(\pi_M, s_0)$ holds for every 2-DFSA M.*

Proposition 10 implies that for every 2-DFSA M the transducer π_M is prefix-free. However, π_M may have infinitely many transitions. Therefore, to make use of Theorem 9 and Proposition 11 for equivalence checking of 2-DFSAs one needs to find a way for representing every transducer π_M as a finite transition system. A key consideration that allows transducers to be represented as finite transition systems is that for every state s and a letter x the set of words $L_M(s, x)$ which transducer π_M outputs in response to x at s is a regular language. Since regular languages admit finitary descriptions (say, deterministic finite state automata), these descriptions themselves can be used for labeling transitions of transducers. A similar technique was used in the model of generalized finite automata [9].

Denote by $PFReg(\Delta)$ the set of prefix-free regular languages over alphabet Δ. Every such language can be specified by the minimal deterministic finite state automaton (1-DFSA). The prefix-free property is manifested in the fact that every such 1-DFSA have a unique accepting state from which no transitions emerge. Concatenation of automata of this kind is defined in the obvious way.

In what follows when dealing with transducers and regexes that involve regular prefix-free languages we will address to these languages by the names of the corresponding 1-DFSAs.

A *generalized prefix-free finite transducer* over languages Σ and Δ is a quadruple $\Pi = \langle Q, q_0, F, \longrightarrow \rangle$, where Q is a finite set of states, q_0 is an initial state, F is a subset of final states, and $\psi : Q \times \Sigma \times Q \to PFReg(\Delta)$ is a finite transition function such that for every state q and a letter x the language $\bigcup_{q' \in Q} \psi(q, x, q')$ is prefix-free. As usual, we write $q \xrightarrow{x/L} q'$ whenever $\psi(q, x, q') = L$.

If $\psi(q, x, q') = \emptyset$ then $q \xrightarrow{x/\emptyset} q'$ is a "seeming" transition which does not contribute to computations of Π. A run of Π is any finite sequence of transitions

$$q \xrightarrow{a_1/L_1} q_1 \xrightarrow{a_2/L_2} \cdots \xrightarrow{a_{n-1}/L_{n-1}} q_{n-1} \xrightarrow{a_n/L_n} q' . \tag{4}$$

When writing $q \xrightarrow{w/L}_* q'$ we mean that Π has a run (4) such that $w = a_1 a_2 \ldots a_n$ and $L = L_1 L_2 \ldots L_n$. A *transduction relation realized by* Π in its state q is the set of pairs $TR(\Pi, q) = \{(w, u) : q \xrightarrow{w/L}_* q', u \in L, q' \in F\}$.

As it follows from Propositions 10 and 11, for every 2-DFSA M there exists a generalized prefix-free finite transducer Π_M such that $TR(M) = TR(\Pi_M, s_0)$.

4 Equivalence Checking of Generalized Prefix-Free Transducers

To check the equivalence of generalized prefix-free finite transducers we adapt appropriately the approach developed for the analysis of ordinary prefix-free finite transducers: for a pair of states q', q'' in a transducer Π build the system of transduction equations $\mathcal{E}_\Pi \cup \{X_{q'} = X_{q''}\}$ and check its solvability. We will discuss only those modifications that need to be made to take into account the specific features of generalized transducers.

Regexes are built of variables $X_1, X_2, \ldots$, constants $0, 1$, and letters from Σ, but instead of Δ we will use prefix-free regular languages from $PFReg$ as constants. Clearly, every such constant $L \in FPReg$ is interpreted as a transduction $\{(\varepsilon, w) : w \in L\}$. Modified Δ-regexes are expressions of the form $L_1 \cdot X_1 + L_2 \cdot X_2 + \cdots + L_n \cdot X_n$, where $L_i \in FPReg$ for every $i, 1 \le i \le n$. The system of equations $\mathcal{E}_1 = \mathcal{E}_\Pi \cup \{X_{q'} = X_{q''}\}$ for a generalized transducer Π is constructed in the same way as for ordinary transducers, and the analogues of Propositions 3–7 also hold for systems of equations with thus modified Δ-regexes. Therefore, the rules (1)–(5) of the solvability checking procedure remain the same for the case of modified system of equations. The only rule which needs an improvement is the rule (6). We modify it as follows.

(6′) *Elimination of nonstandard equations* $E' = E''$. For every equation

$$\sum_{i=1}^{\ell} L_i' \cdot X_i' = \sum_{j=1}^{m} L_j'' \cdot X_j'' \tag{5}$$

check the compatiblity of the languages $L' = \bigcup_{i=1}^{\ell} L'_i$ and $L'' = \bigcup_{j=1}^{m} L''_j$. If the languages are incompatible then complete the procedure and announce the unsolvability of the system. Otherwise, proceed as follows.

6.1 For every $i, 1 \leq i \leq \ell$, such that $L'_i \cap Pref(L'') \neq \emptyset$ find any word $w \in L'_i \cap Pref(L'')$, and add an equation $X'_i = \sum_{j=1}^{m} (w \setminus L''_j) \cdot X''_j$ to the system. Replace all occurrences of X'_i in (5) with the right-hand side of this equation.

6.2 Do the same for every $j, 1 \leq j \leq m$: find any word $u \in L''_j \cap Pref(L')$ in the case when $L''_j \cap Pref(L') \neq \emptyset$, add an equation $X''_j = \sum_{i=1}^{\ell} (u \setminus L'_i) \cdot X'_i$ to the system, and replace all occurrences of X''_j in (5) with the right-hand side of this equation.

6.3 If after applying all the above substitutions, the Eq. (5) does not become an identity, then complete the procedure and announce the unsolvability of the system. Otherwise, remove the resulting identity from the system.

It is easy to verify that the system of equations obtained after the elimination of non-standard equations of the form $E' = E''$ will be equivalent to the original system. Correctness of transformation rule (6') follows from

Proposition 12. *If an equation $L_0 \cdot X_0 = \sum_{i=1}^{n} L_i \cdot X_i$ with a prefix-free Δ-regex at the right-hand side has a prefix-free solution, and $w \in L_0 \cap Pref(\bigcup_{i=1}^{n} L_i)$, then the equation $X_0 = \sum_{i=1}^{n} (w \setminus L_i) \cdot X_i$ has the same solution.*

All the additional equations inserted in the system of equations by the transformation rule (6') can be constructed simultaneously by computing the (descriptions of) languages $L' \cap Pref(L'')$ and $L'' \cap Pref(L')$. When prefix-free regular languages involved in the Eq. (5) are specified by 1-DFSAs, this computation can be performed in time quadratic of their size. Thus, we arrive at

Theorem 13. *The equivalence problem for generalized prefix-free finite transducers is decidable in cubic time.*

Corollary 14. *The equivalence problem for deterministic two-tape finite state automata is decidable in cubic time.*

5   Conclusion

Further development of the results presented in this paper can be carried out in several directions. Perhaps, the most ambitious would be an attempt to extend the approach proposed for checking equivalence of deterministic two-tape automata to n-tape automata. For example, 3-DFSAs can also be translated into generalized prefix-free transducers, but in this case we will have to use prefix-free deterministic 2-DFSAs instead of 1-DFSAs for transition labeling, modify accordingly the notions of Δ- and Σ-regexes, and, finally, adapt the rule $(6')$ of the decision procedure so that it can be applied to work with prefix-free deterministic rational transductions. The latter could be achieved by introducing the notion of left quotient on transductions in such a way as to preserve the analogous of Proposition 12 and by using equivalence checking procedure for deterministic 2-DFSAs. This provides the grounds for the following hypothesis: the equivalence problem for multi-tape automata is decidable in polynomial time but the degree of the polynomial is proportional to the number of tapes.

The author highly appreciates the valuable comments of anonymous referees.

References

1. Bird, M.: The equivalence problem for deterministic two-tape automata. J. Comput. Syst. Sci. **7**, 218–236 (1973)
2. Blattner, M., Head, T.: Single-valued a-transducers. J. Comput. Syst. Sci. **15**, 310–327 (1977)
3. Blattner, M., Head, T.: The decidability of equivalence for deterministic finite transducers. J. Comput. Syst. Sci. **19**, 45–49 (1979)
4. Culik, K., Karhumaki, J.: The equivalence of finite-valued transducers (on HDTOL languages) is decidable. Theor. Comput. Sci. **47**, 71–84 (1986)
5. Fisher, P.S., Rozenberg, A.L.: Multitape one-way nonwriting automata. J. Comput. Syst. Sci. **2**, 88–101 (1968)
6. Friedman, E.P., Greibach, S.A.: A polynomial time algorithm for deciding the equivalence problem for 2-tape deterministic finite state acceptors. SIAM J. Comput. **11**, 166–183 (1982)
7. Griffiths, T.: The unsolvability of the equivalence problem for ε-free nondeterministic generalized machines. J. ACM **15**, 409–413 (1968)
8. Gurari, E., Ibarra, O.: A note on finite-valued and finitely ambiguous transducers. Math. Syst. Theory **16**, 61–66 (1983)
9. Han, Y.-S., Wood, D.: The generalization of generalized automata: expression automata. In: Proceedings of 9th International Conference on Implementation and Application of Automata, pp. 156–166 (2004)
10. Harju, T., Karhumaki, J.: The equivalence problem of multitape finite automata. Theor. Comput. Sci. **78**, 347–355 (1991)
11. Mohri, M.: Minimization algorithms for sequential transducers. Theor. Comput. Sci. **234**, 177–201 (2000)
12. Sakarovitch J., de Souza R.: On the decomposition of k-valued rational relations. In: Proceedings of 25th International Symposium on Theoretical Aspects of Computer Science, pp. 621–632 (2008)
13. Sakarovitch J., de Souza R.: On the decidability of bounded valuedness for transducers. In: Proceedings of the 33rd International Symposium on MFCS, pp. 588–600 (2008)
14. Schutzenberger, M.P.: Sur les relations rationnelles. In: Proceedings of Conference on Automata Theory and Formal Languages, pp. 209–213 (1975)

15. de Souza, R.: On the decidability of the equivalence for k-valued transducers. In: Proceedings of 12th International Conference on Developments in Language Theory, pp. 252–263 (2008)

16. de Souza, R.: On the decidability of the equivalence for a certain class of transducers. In: Proceedings of 13th International Conference on Developments in Language Theory, pp. 478–489 (2009)

17. Valiant, L.G.: The equivalence problem for deterministic finite-turn pushdown automata. Inf. Control **25**, 123–133 (1974)

18. Weber, A.: On the valuedness of finite transducers. Acta Inform. **27**, 749–780 (1989)

19. Weber, A.: On the lengths of values in a finite transducer. Acta Inform. **29**, 663–687 (1992)

20. Weber, A.: Decomposing finite-valued transducers and deciding their equivalence. SIAM J. Comput. **22**, 175–202 (1993)

21. Zakharov, V.A.: Equivalence checking problem for finite state transducers over semigroups. In: Proceedings of the 6th International Conference on Algebraic Informatics, vol. 9270, pp. 208–221 (2015)

Efficient Symmetry Breaking for SAT-Based Minimum DFA Inference

Ilya Zakirzyanov[1,2]($\boxtimes$), Antonio Morgado[3], Alexey Ignatiev[3,4],
Vladimir Ulyantsev[1], and Joao Marques-Silva[3]

[1] ITMO University, St. Petersburg, Russia
{zakirzyanov,ulyantsev}@corp.ifmo.ru
[2] JetBrains Research, St. Petersburg, Russia
[3] Faculty of Science, University of Lisbon, Lisbon, Portugal
{ajmorgado,aignatiev,jpms}@ciencias.ulisboa.pt
[4] ISDCT SB RAS, Irkutsk, Russia

Abstract. Inference of deterministic finite automata (DFA) finds a wide range of important practical applications. In recent years, the use of SAT and SMT solvers for the minimum size DFA inference problem (MinDFA) enabled significant performance improvements. Nevertheless, there are many problems that are simply too difficult to solve to optimality with existing technologies. One fundamental difficulty of the MinDFA problem is the size of the search space. Moreover, another fundamental drawback of these approaches is the encoding size. This paper develops novel compact encodings for Symmetry Breaking of SAT-based approaches to MinDFA. The proposed encodings are shown to perform comparably in practice with the most efficient, but also significantly larger, symmetry breaking encodings.

Keywords: DFA inference · Boolean satisfiability · Symmetry breaking

1 Introduction

The inference of minimum-size deterministic finite automata (DFA) from (positive and negative) examples of their behavior has been investigated since the early days of computing, with continued improvements until the present day. The importance of topic is illustrated not only by recent improvements to tools for computing minimum-size DFAs [27,30], but also by recent and ever growing list of applications [29]. The problem of computing the minimum-size

IZ was supported by RFBR (project 18-37-00425). AM, AI and JMS were supported by FCT grants ABSOLV (PTDC/CCI-COM/28986/2017), FaultLocker (PTDC/CCI-COM/29300/2017), SAFETY (SFRH/BPD/120315/2016), and SAMPLE (CEECIND/04549/2017). VU was supported by the Government of Russia (Grant 08-08).

© Springer Nature Switzerland AG 2019
C. Martín-Vide et al. (Eds.): LATA 2019, LNCS 11417, pp. 159–173, 2019.
https://doi.org/10.1007/978-3-030-13435-8_12

DFA (MinDFA) witnessed seminal work in the early 70s [6]. Moreover, a number of visible contributions were made in the 90s. These include the use of graph coloring [8], constraint programming techniques [9,22], and state merging approaches [17,18]. Approaches based on SAT and SMT were proposed in the last decade, with promising results [12,13,20,21,25]. Nevertheless, the size of existing propositional encodings do not scale for large DFA inference problems. The use of SMT does not represent a clear improvement, since SMT solving approaches for the MinDFA problem will also encode to propositional logic. This paper revisits SAT encodings for the MinDFA problem as well as recent work on exploiting symmetry breaking [25,30], and proposes a (novel) tighter propositional representation of state-of-the-art symmetry breaking predicates, but it also devises new symmetry breaking constraints which serve to achieve more effective pruning of the search space. The new propositional encoding proposed in this paper enables clear performance gains over the state of the art [13,14,26,30].

The paper is organized as follows. Section 2 introduces the definitions used throughout the paper and briefly overviews related work. Section 3 develops new ideas to encode symmetry breaking predicates. Section 4 compares a new tool for the MinDFA problem with the existing state of the art, showing clear performance gains. Section 5 concludes the paper.

2 Background

2.1 Preliminaries

Throughout the paper we assume that automata are defined over some set of symbols Σ, also known as the *alphabet*. The number of symbols in the alphabet is $L = |\Sigma|$. For earlier DFA inference examples, it was often the case that $\Sigma = B = \{0, 1\}$ [18,22]. For more recent DFA inference examples [28], larger alphabets are often considered.

A *deterministic finite automaton* (DFA) is a tuple $\mathcal{D} = (D, \Sigma, \delta, d_1, D^+, D^-)$, where D is a finite set of states, Σ is the (input) alphabet, $\delta : D \times \Sigma \to D$ is the transition function, d_1 is the initial state, D^+ is the set of accepting states and $D^- = D \setminus D^+$ is the set of rejecting sets. For input strings $\pi \in \Sigma^*$ we define $\hat{\delta}(d_1, \pi)$ inductively as follows [16]: (i) $\hat{\delta}(d_1, \epsilon) = d_1$; (ii) If $\pi = \pi'c$, then $\hat{\delta}(d_1, \pi) = \delta(\hat{\delta}(d_1, \pi'), c)$.

We assume the standard setting of inferring a minimum-size DFA given a set of samples of its behavior [7,15], i.e. the training set, each sample represented by an input string that is either accepted or rejected by some DFA $\mathcal{U} = (U, \Sigma, \mu, u_1, U^+, U^-)$, which is not known. This form of learning is often referred to as *passive learning*, as opposed to *active learning* [2,20], which enables a learning algorithm (aiming to create a target DFA) to formulate queries to some teacher (which knows of the unknown DFA).

A *training set* is a set of pairs $\mathbb{T} = \{(\pi_1, o_1), \ldots, (\pi_R, o_R)\}$, where each pair $(\pi_r, o_r) \in \Sigma^* \times \{0, 1\}$ denotes the output o_r observed given input string π_r. If $o_r = 1$ ($o_r = 0$), then π_r is referred to as a *positive* (*negative*) example. Given

Function MINIMUMDFA($\mathcal{T}$)

    Input : $\mathcal{T}$: APTA

    Output: $\mathcal{S}$: minimum size DFA

1    $M \leftarrow$ FindLowerBound($\mathcal{T}$)

2   **while** true **do**

3        $\mathcal{S} \leftarrow$ FindConsistentDFA($\mathcal{T}, M$)

4   **if** $\mathcal{S} \neq \emptyset$ **then** **return** $\mathcal{S}$

5        $M \leftarrow M + 1$

    Algorithm 1: General lower bound refinement algorithm

a training set, we can construct an APTA (*augmented prefix tree acceptor*) [1, 13,24], defined as the DFA $\mathcal{T} = (T, \Sigma, \tau, t_1, T^+, T^-)$, where any input string sharing the same prefix ends up in the same state. Concretely, given input strings $\pi_1 = \pi_a \pi_{b_1}$ and $\pi_2 = \pi_a \pi_{b_2}$ the common prefix π_a will be associated to a unique sequence of states in the APTA. For an APTA $\mathcal{T}$, we have $T^+ \cup T^- \neq T$, and we define $N = |T|$. When clear from the context, the states of $\mathcal{T}$ are referred to by their index, t_i by i, $i = 1, \ldots, N$. In some settings, $\theta(i)$ will be used to denote the distance from the APTA root state t_1 to state t_i.

The *minimum-size DFA* inference problem (MinDFA) is to identify a DFA $\mathcal{S} = (S, \Sigma, \sigma, s_1, S^+, S^-)$, with a minimum number of states, such that for any training pair (π_r, o_r), $\hat{\sigma}(s_1, \pi_r) \in S^+$ iff $o_r = 1$ and $\hat{\sigma}(s_1, \pi_r) \in S^-$ iff $o_r = 0$. For a prospective DFA $\mathcal{S}$, we define $M = |S|$.

Throughout the paper $[R]$ is used to denote the set $\{1, \ldots, R\}$, for some positive integer R. Moreover, we will use integers to refer to either symbols or states. For a given alphabet, by associating states and symbols with integers facilitates imposing a fixed lexicographic order, which will be required later in the paper (see Sect. 3). Additionally, standard SAT definitions are assumed and used [5].

2.2 Minimum Size DFA Inference

This paper focuses on constraint-based exact approaches for the MinDFA problem. Different constraint programming approaches for solving the MinDFA problem have been proposed over the years. More recently, the use of SAT [12–14] and SMT [20,21] has been investigated. A more detailed account of past work is available for example in Neider's PhD thesis [20, Chap. 3].

Algorithm 1 summarizes the most widely used approach for computing a minimum size DFA consistent with a given APTA $\mathcal{T}$ (obtained from the training set). Initially a lower bound on the size of the inferred DFA is computed. An often used heuristic is to compute a maximal clique on states of the APTA that cannot be assigned to the same DFA state [12–14,20–22,26]. Afterwards, starting from the lower bound and for each possible value on the number of states of the DFA, some algorithm decides whether there exists a DFA $\mathcal{S}$ which can be shown consistent with the samples of behavior summarized as the APTA $\mathcal{T}$.

Algorithm 1 is referred to as LSUS (linear-search, UNSAT until SAT) and is used in different settings. Other algorithms can be envisioned. These include binary search, assuming some upper bound is known or can be identified (e.g. with merge-based algorithms). Another alternative is unbounded search with a final binary search step. These algorithms have been used in recent years for solving MaxSAT [19] and for extracting MUSes [4]. The use of propositional encodings can be traced to the work of Grinchtein, Leucker & Piterman [12]. By using two different representations for integers, one in unary and the other in binary, this work proposes two propositional encodings. For the unary representation, the encoding size is in $\mathcal{O}(N \times M^2 + N^2 \times M)$ over $\mathcal{O}(N \times M)$ variables[1]. For the binary representation, the encoding size is in $\mathcal{O}(N \times M \times \log M + N^2 \times M)$ on $\mathcal{O}(N \times \log M)$ variables. More recent work by Heule&Verwer (HV) [13,14] proposed encodings that have been shown effective in practice [28]. The HV encoding builds on the graph coloring analogy proposed in earlier work [8]. The proposed encoding has size $\mathcal{O}(M^3 + N \times M^2)$ over $\mathcal{O}(M^2 + N \times M)$ variables. This encoding is revisited in Sect. 2.3.

2.3 SAT-Based MinDFA

Given an APTA $\mathcal{T}$ and a bound M on the number of states of the inferred DFA $\mathcal{S}$, this subsection provides a derivation of the HV encoding [13,14], based on a different motivation. By careful analysis of this formulation, we achieve a more compact propositional encoding. Instead of relating the MinDFA problem with graph coloring, we formulate it as the problem of matching the N states of the APTA $\mathcal{T}$ to the M states of a target DFA $\mathcal{S}$. The sets of variables of the propositional encoding are as follows:

1. $m_{i,p}$ which is 1 iff state t_i in $\mathcal{T}$ is matched with state s_p in $\mathcal{S}$.
2. $e_{v,p,q}$ which is 1 iff there is a transition from s_p to s_q on symbol l_v in $\mathcal{S}$.
3. a_p which is 1 iff s_p is accepting in $\mathcal{S}$.

The constraints of the proposed encoding are summarized in Table 1. Observe that for encoding the Equals1 constraints, [14] uses a clause to encode an AtLeast1 constraint, and the Pairwise Encoding for encoding an AtMost1 constraint. A simple improvment is to use a more compact encoding, among the many that exist. Concrete examples include sequential counters [23], cardinality networks [3], the ladder encoding [11], sorting networks [10], among several other options. As can be concluded, the proposed encoding grows with $\mathcal{O}(N \times M^2)$. Thus, the encoding is asymptotically (somewhat) tighter than the encoding proposed in [13], in that the encoding of the cardinality constraints changes from $\mathcal{O}(M^3)$ to $\mathcal{O}(M^2)$. This difference can be significant for large values of M. As observed in earlier work [13,14], for some benchmarks [18], the target DFA has hundreds of states, and so an encoding in $\mathcal{O}(M^3)$ is expected to be beyond the memory capacity of existing compute servers. It is straightforward to map the

---

[1] The encoding size shown is adapted from the results in [20], taking into account that both $|T^+|$ and $|T^-|$ can grow with $N = |T|$. The size of $|\Sigma|$ is assumed constant.

Table 1. Constraints of the SAT encoding

Constraint	Range	
$(\sum_{p=1}^{M} m_{i,p}) = 1$	$i \in [N]$	Each state t_i in $\mathcal{T}$ is matched with exactly one state in $\mathcal{S}$
$m_{i,p} \rightarrow a_p$	$i \in [N];\ t_i \in \mathcal{T}^+;$ $p \in [M]$	Each accepting state t_i in $\mathcal{T}$ is matched with an accepting state in $\mathcal{S}$
$m_{i,p} \rightarrow \neg a_p$	$i \in [N];\ t_i \in \mathcal{T}^-;$ $p \in [M]$	Each rejecting state t_i in $\mathcal{T}$ is matched with a rejecting state in $\mathcal{S}$
$(\sum_{q=1}^{M} e_{v,p,q}) = 1$	$v \in [L];\ p \in [M]$	There is exactly one transition from s_p on some symbol l_v in $\mathcal{S}$
$m_{i,p} \wedge m_{k,q} \rightarrow e_{v,p,q}$	$i, k \in [N];\ v \in [L];$ $\sigma(t_i, l_v) = t_k;$ $p, q \in [M]$	A transition between t_i and t_k on l_v in $\mathcal{T}$ forces a transition between its mapped nodes on the same l_v in $\mathcal{S}$
$m_{i,p} \wedge e_{v,p,q} \rightarrow m_{k,q}$	$i, k \in [N];\ v \in [L];$ $\sigma(t_i, l_v) = t_k;$ $p, q \in [M]$	A transition between t_i and t_k on l_v in $\mathcal{T}$, with a transition between the mapped state p and a state q on l_v in $\mathcal{S}$, forces a mapping between t_k and q

sets of clauses in the HV formulation [13,14] into the constraints described above. The main difference is that we explicitly use a tighter encoding for the AtMost1 constraints, which are listed as sets of clauses (capturing the well-known pairwise encoding) in [13]. Additionally, the HV formulation [13] considers different sets of redundant constraints to the basic formulation above. A technique that has been proposed for the SAT formulation is the breaking of symmetries of the DFA constructed [26,30]. Symmetry breaking for the SAT formulation is described in depth in Sect. 3, together with new improvements.

3 Efficient Symmetry Breaking

This section revisits recent symmetry breaking for the MinDFA problem, which imposes an order on the states of the DFA [26,30]. Although effective in practice, the existing propositional encoding is not tight, and so unlikely to scale for larger DFAs. Section 3.2 develops a significantly tighter encoding. Section 3.3 devises novel constraints that serve to furhter prune the search space that a SAT solver needs to explore.

3.1 Propositional Formulation for Breaking Symmetries

This section summarizes the recent work on breaking symmetries of the DFA being constructed, by imposing an ordering on the states of the DFA [26,30]. In this section we follow the original formulation [26]. The approach can be formalized as follows. Assume a target DFA $\mathcal{S} = (S, \Sigma, \sigma, s_1, S^+, S^-)$. The states of the DFA $\mathcal{S}$ are required to be numbered according to the tree induced by a breadth-first search (BFS) of the target DFA. As a result, the formulation of symmetry breaking depends only on the states and transitions of the target DFA $\mathcal{S}$ (independent of the APTA $\mathcal{T}$). In this section we require some fixed (e.g. lexicographic) ordering on the symbols of Σ. Any order of the symbols is valid. The symbols will be numbered from 1 to L, but the numbers respect the fixed ordering.

The propositional variables used in the formulation are as follows:

1. $p_{q,r}$, with $1 \leq r < q \leq M$. $p_{q,r} = 1$ iff state r is the parent of q in the BFS tree.
2. $t_{p,q}$, with $1 \leq p < q \leq M$. $t_{p,q} = 1$ iff there is a transition from p to q in $\mathcal{S}$.
3. $m_{v,p,q}$, with $v \in \Sigma$ and $1 \leq p < q \leq M$. $m_{v,p,q} = 1$ iff there is a transition from state p to state q on symbol l_v and there is no such transition with a lexicographically smaller symbol.

The clauses of the propositional formulation are summarized in Eqs. (1–6).

$$\bigwedge_{2 \leq q \leq M} (p_{q,1} \vee p_{q,2} \vee \cdots \vee p_{q,q-1}) \tag{1}$$

$$\bigwedge_{1 \leq r < s < q < M} (p_{q,s} \rightarrow \neg p_{q+1,r}) \tag{2}$$

$$\bigwedge_{1 \leq r < q \leq M} (t_{r,q} \leftrightarrow e_{1,r,q} \vee \cdots \vee e_{L,r,q}) \tag{3}$$

$$\bigwedge_{1 \leq r < q \leq M} (p_{q,r} \leftrightarrow t_{r,q} \wedge \neg t_{r-1,q} \wedge \cdots \wedge \neg t_{1,q}) \tag{4}$$

$$\bigwedge_{1 \leq r < q \leq M} \bigwedge_{1 \leq v \leq L} (m_{v,r,q} \leftrightarrow e_{v,r,q} \wedge \neg e_{v-1,r,q} \wedge \cdots \wedge \neg e_{1,r,q}) \tag{5}$$

$$\bigwedge_{1 \leq r < q < M} \bigwedge_{1 \leq u < v \leq L} (p_{q,r} \wedge p_{q+1,r} \wedge m_{v,r,q} \rightarrow \neg m_{u,r,q+1}) \tag{6}$$

There are six types of conjunction of clauses considered. (1) relates to the states, and with the exception of the initial state (numbered 1), each clause says that a state must have a parent with smaller number. (2) says that a state q must be enqueued (in the BFS traversal) before the next state $q + 1$, and so the parent r of $q + 1$ cannot be less than the parent s of q. (3) and (4) define the $t_{q,r}$ variables based on the $e_{v,q,r}$ variables and relate them to the parent variables $p_{q,r}$. (5) defines the $m_{v,p,q}$ variables using DFA transitions, and the (6) imposes consecutive states q and $q+1$ with the same parent r to be arranged in the order of the symbols. It is plain to conclude that the size of the encoding grows with $\mathcal{O}(M^3 + M^2 L + M^2 L^2)$. Observe that the contribution of M^3, which dominates the other components assuming $L \ll M$, results from (2) and (4). Moreover, when $|\Sigma| = 2$, [26] proposes to replace (5) and (6) with

$$\bigwedge_{1 \leq r < q < M} (p_{q,r} \wedge p_{q+1,r} \rightarrow e_{1,r,q}) \tag{7}$$

3.2   A Tighter SAT Encoding

A propositional encoding in $\mathcal{O}(M^3 + M^2 L^2 + M^2 L)$ is impractical for the larger DFA inference instances [18,28]. This section shows how to modify the symmetry breaking propositional encoding of Sect. 3.1 such that the encoding size becomes $\mathcal{O}(M^2 L)$. The new encoding develops alternative representations for (2) and the (4), but also for (5) and (6). In addition, one needs to require:

$$\sum_{r=1}^{q-1} p_{q,r} = 1 \quad 1 < q \leq M \tag{8}$$

We first investigate the encoding of (2) and (4). We can view the values of $p_{q,r}$, with $1 \leq r \leq q-1$, as a binary string, with $q-1$ bits, and compare this string with the one of $p_{q+1,r}$, with $1 \leq r \leq q$, and so with q bits. We introduce $p_{q,q} = 0$, and so can also view the values of $p_{q,r}$ as a binary string with q bits (same size).

Observe that (2) encodes the value associated to the binary string of the $p_{q,r}$ variables to be smaller or equal than the value associated to the binary string of the $p_{q+1,r}$ variables. To compare the binary strings, we inspect the bits in order, starting at position q, and moving down to position 1. We consider variables $ng_{q,r}$, such that $ng_{q,r} = 1$ iff the most significant $q - r + 1$ bits of the string associated with $p_{q,r}$ are lexicographically *no greater* than those of $p_{q+1,r}$. The value associated to the binary string of the $p_{q,r}$ variables is smaller or equal than the value associated to the binary string of the $p_{q+1,r}$ variables iff $ng_{q,1} - 1$ holds. Since we enforce $p_{q,q} = 0$, then we must have $ng_{q,q} = 1$. Moreover, we also require $ng_{q,1} \leftrightarrow 1$. Thus we obtain:

$$(ng_{q,1} \leftrightarrow 1) \wedge (ng_{q,q} \leftrightarrow 1) \wedge \bigwedge_{1 \leq r < q} (ng_{q,r} \leftrightarrow ng_{q,r+1} \wedge eq_{q,r} \vee p_{q,r} \wedge \neg p_{q+1,r}) \tag{9}$$

where, $eq_{q,r} \leftrightarrow (p_{q,r} \leftrightarrow p_{q+1,r})$.

Second, a similar approach can be exploited for encoding of (4). We introduce variables $nt_{r,q}$, where $nt_{r,q} = 1$ iff there exists *no* $t_{s,q} = 1$ with $s < r$. Thus, $nt_{r,q}$ can be defined inductively as follows:

$$(nt_{0,q} \leftrightarrow 1) \wedge \bigwedge_{1 < r < q} (nt_{r,q} \leftrightarrow nt_{r-1,q} \wedge \neg t_{r,q}) \tag{10}$$

Thus, (4) can be rewritten, using the $nt_{r,q}$ variables as follows:

$$p_{q,r} \leftrightarrow t_{r,q} \wedge nt_{r-1,q} \tag{11}$$

As can be concluded, by using auxiliary variables $ng_{q,r}$ and $nt_{r,q}$, and Eqs. (9), (10) and (11), we achieve an overall propositional encoding in $\mathcal{O}(M^2 L + M^2 L^2)$.

However, we can tighten further the propositional encoding for breaking symmetries using a BFS tree. This is achieved by devising alternative encodings for (5) and (6). As shown next, this yields a propositional encoding in $\mathcal{O}(M^2 L)$. With respect to (5), we use the additional variables $ne_{v,r,q}$ such that $ne_{v,r,q} = 1$ iff all variables $e_{u,r,q} = 0$ with $u < v$, i.e. there are *no* variables $e_{u,r,q}$ taking value 1, when $u < v$.

$$(ne_{1,r,q} \leftrightarrow \neg e_{1,r,q}) \wedge \bigwedge_{1 < v < L} (ne_{v,r,q} \leftrightarrow \neg e_{v,r,q} \vee ne_{v-1,r,q}) \tag{12}$$

Thus given (12), (5) can be rewritten as follows:

$$\bigwedge_{1\leq r<q\leq M} \bigwedge_{1\leq v\leq L} (m_{v,r,q} \leftrightarrow e_{v,r,q} \wedge ne_{v-1,r,q}) \tag{13}$$

With respect to (6), we use the additional variables $zm_{v,r,q}$ such that $zm_{v,r,q} = 1$ iff all variables $m_{u,r,q}$ are 0-valued, $m_{u,r,q} = 0$, for $u < v$.

$$(zm_{1,r,q} \leftrightarrow \neg m_{1,r,q}) \wedge \bigwedge_{1<v<L} (zm_{v,r,q} \leftrightarrow \neg m_{v,r,q} \wedge zm_{v-1,r,q}) \tag{14}$$

Thus given (14), (6) can be rewritten as follows:

$$\bigwedge_{1\leq r<q\leq M} \bigwedge_{1\leq v\leq L} (p_{q,r} \wedge p_{j+1,r} \wedge m_{v,r,q} \rightarrow zm_{v-1,r,q+1}) \tag{15}$$

One can thus conclude that the resulting propositional encoding size is in $\mathcal{O}(M^2 L)$.

3.3 Exploiting BFS-Based Breaking of Symmetries

This section investigates techniques for developing additional constraints when imposing the ordering of states dictated by a BFS tree of the DFA. Figure 1 shows a possible BFS tree illustrating the *largest* state numbers that can be the children of some other state. The additional constraints proposed in this section will relate with Fig. 1.

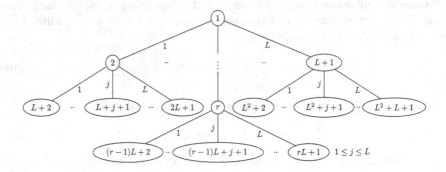

Fig. 1. (Worst case) BFS tree with the largest state numbers that can be the children of some other state. Note that $1 \leq j < L$.

BFS-Induced Properties. Although we have introduced $p_{q,r}$ such that $r < q \leq M$, it is possible to refine the range of q given r.

Property 1. Given a state r, with $1 \leq r \leq M$, in the BFS tree, r can be the parent of states in the range $r + 1$ to $rL + 1$.

Figure 1 illustrates the argument for the upper bound on the number of the children of r. We can conclude that the value of $p_{q,r}$ can be non-zero for $r + 1 \leq q \leq rL + 1$, which also impacts the possible values of some of the $e_{v,r,q}$ and the $t_{r,q}$ variables.

Property 2. For $q > rL + 1$ and $v \in [L]$, then $p_{q,r} = 0$, $e_{v,r,q} = 0$, and $t_{r,q} = 0$.

Given that the BFS tree assumes a fixed ordering not only on the states but also on the input alphabet, it is possible to identify other transitions that must be forced to value 0 (based on the ordering of the symbols). Hence, we have the following.

Property 3. $e_{v,r,rL+2-j} = 0$ for $j \in [L - 1]$ and $v \in [L - j]$.

The above observations enable to devise the additional constraints described in the remainder of this section. The constraints are organized as *shape* or *range*, but also result from information from the APTA and the BFS distance.

Shape Constraints. The possible values of $p_{q,r}$ respect a *continuity* property, dictated by the BFS traversal, in that all children of r are consecutively numbered, and there can be *at most L* of these. This continuity property can be encoded using additional variables. Let $lnp_{q,r}$ be assigned value 1 iff r is the parent of $q + 1$ but not of q (*lnp* stands for *left-no-parent*). Thus,

$$\neg p_{q,r} \wedge p_{q+1,r} \rightarrow lnp_{q,r} \tag{16}$$

Moreover, we have the following:

$$(lnp_{q,r} \rightarrow \neg p_{q,r}) \wedge \bigwedge_{r+1 < q \leq M} (lnp_{q,r} \rightarrow lnp_{q-1,r}) \tag{17}$$

Thus, $lnp_{q,r}$ is 1 from $q = 1$ until the value of q such that $p_{q+1,r}$ holds.

In a similar fashion, let $rnp_{q,r}$ be assigned value 1 if and only if r is the parent of $q - 1$ but not of q (in this case, *rnp* stands for *right-no-parent*). Thus,

$$p_{q-1,r} \wedge \neg p_{q,r} \rightarrow rnp_{q,r} \tag{18}$$

Similarly to the previous case, one can exploit the $rnp_{q,r}$ variables, and derive the following constraints:

$$
\begin{aligned}
rnp_{q,r} &\rightarrow rnp_{q+1,r} \quad r \leq q < M \\
rnp_{q,r} &\rightarrow \neg p_{q,r} \\
rnp_{q,r} &\rightarrow \neg e_{v,q,r} \quad v \in [L]
\end{aligned} \tag{19}
$$

Thus, $rnp_{q,r}$ is 1 from $q = M$ until the value of q such that $p_{q-1,r}$ holds.

Another observation is that r can be the parent of at most L states, due to L outgoing transitions. As a result, we get,

$$
\begin{aligned}
p_{q,r} &\rightarrow rnp_{q+L,r} \quad \text{if } q + L \leq M \\
p_{q,r} &\rightarrow lnp_{q-L,r} \quad \text{if } q - L \geq r + 1
\end{aligned} \tag{20}
$$

The $lnp_{q,r}$ and $rnp_{q,r}$ variables serve to force $p_{q,r}$ variables to be assigned value 0. However, under some circumstances, we can infer that some $p_{q,r}$ variables must be assigned value 1. For example, for the range of values of q for which both $lnp_{q,r}$ and $rnp_{q,r}$ are 0, the value of $p_{q,r}$ must be 1. Thus,

$$\neg lnp_{q_1,r} \wedge \neg rnp_{q_2,r} \rightarrow p_{q',r} \quad \begin{array}{c} q_1 < q' < q_2 \\ q_1 < q_2 \leq \min(q_1 + L - 1, rL + 1, M) \\ r + 1 \leq q_1 < \min(rL + 1, M) \end{array} \quad (21)$$

For any q_1, q_2 can range from $q_1 + 1$ to at most $q_1 + L$. Similarly, we can write,

$$p_{q,r} \wedge p_{s,r} \rightarrow p_{s-1,r} \quad \begin{array}{c} q < s \leq \min(q + L - 1, rL + 1, M) \\ r + 1 \leq q < \min(rL + 1, M) \end{array} \quad (22)$$

As above, for any r, s can range from $r + 1$ to at most $r + L$.

Range Constraints. Given a reference state r, we have shown above that the states of which r can be a parent of range from $r + 1$ until $rL + 1$. Moreover, we also know there is a continuity property, which causes r to be the parent of at most L states, numbered consecutively. This information can be used for constraining the $p_{q,r}$ variables, between states for which r cannot be a parent, as follows,

$$p_{q,r} \rightarrow \neg p_{q+L,r} \quad q \in \{l \,|\, (l \geq r + 1) \wedge (l + L \leq M) \wedge (l + L \leq rL + 1)\} \quad (23)$$

In addition, we get the following stronger condition by directly forcing the value of $e_{v,r,q}$ variables,

$$p_{q,r} \rightarrow \neg e_{v,r,q+L} \quad \begin{array}{c} q \in \{l \,|\, (l \geq r + 1) \wedge (l + L \leq M) \wedge (l + L \leq rL + 1)\} \\ v \in [L] \end{array} \quad (24)$$

Furthermore, we can exploit Property 3, and the imposed ordering of the symbols in the BFS to identify a similar extension to (24) as follows,

$$p_{q,r} \rightarrow \neg e_{v,r,q+j} \quad \begin{array}{c} r + 1 \leq q \leq \min(rL + 1, M) \\ j \in \{l \,|\, l \in [L - 1] \wedge (q + l \leq M) \wedge (q + l \leq rL + 1)\} \\ v \in [j] \end{array} \quad (25)$$

Minimum BFS Distance. Given the way the BFS vertices are visited, one can guarantee a minimum BFS shortest path distance for each state. For state q, the shortest BFS path length is given by $D_{\min}(q) = \lceil \log_L (q(L - 1) + 1) - 1 \rceil$, with $q > 1$, i.e. no matter how the BFS is organized starting at state 1, the shortest path from 1 to q is never less than $D_{\min}(q)$. As a result, if $D_{\min}(q) > \theta(i)$, then $m_{i,q} = 0$. Observe that, under any possible setting in the DFA, the shortest path to q is larger than the distance to state i in the APTA. Thus, to get to q it would require more transitions that those allowed to get from inital state to i.

Exploiting APTA Information. By exploiting the variables and constraints used for breaking symmetries and using a BFS tree on the target DFA, we can devise additional constraints. Observe that, if the depth of a state i in the APTA is some value K, then in the DFA, we *must* be able to move from 1 to q in K of fewer transitions. However, if the shortest path from 1 to q in the DFA exceeds the depth K of vertex of i in the APTA, then it would be impossible to move from state 1 to state q in K or fewer transitions.

We consider the propositional variables $d_{q,j}$, with $q \in [M]$ and $1 \le j < q$, such that $d_{q,j} = 1$ iff the length of the shortest path in the BFS tree from state 1 to q is j. Moreover, we consider propositional variables $se_{q,j}$, with $q \in [M]$ and $1 \le j < q$, such that $se_{q,j} = 1$ iff the length of the shortest path in the BFS tree from state 1 to q is *smaller* than or *equal* to j. We can use an inductive definition for $se_{q,j}$ as follows:

$$se_{q,0} \leftrightarrow 0 \quad \text{and} \quad se_{q,j} \leftrightarrow se_{q,j-1} \vee d_{q,j} \tag{26}$$

Similarly to Sect. 3.2, we devise a tight encoding for the definition of the $d_{q,j}$ variables, suitable for larger problem instances. The insight is to introduce additional variables, which are inductively defined. Let $er_{q,r,j}$ be such that $er_{q,r,j} = 1$ iff there *exists* some index $r < q$ such that $p_{q,r} = 1$ and $d_{r,j} = 1$.

$$er_{q,r,j} \leftrightarrow p_{q,r} \wedge d_{r,j} \vee er_{q,r+1,j} \quad j < r < q - 1 \tag{27}$$
$$er_{q,q-1,j} \leftrightarrow p_{q,q-1} \wedge d_{q-1,j}$$

we can now derive constraints on the $m_{i,p}$ variables. Let t_i be a state of the APTA such that the depth of t_i is I. We can define $d_{q,j}$ as follows:

$$d_{q,j} \leftrightarrow \neg se_{q,j-1} \wedge er_{q,j,j-1} \tag{28}$$
$$\neg se_{q,I} \rightarrow \neg m_{i,q} \tag{29}$$

One can conclude that the modified constraints have an encoding size in $\mathcal{O}(N \times M^2)$.

4   Experimental Results

This section evaluates the ideas described above, namely a compact SAT encoding and symmetry breaking predicates for solving the MinDFA problem. For this, the ideas were implemented on top of a known MinDFA solver called *DFA-Inductor* [26, 30] written in Java[2]. The new prototype is referred to as *DFA-Inductor 2*. For comparison, two competitors were considered: the original DFA-Inductor and also *dfasat* [13]. All the selected tools apply the Glucose 4.1[3] SAT solver iteratively and *non-incrementally*, i.e. each call to the oracle is made *from scratch*. All the conducted experiments were performed in Ubuntu Linux on an AMD Opteron 6378 2.40 GHz processor with 496GByte of memory. For

---

[2] https://github.com/ctlab/DFA-Inductor.
[3] http://www.labri.fr/perso/lsimon/glucose/.

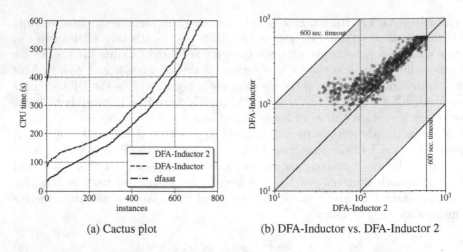

(a) Cactus plot   (b) DFA-Inductor vs. DFA-Inductor 2

Fig. 2. Detailed performance of dfasat, DFA-Inductor, and DFA-Inductor 2

each individual process, the time limit was set to 600 s and the memory limit to
1 GByte. For the comparison, a number of benchmark instances were randomly
generated, following the procedure described in [30]. Concretely, starting from a
randomly generated APTA of *even* size N, $N \in [20, 36]$, $50 \times N$ samples were
generated. The size of the Σ is two. For each even number $N \in [20, 36]$, exactly
100 benchmark instances were created such that given value N, the resulting
DFA for each of the corresponding 100 instances is guaranteed to be N. This
way, the number of benchmark families defined by values N is 9. Thus, the total
number of instances considered is 900. Figure 2a shows a cactus plot depicting the
performance of all the selected solvers. As one can observe, dfasat is significantly
outperformed by the compact encoding implemented in DFA-Inductor. In total,
dfasat is able to solve only 51 benchmark instances (out of 900). Also observe that
the symmetry breaking predicates described above further improve the performance of DFA-Inductor (see DFA-Inductor 2 compared to DFA-Inductor in the
Fig. 2a). A comparison between DFA-Inductor and DFA-Inductor 2 is detailed
in Fig. 2b and also in Table 2. Except for a few outliers, the symmetry breaking
predicates of DFA-Inductor 2 are responsible for 20–40% performance improvement on average. Also it is important to note that the harder the problems are,
the smaller is the performance gap between the two configurations. Although
this can be seen as a drawback, the phenomenon requires further investigation
on the use of symmetry breaking with various SAT solvers and a multitude
of families benchmark sets. In total, the number of instances solved by DFA-
Inductor and DFA-Inductor 2 is 678 and 731, respectively, thus, comprising a
gap of 53 benchmark instances. Therefore, symmetry breaking brings more 7.2%
instances solved.

Table 2. The effect of applying the symmetry breaking predicates described above. The solver configuration using the proposed symmetry breaking is referred to as *DFA-Inductor 2* and compared to the base configuration, i.e. *DFA-Inductor*. If an instance is timed out, its contribution to the average time of the corresponding benchmark family is assumed to be 600 s. The corresponding values are written in *italic*.

N	DFA-Inductor				DFA-Inductor 2			
	min	*avg*	*max*	*# solved*	*min*	*avg*	*max*	*# solved*
20	86.8	148.3	221.0	**100**	33.3	91.9	228.4	**100**
22	85.5	*147.1*	—	**99**	49.2	*100.4*	—	**99**
24	128.6	181.5	287.8	**100**	80.4	136.8	262.5	**100**
26	158.1	*251.8*	—	**99**	114.8	*209.3*	—	**99**
28	223.4	317.9	534.5	**100**	164.2	*268.9*	—	**99**
30	307.2	*443.8*	—	**91**	227.1	*389.2*	—	**95**
32	326.0	*506.5*	—	**76**	249.2	*447.4*	—	**86**
34	414.5	*591.1*	—	**13**	392.1	*569.9*	—	**41**
36	—	*600.0*	—	**0**	448.4	*594.8*	—	**12**

5 Conclusions

This paper proposes a number of novel techniques for encoding and reasoning about symmetries when exploiting SAT oracles for inferring minimum-size DFAs. The experimental results provide evidence of the improvements that can be achieved when compared with the state of the art [26,30], also enabling significant gains over the best exact methods proposed in recent years [13]. The novel symmetry-breaking ideas described in the paper can be applied to other approaches for inferring minimum-size DFAs, including the use of SMT solvers [20], and also in other settings.

References

1. Abela, J., Coste, F., Spina, S.: Mutually compatible and incompatible merges for the search of the smallest consistent DFA. In: ICGI, pp. 28–39 (2004)
2. Angluin, D.: Learning regular sets from queries and counterexamples. Inf. Comput. **75**(2), 87–106 (1987)
3. Asín, R., Nieuwenhuis, R., Oliveras, A., Rodríguez-Carbonell, E.: Cardinality networks: a theoretical and empirical study. Constraints **16**(2), 195–221 (2011)
4. Belov, A., Lynce, I., Marques-Silva, J.: Towards efficient MUS extraction. AI Commun. **25**(2), 97–116 (2012)
5. Biere, A., Heule, M., van Maaren, H., Walsh, T. (eds.): Handbook of Satisfiability, Frontiers in Artificial Intelligence and Applications, vol. 185. IOS Press, Amsterdam (2009)

6. Biermann, A.W., Feldman, J.A.: On the synthesis of finite-state machines from samples of their behavior. IEEE Trans. Comput. **21**(6), 592–597 (1972)
7. Bugalho, M.M.F., Oliveira, A.L.: Inference of regular languages using state merging algorithms with search. Pattern Recognit. **38**(9), 1457–1467 (2005)
8. Coste, F., Nicolas, J.: Regular inference as a graph coloring problem. In: IWGI (1997)
9. Coste, F., Nicolas, J.: How considering incompatible state mergings may reduce the DFA induction search tree. In: ICGI, pp. 199–210 (1998)
10. Eén, N., Sörensson, N.: Translating Pseudo-Boolean constraints into SAT. JSAT **2**(1–4), 1–26 (2006)
11. Gent, I.P., Nightingale, P.: A new encoding of all different into SAT. In: Workshop on Modelling and Reformulating Constraint Satisfaction Problems, pp. 95–110 (2004)
12. Grinchtein, O., Leucker, M., Piterman, N.: Inferring network invariants automatically. In: Furbach, U., Shankar, N. (eds.) IJCAR 2006. LNCS (LNAI), vol. 4130, pp. 483–497. Springer, Heidelberg (2006). https://doi.org/10.1007/11814771_40
13. Heule, M.J.H., Verwer, S.: Exact DFA identification using SAT solvers. In: Sempere, J.M., García, P. (eds.) ICGI 2010. LNCS (LNAI), vol. 6339, pp. 66–79. Springer, Heidelberg (2010). https://doi.org/10.1007/978-3-642-15488-1_7
14. Heule, M., Verwer, S.: Software model synthesis using satisfiability solvers. Empir. Softw. Eng. **18**(4), 825–856 (2013)
15. de la Higuera, C.: A bibliographical study of grammatical inference. Pattern Recognit. **38**(9), 1332–1348 (2005)
16. Hopcroft, J.E., Motwani, R., Ullman, J.D.: Introduction to Automata Theory, Languages, and Computation - International Edition, 2nd edn. Addison-Wesley, Boston (2003)
17. Lang, K.J.: Faster algorithms for finding minimal consistent DFAs. Technical report, NEC Research Institute (1999)
18. Lang, K.J., Pearlmutter, B.A., Price, R.A.: Results of the Abbadingo one DFA learning competition and a new evidence-driven state merging algorithm. In: Honavar, V., Slutzki, G. (eds.) ICGI 1998. LNCS, vol. 1433, pp. 1–12. Springer, Heidelberg (1998). https://doi.org/10.1007/BFb0054059
19. Morgado, A., Heras, F., Liffiton, M.H., Planes, J., Marques-Silva, J.: Iterative and core-guided MaxSAT solving: a survey and assessment. Constraints **18**(4), 478–534 (2013)
20. Neider, D.: Applications of automata learning in verification and synthesis. Ph.D. thesis, RWTH Aachen University (2014)
21. Neider, D., Jansen, N.: Regular model checking using solver technologies and automata learning. In: NFM, pp. 16–31 (2013)
22. Oliveira, A.L., Marques-Silva, J.: Efficient algorithms for the inference of minimum size DFAs. Mach. Learn. **44**(1/2), 93–119 (2001)
23. Sinz, C.: Towards an optimal CNF encoding of Boolean cardinality constraints. In: van Beek, P. (ed.) CP 2005. LNCS, vol. 3709, pp. 827–831. Springer, Heidelberg (2005). https://doi.org/10.1007/11564751_73
24. Trakhtenbrot, B.A., Barzdin, Y.M.: Finite Automata: Behavior and Synthesis. North-Holland Publishing Company, Amsterdam (1973)
25. Ulyantsev, V., Tsarev, F.: Extended finite-state machine induction using SAT-solver. In: ICMLA, pp. 346–349 (2011)

26. Ulyantsev, V., Zakirzyanov, I., Shalyto, A.: BFS-based symmetry breaking predicates for DFA identification. In: Dediu, A.-H., Formenti, E., Martín-Vide, C., Truthe, B. (eds.) LATA 2015. LNCS, vol. 8977, pp. 611–622. Springer, Cham (2015). https://doi.org/10.1007/978-3-319-15579-1_48
27. Verwer, S., Hammerschmidt, C.A.: flexfringe: a passive automaton learning package. In: ICSME, pp. 638–642 (2017)
28. Walkinshaw, N., Lambeau, B., Damas, C., Bogdanov, K., Dupont, P.: STAMINA: a competition to encourage the development and assessment of software model inference techniques. Empir. Softw. Eng. **18**(4), 791–824 (2013)
29. Wieman, R., Aniche, M.F., Lobbezoo, W., Verwer, S., van Deursen, A.: An experience report on applying passive learning in a large-scale payment company. In: ICSME, pp. 564–573 (2017)
30. Zakirzyanov, I., Shalyto, A., Ulyantsev, V.: Finding all minimum-size DFA consistent with given examples: SAT-based approach. In: Cerone, A., Roveri, M. (eds.) SEFM 2017. LNCS, vol. 10729, pp. 117–131. Springer, Cham (2018). https://doi.org/10.1007/978-3-319-74781-1_9

Complexity

Closure and Nonclosure Properties of the Compressible and Rankable Sets

Jackson Abascal$^{(\boxtimes)}$, Lane A. Hemaspaandra, Shir Maimon, and Daniel Rubery

Department of Computer Science, University of Rochester,
Rochester, NY 14627, USA
jabascal@u.rochester.edu

Abstract. The rankable and compressible sets have been studied for more than a quarter of a century, ever since Allender [2] and Goldberg and Sipser [7] introduced the formal study of polynomial-time ranking. Yet even after all that time, whether the rankable and compressible sets are closed under the most important boolean and other operations remains essentially unexplored. The present paper studies these questions for both polynomial-time and recursion-theoretic compression and ranking, and for almost every case arrives at a Closed, a Not-Closed, or a Closed-Iff-Well-Known-Complexity-Classes-Collapse result for the given operation. Even though compression and ranking classes are capturing something quite natural about the structure of sets, it turns out that they are quite fragile with respect to closure properties, and many fail to possess even the most basic of closure properties. For example, we show that with respect to the join (aka disjoint union) operation: the P-rankable sets are not closed, whether the semistrongly P-rankable sets are closed is closely linked to whether $P = UP \cap coUP$, and the strongly P-rankable sets are closed.

Keywords: Complexity theory · Closure properties · Compression ·
Ranking · Computability

1 Introduction

A compression function f for a set A is a function over the domain Σ^* such that (a) $f(A) = \Sigma^*$ and (b) $(\forall a, b \in A : a \neq b)[f(a) \neq f(b)]$. That is, f puts A in 1-to-1 correspondence with Σ^*. This is sometimes described as providing a minimal perfect hash function for A: It is perfect since there are no collisions (among elements of A), and it is minimal since not a single element of the codomain is missed. Note that the above does not put any constraints on what strings the elements of $\overline{A}$ are mapped to, or even about whether the compression function needs to be defined on such strings. A ranking function is similar, yet stronger, in that a ranking function sends the ith string in A to the integer i; it respects the ordering of the members of A.

Supported in part by a CRA-W Collaborative Research Experiences for Undergraduates (CREU) grant. S. Maimon's current affiliation: Cornell CS.

© Springer Nature Switzerland AG 2019
C. Martín-Vide et al. (Eds.): LATA 2019, LNCS 11417, pp. 177–189, 2019.
https://doi.org/10.1007/978-3-030-13435-8_13

The study of ranking was started by Allender [2] and Goldberg and Sipser [7], and has been pursued in many papers since, especially in the early 1990 s, e.g., [3,6,10,14]. The study of ranking led to the study of compression, which was started—in its current form, though already foreshadowed in a notion of [7]— by Goldsmith, Hemachandra, and Kunen [8] (see also [9]). The abovementioned work focused on polynomial-time or logarithmic-space ranking or compression functions. More recently, both compression and ranking have also been studied in the recursion-theoretic context ([12], and see the discussion therein for precursors in classic recursive function theory), in particular for both the case of (total) recursive compression/ranking functions (which of course must be defined on all inputs in Σ^*) and the case of partial-recursive compression/ranking functions (i.e., functions that on some or all elements of the complement of the set being compressed/ranked are allowed to be undefined).

In the present paper, we continue the study of both complexity-theoretic and recursion-theoretic compression and ranking functions. In particular, the earlier papers often viewed the compressible sets or the rankable sets as a *class*. Our main contributions can be seen in Table 1, where we obtain closure and nonclosure results for many previously studied variations of compressible and rankable

Table 1. Overview of results for closure of these classes under boolean operations. If an entry does not contain "No" or "Yes" then the class is closed under the operation if and only if the entry holds. A special case is semistrong-P-rankable and semistrong-P-rankable[C], in which we deliberately use the $\approx$ symbol to indicate that the implication is true in one direction and in the other direction currently is known to be true only for a broad subclass of these sets. Specifically, if $P = UP \cap coUP$ then the complements of all "nongappy" semistrong-P-rankable sets are themselves semistrong-P-rankable.

Class	$\cap$	$\cup$	Complement
strong-P-rankable	$P = P^{\#P}$ (Theorem 4.2)	$P = P^{\#P}$ (Theorem 4.2)	Yes (Proposition 4.3)
semistrong-P-rankable	$P = P^{\#P}$ (Theorem 4.2)	$P = P^{\#P}$ (Theorem 4.2)	$\approx P = UP \cap coUP$ (Theorem 4.5, Corollary 4.8)
P-rankable, P-compressible', F_{REC}-rankable, F_{REC}-compressible, F_{PR}-rankable, and F_{PR}-compressible	No (Theorem 4.9)	No (Theorem 4.10)	No (Theorem 4.11)
strong-P-rankable[C]	No (Theorem 4.12)	No (Theorem 4.12)	Yes (Proposition 4.3)
semistrong-P-rankable[C]	No (Theorem 4.12)	No (Theorem 4.12)	$\approx P = UP \cap coUP$ (Theorem 4.5, Corollary 4.8)
P-rankable[C], P-compressible[C], F_{REC}-rankable[C], F_{REC}-compressible[C], F_{PR}-rankable[C], and F_{PR}-compressible[C]	No (Theorem 4.12)	No (Theorem 4.12)	No (Theorem 4.11)

sets under boolean operations (Sect. 4). We also study the closure of these sets under additional operations, such as the join, aka disjoint union (Sect. 5). And we introduce the notion of compression *onto a set* and characterize the robustness of compression under this notion. In particular, by a finite-injury priority argument with some interesting features we show that there exist RE sets that each compress to the other, yet that nonetheless are not recursively isomorphic (Sect. 3).

2 Definitions

Throughout this paper, "P" when used in a function context (e.g., the P-rankable sets) will denote the class of total, polynomial-time computable functions from Σ^* to Σ^*. Additionally, throughout this paper, $\Sigma = \{0, 1\}$. F_{REC} will denote the class of total, recursive functions from Σ^* to Σ^*. F_{PR} will denote the class of partial recursive functions from Σ^* to Σ^*. ϵ will denote the empty string. We define the function shift(x, n) for $n \in \mathbb{Z}$. If $n \geq 0$, then shift(x, n) is the string n spots after x in lexicographical order, e.g., shift$(\epsilon, 4) = 01$. For $n > 0$, define shift$(x, -n)$ as the string n spots before x in lexicographical order, or ϵ if no such string exists. The symbol $\mathbb{N}$ will denote the natural numbers $\{0, 1, 2, 3, \dots\}$.

We now define the notion of compression onto a set, and subsequently use it to define the classical notion of compressible sets from [12].

Definition 2.1 (Compressible to B)

1. *Given sets $A \subseteq \Sigma^*$ and $B \subseteq \Sigma^*$, a (possibly partial) function f is a com-pression function for A to B exactly if*
   (a) *domain$(f) \supseteq A$,*
   (b) *$f(A) = B$, and*
   (c) *for all a and b in A, if $a \neq b$ then $f(a) \neq f(b)$.*
2. *Let $\mathcal{F}$ be any class of (possibly partial) functions mapping from Σ^* to Σ^*. A set A is $\mathcal{F}$-compressible to B if some $f \in \mathcal{F}$ is a compression function for A to B.*

Note that a compression function for A to B may have any behavior on elements not in A. In particular, for the case of F_{PR}, the function need not even be defined outside of A. We now use the general notion of compressibility to B to define compressible sets.

A set is *compressible* if it is compressible to Σ^*. A compression function for A to Σ^* is simply called a *compression function* for A. We define the class $\mathcal{F}$-compressible $= \{A \mid A \text{ is } \mathcal{F}\text{-compressible}\}$. Note that no finite sets are compressible, as they do not contain enough elements to map onto Σ^*. So that finite sets do not affect whether we consider a class of sets to be compressible, we define $\mathcal{F}$-compressible' as the union of $\mathcal{F}$-compressible with the class of finite sets and say that a class of sets $\mathcal{C}$ is $\mathcal{F}$-compressible if $\mathcal{C} \subseteq \mathcal{F}$-compressible'.

Ranking can be informally thought of as a sibling of compression that preserves lexicographical order within the set. We consider three classes of rankable functions that differ in how they are allowed to behave on the complement of the set they rank. Although ever since the paper of Hemachandra and Rudich [10], which introduced two of the three types, there have been those three types of ranking classes, different papers have used different (and sometimes conflicting) terminology for these types. Here, we use the (without modifying adjective) terms "ranking function" and "rankable" in the same way as Hemaspaandra and Rubery [12] do, for the least restrictive form of ranking (the one that can even "lie" on the complement). That is the form of ranking that is most naturally analogous with compression, and so it is natural that both terms should lack a modifying adjective. For the most restrictive form of ranking, which even for strings x in the complement of the set A being ranked must determine the number of strings up to x that are in A, like Hemachandra and Rudich [10] we use the terms "strong ranking function" and "strong(ly) rankable." And for the version of ranking that falls between those two, since for strings in the complement it need only detect that they are in the complement, we use the terms "semistrong ranking function" and "semistrong(ly) rankable."

For a set $A \subseteq \Sigma^*$, we define $\mathrm{rank}_A(y) = \|\{z \mid z \leq y \land z \in A\}\|$ [2,7]. A ranking function for A is a (possibly partial) compression function for A which preserves lexicographical order of elements in A. In other words, it sends the ith string in A to the ith string in Σ^*, ordered lexicographically. If we identify the ith string in Σ^* with the natural number i, we may alternatively define a ranking function f for A as any (possibly partial) function such that $f(x) = \mathrm{rank}_A(x)$ for $x \in A$. We adopt this perspective, and for the remainder of this paper identify the codomain of ranking functions with $\mathbb{N}$, allowing arithmetic operations to be performed. A semistrong ranking f function for A is a ranking function for A such that for $x \notin A$, $f(x)$ indicates "not in set" (e.g., via the machine computing f halting in a special state; we still view this as a case where x belongs to domain(f)). A strong ranking function f is a ranking function such such that $f(x) = \mathrm{rank}_A(x)$ for all inputs x, not just those in A.

Let $\mathcal{F}$ be any class of (possibly partial) functions from $\Sigma^* \to \Sigma^*$. A set A is $\mathcal{F}$-rankable if some $f \in \mathcal{F}$ is a ranking function for A. We define $\mathcal{F}$-rankable as $\{A \mid A \text{ is } \mathcal{F}\text{-rankable}\}$. A class of sets $\mathcal{C}$ is $\mathcal{F}$-rankable if all sets in $\mathcal{C}$ are $\mathcal{F}$-rankable. Analogous definitions hold for both the strong and semistrong ranking cases: A set A is (semi)strong-$\mathcal{F}$-rankable if some $f \in \mathcal{F}$ is a (semi)strong ranking function for A. We define (semi)strong-$\mathcal{F}$-rankable as $\{A \mid A \text{ is (semi)strong-}\mathcal{F}\text{-rankable}\}$. A class of sets $\mathcal{C}$ is (semi)strong-$\mathcal{F}$-rankable if all sets in $\mathcal{C}$ are (semi)strong-$\mathcal{F}$-rankable.

For almost any natural class of functions, $\mathcal{F}$, we will have that $\mathcal{F}$-rankable $\subseteq \mathcal{F}$-compressible'. In particular, P, $\mathrm{F_{PR}}$, and $\mathrm{F_{REC}}$ each have this property. If f is a ranking function for A, for our same-class compression function for A we can map $x \in \Sigma^*$ to the $f(x)$-th string in Σ^* (where we consider ϵ to be the first string in Σ^*) if $f(x) > 0$, and if $f(x) = 0$ what we map to is irrelevant so map to any particular fixed string (for concreteness, ϵ). For each class $\mathcal{C} \subseteq 2^{\Sigma^*}$, $\mathcal{C}^\complement$ will denote the complement of $\mathcal{C}$, i.e., $2^{\Sigma^*} - \mathcal{C}$. For example, P-rankable$^\complement$ is the class of non-P-rankable sets.

The class semistrong-P-rankable is a subset of P (indeed, a strict subset unless $P = P^{\#P}$ [10]), but there exist undecidable sets that are P-rankable. Clearly, semistrong-REC-rankable = strong-REC-rankable.

3 Compression onto B: Robustness with Respect to Target Set

A natural question to ask is whether compression to B is a new notion, or whether it coincides with our existing notion of compression to Σ^*, at least for sets B from common classes such as REC and RE. The following result shows that for REC and RE this new notion does coincide with our existing one.

Theorem 3.1. *Let A and B be infinite sets.*

1. *If $B \in$ REC, then A is F_{REC}-compressible to B if and only if A is F_{REC}-compressible to Σ^*.*
2. *If $B \in$ RE, then A is F_{PR}-compressible to B if and only if A is F_{PR}-compressible to Σ^*.*

Theorem 3.1 covers the two most natural pairings of set classes with function classes: recursive sets B with F_{REC} compression, and RE sets B with F_{PR} compression. What about pairing recursive sets under F_{PR} compression, or RE sets under recursive compression? We note as the following theorem that one and a half of the analogous statements hold, but the remaining direction fails.

Theorem 3.2. *1. Let A and B be infinite sets and suppose that $B \in$ REC. Then A is F_{PR}-compressible to B if and only if A is F_{PR}-compressible to Σ^*.*
2. *Let A and B be infinite sets with $B \in$ RE. If A is F_{PR}-compressible to Σ^*, then A is F_{PR}-compressible to B. In fact, we may even require that the compression function for A to B satisfies $f(\Sigma^*) = B$.*
3. *There are infinite sets A and B with $B \in$ RE such that A is F_{REC}-compressible to B but A is not F_{REC}-compressible to Σ^*.*

Proof. The first part follows immediately from Theorem 3.1, part 2. The second part follows as a corollary to the proof of Theorem 3.1, part 2. In particular, the proof of the "$\Leftarrow$" direction proves the second part, since it is clear that if f is a recursive function the f' defined there is also recursive.

The third part follows from [12] in which it is shown that any set in RE $-$ REC is not F_{REC}-compressible to Σ^*. Thus if we let $A = B$ be any set in RE $-$ REC, then A is F_{REC}-compressible to B by the function $f(x) = x$ but B is not F_{REC}-compressible to Σ^*.   □

Another interesting question is how recursive compressibility to B is, or is not, linked to recursive isomorphism. Recall that two sets A and B are recursively isomorphic (notated $A \equiv_{iso} B$) if there exists a recursive bijection $f : \Sigma^* \to \Sigma^*$ with $f(A) = B$. Although recursive isomorphism of sets implies mutual compressibility to each other, we prove via a finite-injury priority argument that the converse does not hold (even when restricted to the RE sets). The argument has an interesting graph-theoretic flavor, and involves queuing infinitely many strings to be added to a set at once.

Theorem 3.3. *If $A \equiv_{iso} B$, then A is F_{REC}-compressible to B and B is F_{REC}-compressible to A.*

Theorem 3.4. *There exist RE sets A and B such that A is F_{REC}-compressible to B and B is F_{REC}-compressible to A, yet $A \not\equiv_{iso} B$.*

Proof of Theorem 3.3. Now A is F_{REC}-compressible to B by simply letting our F_{REC}-compression function be the recursive isomorphism function f. Since each recursive isomorphism has a recursive inverse, B is F_{REC}-compressible to A by letting our F_{REC}-compression function be the inverse of f.   □

Proof of Theorem 3.4. Before defining A and B, we will define a function f which will serve as both a compression function from A to B and a compression function from B to A. First, fix a recursive isomorphism between Σ^* and $\{\langle t, j, k \rangle \mid t \in \{0, 1, 2, 3\} \wedge j, k \in \mathbb{N}\}$. Now we will define f as follows. For each $j, k \in \mathbb{N}$, let $f(\langle 3, j, k \rangle) = \langle 3, j + 1, k \rangle$. For each $j, k \in \mathbb{N}$, $j > 0$, and $t \in \{0, 1, 2\}$, let $f(\langle t, j, k \rangle) = \langle t, j - 1, k \rangle$. Finally, for each $k \in \mathbb{N}$, let $f(\langle 0, 0, k \rangle) = \langle 3, 0, k \rangle$, $f(\langle 1, 0, k \rangle) = \langle 0, 0, k \rangle$, and $f(\langle 2, 0, k \rangle) = \langle 3, 0, k \rangle$. Let $\ell : \Sigma^* \to \{0, 1\}$ be the unique function such that $\ell(\langle 0, 0, k \rangle) = 0$ for all $k \in \mathbb{N}$ and $\ell(f(x)) = 1 - \ell(x)$. Let D_f be the directed graph with edges $(x, f(x))$. Note that ℓ is a 2-coloring of D_f if we treat the edges as being undirected. See Fig. 1.

Call a set C a *path set* if for all $x \in C$, $f(x) \in C$ and there is exactly one $y \in C$ such that $f(y) = x$. Suppose C is a path set. Let $C_i = \{x \in C \mid \ell(x) = i\}$ for $i \in \{0, 1\}$. By the assumed property of C, we have C_0 and C_1 are F_{REC}-compressible to each other by f. Furthermore, if C is RE then so are C_0 and C_1 since $C_i = C \cap \{x \mid \ell(x) = i\}$ is the intersection of an RE set with a recursive set. Thus if we provide an enumerator for a path set C such that $C_0 \not\equiv_{iso} C_1$, we may let $A = C_0$ and $B = C_1$ and be done.

Our enumerator for C proceeds in two interleaved types of stages: printing stages P_i and evaluation stages E_i. More formally, we proceed in stages labeled E_i and P_i for $i \geq 1$, interleaved as $E_1, P_1, E_2, P_2, \ldots, E_n, P_n, \ldots$ when running. We also maintain a set Q of elements of the form $\langle t, k \rangle$, where $t \in \{0, 1, 2\}$ and $k \in \mathbb{N}$. This set Q will only ever be added to as the procedure runs.

In the printing stage P_i, we do the following for every $\langle t, k \rangle$ in Q. Enumerate $\langle 3, j, k \rangle$ and $\langle t, j, k \rangle$ for all $j \leq i$. If $t = 1$, additionally enumerate $\langle 0, 0, k \rangle$.

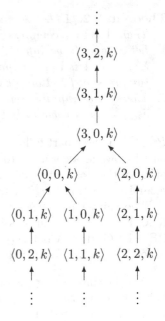

Fig. 1. A diagram of D_f, for fixed k.

Adding an element $\langle t, k \rangle$ to Q in some evaluation stage E_i is essentially adding an infinite path of nodes in D_f to C.

In addition to Q, we also maintain an integer b and a set R of elements $\langle n, k \rangle$ where $n, k \in \mathbb{N}$. If $\langle n, k \rangle \in R$ after stage i, it signifies that we have not yet satisfied the condition that φ_n, the nth partial recursive function, is not an isomorphism function between C_0 and C_1. In stage E_i we perform the following. Add $\langle i, b \rangle$ to R. Increment b by one. For each $\langle n, k \rangle \in R$, run φ_n, the nth partial recursive function, on $\langle 0, 0, k \rangle$ for i steps. If none of these machines halt in their allotted time, end the stage. Otherwise, let n_i be the smallest number such that φ_{n_i} produced an output $w_i = \langle x_i, y_i, z_i \rangle$ on its respective input $\langle 0, 0, k_i \rangle$. We now break into cases:

1. If $\ell(w_i) = 0$ add $\langle 0, k_i \rangle$ to Q.
2. If $z_i \neq k_i$ and $\ell(w_i) = 1$ and as it stands w_i would not be printed eventually if there were only type P stages from now on, add $\langle 0, k_i, \rangle$ to Q.
3. If $z_i \neq k_i$ and $\ell(w_i) = 1$ and as it stands w_i would be printed eventually if there were only type P stages from now on, do nothing.
4. If $z_i = k_i$ and $\ell(w_i) = 1$ and $x_i = 0$, add $\langle 1, k_i \rangle$ to Q.
5. If $z_i = k_i$ and $\ell(w_i) = 1$ and either $x_i = 1$ or $x_i = 2$, add $\langle 0, k_i \rangle$ to Q.
6. If $z_i = k_i$ and $\ell(w_i) = 1$ and $x_i = 3$, add $\langle 2, k_i \rangle$ to Q.

Set $b = \max(k_i, z_i) + 1$. Remove all pairs $\langle n, k \rangle$ with $n \geq n_i$ from R. Then for each n from $n_i + 1$ to i, first add $\langle n, b \rangle$ and subsequently increment b by 1.

We will first prove that C is a path set. If $x \in C$, then it is printed in some printing stage P_i. By tracing the definition of f and the procedure for printing stages, one can verify that both $f(x)$ and exactly one y such that $f(y) = x$ will be printed in stage P_j for $j \geq i$. This string y will be the only one ever printed, since no two elements with the same second coordinate will ever be added to Q, as every element added to Q has the current state of b as its second coordinate, and b only ever strictly increases between additions to Q.

Let F_n be the condition that φ_n fails to be a recursive isomorphism of C_0 onto C_1. Fix n. Say during E_i we have $n_i = n$. In cases 1, 2, 4, and 5, we force φ_n to map $\langle 0, 0, k_i \rangle \in C_0$ to something out of C_1. In cases 3 and 6, we force φ_n to map $\langle 0, 0, k_i \rangle \notin C_0$ to something in C_1. Thus whenever at stage i we have $n_i = n$, condition F_n becomes satisfied, though perhaps not permanently. Specifically, in case 2, w could be printed later to satisfy some other F_m and in doing so "injure" F_n. However, note that during E_i the variable b is set to $\max(k_i, z_i)$, thus F_n can only be injured when satisfying conditions F_m for $m < n$. Pairs with first coordinate n will only ever be added to R when after satisfying some such F_m, in addition to once initially, so in total only a finite number of times. If φ_n always halts, F_n will eventually be satisfied and never injured again.

This proves that C is a path set such that $C_0 \neq_{iso} C_1$. Thus C_0 and C_1 are RE sets that are F_{REC}-compressible to each other by f, but are not recursively isomorphic. □

For those interested in the issue of isomorphism in the context of complexity-theoretic functions, which was not the focus above, we mention that: Hemaspaandra, Zaki, and Zimand [13] prove that the P-rankable sets are not closed under $\equiv_{iso}^p$; Goldsmith and Homer [9] prove that the strong-P-rankable sets are closed under $\equiv_{iso}^p$ if and only if $P = P^{\#P}$; and [13] notes that the semistrong-P-rankable sets similarly are closed under $\equiv_{iso}^p$ if and only if $P = P^{\#P}$.

4   Closures and Nonclosures Under Boolean Operations

We now move on to a main focus of this paper, the closure properties of the compressible and the rankable sets. We explore these properties both in the complexity-theoretic and the recursion-theoretic domains. Table 1 (which appears in Sect. 1) summarizes our findings.

Lemma 4.1. *Let A and B be strong-P-rankable. Then $A \cup B$ is strong-P-rankable if and only if $A \cap B$ is.*

Proof. The identity $\mathrm{rank}_{A \cap B}(x) + \mathrm{rank}_{A \cup B} = \mathrm{rank}_A(x) + \mathrm{rank}_B(x)$ allows us to compute either of $\mathrm{rank}_{A \cap B}(x)$ or $\mathrm{rank}_{A \cup B}(x)$ from the other.   □

Theorem 4.2. *The following conditions are equivalent:*

1. *the classes strong-P-rankable and semistrong-P-rankable are closed under intersection,*
2. *the classes strong-P-rankable and semistrong-P-rankable are closed under union, and*
3. $P = P^{\#P}$.

Proof. It was proven in [10] by Hemachandra and Rudich that $P = P^{\#P}$ implies $P = $ strong-P-rankable = semistrong-P-rankable. Since P is closed under intersection and union, this shows that 3 implies 1 and 2. To show, in light of Lemma 4.1, that either 1 or 2 would imply 3, we will construct two strong-P-rankable sets whose intersection is not P-rankable unless $P = P^{\#P}$.

Let A_1 be the set of $x1y1$ such that $|x| = |y|$, x encodes a boolean formula, and y (padded with 0s so that it has length $|x|$) encodes a satisfying assignment for the formula x. Let A_0 be the set of $x1y0$ such that $|x| = |y|$, and $x1y1 \notin A_1$. Let A_2 be the set of strings $x0^{|x|+1}1$. Let $A = A_0 \cup A_1 \cup A_2$. For every x, and every y such that $|x| = |y|$, exactly one of $x1y0$ and $x1y1$ is in A. Thus, for any x, we can find $\mathrm{rank}_{A_0 \cup A_1}(x)$ in polynomial time. Clearly A_2 is strong-P-rankable. Since $A_0 \cup A_1$ and A_2 are disjoint, $\mathrm{rank}_{A_0 \cup A_1 \cup A_2}(x) = \mathrm{rank}_{A_0 \cup A_1}(x) + \mathrm{rank}_{A_2}(x)$, so A is strong-P-rankable.

Let $B = \Sigma^*1$. Then $A \cap B = A_1 \cup A_2$ is the set of $x1y1$ such that y encodes a satisfying assignment for x, along with all strings $x0^{|x|+1}1$. If $A_1 \cup A_2$ were P-rankable, then we could count satisfying assignments of a formula x in polynomial time by computing $\mathrm{rank}_{A \cap B}(\mathrm{shift}(x,1)0^{|\mathrm{shift}(x,1)|+1}1) - \mathrm{rank}_{A \cap B}(x0^{|x|+1}1) - 1$. Thus #SAT is polynomial-time computable and so $P = P^{\#P}$.   □

Proposition 4.3. *strong-P-rankable and strong-P-rankable$^{\complement}$ are closed under complementation.*

Proof. If A is strong-P-rankable, the identity $\text{rank}_A(x) + \text{rank}_{\overline{A}}(x) = \text{rank}_{\Sigma^*}(x)$ allows us to compute either of $\text{rank}_A(x)$ or $\text{rank}_{\overline{A}}(x)$ from the other.   □

Lemma 4.4. *The class semistrong-P-rankable is closed under complementation if and only if semistrong-P-rankable = strong-P-rankable.*

Proof. The "if" direction follows directly from Proposition 4.3. For the "only if" direction, let A be a semistrong-P-rankable set with ranking function r_A, and suppose $\overline{A}$ is semistrong-P-rankable with semistrong ranking function $r_{\overline{A}}$. Then $\text{rank}_A(x) = r_A(x)$ if $x \in A$, and equals $\text{rank}_{\Sigma^*}(x) - r_{\overline{A}}(x)$ otherwise. The function r_A decides membership in A, so we can compute $\text{rank}_A(x)$ in polynomial time.   □

Theorem 4.5. *If semistrong-P-rankable is closed under complementation, then* P = *UP∩ coUP.*

Proof. Suppose semistrong-P-rankable is closed under complementation. Let A be in UP $\cap$ coUP. Then there exists a UP machine U recognizing A, and a UP machine $\hat{U}$ recognizing $\overline{A}$. If $x \in A$, let $f(x)$ be the unique accepting path for x in U. Otherwise, let $f(x)$ be the unique accepting path for x in $\hat{U}$. Choose a polynomial p such that, without loss of generality, $p(x)$ is monotonically increasing and $|f(x)| = p(|x|)$ (we may pad accepting paths with 0s to make this true).

The language $B = \{xf(x)1 \mid x \in \Sigma^*\} \cup \{x0^{p(|x|)+1} \mid x \in \Sigma^*\}$ is semistrong-P-rankable since $\text{rank}_B(x0^{p(|x|)+1}) = 2\text{rank}_{\Sigma^*}(x) - 1$ and $\text{rank}_B(xf(x)1) = 2\text{rank}_{\Sigma^*}(x)$. Since semistrong-P-rankable is closed under complementation, and B is semistrong-P-rankable, B is also strong-P-rankable by Lemma 4.4. Let x be a string, and let $y = \text{shift}(x, 1)$. We can binary search on the value of rank_B in the range from $x0^{p(|x|)+1}$ to $y0^{p(|y|)+1}$ to find the first value xz where $|z| = p(|x|)+1$ and $\text{rank}_B(xz) = 2\text{rank}_{\Sigma^*}(x)$. See that $f(x)$ must equal z. We then simulate U on the path z and $\hat{U}$ on the path z. Now z must be an accepting path for one of these machines, so either U accepts and $x \in A$, or $\hat{U}$ accepts and $x \notin A$.   □

Definition 4.6. *A set is* nongappy *if there exists a polynomial p such that, for each $n \in \mathbb{N}$, there is some element $y \in A$ such that $n \leq |y| \leq p(n)$.*

Theorem 4.7. *If* P = *UP∩ coUP then each nongappy semistrong-P-rankable set is strong-P-rankable.*

Proof. Let A be a nongappy semistrong-P-rankable set, and let p be a polynomial such that, for each $n \in \mathbb{N}$, there is y in A such that $n \leq |y| \leq p(y)$. Let r be a polynomial-time semistrong ranking function for A. The coming string comparisons of course will be lexicographical. Let L be the set of $\langle x, b \rangle$ such that there exists at least one string in A that is less than or equal to x and b a prefix of the greatest string in A that is lexicographically less than or equal to x. L is in UP∩coUP by the following procedure. Let x_0 be the lexicographically first string in A. If $x < x_0$ output 0. Otherwise, guess a string $z > x$ such that $|z| \leq p(|x|+1)$. Then guess a $y \leq x$. If y and z are in A and $r(y)+1 = r(z)$, then we know that and y and z are the (unique) strings in A that most tightly bracket x in the $\leq$ and the $>$ directions. We can in our current case build the greatest string less than or equal to x that is in A bit by bit, querying potential prefixes, in polynomial time. Since $\text{rank}_A(x) = \text{rank}_A(y)$, we can compute $\text{rank}_A(x)$ in polynomial time for arbitrary x.   □

From Proposition 4.3 and Theorem 4.7, we obtain the following corollary.

Corollary 4.8. *If* P $=$ UP $\cap$ coUP *then the complement of each nongappy semistrong-P-rankable set is strong-P-rankable (and so certainly is semistrong-P-rankable).*

The proofs of the following four theorems can be found in the technical report version of this paper [1].

Theorem 4.9. *There exist P-rankable sets A and B such that $A \cap B$ is infinite but not F_{PR}-compressible.*

Theorem 4.10. *There exist infinite P-rankable sets A and B such that $A \cup B$ is not F_{PR}-compressible.*

Theorem 4.11. *There exists an infinite P-rankable set whose complement is infinite but not F_{PR}-compressible.*

Theorem 4.12. *There exist sets A and B that are not F_{PR}-compressible, yet $A \cup B$ is strong-P-rankable. In addition, there exist sets A and B that are not F_{PR}-compressible, yet $A \cap B$ is strong-P-rankable.*

5   Additional Closure and Nonclosure Properties

We focus on the join (aka disjoint union), giving a full classification of the closure properties of the P-rankable, semistrong-P-rankable, and strong-P-rankable sets, as well as their complements, under this operation. The literature is inconsistent as to whether the low-order or high-order bit is the "marking" bit for the join. Here, we follow the classic computability texts of Rogers [15] and Soare [17] and the classic structural-complexity text of Balcázar, Díaz, Gabarró [5], and define the join using low-order-bit marking: The join of A and B, denoted $A \oplus B$, is $A0 \cup B1 = \{x0 \mid x \in A\} \cup \{x1 \mid x \in B\}$. For classes invariant under reversal, which end is used for the marking bit is not important (in the sense that the class itself is closed under upper-bit-marked join if and only if it is closed under lower-bit-marked join). However, the placement of the marking bit potentially matters for ranking-based classes, as those classes rely on lexicographical order.

The join is such a basic operation that it seems very surprising that any class would not be closed under it, and it would be even more surprising if the join of two sets that lack some nice organizational property (such as being P-rankable) can itself have that property (can be P-rankable, and we indeed show in this section that that happens)—i.e., the join of two sets can be "simpler" than either of them (despite the fact that the join of two sets is the least upper bound for them with respect to $\leq_m^p$ [16], and in the sense of reductions captures the power-as-a-target of both sets). However, there is in the literature a precedent for the just-mentioned surprising behavior. It is known that $(\mathrm{EL}_2)^\complement$ is not closed under the join [11], where EL_2 is the second level of the extended low hierarchy [4].

Theorem 5.1. *If* $\mathrm{P} \neq \mathrm{P}^{\#P}$ *then there exist sets* $A \in \mathrm{P}$ *and* $B \in \mathrm{P}$ *that are not* P-*rankable yet* $A \cap B$, $A \cup B$, *and* $A \oplus B$ *are strong-*P-*rankable.*

Proof. We construct a set A_1 whose members represent satisfying assignments of boolean formulas. We force certain elements, or beacons, into A_1 to obtain a set A such that if A were rankable, we could count the number of satisfying assignments to a boolean formula using the ranks of the beacons. The set B is built similarly, but so that $A \cup B$, $A \cap B$, and $A \oplus B$ are strong-P-rankable.

Let $A_1 = \{\alpha 01\beta \mid \alpha, \beta \in \Sigma^* \wedge |\alpha| = |\beta| \wedge \alpha$ encodes a boolean formula F with (without loss of generality) $k \leq |\alpha|$ variables, the first k bits of β encode a satisfying assignment of F, and the remaining $|\beta| - k$ bits of β are 0$\}$. Given a string $x = \alpha 01\beta \in A_1$, we can unambiguously extract α and β because they have length $(|x| - 2)/2$. Let $B_1 = \{\alpha 01\beta \mid \alpha, \beta \in \Sigma^* \wedge |\alpha| = |\beta| \wedge \alpha 01\beta \notin A_1\}$. Let $Beacons = \{\alpha 000^{|\alpha|} \mid \alpha \in \Sigma^*\} \cup \{\alpha 110^{|\alpha|} \mid \alpha \in \Sigma^*\}$. Similarly to A_1, strings in B_1 and $Beacons$ can be parsed unambiguously. Let $A = A_1 \cup Beacons$ and $B = B_1 \cup Beacons$. Note A and B are both in P as A_1, B_1, and $Beacons$ are.

We will now demonstrate that if either A or B are P-rankable, then #SAT is polynomial-time computable. Suppose that A is P-rankable and let f be a polynomial-time ranking function for A. Let α be a string encoding a boolean formula F with k variables. Then we can compute $j = f(\alpha 110^{|\alpha|}) - f(\alpha 000^{|\alpha|})$ in polynomial time. Both $\alpha 110^{|\alpha|}$ and $\alpha 000^{|\alpha|}$ are in $Beacons$ and thus in A, so f gives a true ranking for these values. Every string in A strictly between these $Beacons$ strings is from A_1 and so represents a satisfying assignment for F, and every satisfying assignment for F is represented by a string between these $Beacons$ strings. The last $|\beta| - k$ bits of β are 0, so each satisfying assignment for F is represented exactly once between the two $Beacons$ strings. Thus F has $j - 1$ satisfying assignments. We can find j in polynomial time, so #SAT is polynomial-time computable and $\mathrm{P} = \mathrm{P}^{\#P}$, contrary to our $\mathrm{P} \neq \mathrm{P}^{\#P}$ hypothesis.

Now suppose that B is P-rankable and let f be a P-time ranking function for it. Let α be as before and $j = f(\alpha 110^{|\alpha|}) - f(\alpha 000^{|\alpha|})$. The strings in B between $\alpha 110^{|\alpha|}$ and $\alpha 000^{|\alpha|}$ are the strings of the form $\alpha 01\Sigma^{|\alpha|}$ except for those that are in A_1 (and recall that those that are in A_1 are precisely the padded-with-0s satisfying assignments for F). We know the number of strings of the form $\alpha 01\Sigma^{|\alpha|}$, so we can find the number of satisfying assignments for F. Namely, we have that $j = 1 + 2^{|\alpha|} - s$, where s is the number of satisfying assignments of F. Thus if B is P-rankable then #SAT is polynomial-time computable contrary to our $\mathrm{P} \neq \mathrm{P}^{\#P}$ hypothesis.

We omit the construction of strong-ranking functions for $A \cup B$, $A \cap B$, and $A \oplus B$, which can be found in the technical report version of this paper [1]. $\square$

Theorem 5.2. *The following are equivalent: (1) strong-P-rankable^C is closed under join, (2) semistrong-P-rankable^C is closed under join, and (3) $P = P^{\#P}$.*

Theorem 5.3. *(1) The class P-rankable^C is not closed under join.*
(2) The class P-rankable is not closed under join.
(3) The class strong-P-rankable is closed under join.
(4) The class semistrong-P-rankable is closed under complement if and only if it is closed under join.
(5) The class P-compressible′ is closed under join.

Proofs of Theorems 5.2 and 5.3 can be found in our technical report version [1].

6   Conclusions

Taking to heart the work in earlier papers that views as classes the collections of sets that have (or lack) rankability/compressibility properties, we have studied whether those classes are closed under the most important boolean and other operations. For the studied classes, we in almost every case prove that they are closed under the operation, or prove that they are not closed under the operation, or prove that whether they are closed depends on well-known questions about standard complexity classes. Additionally, we introduced compression onto a set and showed the robustness of compression under this notion, as well as the limits of that robustness.

Acknowledgments. We thank the anonymous referees for helpful comments.

References

1. Abascal, J., Hemaspaandra, L., Maimon, S., Rubery, D.: Closure and non-closure properties of the compressible and rankable sets. Technical report. arXiv:1611.01696 [cs.LO], Computing Research Repository, arXiv.org/corr/, November 2016. Revised, October 2018
2. Allender, E.: Invertible functions. Ph.D. thesis, Georgia Institute of Technology (1985)
3. Álvarez, C., Jenner, B.: A very hard log-space counting class. Theoret. Comput. Sci. **107**, 3–30 (1993)
4. Balcázar, J., Book, R., Schöning, U.: Sparse sets, lowness and highness. SIAM J. Comput. **15**(3), 739–746 (1986)
5. Balcázar, J., Díaz, J., Gabarró, J.: Structural Complexity I. EATCS Texts in Theoretical Computer Science, 2nd edn. Springer, Heidelberg (1995). https://doi.org/10.1007/978-3-642-79235-9
6. Bertoni, A., Goldwurm, M., Sabadini, N.: The complexity of computing the number of strings of given length in context-free languages. Theoret. Comput. Sci. **86**(2), 325–342 (1991)
7. Goldberg, A., Sipser, M.: Compression and ranking. SIAM J. Comput. **20**(3), 524–536 (1991)

8. Goldsmith, J., Hemachandra, L., Kunen, K.: Polynomial-time compression. Comput. Complex. **2**(1), 18–39 (1992)
9. Goldsmith, J., Homer, S.: Scalability and the isomorphism problem. Inf. Process. Lett. **57**(3), 137–143 (1996)
10. Hemachandra, L., Rudich, S.: On the complexity of ranking. J. Comput. Syst. Sci. **41**(2), 251–271 (1990)
11. Hemaspaandra, L., Jiang, Z., Rothe, J., Watanabe, O.: Boolean operations, joins, and the extended low hierarchy. Theoret. Comput. Sci. **205**(1–2), 317–327 (1998)
12. Hemaspaandra, L., Rubery, D.: Recursion-theoretic ranking and compression. J. Comput. Syst. Sci. **101**, 31–41 (2019)
13. Hemaspaandra, L., Zaki, M., Zimand, M.: Polynomial-time semi-rankable sets. J. Comput. Inf. **2**(1), 50–67 (1996). Special Issue: Proceedings of the 8th International Conference on Computing and Information
14. Huynh, D.: The complexity of ranking simple languages. Math. Syst. Theory **23**(1), 1–20 (1990)
15. Rogers Jr., H.: The Theory of Recursive Functions and Effective Computability. McGraw-Hill, New York (1967)
16. Schöning, U. (ed.): Complexity and Structure. LNCS, vol. 211. Springer, Heidelberg (1986). https://doi.org/10.1007/3-540-16079-5
17. Soare, R.: Recursively Enumerable Sets and Degrees: A Study of Computable Functions and Computably Generated Sets. Springer, Heidelberg (1987). Perspectives in Mathematical Logic

The Range of State Complexities
of Languages Resulting from the
Cut Operation

Markus Holzer[1] and Michal Hospodár[2](✉)

[1] Institut für Informatik, Universität Giessen,
Arndtstr. 2, 35392 Giessen, Germany
holzer@informatik.uni-giessen.de
[2] Mathematical Institute, Slovak Academy of Sciences,
Grešákova 6, 040 01 Košice, Slovakia
hosmich@gmail.com

Abstract. We investigate the state complexity of languages resulting from the cut operation of two regular languages represented by minimal deterministic finite automata with m and n states. We show that the entire range of complexities, up to the known upper bound, can be produced in the case when the input alphabet has at least two symbols. Moreover, we prove that in the unary case, only complexities up to $2m-1$ and between n and $m+n-2$ can be produced, while if $2m \leq n-1$, then the complexities from $2m$ up to $n-1$ cannot be produced.

1 Introduction

It is well known that for every n-state nondeterministic finite automaton (NFA), there exists a language-equivalent deterministic finite automaton (DFA) with at most 2^n states [21]. This bound is tight in the sense that for an arbitrary integer n there is always some n-state NFA which cannot be simulated by any DFA with less than 2^n states [17–19,23].

Nearly two decades ago a very fundamental question on determinization was raised by Iwama, Kambayashi, and Takaki [9]: does there always exist a minimal n-state NFA whose equivalent minimal DFA has α states for all n and α with $n \leq \alpha \leq 2^n$? Iwama, Matsuura, and Paterson [10] called a number α in the range from n to 2^n *magic* if no minimal n-state NFA has an equivalent minimal α-state DFA. The simple question whether for every n no number is magic turned out to be harder than expected. In a series of papers, non-magic (attainable) numbers were identified [6,11,12] until the problem was solved in [14] showing that for ternary languages *no* magic numbers exist. On the contrary, Geffert [5] proved that most of the numbers in the range from n up to $F(n) + n^2$, where $F(n)$

M. Hospodár—Research supported by VEGA grant 2/0132/19 and by grant APVV-15-0091. This work was conducted during a research visit at the Institut für Informatik, Universität Giessen, Germany, funded by the DAAD short-term grant ID 57314022.

C. Martín-Vide et al. (Eds.): LATA 2019, LNCS 11417, pp. 190–202, 2019.
https://doi.org/10.1007/978-3-030-13435-8_14

is the Landau function, is not attainable as the state complexity of a language accepted by a minimal unary n-state NFA. However, his proof is existential, and no specific value is known to be unattainable. For binary languages, the original problem from [9] is still open.

The idea behind the magic number problem is not limited to the determinization of NFAs. In fact every (regularity preserving) formal language operation can be used to define a magic number problem for the operation in question. For instance, consider the intersection operation on languages. Let A and B be minimal finite automata with m and n states, respectively. Then the size of the minimal automaton for the intersection of $L(A)$ and $L(B)$ is between 1 and mn. The value one is induced by the intersection of disjoint languages and the value mn by the standard cross-product construction for the intersection operation. Thus, in a similar way as for the determinization, one may now ask, whether every α within the range between 1 and mn can be attained by the size of minimal automaton for intersection of languages given by two minimal automata with m and n states, respectively? In other words, is the outcome of the intersection operation in terms of the number of states contiguous or are there any gaps, hence magic numbers? In [8] it was shown that for the intersection on DFAs *no* number from 1 up to mn is magic—this already holds for binary automata. Besides intersection, also other formal language operations were investigated from the "magic number" perspective. It turned out that magic numbers are quite rare, and most of them occur in the unary case. For example, Čevorová [2] studied the complexity of languages resulting from the Kleene star operation in the unary case. In such a case, the known upper bound is $(n-1)^2 + 1$ [24]. She proved that the values from 1 to n, as well as the values $n^2 - 2n + 2$ and $n^2 - 3n + 3$, are attainable, while the value $n^2 - 3n + 2$ is attainable if n is odd and it is not attainable otherwise. Moreover, she showed that all the values from $n^2 - 3n + 4$ up to $n^2 - 2n + 1$ and from $n^2 - 4n + 7$ up to $n^2 - 3n + 1$ cannot be attained by the state complexity of the Kleene star of any language accepted by minimal unary DFA with n states. The magic number problem was also examined for concatenation [13, 16], square [3], star on general alphabet [15], and reversal [22].

We contribute to the list of magic number problems for formal language operations by studying the cut operation. The cut operation was introduced in [1] as a machine implementation of "concatenation" on UNIX text processors which behaves greedy-like in its left term of concatenation. Tight upper bounds for the state complexity of the cut and iterated cut operations on DFAs were obtained in [4]. While the state complexity of concatenation is growing linearly with the first parameter (the number of states of the left automaton) and exponentially with the second parameter (the number of states of the right automaton), the state complexity of the cut operation is only linearly growing with both parameters. In the general case, the known tight upper bound is given by the function $f(m, n)$ such that $f(m, 1) = m$ and $f(m, n) = (m-1)n + m$ if $n \geq 2$. In the unary case, the known tight upper bound is given by the function

$f_1(m,n)$ such that $f_1(1,n) = 1$, $f_1(m,1) = m$, $f_1(m,n) = 2m-1$ if $m,n \geq 2$ and $m \geq n$, and finally let $f_1(m,n) = m+n-2$ if $m,n \geq 2$ and $m < n$ [4].

In this paper, we show for every value from 1 up to $f_1(m,n)$ whether or not it can be attained by the state complexity of the cut of two languages accepted by minimal unary DFAs with m and n states. We show that only complexities up to $2m-1$ and between n and $m+n-2$ can be attained, while complexities from $2m$ up to $n-1$ turn out to be magic. To get these results, the tail-loop structure of minimal unary DFAs is very valuable in the proofs.

On the other hand, we show that the entire range of complexities, up to the known upper bound $f(m,n)$, can be produced by the cut operation on minimal DFAs with m and n states, respectively, in case when the input alphabet consists of at least two symbols. The proof of this result resembles some ideas used in [8] for the magic number problem of the intersection and union operations on DFAs.

To the best of our knowledge, this is the first operation where for every alphabet, every value in the range of possible complexities is known to be either attainable or not, and not all values are attainable in the unary case. However, all values are attainable in every other alphabet size. Hence, the magic number problem for the cut operation is completely solved in this paper.

2 Preliminaries

We recall some definitions on finite automata as contained in [7]. Let Σ^* denote the set of all words over a finite alphabet Σ. The *empty word* is the word with length zero. If u,v,w are words over Σ such that $w = uv$, then u is a *prefix* of w. Further, we denote the set $\{i, i+1, \ldots, j\}$ by $[i,j]$ if i and j are integers.

A *deterministic finite automaton* (DFA) is a quintuple $A = (Q, \Sigma, \delta, s, F)$ where Q is a finite nonempty set of *states*, Σ is a finite nonempty set of *input symbols*, $s \in Q$ is the *initial* state, $F \subseteq Q$ is the set of *final* (or *accepting*) states, and $\delta \colon Q \times \Sigma \to Q$ is the *transition function* which can be extended to the domain $Q \times \Sigma^*$ in the natural way. The *language accepted* (or *recognized*) by the DFA A is defined as $L(A) = \{ w \in \Sigma^* \mid \delta(s,w) \in F \}$.

Two DFAs A and B are *equivalent* if they accept the same language, that is, if $L(A) = L(B)$. An automaton is *minimal* if it admits no smaller equivalent automaton with respect to the number of states. For DFAs this property can be verified by showing that all states are reachable from the initial state and all states are pairwise distinguishable. It is well known that every regular language has a unique, up to isomorphism, minimal DFA.

The *state complexity* of a regular language is the number of states in the minimal DFA recognizing this language.

In [1] the *cut operation* on languages K and L, denoted by $K\,!\,L$, is defined as

$$K\,!\,L = \{\, uv \mid u \in K, v \in L, \text{ and } uv' \notin K \text{ for every nonempty prefix } v' \text{ of } v \}.$$

The above defined cut operation preserves regularity as shown in [1]. Since we are interested in the descriptional complexity of this operation we briefly recall the construction of a DFA for the cut operation; we slightly deviate from the presentation of the construction given in [4].

Let $A = (Q_A, \Sigma, \delta_A, s_A, F_A)$ and $B = (Q_B, \Sigma, \delta_B, s_B, F_B)$ be two DFAs. Let $\bot \notin Q_B$. Define the cut automaton $A!B = (Q, \Sigma, \delta, s, F)$ with the state set $Q = (Q_A \times \{\bot\}) \cup (Q_A \times Q_B)$, the initial state $s = (s_A, \bot)$ if the empty word is not in $L(A)$ and $s = (s_A, s_B)$ otherwise, the set of final states $F = Q_A \times F_B$, and for each state (p, q) in Q and each input a in Σ we have

$$\delta((p, \bot), a) = \begin{cases} (\delta_A(p, a), \bot), & \text{if } \delta_A(p, a) \notin F_A; \\ (\delta_A(p, a), s_B), & \text{otherwise}; \end{cases}$$

and

$$\delta((p, q), a) = \begin{cases} (\delta_A(p, a), \delta_B(q, a)), & \text{if } \delta_A(p, a) \notin F_A; \\ (\delta_A(p, a), s_B), & \text{otherwise}. \end{cases}$$

Then $L(A!B) = L(A)!L(B)$.

In [4], the following functions were introduced.

$$f(m, n) = \begin{cases} m, & \text{if } n = 1; \\ (m - 1)n + m, & \text{if } n \geq 2 \end{cases} \tag{1}$$

and

$$f_1(m, n) = \begin{cases} 1, & \text{if } m = 1; \\ m, & \text{if } m \geq 2 \text{ and } n = 1; \\ 2m - 1, & \text{if } m, n \geq 2 \text{ and } m \geq n; \\ m + n - 2, & \text{if } m, n \geq 2 \text{ and } m < n \end{cases} \tag{2}$$

It was proven in [4, Theorems 3.1 and 3.2] that if A and B are DFAs with m and n states, respectively, then $f(m, n)$ states, resp. $f_1(m, n)$ states if A and B are unary, are sufficient and necessary in the worst case for any DFA accepting the language $L(A)!L(B)$.

3   The Descriptional Complexity of the Cut Operation

In this section we investigate the range of attainable complexities for the cut operation. In the first subsection we investigate the unary case and we show that some values may be unattainable. In the second subsection we study this problem for regular languages over an arbitrary alphabet, and we obtain a contiguous range of complexities from one up to the known upper bound already in the binary case.

3.1   The Cut Operation on Unary Regular Languages

When working with unary DFAs, we use the notational convention proposed by Nicaud in [20].

Every unary DFA consists of a tail path, which starts from the initial state, followed by a loop of one or more states. Let $A = (Q, \{a\}, \delta, q_0, F)$ be a unary DFA with $|Q| = n$. We can identify the states of A with integers from $[0, n-1]$ via $q \mapsto \min\{i \mid \delta(q_0, a^i) = q\}$. In particular the initial state q_0 is mapped to 0. Let $\ell = \delta(q_0, a^n)$. Then the unary DFA A with n states, loop number ℓ ($0 \leq \ell \leq n-1$), and set of final states F ($F \subseteq [0, n-1]$) is referred to as $A = (n, \ell, F)$. The following characterization of minimal unary DFAs is known.

Lemma 1 ([20]). *A unary DFA $A = (n, \ell, F)$ is minimal if and only if*

1. *its loop is minimal, and*
2. *if $\ell \neq 0$, then states $n-1$ and $\ell - 1$ do not have the same finality, that is, exactly one of them is final.*

Now we are ready for our first result on the cut operation of unary regular languages represented by DFAs. In a series of lemmata we consider the state complexity α of the resulting language in increasing order of α. The first interval we are going to discuss is $[1, m]$.

Lemma 2. *Let $m, n \geq 1$ and $1 \leq \alpha \leq m$. There exist a minimal unary m-state DFA A and a minimal unary n-state DFA B such that the minimal DFA for $L(A) \,! \, L(B)$ has α states.*

Proof. The proof has five cases:

1. Let $m = 1$, so we must have $\alpha = 1$. Let A be the one-state DFA accepting the empty language and B be the minimal n-state DFA for $a^{n-1}a^*$. Then $L(A) \,! \, L(B) = \emptyset$ which is accepted by a minimal one-state DFA.
2. Let $m \geq 2$ and $n = 1$. Let A be the minimal m-state DFA for $a^{\alpha-1}(a^m)^*$ and B be the one-state DFA for a^*. The reachable part of the cut automaton $A \,! \, B$ consists of the tail of non-final states $(i, \perp)$ with $0 \leq i \leq \alpha - 2$ and the loop of final states $(i, 0)$ with $0 \leq i \leq m - 1$. Since all the final states are equivalent, the minimal DFA for $L(A) \,! \, L(B)$ has α states.
3. Let $m, n \geq 2$ and $\alpha = 1$. Consider the unary languages $a^{m-1}a^*$ and $a^{n-1}a^*$ accepted by minimal DFAs A and B of m and n states, respectively. Then the reachable part of the cut automaton $A \,! \, B$ consists of the tail of non-final states $(i, \perp)$ with $1 \leq i \leq m - 2$, and the loop consisting of a single non-final state $(m-1, 0)$; notice that 0 is a non-final state in B. Hence $L(A) \,! \, L(B)$ is the empty language accepted by a one-state DFA.
4. Let $m \geq 2$, $n = 2$, and $2 \leq \alpha \leq m$. Consider the unary languages K and L defined as follows. If $m - \alpha$ is even, then $K = \{a^{\alpha-2}, a^{m-2}\}$ and $L = a(aa)^*$, otherwise, $K = \{a^{\alpha-1}, a^{m-2}\}$ and $L = (aa)^*$. The minimal DFAs for K and L have m and 2 states, respectively. We have $K \,! \, L = a^{\alpha-1}(aa)^*$, which is accepted by a minimal α-state DFA.

5. Let $m \geq 2$, $n \geq 3$, and $2 \leq \alpha \leq m$. Consider the unary deterministic finite automata $A = (m, \alpha-2, [\alpha - 1, m - 1])$ and $B = (n, n - 1, [0, n - 2])$. By Lemma 1, the DFAs A and B are minimal. The reachable part of the cut automaton consists of the tail of $\alpha - 1$ non-final states and of the loop of $m - \alpha + 2$ final states. Hence the minimal DFA $(\alpha, \alpha - 1, \{\alpha - 1\})$ for $L(A)\,!\,L(B)$ has α states. □

Our next interval is $[m + 1, 2m - 1]$; cf. $f_1(m, n)$ defined by (2) on page 4.

Lemma 3. *Let $m, n \geq 2$ and $m + 1 \leq \alpha \leq 2m - 1$. There exist a minimal unary m-state DFA A and a minimal unary n-state DFA B such that the minimal DFA for $L(A)\,!\,L(B)$ has α states.* □

The last interval we are considering in this series of lemmata is $[n, m + n - 2]$.

Lemma 4. *Let $m, n \geq 2$, $\alpha \geq m$, and $n \leq \alpha \leq m + n - 2$. There exist a minimal unary m-state DFA A and a minimal unary n-state DFA B such that the minimal DFA for $L(A)\,!\,L(B)$ has α states.* □

For certain values of m and n the intervals stated in the previous lemmata may not be contiguous. For instance, if we choose $m = 2$ and $n = 5$, then the intervals from Lemmata 2, 3, and 4 cover $\{1, 2, 3, 5\}$. Hence the value 4, which comes from the interval $[2m, n - 1]$, is missing. In fact, we show that whenever this interval is nonempty, these values cannot be obtained by an application of the cut operation on minimal DFAs with an appropriate number of states.

Lemma 5. *Let $m, n \geq 2$ be numbers satisfying $2m \leq n - 1$. Then for every α with $2m \leq \alpha \leq n - 1$, there exist no minimal unary m-state DFA A and minimal unary n-state DFA B such that the minimal DFA for $L(A)\,!\,L(B)$ has α states.*

Proof. We discuss two cases depending on whether $L(A)$ is infinite or finite.

If $L(A)$ is infinite, then A must have a final state in its loop. Denote the size of loop in A by ℓ and the smallest final state in the loop of A by j. Consider the cut automaton $A\,!\,B$. Notice that its initial state is sent to the state $(j, 0)$ by the word a^j. Next, the state $(j, 0)$ is sent to itself by the word a^ℓ. It follows that $A\,!\,B$ is equivalent to a DFA $(j + \ell, j, F)$ for some set $F \subseteq [0, j + l - 1]$. Since $j \leq m - 1$ and $\ell \leq m$, the DFA for $L(A)\,!\,L(B)$ has at most $2m - 1$ states.

If $L(A)$ is finite, then A has a loop in the non-final state $m - 1$ and the state $m - 2$ is final. Let $A = (m, m - 1, F)$ and $B = (n, k, F')$ be minimal unary DFAs for some sets $F \subseteq [0, m - 1]$ and $F' \subseteq [0, n - 1]$. It follows that in the cut automaton $A\,!\,B$, the state $(m - 2, 0)$ and the states $(m - 1, j)$ with $1 \leq j \leq n - 1$ are reachable. Two distinct states $(m - 1, j)$ and $(m - 1, j')$ are distinguishable by the same word as the states j and j' in B, and the state $(m - 2, 0)$ and a state $(m - 1, j)$ are distinguishable by the same word as 0 and j are distinguishable in B. It follows that the cut automaton has at least n reachable and pairwise distinguishable states, and the theorem follows. □

Now let us summarize the results of this subsection; recall that the state complexity of the cut operation on unary languages is given by the function $f_1(m, n)$ defined by (2) on page 4 such that $f_1(1, n) = 1$, $f_1(m, 1) = m$, $f_1(m, n) = 2m - 1$ if $m, n \geq 2$ and $m \geq n$, and $f_1(m, n) = m + n - 2$ if $m, n \geq 2$ and $m < n$.

Theorem 6 (Unary Case). *For every* $m, n, \alpha \geq 1$ *such that*

(i) $\alpha = 1$ *if* $m = 1$,
(ii) $1 \leq \alpha \leq m$ *if* $m \geq 2$ *and* $n = 1$, *or*
(iii) $1 \leq \alpha \leq 2m - 1$ *or* $n \leq \alpha \leq m + n - 2$ *if* $m, n \geq 2$,

there exist a minimal unary m-*state DFA* A *and a minimal unary* n-*state DFA* B *such that the minimal DFA for* $L(A) ! L(B)$ *has* α *states. In the case of* $m, n \geq 2$ *and* $2m \leq \alpha \leq n - 1$, *there do not exist minimal unary* m-*state and* n-*state DFAs* A *and* B *such that the minimal DFA for* $L(A) ! L(B)$ *has* α *states.* □

3.2   The Cut Operation on Binary Regular Languages

Next we consider the range of state complexities of languages resulting from the cut operation on regular languages over an arbitrary alphabet. The aim of this subsection is to show that the entire range of complexities up to the known upper bound can be produced in this case, even for languages over a binary alphabet. First, we show that the numbers in $[1, m + n - 2]$ are attainable in the binary case. The values in $[1, 2m - 1]$ as well as the cases of $m = 1$ or $n = 1$ are covered by Theorem 6 since duplicating the symbols does not change the state complexity.

Lemma 7. *Let* $m, n \geq 2$ *and* $2m \leq \alpha \leq m + n - 2$. *There exist a minimal binary* m-*state DFA* A *and a minimal binary* n-*state DFA* B *such that the minimal DFA for* $L(A) ! L(B)$ *has* α *states.*

Proof. Notice that in this case we must have and $m < n$. Consider the binary DFA $A = ([0, m - 1], \{a, b\}, \delta_A, 0, \{m - 1\})$, where

$$\delta_A(i, a) = (i + 1) \bmod m \quad \text{and} \quad \delta_A(i, b) = \begin{cases} (i + 1) \bmod m, & \text{if } i \neq m - 2, \\ m - 2, & \text{otherwise.} \end{cases}$$

Next, consider the binary DFA $B = ([0, n - 1], \{a, b\}, \delta_B, 0, \{m - 1\})$, where

$$\delta_B(j, a) = (j + 1) \bmod n \quad \text{and} \quad \delta_B(j, b) = \begin{cases} (j + 1) \bmod n, & \text{if } j \neq \alpha - m, \\ m - 1, & \text{otherwise.} \end{cases}$$

Both automata A and B are depicted in Fig. 1.

In the cut automaton $A ! B$ we consider the following sets of states:

$$\mathcal{R}_1 = \{ (i, \bot) \mid 0 \leq i \leq m - 2 \} \cup \{(m - 1, 0)\} \cup \{ (i, i + 1) \mid 0 \leq i \leq m - 3 \},$$
$$\mathcal{R}_2 = \{ (m - 2, j) \mid m - 1 \leq j \leq \alpha - m \}.$$

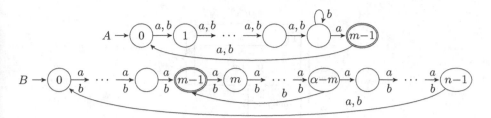

Fig. 1. The DFAs A (top) and B (bottom) for the case $m < n$ and $2m \leq \alpha \leq m+n-2$

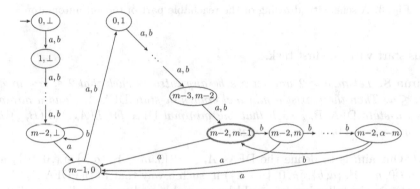

Fig. 2. The cut automaton for the DFAs in Fig. 1

Each state in $\mathcal{R}_1 \cup \{(m-2, m-1)\}$ is reached from $(0, \perp)$ by a word in a^*, and each state in $\mathcal{R}_2$ is reached from $(m-2, m-1)$ by a word in b^*. Figure 2 shows that no other state is reachable in the cut automaton.

To prove distinguishability, notice that two distinct states in $\mathcal{R}_1$ are distinguishable by a word in a^* and two distinct states in $\mathcal{R}_2$ are distinguishable by a word in b^*. The states $(i, \perp)$ in $\mathcal{R}_1$ are distinguishable from each state in $\mathcal{R}_2$ by a word in b^*. Every other state in $\mathcal{R}_1$ is distinguishable from each state in the set $\mathcal{R}_2$ by a word in a^*. Since $|\mathcal{R}_1 \cup \mathcal{R}_2| = \alpha$, our proof is complete.   □

Since the state complexity of the cut operation for regular languages in general is higher than those for unary languages, we have to consider the remaining interval $[m+n-1, (m-1)n+m]$. This is done in the following steps (cf. [8]):

1. First we show that some special values of α, corresponding to the number of states of the cut automaton in the first r rows and the first s columns, see Fig. 3, are attainable, namely $\alpha = 1 + (r-1)n + (m-r)s$ for some r, s with $2 \leq r \leq m$ and $1 \leq s \leq n$.
2. Then we show that all the remaining values of α in $[m+n-1, (m-1)n+1]$ are attainable.
3. Finally, we show that all the values of α in $[(m-1)n+2, (m-1)n+m]$ are attainable.

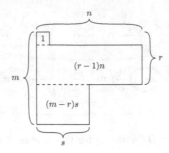

Fig. 3. A schematic drawing of the reachable part of the cut automaton

Let us start with the first task.

Lemma 8. *Let $m, n \geq 2$ and let r, s be any integers such that $2 \leq r \leq m$ and $1 \leq s \leq n$. Then there exist a minimal binary m-state DFA $A_{r,s}$ and a minimal binary n-state DFA $B_{r,s}$ such that the minimal DFA for $L(A_{r,s}) \,!\, L(B_{r,s})$ has exactly $1 + (r-1)n + (m-r)s$ states.*

Proof. Our aim is to define the DFAs $A_{r,s} = ([0, m-1], \{a, b\}, \delta_A, 0, \{0\})$ and $B_{r,s} = ([0, n-1], \{a, b\}, \delta_B, 0, \{n-1\})$ in such a way that in the DFA $A_{r,s} \,!\, B_{r,s}$ the states in the following set would be reachable and pairwise distinguishable:

$$\mathcal{R} = \{(0,0)\} \cup \{(i,j) \mid 1 \leq i \leq r-1 \text{ and } 0 \leq j \leq n-1\}$$
$$\cup \{(i,j) \mid r \leq i \leq m-1 \text{ and } 0 \leq j \leq s-1\}.$$

Moreover, we have to assure that no other state of the cut automaton is reachable. Because $|\mathcal{R}| = 1 + (r-1)n + (m-r)s$, the DFAs $A_{r,s}$ and $B_{r,s}$ will be the desired DFAs. To this aim, we define δ_A and δ_B as follows:

$$\delta_A(i,a) = (i+1) \bmod m \quad \text{and} \quad \delta_A(i,b) = \begin{cases} i, & \text{if } i \leq r-1; \\ r, & \text{if } i \geq r; \end{cases}$$

and

$$\delta_B(j,b) = (j+1) \bmod n \quad \text{and} \quad \delta_B(j,a) = \begin{cases} j, & \text{if } j \leq s-1; \\ s-1, & \text{if } j \geq s. \end{cases}$$

In the cut automaton $A_{r,s} \,!\, B_{r,s}$, the state $(0,0)$ is the initial state, and each state (i,j) in $\mathcal{R}$ is reached from $(0,0)$ by $a^i b^j$. To show that no other state is reachable, notice that each state (i,j) in $\mathcal{R}$ goes on a to a state (i',j') where $j' \leq s-1$, and it goes on b to a state (i'',j'') where $i'' \leq r-1$. Since both resulting states are in $\mathcal{R}$, no other state is reachable in the cut automaton.

It remains to prove the distinguishability of states in $\mathcal{R}$. The state $(0,0)$ and any other state in $\mathcal{R}$ are distinguishable by a word in b^*. Two states in different columns are distinguishable by a word in b^* since exactly one of them can be moved to the last column containing the final states of the cut automaton. Two states in different rows are distinguishable by a word in a^* since exactly one of them can be moved to the state $(0,0)$. This proves distinguishability and concludes the proof.   □

In the above lemma we obtained the values $\alpha_{r,s} = 1 + (r-1)n + (m-r)s$ in $[m+n-1, (m-1)n+1]$. We still need to get the values between $\alpha_{r,s}$ and $\alpha_{r+1,s}$ resp. $\alpha_{r,s+1}$. We have $\alpha_{r+1,s} - \alpha_{r,s} = n - s$ and $\alpha_{r,s+1} - \alpha_{r,s} = m - r$, so we need to obtain the complexities $\alpha_{r,s} + t$, where $1 \leq t \leq \min\{n-s, m-r\} - 1$. The next lemma produces these complexities.

Lemma 9. *Let $m, n \geq 2$ and let r, s be any integers such that $2 \leq r \leq m$ and $1 \leq s \leq n$. Moreover let t satisfy $1 \leq t \leq \min\{n-s, m-r\} - 1$. Then there exist a minimal binary m-state DFA $A_{r,s,t}$ and a minimal binary n-state DFA $B_{r,s,t}$ such that the minimal DFA for the language $L(A_{r,s,t}) ! L(B_{r,s,t})$ has exactly $1 + (r-1)n + (m-r)s + t$ states.*

Proof. Let $\alpha_{r,s} = 1 + (r-1)n + (m-r)s$. Then in the cut automaton $A_{r,s} ! B_{r,s}$ described in the previous proof, exactly $\alpha_{r,s}$ states are reachable and distinguishable. Our aim is to modify both automata in such a way that the resulting cut automaton has t more reachable states. To achieve this goal, we modify DFAs $A_{r,s}$ and $B_{r,s}$ as follows.

In $A_{r,s}$ we replace each transition $(r+i, b, r-1)$ by $(r+i, b, r+i-1)$, if $2 \leq i \leq t$ and i is even. Since $i \leq t \leq (m-r) - 1$, we have $r+i \leq m-1$. In $B_{r,s}$ we replace each transition $(s+i, a, s-1)$ by $(s+i, a, s+i-1)$, if $1 \leq i \leq t$ and i is odd. Since $i \leq t \leq (n-s) - 1$, we have $s+i \leq n-1$. Denote the resulting DFAs by $A_{r,s,t}$ and $B_{r,s,t}$, respectively. Consider the cut automaton $A_{r,s,t} ! B_{r,s,t}$. Let $\mathcal{R}$ be the same set as in the previous proof. Then each state (i,j) in $\mathcal{R}$ is reachable from $(0,0)$ by $a^i b^j$. Next, if i is odd, then each state $q_i = (r, s+i-1)$ is reached from $(r-1, s+i)$ by a, and otherwise, each state $q_i = (r+i-1, s)$ is reached from $(r+i, s-1)$ by b.

Now, let us show that no other state is reachable. Notice that each state in $\mathcal{R}$ goes either to a state in $\mathcal{R}$ or to a state in $\{q_1, q_2, \ldots, q_t\}$ on a and b; each state $(r-1, s+i)$ with i even goes to $(r, s-1)$ on a, and each state $(r+i, s-1)$ with $i = 0$ or i odd goes to $(r-1, s)$ on b. Next, each state q_i with i odd goes to the state $(r+1, s-1)$ on a and to a state in row $r-1$ on b. Finally, each state q_i with i even goes to a state in column $s-1$ on a and to the state $(r-1, s+1)$ on b. Since all the resulting states are in $\mathcal{R} \cup \{q_1, q_2, \ldots, q_t\}$, no other state is reachable in the cut automaton.

The proof of distinguishability is exactly the same as in Lemma 8.   □

In the two lemmata above, we have produced all the complexities in the range from $m+n-1$ to $(m-1)n+1$. It remains to show that the complexities in $[(m-1)n+2, (m-1)n+m]$ are attainable.

Lemma 10. *Let $m, n \geq 2$ and $(m-1)n+2 \leq \alpha \leq (m-1)n+m$. There exist a minimal binary m-state DFA A and a minimal binary n-state DFA B such that the minimal DFA for $L(A) ! L(B)$ has exactly α states.*

Proof. We have $\alpha = (m-1)n + 1 + \beta$ for some β with $1 \le \beta \le m-1$. Let A be a minimal m-state DFA over $\{a, b\}$ that accepts the words in which the number of a's modulo m is β. Let B be a minimal n-state DFA over $\{a, b\}$ that accepts the words in which the number of b's modulo n is $n-1$.

Consider the cut automaton $A \mathbin{!} B$. Denote

$$\mathcal{R}_1 = \{\, (i, \bot) \mid i \in [0, \beta-1] \,\} \cup \{(\beta, 0)\},$$

and

$$\mathcal{R}_2 = \{\, (i, j) \mid i \in [0, \beta-1] \cup [\beta+1, m-1] \text{ and } j \in [0, n-1] \,\}.$$

Notice that each state $(i, \bot)$ in $\mathcal{R}_1$ is reachable from the initial state $(0, \bot)$ by a^i, and each state $(i, 0)$ is reachable by a^{m+i}. Each state (i, j) in $\mathcal{R}_2$ is reached from $(0, 0)$ by $a^i b^j$. Since the state β is a final state in A, it follows from the construction of the cut automaton that no state $(i, \bot)$ with $i \ge \beta$ and no state in row β except for $(\beta, 0)$ is reachable.

To prove distinguishability, let p and q be two different states in $\mathcal{R}_1 \cup \mathcal{R}_2$. If $p \in \mathcal{R}_1$ and $q \in \mathcal{R}_2$, then p is a non-final state with a loop on b, while a word in b^* is accepted from q. If both p and q are in $\mathcal{R}_1$, then a word in a^* leads one of them to the state $((\beta+1) \bmod m, 0)$ in $\mathcal{R}_2$, while it leads the second one to a state in $\mathcal{R}_1$, and the resulting states are distinguishable as shown above. Finally, let p and q be two states in $\mathcal{R}_2$. If they are in different columns, then a word in b^* distinguishes them. If p and q are in different rows, then a word in a^* leads one of them to the state $(\beta, 0)$ in $\mathcal{R}_1$, and it leads the second one to a state in $\mathcal{R}_2$. □

The next theorem summarizes the results of this section; recall that the state complexity of the cut operation is given by the function $f(m, n)$ defined by (1) on page 4 such that $f(m, 1) = m$ and $f(m, n) = (m-1)n + m$ if $n \ge 2$.

Theorem 11 (General Case). *Let $m, n \ge 1$ and $f(m, n)$ be the state complexity of the cut operation. For each α such that $1 \le \alpha \le f(m, n)$, there exist a minimal binary m-state DFA A and a minimal binary n-state DFA B such that the minimal DFA for $L(A) \mathbin{!} L(B)$ has α states.* □

Observe that this theorem solves the magic number problem for the cut operation for every alphabets of size at least two by duplicating input symbols.

4　Conclusions

We examined the state complexity of languages resulting from the cut operation on minimal DFAs with m and n states. We showed that the range of state complexities of languages resulting from the cut operation is contiguous from one up to the known upper bound for every alphabet of size at least two. Our results in the unary case are different. We proved that no value from $2m$ up to $n-1$ is attainable by the state complexity of the cut of two unary languages represented by minimal deterministic finite automata with m and n states. All the remaining values up to the known upper bound are attainable. This means that the problem of finding all attainable complexities for the cut operation is completely solved for every size of alphabet. To the best of our knowledge, the cut operation is the first operation where this is the case.

Acknowledgments. We thank Juraj Šebej and Jozef Jirásek Jr. for their help on border values in our theorems. Moreover, also thanks to Galina Jirásková for her support and to all who helped us to improve the presentation of the paper.

References

1. Berglund, M., Björklund, H., Drewes, F., van der Merwe, B., Watson, B.: Cuts in regular expressions. In: Béal, M.-P., Carton, O. (eds.) DLT 2013. LNCS, vol. 7907, pp. 70–81. Springer, Heidelberg (2013). https://doi.org/10.1007/978-3-642-38771-5_8
2. Čevorová, K.: Kleene star on unary regular languages. In: Jurgensen, H., Reis, R. (eds.) DCFS 2013. LNCS, vol. 8031, pp. 277–288. Springer, Heidelberg (2013). https://doi.org/10.1007/978-3-642-39310-5_26
3. Čevorová, K., Jirásková, G., Krajňáková, I.: On the square of regular languages. In: Holzer, M., Kutrib, M. (eds.) CIAA 2014. LNCS, vol. 8587, pp. 136–147. Springer, Cham (2014). https://doi.org/10.1007/978-3-319-08846-4_10
4. Drewes, F., Holzer, M., Jakobi, S., van der Merwe, B.: Tight bounds for cut-operations on deterministic finite automata. Fundam. Inform. **155**(1–2), 89–110 (2017)
5. Geffert, V.: Magic numbers in the state hierarchy of finite automata. Inf. Comput. **205**(11), 1652–1670 (2007)
6. Geffert, V.: State hierarchy for one-way finite automata. J. Autom. Lang. Comb. **12**(1–2), 139–145 (2007)
7. Harrison, M.A.: Introduction to Formal Language Theory. Addison-Wesley, Boston (1978)
8. Hricko, M., Jirásková, G., Szabari, A.: Union and intersection of regular languages and descriptional complexity. In: DCFS 2005, pp. 170–181. Università degli Studi di Milano, Italy (2005)
9. Iwama, K., Kambayashi, Y., Takaki, K.: Tight bounds on the number of states of DFAs that are equivalent to n-state NFAs. Theoret. Comput. Sci. **237**(1–2), 485–494 (2000)
10. Iwama, K., Matsuura, A., Paterson, M.: A family of NFA's which need $2^n - \alpha$ deterministic states. In: Nielsen, M., Rovan, B. (eds.) MFCS 2000. LNCS, vol. 1893, pp. 436–445. Springer, Heidelberg (2000). https://doi.org/10.1007/3-540-44612-5_39
11. Jirásek, J., Jirásková, G., Szabari, A.: Deterministic blow-ups of minimal nondeterministic finite automata over a fixed alphabet. Internat. J. Found. Comput. Sci. **19**(3), 617–631 (2008)
12. Jirásková, G.: Deterministic blow-ups of minimal NFA's. RAIRO-ITA **40**(3), 485–499 (2006)
13. Jirásková, G.: Concatenation of regular languages and descriptional complexity. Theory Comput. Syst. **49**(2), 306–318 (2011)

14. Jirásková, G.: Magic numbers and ternary alphabet. Int. J. Found. Comput. Sci. **22**(2), 331–344 (2011)
15. Jirásková, G., Palmovský, M., Šebej, J.: Kleene closure on regular and prefix-free languages. In: Holzer, M., Kutrib, M. (eds.) CIAA 2014. LNCS, vol. 8587, pp. 226–237. Springer, Cham (2014). https://doi.org/10.1007/978-3-319-08846-4_17
16. Jirásková, G., Szabari, A., Šebej, J.: The complexity of languages resulting from the concatenation operation. In: Câmpeanu, C., Manea, F., Shallit, J. (eds.) DCFS 2016. LNCS, vol. 9777, pp. 153–167. Springer, Cham (2016). https://doi.org/10.1007/978-3-319-41114-9_12
17. Lupanov, O.B.: A comparison of two types of finite automata. Problemy Kibernetiki **9**, 321–326 (1963). (in Russian). German translation: Über den Vergleich zweier Typen endlicher Quellen. Probleme der Kybernetik 6, 328–335 (1966)
18. Meyer, A.R., Fischer, M.J.: Economy of description by automata, grammars, and formal systems. In: Proceedings of the 12th Annual Symposium on Switching and Automata Theory, pp. 188–191. IEEE Computer Society Press (1971)
19. Moore, F.R.: On the bounds for state-set size in the proofs of equivalence between deterministic, nondeterministic, and two-way finite automata. IEEE Trans. Comput. **C–20**, 1211–1219 (1971)
20. Nicaud, C.: Average state complexity of operations on unary automata. In: Kutyłowski, M., Pacholski, L., Wierzbicki, T. (eds.) MFCS 1999. LNCS, vol. 1672, pp. 231–240. Springer, Heidelberg (1999). https://doi.org/10.1007/3-540-48340-3_21
21. Rabin, M.O., Scott, D.: Finite automata and their decision problems. IBM J. Res. Dev. **3**, 114–125 (1959)
22. Šebej, J.: Reversal on regular languages and descriptional complexity. In: Jurgensen, H., Reis, R. (eds.) DCFS 2013. LNCS, vol. 8031, pp. 265–276. Springer, Heidelberg (2013). https://doi.org/10.1007/978-3-642-39310-5_25
23. Yershov, Y.L.: On a conjecture of V. A. Uspenskii. Algebra i logika **1**, 45–48 (1962). (in Russian)
24. Yu, S., Zhuang, Q., Salomaa, K.: The state complexities of some basic operations on regular languages. Theoret. Comput. Sci. **125**(2), 315–328 (1994)

State Complexity of Pseudocatenation

Lila Kari and Timothy Ng[✉]

School of Computer Science, University of Waterloo,
Waterloo, ON N2L 3G1, Canada
{lila.kari,tim.ng}@uwaterloo.ca

Abstract. The state complexity of a regular language L_m is the number m of states in a minimal deterministic finite automaton (DFA) accepting L_m. The state complexity of a regularity-preserving binary operation on regular languages is defined as the maximal state complexity of the result of the operation, where the two operands range over all languages of state complexities $\leq m$ and $\leq n$, respectively. We consider the deterministic and nondeterministic state complexity of pseudocatenation. The pseudocatenation of two words x and y with respect to an antimorphic involution θ is the set $\{xy, x\theta(y)\}$. This operation was introduced in the context of DNA computing as the generator of pseudopowers of words (a pseudopower of a word u is a word in $u\{u, \theta(u)\}^*$). We prove that the state complexity of the pseudocatenation of languages L_m and L_n, where $m, n \geq 3$, is at most $(m-1)(2^{2n} - 2^{n+1} + 2) + 2^{2n-2} - 2^{n-1} + 1$. Moreover, for $m, n \geq 3$ there exist languages L_m and L_n over an alphabet of size 4, whose pseudocatenation meets the upper bound. We also prove that the state complexity of the positive pseudocatenation closure of a regular language L_n has an upper bound of $2^{2n-1} - 2^n + 1$, and that this bound can be reached, with the witness being a language over an alphabet of size 4.

1 Introduction

In the context of DNA computing, the fact that one can consider a DNA strand and its Watson-Crick complement "equivalent" from the point of view of their information content led to several natural, as well as theoretically interesting, extensions of notions in combinatorics of words and formal language theory such as the pseudo-palindrome [21], pseudo-commutativity [18], or pseudoknot-bordered words [19]. In this context, Watson-Crick complementarity has been modelled mathematically by an antimorphic involution θ, i.e., a function that is an antimorphism, $\theta(uv) = \theta(v)\theta(u)$, $\forall u, v \in \Sigma^*$, and an involution, $\theta(\theta(x)) = x$, $\forall x \in \Sigma^*$. For example, in [10], a word w is called a θ-*power* or *pseudopower* if it is of the form $w \in u\{u, \theta(u)\}^*$, and the related notions of θ-periodicity and θ-primitivity can be analogously defined. The static notions of the power

This research was supported by the Natural Sciences and Engineering Research Council of Canada Discovery Grant R2824A01, and a University of Waterloo School of Computer Science Grant to L.K.

C. Martín-Vide et al. (Eds.): LATA 2019, LNCS 11417, pp. 203–214, 2019.
https://doi.org/10.1007/978-3-030-13435-8_15

and period of a word are intrinsically connected to the word operation that dynamically generates that power. In the case of the classical notion of power and period of a word, that operation is catenation, and in the case of θ-power and θ-periodicity, that operation is θ-catenation, defined and studied in [17]. Here we continue the investigation of θ-catenation, defined by $x \odot^\theta y = \{xy, x\theta(y)\}$, by studying its state complexity.

The state complexity of a language operation is a complexity measure based on the number of states of the machine that recognizes the result of the language operation, expressed as a function of the size of the machines recognizing the operand languages. Operational state complexity has been studied since the early 90s and continues to be an active area of research [12,26]. Recently, there have been several investigations of state complexity for operations modelling biological phenomena, such as hairpin completion [16], inversion [6], duplication [5], and overlap assembly [3].

The state complexity of combinations of operations has also been studied extensively, as many language operations can be expressed as a combination of several basic operations. While one can obtain an upper bound for the state complexity of multiple operations by simply composing the state complexities of each operation, in many cases, the exact state complexity of the combination of operations is much lower than the bound obtained in this fashion [23]. Furthermore, the exact state complexity of a combination of operations is undecidable [24], thus motivating further study in this direction [1,2,7–9,13,14,20].

In this paper, we consider the deterministic and nondeterministic state complexity of the pseudocatenation and positive pseudocatenation closure operations with respect to an antimorphism θ. We note that for our constructions, θ need not be an involution. We fix notation and definitions in Sect. 2. In Sect. 3, we consider the state complexity of the pseudocatenation operation. In Sect. 4, we consider the positive closure of a language with respect to pseudocatenation. We conclude in Sect. 5.

2   Preliminaries

Let Σ be a finite alphabet. We denote by Σ^* the set of all finite words over Σ, including the empty word, which we denote by ε. We denote the length of a word $w = a_1 a_2 \cdots a_n$ by $|w| = n$. The reversal of a word $w = a_1 a_2 \cdots a_n$ is denoted by $w^R = a_n \cdots a_2 a_1$. If $w = xyz$, then we say that x is a prefix of w, y is a factor or subword of w, and z is a suffix of w. For a word $u \in \Sigma^*$, we denote the number of occurrences of u as a factor of w by $|w|_u$.

A deterministic finite automaton (DFA) is a tuple $A = (Q, \Sigma, \delta, s, F)$ where Q is a finite set of states, Σ is an alphabet, δ is a function $\delta : Q \times \Sigma \to Q$, $s \in Q$ is the initial state, and $F \subset Q$ is a set of final states. We extend the transition function δ to a function $Q \times \Sigma^* \to Q$ in the usual way. A DFA A is complete if δ is defined for all $q \in Q$ and $a \in \Sigma$. We will also make use of the notation $q \xrightarrow{w} q'$ for $\delta(q, w) = q'$ whenever convenient.

A word $w \in \Sigma^*$ is accepted by A if $\delta(s, w) \in F$. The language recognized by A is $L(A) = \{w \in \Sigma^* \mid \delta(s, w) \in F\}$. A state q is reachable if there exists a string $w \in \Sigma^*$ such that $\delta(s, w) = q$. Two states p and q of A are equivalent if $\delta(p, w) \in F$ if and only if $\delta(q, w) \in F$ for every word $w \in \Sigma^*$. A DFA A is minimal if each state $q \in Q$ is reachable from the initial state and no two states are equivalent. The state complexity of a regular language L, denoted $\mathrm{sc}(L)$ is the number of states of the minimal complete DFA recognizing L [25].

A nondeterministic finite automaton (NFA) is a tuple $A = (Q, \Sigma, \delta, I, F)$ where Q is a finite set of states, Σ is an alphabet, δ is a function $\delta : Q \times \Sigma \to 2^Q$, $I \subseteq Q$ is a set of initial states, and F is a set of final states. The language recognized by an NFA A is $L(A) = \{w \in \Sigma^* \mid \bigcup_{q \in I} \delta(q, w) \cap F \neq \emptyset\}$. The nondeterministic state complexity of a regular language is the minimum number of states for any NFA which accepts L. We denote the nondeterministic state complexity of L by $\mathrm{nsc}(L)$.

A set of pair of strings $S = \{(x_1, y_1), \ldots, (x_m, y_m)\}$ with $x_i, y_i \in \Sigma^*$ for $1 \leq i \leq m$ is a fooling set for a regular language L if $x_i y_i \in L$ for $1 \leq i \leq m$ and for all $1 \leq i < j \leq m$, either $x_i y_j \notin L$ or $x_j y_i \notin L$. If L has a fooling set S, then $\mathrm{nsc}(L) \geq |S|$ [15].

Let $\theta : \Sigma^* \to \Sigma^*$ be a mapping. We say θ is a morphism if for $u, v \in \Sigma^*$, we have $\theta(uv) = \theta(u)\theta(v)$. We say θ is an antimorphism if we have $\theta(uv) = \theta(v)\theta(u)$. The mapping θ is an involution if for all words $u \in \Sigma^*$, we have $\theta(\theta(u)) = u$. For example, if $\Sigma = \{A, C, G, T\}$ we can define Watson-Crick complementarity for DNA as an antimorphic involution θ by $\theta(A) = T$, $\theta(C) = G$, $\theta(G) = C$, and $\theta(T) = A$. Then the Watson-Crick complement of a DNA string w is given by $\theta(w)$.

Definition 1. *Let θ be an antimorphic involution and $x, y \in \Sigma^*$. We define the θ-catenation operation $\odot^\theta$, also called pseudocatenation with respect to θ, by*

$$x \odot^\theta y = \{xy, x\theta(y)\}.$$

We can define θ-catenation for languages by

$$L_1 \odot^\theta L_2 = \{xy, x\theta(y) \mid x \in L_1, y \in L_2\}.$$

This operation can be extended to an iterated variant by $L^{\odot^\theta_0} = \{\varepsilon\}$, $L^{\odot^\theta_1} = L$, and $L^{\odot^\theta_n} = L^{\odot^\theta_{n-1}} \odot^\theta L$. Then we can take the positive θ-catenation closure by

$$L^{\odot^\theta_+} = \bigcup_{i \geq 1} L^{\odot^\theta_i}.$$

Although θ-catenation is defined for both morphisms and antimorphisms, we will consider only the state complexity for antimorphisms. For morphic θ, many state complexity results are the same as the state complexity of combined operations studied previously. Furthermore, we note that the condition that θ be involutive is not strictly necessary in our constructions.

We will make use of the following notation for the NFA recognizing $\theta(L(A))$ for a given DFA A and antimorphism θ. Let $A = (Q, \Sigma, \delta, s, F)$ be a DFA. Let $P \subseteq Q$ be a set of states of Q. We denote by $\overline{P} = \{\overline{q} \mid q \in P\}$. Then define the

DFA $\overline{A} = (\overline{Q}, \Sigma, \delta^{-1}, \overline{F}, \{\overline{s}\})$, where the transition function $\delta^{-1} : \overline{Q} \times \Sigma \to 2^{\overline{Q}}$ is defined for $\overline{q} \in \overline{Q}$ and $a \in \Sigma$ by $\delta^{-1}(\overline{q}, a) = \{\overline{q}' \mid \delta(q', \theta(a)) = q\}$. In other words, every transition of A is reversed and relabeled according to θ in $\overline{A}$. Then $L(\overline{A}) = \theta(L(A))$.

3　State Complexity of θ-Catenation

We will consider the state complexity of the θ-catenation of two regular languages. It was shown in [17] that the class of regular languages is closed under θ-catenation. This is easy to see from the following expression for the θ-catenation of L_1 and L_2, which follows directly from the definition.

Proposition 2. *Let $L_1, L_2 \subseteq \Sigma^*$ be languages and θ an antimorphism. Then $L_1 \odot^\theta L_2 = L_1(L_2 \cup \theta(L_2))$.*

First, we consider an NFA for recognizing $L_1 \odot^\theta L_2$ and its nondeterministic state complexity.

Proposition 3. *For $m, n \geq 1$, let A and B be NFAs defined over an alphabet Σ with m and n states and let θ be an antimorphism. Then there exists an NFA that recognizes $L(A) \odot^\theta L(B)$ with at most $m + 2n$ states and this bound can be reached.*

The proof of Proposition 3 makes use of the following construction for an NFA C that recognizes $L(A) \odot^\theta L(B)$. Let $A = (Q_A, \Sigma, \delta_A, I_A, F_A)$ and $B = (Q_B, \Sigma, \delta_B, I_B, F_B)$. We denote by $\overline{B}$ the NFA for $\theta(L(B))$, defined $\overline{B} = (\overline{Q_B}, \Sigma, \delta_B^{-1}, \overline{F_B}, \overline{I_B})$. We define an NFA $C = (Q_C, \Sigma, \delta_C, I_C, F_C)$ where $Q_C = Q_A \cup Q_B \cup \overline{Q_B}$, $I_C = I_A$, $F_C = F_B \cup \overline{I_B}$, and the transition function $\delta_C : Q_C \times \Sigma \to 2^{Q_C}$ is defined for $q \in Q_C$ and $a \in \Sigma$ by

$$\delta_C(q, a) = \begin{cases} \delta_A(q, a) & \text{if } q \in Q_A, \\ \delta_B(q, a) & \text{if } q \in Q_B, \\ \delta_B^{-1}(q, a) & \text{if } q \in \overline{Q_B}, \\ \delta_A(q, a) \cup I_B \cup \overline{F_B} & \text{if } (\delta_A(q, a) \cap F_A) \neq \emptyset. \end{cases}$$

From this construction, it follows that C has at most $m + 2n$ states, and this bound is also reachable.

We will now consider the deterministic state complexity of θ-catenation. We note again that $L(A) \odot^\theta L(B) = L(A)(L(B) \cup \theta(L(B)))$. By directly computing the state complexity of the union $L(B) \cup \theta(L(B))$ and composing it with the state complexity for the catenation $L(A)(L(B) \cup \theta(L(B)))$, we obtain an upper bound of $m2^{n2^n} - 2^{n2^n-1}$ states for $L(A) \odot^\theta L(B)$. This is clearly incorrect, since determinizing the NFA from Proposition 3 gives at most 2^{m+2n} states. Instead, we apply a construction similar to the one from [7] to $L(A)(L(B) \cup \theta(L(B)))$.

Proposition 4. *Let $m, n \geq 3$, θ be an antimorphism, and A and B be DFAs defined over an alphabet Σ with m and n states, respectively. Then there exists a DFA that recognizes $L(A) \odot^\theta L(B)$ with at most $(m-1)(2^{2n} - 2^{n+1} + 2) + 2^{2n-2} - 2^{n-1} + 1$ states.*

Proof. We will define a DFA C that recognizes $L(A) \odot^\theta L(B)$ given two DFAs A and B. Let $A = (Q_A, \Sigma, \delta_A, s_A, F_A)$ and $B = (Q_B, \Sigma, \delta_B, s_B, F_B)$. We define the DFA $C = (Q_C, \Sigma, \delta_C, s_C, F_C)$ by the set of states

$$Q_C = \{\langle q, P, \overline{R}\rangle \mid q \in Q_A - F_A, P \in 2^{Q_B} - \{\emptyset\}, \overline{R} \in 2^{\overline{Q_B}} - \{\emptyset\}\}$$
$$\cup \{\langle q, \emptyset, \emptyset\rangle \mid q \in Q_A - F_A\}$$
$$\cup \{\langle q, P \cup \{s_B\}, \overline{R} \cup \overline{F_B}\rangle \mid q \in F_A, P \in 2^{Q_B - \{s_B\}}, \overline{R} \in 2^{\overline{Q_B} - \overline{F_B}}\},$$

the initial state

$$s_C = \begin{cases} \langle s_A, \emptyset, \emptyset\rangle & \text{if } s_A \notin F_A, \\ \langle s_A, \{s_B\}, \overline{F_B}\rangle & \text{otherwise,} \end{cases}$$

the set of final states $F_C = \{\langle q, P, \overline{R}\rangle \in Q_C \mid (P \cup \overline{R}) \cap (F_B \cup \{\overline{s_B}\}) \neq \emptyset\}$, and the transition function $\delta_C(\langle q, P, \overline{R}\rangle, a) = \langle q', P', \overline{R}'\rangle$ for $a \in \Sigma$ where $q' = \delta_A(q, a)$,

$$P' = \begin{cases} \bigcup_{p \in P} \delta_B(p, a) \cup \{s_B\} & \text{if } q' \in F_A, \\ \bigcup_{p \in P} \delta_B(p, a) & \text{otherwise,} \end{cases}$$

$$\overline{R}' = \begin{cases} \delta_B^{-1}(\overline{R}, a) \cup \overline{F_B} & \text{if } q' \in F_A, \\ \delta_B^{-1}(\overline{R}, a) & \text{otherwise.} \end{cases}$$

Informally, the DFA C operates as follows. The states of C are 3-tuples $\langle q, P, \overline{R}\rangle$, where q is a state of A, and P and $\overline{R}$ are subsets of states of B. The first component q denotes the current state of a computation on A, the second component P denotes a set of states corresponding to the current states of computations on B, and the third component $\overline{R}$ denotes a set of states corresponding to the current states of computations on $\overline{B}$, the NFA recognizing $\theta(L(B))$.

Upon reading a symbol $a \in \Sigma$, the computations advance one step to $\langle q', P', R'\rangle$. If q' is a final state of A, then in addition to updating the sets P and R' to advance one step in computation, the initial state s_B of B is added to P' and the set of initial states $\overline{F_B}$ of $\overline{B}$, the NFA recognizing $\theta(L(B))$, is added to $\overline{R}'$.

We will now consider the size of Q_C, the state set of C. Let $k_A = |F_A|$ and $k_B = |F_B|$. We have

$$|Q_C| = (m - k_A)(2^n - 1)(2^n - 1) + (m - k_A) + k_A(2^{n-1})(2^{n-k_b}).$$

However, note that since B is a complete DFA, we have $\delta_B^{-1}(\overline{Q_B}, \sigma) = \overline{Q_B}$ for all $\sigma \in \Sigma$. Then for all states $q \in Q_A$, $P \subseteq Q_B$, and symbols $\sigma \in \Sigma$, we have $\delta_C(\langle q, P, \overline{Q_B}\rangle, \sigma) = \langle q', P', \overline{Q_B}\rangle$. Since $\overline{s} \in \overline{Q_B}$, any state of the form $\langle q, P, \overline{Q_B}\rangle$ is a final state. Thus, for all states $q \in Q_A$, $P \subseteq Q_B$, and words $w \in \Sigma^*$, we have

$\delta_C(\langle q, P, \overline{Q_B} \rangle, w) \in F_C$. Therefore, all states with the third component $\overline{R} = \overline{Q_B}$ are equivalent and indistinguishable and we revise our upper bound down to

$$(m - k_A)(2^n - 1)(2^n - 1) + (m - k_A) + k_A(2^{n-1})(2^{n-k_B} - 1) + 1.$$

This value is maximized when $k_A = 1$ and $k_B = 1$, giving a total of $(m-1)(2^{2n} - 2^{n+1} + 2) + 2^{2n-2} - 2^{n-1} + 1$ states.   □

We will now show that this bound is reachable.

Lemma 5. *For $m, n \geq 3$, there exist an m-state DFA A, an n-state DFA B, and an antimorphism θ over an alphabet of size 4 such that*

$$sc(L(A) \odot^\theta L(B)) \geq (m - 1)(2^{2n} - 2^{n+1} + 2) + 2^{2n-2} - 2^{n-1} + 1.$$

The main idea of the proof of Lemma 5 is to demonstrate that the bound from Proposition 4 is reachable by using the witness $\mathcal{W}_n(a, b, c, d)$ defined by Brzozowski [2]. Let $\Sigma = \{a, b, c, d\}$ and let $\theta : \Sigma^* \to \Sigma^*$ be the Watson-Crick antimorphism defined by

$$\theta(a) = d \quad \theta(b) = c \quad \theta(c) = b \quad \theta(d) = a.$$

We set $A = \mathcal{W}_m(a, b, c, d)$ with m states and $B = \mathcal{W}_n(a, b, c, d)$ with n states. Then we define $\overline{B} = \overline{\mathcal{W}_n(a, b, c, d)}$. That is, $L(\overline{B}) = \theta(L(\mathcal{W}_n(a, b, c, d))) = L(\mathcal{W}_n(d, c, b, a))^R$. The DFA $\mathcal{W}_3(a, b, c, d)$ is shown in Fig. 1 and the DFA B and the NFA $\overline{B}$ are shown in Fig. 2.

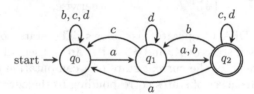

Fig. 1. The DFA $\mathcal{W}_3(a, b, c, d)$

Proposition 4 and Lemma 5 are summarized in the following theorem.

Theorem 6. *For $m, n \geq 3$, regular languages L_m and L_n with $sc(L_m) = m$ and $sc(L_n) = n$, and antimorphism θ,*

$$sc(L_m \odot^\theta L_n) \leq (m - 1)(2^{2n} - 2^{n+1} + 2) + 2^{2n-2} - 2^{n-1} + 1$$

and this bound can be reached in the worst case.

Furthermore, observe that the witnesses used in Lemma 5 belong to the same family of DFAs $\mathcal{W}_n(a, b, c, d)$. Setting $m = n$ gives us the same DFA and we obtain a tight bound for the state complexity of the pseudosquare of L, $L^{\odot^\theta}_2$, via Lemma 5.

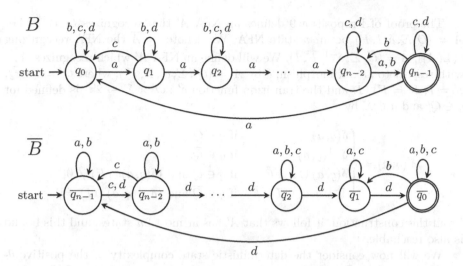

Fig. 2. The DFA $B = \mathcal{W}_n(a, b, c, d)$ and the NFA $\overline{B} = \overline{\mathcal{W}_n(a, b, c, d)}$

Corollary 7. *For $n \geq 3$, let L_n be a regular language with $sc(L_n) = n$ and let θ be an antimorphism. Then*

$$sc(L_n^{\odot_2^\theta}) \leq (n-1)(2^{2n} - 2^{n+1} + 2) + 2^{2n-2} - 2^{n-1} + 1$$

and this bound can be reached in the worst case.

4 State Complexity of θ-Catenation Closure

In this section, we consider the θ-catenation closure of a regular language. This is analogous to the positive Kleene closure, but with respect to θ-catenation. It was shown in [17] that the positive closure of a regular language with respect to θ-catenation is also regular. The following equality follows from the definition.

Proposition 8. *Let L be a language and let θ be an antimorphism. Then the positive θ-catenation closure is $L^{\odot_+^\theta} = L(L \cup \theta(L))^*$.*

It is important to note that the positive closure with respect to θ-catenation is not $(L \cup \theta(L))^+$, as words $w \in L^{\odot_+^\theta}$ have the form $w = uv_1v_2 \cdots v_{k-1}$ where $u \in L$ and $v_i \in L \cup \theta(L)$ for $1 \leq i \leq k$ [17], whereas $(L \cup \theta(L))^+$ also contains words of the form $\theta(u)v_1v_2 \cdots v_{k-1}$.

We will first consider an NFA for recognizing $L^{\odot_+^\theta}$ and its nondeterministic state complexity.

Proposition 9. *For $n \geq 1$, let A be an NFA with n states defined over an alphabet Σ and let θ be an antimorphism. Then there exists an NFA that recognizes $L(A)^{\odot_+^\theta}$ with at most $2n$ states. Furthermore, this bound can be reached in the worst case.*

The proof of Proposition 9 defines an NFA A' that recognizes $L(A)^{\odot_+^\theta}$. Let $A = (Q, \Sigma, \delta, I, F)$ be an n-state NFA. We denote by $\overline{A}$ the NFA recognizing $\theta(L(A))$, $\overline{A} = (\overline{Q}, \Sigma, \delta^{-1}, \overline{F}, \overline{I})$. We will define an NFA A' which recognizes $A^{\odot_+^\theta}$ with respect to an antimorphism θ by $A' = (Q', \Sigma, \delta', I', F')$, where $Q' = Q \cup \overline{Q}$, $I' = I$, $F' = F \cup \overline{I}$, and the transition function $\delta' : Q' \times \Sigma \to 2^{Q'}$ is defined for $q \in Q'$ and $a \in \Sigma$ by

$$\delta'(q, a) = \begin{cases} \delta(q, a) & \text{if } q \in Q, \\ \delta^{-1}(q, a) & \text{if } q \in \overline{Q}, \\ \delta(q, a) \cup I \cup \overline{F} & \text{if } q \in Q \text{ and } (\delta(q, a) \cap F) \neq \emptyset, \\ \delta^{-1}(q, a) \cup I \cup \overline{F} & \text{if } q \in \overline{Q} \text{ and } (\delta^{-1}(q, a) \cap \overline{I}) \neq \emptyset. \end{cases}$$

From this construction, it follows that A' has at most $2n$ states, and this bound is also reachable.

We will now consider the deterministic state complexity of the positive θ-catenation closure. In [23], it was shown that the state complexity of $(L_1 \cup L_2)^*$ was much lower than the straightforward upper bound of $2^{mn-1} + 2^{mn-2}$. Indeed, the bound obtained from the NFA of Proposition 9 is already 2^{2n} states. We will show that the state complexity of θ-catenation closure is still lower than this.

Proposition 10. *For $n \geq 3$, let A be a DFA defined over an alphabet Σ with n states and let θ be an antimorphism. Then there exists a DFA that recognizes $L(A)^{\odot_+^\theta}$ with at most $2^{2n-1} - 2^n + 1$ states.*

Proof. We define a DFA A' that recognizes $L(A)^{\odot_+^\theta}$ given a DFA A. Let $A = (Q, \Sigma, \delta, s, F)$. We define the DFA $A' = (Q', \Sigma, \delta', s', F')$ with the set of states

$$Q' = \{\langle P, \overline{R} \rangle \mid \emptyset \neq P \subseteq Q - F, \overline{R} \subseteq \overline{Q} - \{\overline{s}\}\}$$
$$\cup \{\langle P \cup \{s\}, \overline{R} \cup \overline{F} \rangle \subseteq Q \times \overline{Q} \mid (P \cup \overline{R}) \cap (F \cup \{\overline{s}\}) \neq \emptyset\},$$

the initial state

$$s' = \begin{cases} \langle \{s\}, \emptyset \rangle & \text{if } s \notin F, \\ \langle \{s\}, \overline{F} \rangle & \text{if } s \in F, \end{cases}$$

the set of final states $F' = \{\langle P, \overline{R} \rangle \subseteq Q \times \overline{Q} \mid (P \cup \overline{R}) \cap (F \cup \{\overline{s}\}) \neq \emptyset\}$, and the transition function for a state $\langle P, \overline{R} \rangle$ and symbol $a \in \Sigma$ with $P' = \delta(P, a)$ and $\overline{R}' = \delta^{-1}(\overline{R}, a)$ is defined by

$$\delta'(\langle P, \overline{R} \rangle, a) = \begin{cases} \langle P' \cup \{s\}, \overline{R}' \cup \overline{F} \rangle & \text{if } (P' \cup \overline{R}') \cap (F \cup \{\overline{s}\}) \neq \emptyset, \\ \langle P', \overline{R}' \rangle & \text{otherwise.} \end{cases}$$

Informally, DFA A' operates by first simulating a computation of A, since by definition, we have $L(A)^{\odot_+^\theta} = L(A)(L(A) \cup \theta(L(A)))^*$. Once the computation reaches a final state of A, an initial state for A and $\overline{A}$ is added to the current state set and the computation continues. Whenever the current state of A' contains a final state of A or $\overline{A}$, the initial states of both machines are added. The computation continues until the input is read and accepts if and only if a final state of A or $\overline{A}$ is contained in the state of A' when the input has been read.

Now let us consider the state set Q' of A',

$$Q' = Q_1 \cup Q_2,$$
$$Q_1 = \{\langle P, \overline{R} \rangle \mid \emptyset \neq P \subseteq Q - F, \overline{R} \subseteq \overline{Q} - \{\overline{s}\}\},$$
$$Q_2 = \{\langle P \cup \{s\}, \overline{R} \cup \overline{F} \rangle \subseteq Q \times \overline{Q} \mid (P \cup \overline{R}) \cap (F \cup \{\overline{s}\}) \neq \emptyset\}.$$

The size of Q' will depend on k and whether or not $s \in F$. We will consider each term. Let $k = |F|$.

First, Q_1 is the set of states with components that do not contain any final state of A or $\overline{A}$. There are $2^{n-k} - 1$ nonempty subsets of $Q - F$ and there are 2^{n-1} subsets of $\overline{Q} - \{\overline{s}\}$. This gives us $|Q_1| = (2^{n-k} - 1)(2^{n-1})$.

Then, Q_2 is the set of states of the form $\langle P, \overline{R} \rangle$ with $P \subseteq Q$ and $\overline{R} \subseteq \overline{Q}$ such that $(P \cup \overline{R}) \cap (F \cup \{\overline{s}\}) \neq \emptyset$. That is at least one of a final state of A is in P or $\overline{s}$ is in $\overline{R}$. Then $s \in P$ and $\overline{F} \subseteq \overline{R}$. This gives up to $(2^{n-1})(2^{n-k})$ states.

This count may include states such that $s \in P$ and $\overline{F} \subseteq \overline{R}$ but $(P \cup \overline{R}) \cap (F \cup \{\overline{s}\}) = \emptyset$, depending on whether or not $s \in F$. If $s \in F$, then there are no such states, since $s \in F$ and $s \in P$ implies that $(P \cup \overline{R}) \cap (F \cup \{\overline{s}\}) \neq \emptyset$.

However, if $s \notin F$, there are up to $(2^{n-1-k})^2$ states $\langle P, \overline{R} \rangle$ such that $s \in P$, $\overline{F} \subseteq \overline{R}$, and $(P \cup \overline{R}) \cap (F \cup \{\overline{s}\}) = \emptyset$ which must be removed from the total, resulting in at most $(2^{n-1})(2^{n-k}) - (2^{n-1-k})^2$ states when $s \notin F$.

Finally, we must account for states of the form $\langle P, \overline{Q} \rangle$. Since A is a complete DFA, we have $\delta^{-1}(\overline{Q}, \sigma) = \overline{Q}$ for all $\sigma \in \Sigma$. Since $\overline{s} \in \overline{Q}$, we have $\langle P, \overline{Q} \rangle$ for all $P \subseteq Q$ and therefore $\delta'(\langle P, \overline{Q} \rangle, w) \in F'$ for all $P \subseteq Q$ and $w \in \Sigma^*$. Thus, all such states are equivalent and indistinguishable. Since $\overline{s} \in \overline{Q}$, for all states $\langle P, \overline{Q} \rangle$, we have $s \in P$ and thus there are 2^{n-1} such states to be merged into a single state.

Thus, in total, we have

$$|Q'| \leq \begin{cases} (2^{n-k} - 1)(2^{n-1}) + (2^{n-1})(2^{n-k}) - 2^{n-1} + 1 & \text{if } s \in F, \\ (2^{n-k} - 1)(2^{n-1}) + (2^{n-1})(2^{n-k}) - (2^{n-1-k})^2 - 2^{n-1} + 1 & \text{if } s \notin F. \end{cases}$$

From this, we can see that the size of Q' is maximized when $k = 1$ and $s \in F$. Thus, Q' has size at most $(2^{n-1} - 1)2^{n-1} + (2^{n-1})^2 - 2^{n-1} + 1 = 2^{2n-1} - 2^n + 1$. $\square$

Lemma 11. *Let $n \geq 3$. Then there exists an n-state DFA A and an antimorphism θ over an alphabet of size 4 such that*

$$sc(L(A)^{\odot^{\theta}_+}) \geq 2^{2n-1} - 2^n + 1.$$

To prove Lemma 11, we demonstrate that the upper bound from Proposition 10 is reachable via the following witness. Let $\Sigma = \{a, b, c, d\}$ and let $\theta : \Sigma^* \to \Sigma^*$ be the antimorphism defined by

$$\theta(a) = b \quad \theta(b) = a \quad \theta(c) = d \quad \theta(d) = c.$$

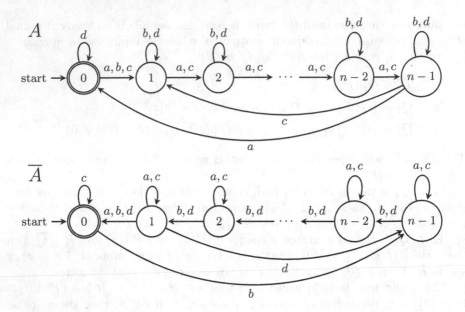

Fig. 3. The DFA A and the NFA $\overline{A}$

We define a DFA A, shown in Fig. 3 together with the NFA $\overline{A}$ which recognizes the language $\theta(L(A))$.

From Proposition 10 and Lemma 11, we can summarize our results in the following theorem.

Theorem 12. *For $n \geq 3$, a regular language L with $sc(L) = n$, and an anti-morphism θ,*

$$sc(L^{\odot^{\theta}}_{+}) \leq 2^{2n-1} - 2^n + 1$$

and this bound can be reached in the worst case.

5   Conclusion

We have given tight bounds for the deterministic and nondeterministic state complexity of pseudocatenation and positive pseudocatenation closure. The deterministic state complexity bounds for each operation differ from those for the corresponding classical operations, catenation and star, and the bounds derived from combined operations. A comparison between the bounds is given in Table 1.

One question that arises is to consider variants of the pseudocatenation operation. The definition of $\odot^{\theta}$ on two words u and v was defined by Kari and Kulkarni [17] to be the set comprising uv and $u\theta(v)$. This definition coincides with θ-powers and θ-primitivity as defined by Czeizler et al. [10]. However, a definition of θ-catenation that incorporates $\theta(u)v$ also makes sense to consider from the biological point of view, since it is the Watson-Crick complement of $u\theta(v)$.

Table 1. A comparison of the deterministic state complexity for each operation

Operation	State complexity	
$L_m \odot^\theta L_n$	$(m-1)(2^{2n} - 2^{n+1} + 2) + 2^{2n-2} - 2^{n-1} + 1$	Theorem 6
$L_m L_n$	$m2^n - 2^{n-1}$	[26]
$L_m(L_n \cup L_p)$	$(m-1)(2^{n+p} - 2^n - 2^p + 2) + 2^{n+p-2}$	[7]
$L_n^{\odot^\theta_2}$	$(n-1)(2^{2n} - 2^{n+1} + 2) + 2^{2n-2} - 2^{n-1} + 1$	Corollary 7
L_n^2	$n2^n - 2^{n-1}$	[22]
$L_n^{\odot^\theta_+}$	$2^{2n-1} - 2^n + 1$	Theorem 12
L_n^*	$2^{n-1} + 2^{n-2}$	[26]
$(L_m \cup L_n)^*$	$2^{m+n-1} - 2^{m-1} - 2^{n-1} + 1$	[23]

There are also further questions considering the state complexity of the current pseudocatenation operation $\odot^\theta$. We can consider the state complexity of pseudocatenation for sub-regular language classes, such as finite languages. We also noted earlier that as a result of our choice of witnesses in Lemma 5, we were also able to obtain the state complexity for the pseudosquare $L^{\odot^\theta_2}$ (Corollary 7). Domaratzki and Okhotin [11] gave a tight state complexity bound for the cube of a language L^3, which was improved by Caron et al. [4]. Asymptotic state complexity bounds for the kth power of a language L^k are also given in [11]. A natural next question to consider is the state complexity of pseudocubes and pseudopowers with respect to θ.

References

1. Brzozowski, J., Liu, D.: Universal witnesses for state complexity of basic operations combined with reversal. In: Konstantinidis, S. (ed.) CIAA 2013. LNCS, vol. 7982, pp. 72–83. Springer, Heidelberg (2013). https://doi.org/10.1007/978-3-642-39274-0_8

2. Brzozowski, J., Liu, D.: Universal witnesses for state complexity of boolean operations and concatenation combined with star. In: Jurgensen, H., Reis, R. (eds.) DCFS 2013. LNCS, vol. 8031, pp. 30–41. Springer, Heidelberg (2013). https://doi.org/10.1007/978-3-642-39310-5_5

3. Brzozowski, J.A., Kari, L., Li, B., Szykuła, M.: State complexity of overlap assembly. In: Câmpeanu, C. (ed.) CIAA 2018. LNCS, vol. 10977, pp. 109–120. Springer, Cham (2018). https://doi.org/10.1007/978-3-319-94812-6_10

4. Caron, P., Luque, J.G., Patrou, B.: State complexity of multiple catenation. arXiv:1607.04031 (2016)

5. Cho, D.J., Han, Y.S., Kim, H., Palioudakis, A., Salomaa, K.: Duplications and pseudo-duplications. Int. J. Unconv. Comput. **12**(2–3), 157–168 (2016)

6. Cho, D.J., Han, Y.S., Ko, S.K., Salomaa, K.: State complexity of inversion operations. Theoret. Comput. Sci. **610**, 2–12 (2016)

7. Cui, B., Gao, Y., Kari, L., Yu, S.: State complexity of two combined operations: catenation-union and catenation-intersection. Int. J. Found. Comput. Sci. **22**(08), 1797–1812 (2011)

8. Cui, B., Gao, Y., Kari, L., Yu, S.: State complexity of combined operations with two basic operations. Theoret. Comput. Sci. **437**, 82–102 (2012)

9. Cui, B., Gao, Y., Kari, L., Yu, S.: State complexity of two combined operations: catenation-star and catenation-reversal. Int. J. Found. Comput. Sci. **23**(01), 51–66 (2012)

10. Czeizler, E., Kari, L., Seki, S.: On a special class of primitive words. Theoret. Comput. Sci. **411**(3), 617–630 (2010)

11. Domaratzki, M., Okhotin, A.: State complexity of power. Theoret. Comput. Sci. **410**(24–25), 2377–2392 (2009)

12. Gao, Y., Moreira, N., Reis, R., Yu, S.: A survey on operational state complexity. J. Automata Lang. Comb. **21**(4), 251–310 (2016)

13. Gao, Y., Yu, S.: State complexity of four combined operations composed of union, intersection, star and reversal. In: Holzer, M., Kutrib, M., Pighizzini, G. (eds.) DCFS 2011. LNCS, vol. 6808, pp. 158–171. Springer, Heidelberg (2011). https://doi.org/10.1007/978-3-642-22600-7_13

14. Gao, Y., Yu, S.: State complexity of combined operations with union, intersection, star and reversal. Fundamenta Informaticae **116**, 79–92 (2012)

15. Glaister, I., Shallit, J.: A lower bound technique for the size of nondeterministic finite automata. Inf. Process. Lett. **59**(2), 75–77 (1996)

16. Kari, L., Konstantinidis, S., Losseva, E., Sosik, P., Thierrin, G.: A formal language analysis of DNA hairpin structures. Fundamenta Informaticae **71**, 453–475 (2006)

17. Kari, L., Kulkarni, M.: Generating the pseudo-powers of a word. J. Automata Lang. Comb. **19**(1–4), 157–171 (2014)

18. Kari, L., Mahalingam, K.: Watson-Crick conjugate and commutative words. In: Garzon, M.H., Yan, H. (eds.) DNA 2007. LNCS, vol. 4848, pp. 273–283. Springer, Heidelberg (2008). https://doi.org/10.1007/978-3-540-77962-9_29

19. Kari, L., Seki, S.: On pseudoknot-bordered words and their properties. J. Comput. Syst. Sci. **75**, 113–121 (2009)

20. Liu, G., Martin-Vide, C., Salomaa, A., Yu, S.: State complexity of basic language operations combined with reversal. Inf. Comput. **206**(9–10), 1178–1186 (2008)

21. de Luca, A., Luca, A.D.: Pseudopalindrome closure operators in free monoids. Theoret. Comput. Sci. **362**(1–3), 282–300 (2006)

22. Rampersad, N.: The state complexity of L^2 and L^k. Inf. Process. Lett. **98**(6), 231–234 (2006)

23. Salomaa, A., Salomaa, K., Yu, S.: State complexity of combined operations. Theoret. Comput. Sci. **383**(2–3), 140–152 (2007)

24. Salomaa, A., Salomaa, K., Yu, S.: Undecidability of state complexity. Int. J. Comput. Math. **90**(6), 1310–1320 (2013)

25. Yu, S.: Regular languages. In: Rozenberg, G., Salomaa, A. (eds.) Handbook of Formal Languages, pp. 41–110. Springer, Heidelberg (1997). https://doi.org/10.1007/978-3-642-59136-5_2

26. Yu, S., Salomaa, K., Zhuang, Q.: The state complexities of some basic operations on regular languages. Theoret. Comput. Sci. **125**(2), 315–328 (1994)

Complexity of Regex Crosswords

Stephen Fenner[(✉)] and Daniel Padé[(✉)]

University of South Carolina, Columbia, SC 29201, USA
fenner@cse.sc.edu, pade@email.sc.edu

Abstract. In a regular expression crossword puzzle, one is given two non-empty lists $\langle\langle R_1, \ldots, R_m\rangle$ and $\langle C_1, \ldots, C_n\rangle\rangle$ over some alphabet, and the challenge is to fill in an $m \times n$ grid of characters such that the string formed by the ith row is in $L(R_i)$ and the string in the jth column is in $L(C_j)$. We consider a restriction of this puzzle where all the R_i are equal to one another and similarly the C_j. We consider a 2-player version of this puzzle, showing it to be **PSPACE**-complete. Using a reduction from **3SAT**, we also give a new, simple proof of the known result that the existence problem of a solution for the restricted (1-player) puzzle is **NP**-complete.

Keywords: Complexity · Regular expressions · Regex crossword · Picture language · NP-complete

1 Introduction

Regular expression crossword puzzles (regex crosswords, for short) share some traits in common with traditional crossword puzzles and with sudoku. One is typically given two lists $R_1, \ldots, R_m$ and $C_1, \ldots, C_n$ of regular expressions labeling the rows and columns, respectively, of an $m \times n$ grid of blank squares. The object is to fill in the squares with letters so that each row, read left to right as a string, *matches* (i.e., is in the language denoted by) the corresponding regular expression, and similarly for each column, read top to bottom. The solution itself may have some additional property, e.g., spelling out a phrase or sentence in row major order.

Regex crosswords have enjoyed some recent popularity, having been discussed in several popular media sources [5,7], and thanks to some websites where people can solve the puzzles online [1,2]. Some variants of the basic puzzle have also been posed [3].

A natural complexity theoretic question to ask is: How hard is it to solve a regex crossword in general?[1] The folklore answer—easy to show and apparently found by several people independently—is that it is **NP**-hard, and the corresponding decision problem ("Does a solution exist?") is **NP**-complete.

---

[1] Glen Takahashi posted this question to Stack Exchange in 2012 [13], but it has been asked by others independently.

© Springer Nature Switzerland AG 2019
C. Martín-Vide et al. (Eds.): LATA 2019, LNCS 11417, pp. 215–230, 2019.
https://doi.org/10.1007/978-3-030-13435-8_16

In this paper, we consider two variations on the basic regex crossword puzzle: (1) a restriction of the puzzle where all the row regexes $R_1, \ldots, R_m$ are equal and all the column regexes $C_1, \ldots, C_n$ are equal; and (2) a 2-player game where players take turns attempting to fill in successive rows and columns of the grid. Variation (2) can also be restricted to having equal row regexes and equal column regexes for the two players. These variants have corresponding decision problems: Let **RC** be the solution existence problem for variation (1), **RCG'** the first-player-win problem for variation (2), and **RCG** the first-player-win problem for the restricted version of (2) (see Sects. 3 and 4 for precise definitions). Our main result is that **RCG'** and **RCG** are both **PSPACE**-complete (see Sect. 4, below). We give explicit polynomial reductions from **TQBF** to **RCG'** and from **RCG'** to **RCG**.

The **NP**-completeness of **RC** was shown in [8],[2] but the polynomial reduction used there was indirect and needlessly complicated for its purpose. As a warm-up to our main result, we give a simple, straightforward polynomial reduction from **3SAT** to **RC**.

In the spirit of the Post Correspondence Problem in computability, our results have the pedagogical benefit of showing the hardness of some decision problems in automata theory that are simply stated and accessible to any undergraduate theory student. The proofs given here are similarly accessible.

1.1 Connections to Other Work

Regex crossword techniques bear some similarity to results in cellular automata, to the Cook-Levin theorem, and to results of Berger from the 1960s showing the undecidability of tiling the plane with Wang tiles (the so-called "domino problem" [6], which was the first proof that there exist finite tile sets that tile the whole plane but only aperiodically).

The particular problems we study here are perhaps chiefly inspired by results in the theory of two-dimensional languages (picture languages) from formal language theory [10]. Given two regexes R and C for the rows and columns, respectively, the *unbounded* (R, C)-crossword problem asks whether a solution grid exists of *any size*. One can show that the recognizable picture languages coincide exactly with the letter-to-letter projections of (R, C)-crossword solutions [10, Theorem 8.6] (except that the empty picture may also be included in the language). Recognizable picture languages can be defined in terms of finite objects known as tiling systems [9] (cf. [10, Definition 7.2]), and given a tiling system T, it is not hard to show that one can effectively find two regular expressions R and C (over some alphabet) and a projection π that defines the same picture language as T. The existence problem for recognizable picture languages ("Given a tiling system, does it define a nonempty language?") is known to be undecidable ([10, Theorem 9.1]), and so, putting these results together, we get that the existence problem for unbounded (R, C)-crosswords is undecidable as

---

[2] In the same paper, a restriction of **RC** where the unique row and column regexes are equal to *each other* was also shown **NP**-complete.

well. A much more direct reduction from the halting problem to unbounded (R, C)-crossword existence was given in [8], where it was also shown that one could even fix the column regex C once and for all, as well as restricting R and C to be over a binary alphabet.

The unbounded regex crossword problem naturally assumes one regex R for all rows and one regex C for all columns, since the number of rows and columns is unspecified. This directly motivates us to impose similar restrictions on the bounded regex crossword problems we study here, where the dimensions of the grid are given as part of the input.

We give some basic concepts and definitions in Sect. 2. Section 3 gives our polynomial reduction from **3SAT** to **RC**. This reduction suggests the technique we use to show our main results about 2-player crossword games in Sect. 4. We give open problems in Sect. 5.

2   Preliminaries

We fix an alphabet Σ once and for all and assume it contains the symbols 0 and 1 at least. For the **NP**-completeness result of Sect. 3, one can assume that $\Sigma = \{0, 1\}$. For the **PSPACE**-completeness result of Sect. 4, it suffices that $\Sigma = \{0, 1, 2\}$.

2.1   3SAT

An instance of **3SAT** is described by a Boolean formula φ over k variables $x_1, \ldots, x_k$, given in conjunctive normal form:

$$\varphi := C_i \wedge \cdots \wedge C_d$$

where each C_i is a clause of three literals (each a variable or its negation) connected by disjunctions:

$$C_i := \ell_{i,1} \vee \ell_{i,2} \vee \ell_{i,3}$$

The question is, is φ true (is it *satisfied*) for some assignment of the variables. This is the canonical complete problem for **NP**. In Sect. 3 we show that the language **RC**—the language of (R, C)-crosswords—is **NP**-complete by giving reduction from **3SAT**.

2.2   TQBF

An instance of **TQBF** is described by a closed Boolean formula φ, given in prenex normal form:

$$\varphi := \exists x_0 \forall y_0 \cdots \exists x_{k-1} \forall y_{k-1} \exists x_k \tilde{\varphi}(x_0, y_0, \ldots, x_{k-1}, y_{k-1}, x_k) \tag{1}$$

where $\tilde{\varphi}$ is a quantifier-free Boolean formula which can be assumed to be in conjunctive normal form with c clauses and $2k + 1$ variables, for some positive c and k.

The sentence φ is naturally viewed as a two-player game, where the players alternate choosing truth values for the variables in order, the first player wishing to make the formula $\tilde{\varphi}$ true and second player wishing to make it false. The question to be answered is whether φ is true when the quantified variables range over the Boolean values FALSE and TRUE.[3] That is, whether the first player has a winning strategy in the corresponding game.

As **3SAT** is for **NP**, **TQBF** is the canonical complete problem for **PSPACE**. In Sect. 4, **RCG**—the language of (R, C)-crossword games (defined below) with a winning strategy for the first player—is **PSPACE**-complete by reduction from **TQBF**.

3   (R, C)-Crosswords

For two given regexes R and C over Σ, an (R, C)-*crossword solution* is a two-dimensional m by n grid of symbols from the alphabet. Interpreting rows and columns as strings, each row must match R and each column must match C.

An (R, C)-*crossword* is represented as a 4-tuple $\langle 0^m, 0^n, R, C \rangle$ where the number of rows and columns are given in unary as m and n, and R and C are row and column regexes over Σ (defined in the usual way, using the operators $\cup$, $\|$, $*$, where $\|$ or juxtaposition both indicate concatenation).

Definition 1. *The language* **RC** *is the set of all* (R, C)-*crosswords for which there exists an* (R, C)-*crossword solution of the given dimensions.*

RC was shown to be **NP**-complete in [8] via an indirect, complicated reduction. In this section, we give a much more straightforward polynomial reduction from **3SAT** to **RC**.

3.1   The Reduction

Given a Boolean formula φ with $k \geq 1$ variables and d clauses as defined in Sect. 2.1 above (where we can assume $d \geq 3$), we construct an instance $\langle 0^{d+1}, 0^{k+d}, R, C \rangle$ of **RC** as follows: For $1 \leq i \leq d$, we define t_i to be the regex

$$t_i = 0^{i-1} 1 0^{d-i} = \underbrace{0 \cdots 0}_{i-1} 1 \underbrace{0 \cdots 0}_{d-i} \ .$$

---

[3] More precisely, the question is whether the sentence $\exists x_0 \forall y_0 \cdots \exists x_{k-1} \forall y_{k-1} \exists x_k$ $[\tilde{\varphi}(x_0, y_0, \ldots, x_{k-1}, y_{k-1}, x_k) = \text{TRUE}]$ holds in the two-element Boolean algebra $(\{\text{FALSE}, \text{TRUE}\}, \wedge, \vee, \neg)$.

Then we define

$$S = 1^d 0^*$$

$$R = \left(\bigcup_{i=1}^{d} t_i R_i \right) \cup S$$

$$C = 1 \, (0^* 10^*) \cup 0 (0^* \cup 1^*)$$

where S is called the 'spine,' and for $1 \le i \le d$, R_i is derived from the formula φ as follows:

$$R_i = (a_{i,1} \cdots a_{i,k}) \cup (b_{i,1} \cdots b_{i,k}) \cup (c_{i,1} \cdots c_{i,k})$$

where, for $1 \le j \le k$,

$$a_{i,j} = \begin{cases} 1 & \text{if the first literal in the } i^{\text{th}} \text{ clause is } x_j \\ 0 & \text{if the first literal in the } i^{\text{th}} \text{ clause is } \overline{x_j} \\ (1 \cup 0) & \text{otherwise} \end{cases}$$

and $b_{i,j}$, $c_{i,j}$ are set similarly according to the second and third literals in each clause.

We show that φ is satisfiable iff an (R, C)-crossword solution exists.

First, assuming that φ is satisfiable, where $\langle z_1, \ldots, z_k \rangle$ is a satisfying assignment, then this sets up a $d + 1$ by $d + k$ crossword solution of the following form:

	c_1	c_2	c_3	$\cdots$	c_d	c_{d+1}	$\cdots$	c_{d+k}
r_0	1	1	1	$\cdots$	1	0	$\cdots$	0
r_1	1	0	0	$\cdots$	0	z_1	$\cdots$	z_k
r_2	0	1	0	$\cdots$	0	z_1	$\cdots$	z_k
$\vdots$	$\vdots$	$\vdots$	$\vdots$	$\ddots$	$\vdots$	z_1	$\cdots$	z_k
r_c	0	0	0	$\cdots$	1	z_1	$\cdots$	z_k

Fig. 1. Solution

Here, the first row is the spine (matching S); the block on the left below the spine is akin to an identity matrix; and the block on the right consists of columns where each column is either all 1's or all 0's (save the first element, which is always 0), according to each z_i. An overview representation is shown below:

Spine	
Calibration Region	Clause Verification

Where the spine is the string that matches S. The 'clause verification region' is determined by the satisfying assignment to φ, i.e., if z_j is true in the satisfying assignment, then column c_{d+j} will match the regex $\mathbf{01}^*$; otherwise it will match $\mathbf{00}^*$.

By construction, it is clear that if φ is satisfiable, then the (R, C)-crossword constructed above is solvable. In other words, there is a way to fill in the crossword such that all rows match the regular expression R, and all columns match the regular expression C.

In fact, since the calibration region requires only one 1 per row and column, the solution given in Fig. 1 is not the only valid one. It is easy to see that once any solution is given, any rearranging of the (non-spine) rows gives another valid solution. Due to this fact it is guaranteed that for each i, some row matches $t_i R_i$, which is important for the converse below.

3.2   An (R, C)-Crossword Solution Guarantees φ Is Satisfiable

To complete the proof, it must be shown that if the crossword is solvable, this implies that φ is satisfiable. We do this via a series of lemmas.

Here we assume an (R, C)-crossword solution exists with rows $\langle r_0, \ldots, r_d \rangle$ and columns $\langle c_1, \ldots, c_{d+k} \rangle$.

Observe that since each r_j matches R, it must either start with d many 1's or else have exactly one 1 among its first d symbols.

Lemma 1. *The string r_0 matches S.*

Proof. Assume not. Then r_0 must match $t_i R_i$ for some $1 \leq i \leq d$. Fix such an i. The picture below shows the case where r_0 matches $t_2 R_2$, i.e., $r_0 = 010 \cdots$:

	c_1	c_2	c_3	c_4	$\cdot$	c_d	$\cdot$	$\cdot$
r_0	0	1	0	0	$\cdot$	0		
$\vdots$								

From the definition of C, we see that c_i must match $\mathbf{1}(\mathbf{0}^*\mathbf{10}^*)$, that is, $c_i = 10^{j-1}10^{d-j}$ for some $1 \leq j \leq d$. The picture below shows the case where $i = 2$ and $j = 2$, that is, where $c_i = c_2 = 10100 \cdots 0$:

	c_1	c_2	c_3	c_4	$\cdot$	c_d		$\cdot$	$\cdot$
r_0	0	1	0	0	$\cdot$	0			
r_1		0							
r_2		1							
r_3		0							
$\vdots$		$\vdots$							

For r_j, we have two cases, both leading to contradiction:

r_j **matches** S: This requires that all of the first d columns other than c_i match **01***, which means $r_{j'}$ starts with $1^{i-1}01^{d-i}\cdots$ for all $j' \geq 1$ such that $j' \neq j$. These rows do not match R.

r_j **matches** $t_i R_i$, **that is,** $r_j = 0^{i-1}10^{d-i}\cdots$: This requires that all of the first d columns other than c_i match **0***, which means no rows other than r_j and r_0 will match R, since they all start with 0^d.

This proves the lemma.

By Lemma 1, the first d columns must match $\mathbf{1(0^*10^*)}$; we call such columns *calibration columns*.

Lemma 2. *No row other than r_0 matches S.*

Proof. Again assume this not the case. By the previous lemma, r_0 must match S. Suppose r_j also matches S for some $j \geq 1$. Then C forces $r_{j'}$ to start with d many 0's for all $1 \leq j' \neq j$, because the calibration columns are only allowed a single 1 below the spine. Thus none of these $r_{j'}$ matches R.

Lemma 3. *For any i, $1 \leq i \leq d$, some row matches $t_i R_i$*

Proof. By Lemmas 1 and 2, we have that r_0 is the only row to match the spine S. Since $R = (\bigcup_{i=1}^{d} t_i R_i) \cup S$, it follows that each of the other rows matches $t_i R_i$ for some i. For the purposes of contradiction, assume that there is some $t_i R_i$ not matched by any row. Then by the pigeonhole principle, there must be two distinct rows r_n and r_m both matching $t_\ell R_\ell$ for the same ℓ. By the definition of t_ℓ, the column c_ℓ will thus have at least two 1's:

	c_1	·	$c_{\ell-1}$	c_ℓ	$c_{\ell+1}$	·	c_d	c_{d+1}	·
r_0	1	·	1	1	1	·	1		·
⋮			⋮						
r_n	0	·	0	1	0	·	0		·
⋮			⋮						
r_m	0	·	0	1	0	·	0		·
⋮									

But then column c_ℓ does not match C. This completes the proof.

Lemma 4. φ *is satisfiable.*

Proof. Because of the spine in the first row, note that for $1 \le j \le k$, c_{d+j} matches either $\mathbf{01^*}$ or $\mathbf{00^*}$. Set

$$z_j = \begin{cases} 1 \text{ if } c_{d+j} \text{ matches } \mathbf{01^*}, \\ 0 \text{ if } c_{d+j} \text{ matches } \mathbf{00^*}. \end{cases}$$

We show that $\langle z_1, \ldots, z_k \rangle$ is a satisfying truth assignment for φ. Consider the ith clause C_i of φ. By Lemma 3, some non-spine row matches $t_i R_i$. Let r be the suffix of that row obtained by removing its first d symbols. Then r matches either $a_{i,1} \cdots a_{i,k}$, $b_{i,1} \cdots b_{i,k}$, or $c_{i,1} \cdots c_{i,k}$. Suppose r matches $a_{i,1} \cdots a_{i,k}$ (the other two cases are handled similarly). Let x_j be the variable mentioned by the first literal $\ell_{i,1}$ of C_i. If $\ell_{i,1} = x_j$, then $a_{i,j} = 1$, whence r has a 1 as its jth symbol, whence c_{d+j} matches $\mathbf{01^*}$, whence $z_j = 1$, which makes $\ell_{i,1}$ true, satisfying C_i. Similarly, if $\ell_{i,1} = \overline{x_j}$, then $z_j = 0$, also satisfying C_i.

Since i was arbitrary, we have that φ is satisfied by $\langle z_1, \ldots, z_k \rangle$. ∎

4   (R, C)-Crossword Games

For two given regexes R and C over Σ, an (R, C)-*game* is a two-player combinatorial game that can be thought of as follows: we start with a two-dimensional grid X with m rows and n columns (m and n are positive integers). X is initially empty. Player 1, who we call *Rose*, fills in the first row of X with symbols from Σ to form a string matching R.

Player 2, who we call *Colin*, responds by filling the remainder of the first column of X with symbols from Σ so that the entire column matches C. Rose then fills the remainder of the second row so that it matches R, then Colin the remainder of the second column to match C, etc. The first player unable to fill a row (respectively, column) in this way loses, and the other player wins.[4]

We represent an (R, C)-game as a 4-tuple $\langle 0^m, 0^n, R, C \rangle$, where m and n are positive integers (the number of rows and columns of the grid, respectively), and R and C are the corresponding regexes over Σ (defined in the usual way, using the operators $\cup, \|, *$).

Note that the numbers m and n are given in *unary*.

Definition 2. *The language* **RCG** *is the set of all* (R, C)*-games where Rose has a winning strategy.*

4.1 RCG $\in$ PSPACE

It is straightforward to observe that **RCG** $\in$ **PSPACE**. This follows from the properties of (R, C)-games: Given an instance of **RCG** of size $N = m \cdot n$,

- all game positions are representable by strings of polynomial length (in N),
- any play of the game lasts for at most polynomially many turns, and
- given any game position, whether a given next move is legal can be determined in polynomial space (polynomial time, in fact).

For this it is crucial that the dimensions of the board be given in unary. If the dimensions were given in binary, then we conjecture that the corresponding language would be complete for **EXPSPACE**. Also note that the regex matching problem ("Given a regex E and string w, does w match E?") is in **P**.

4.2 Hardness of RCG

Here is the main result of our paper.

Theorem 1. TQBF $\leq_p$ RCG.

To prove Theorem 1, our main result, we first consider a variant of **RCG**, where each row and each column may correspond to a different regex, that is, the input is a pair $\langle \langle R_1, \ldots, R_m \rangle, \langle C_1, \ldots, C_n \rangle \rangle$ of lists of regexes. Rose and Colin alternate turns as before, but on her i^{th} turn, Rose must fill the remainder of the ith row so that it matches R_i, and similarly, on his j^{th} turn, Colin must fill the remainder of the jth column so that it matches C_j. Call this variant **RCG'**.

We show our main result in two steps: in Lemma 5 we show how to polynomially reduce **TQBF** to **RCG'**; then we give a polynomial reduction from **RCG'** to **RCG** (Theorem 2 below). In using **RCG'**, the goal is to first consider a 'simpler' game to verify that there is a correspondence between the formulæ in **TQBF** and the possible games in **RCG**.

---

[4] For the last move of the game, Rose or Colin may encounter a row or column, respectively, that is already completely filled in. In this case, she or he wins if and only if the row or column matches the corresponding regular expression.

Lemma 5. $\textbf{TQBF} \leq_p \textbf{RCG}'$.

Proof. Given an instance φ of **TQBF** as in Eq. (1) of Sect. 2.2 with c clauses and $2k+1$ variables, we construct an equivalent instance of **RCG**$'$ with $m := k+c+1$ rows and $n := k + c$ columns. The intersection of the first $k + 1$ rows and first $k+1$ columns we will call the *variable region*. There are c rows below this region, one for each clause of $\bar{\varphi}$, which we collectively call the *clause region*. The regular expressions for each player in **RCG**$'$ are over the alphabet $\{0,1\}$ and are defined as follows (with an explanation afterward): for $1 \leq i \leq m$, we let

$$R_i := \begin{cases} (\mathbf{0} \cup \mathbf{1})^* \mathbf{0}^{c-1} & \text{if } 1 \leq i \leq k+1, \\ (\mathbf{0} \cup \mathbf{1})^* \mathbf{1}(\mathbf{0} \cup \mathbf{1})^* \mathbf{0}^{c-1} & \text{if } k+2 \leq i \leq m, \end{cases}$$

and for all $1 \leq i \leq n$, we let

$$C_i := \begin{cases} \displaystyle\bigcup_{a,b \in \{0,1\}} (\mathbf{0} \cup \mathbf{1})^{i-1} ab (\mathbf{0} \cup \mathbf{1})^{k-i} \| S_{iab} & \text{if } 1 \leq i \leq k, \\ \displaystyle\bigcup_{a \in \{0,1\}} (\mathbf{0} \cup \mathbf{1})^k a \| T_a & \text{if } i = k+1, \\ \mathbf{0}^* & \text{if } k+2 \leq i \leq n, \end{cases}$$

where given $1 \leq i \leq k+1$, and $a, b \in \{0,1\}$, the regexes S_{iab} and T_a are defined as follows: First, for $1 \leq j \leq c$ let

$$u_j := \begin{cases} \mathbf{0} & \text{if } x_{i-1} \text{ occurs negatively in the } j\text{th clause,} \\ \mathbf{1} & \text{if } x_{i-1} \text{ occurs positively in the } j\text{th clause,} \\ \bot & \text{if } x_{i-1} \text{ does not occur in the } j\text{th clause.} \end{cases}$$

$$v_j := \begin{cases} \mathbf{0} & \text{if } y_{i-1} \text{ occurs negatively in the } j\text{th clause,} \\ \mathbf{1} & \text{if } y_{i-1} \text{ occurs positively in the } j\text{th clause,} \\ \bot & \text{if } y_{i-1} \text{ does not occur in the } j\text{th clause.} \end{cases}$$

Now for $1 \leq j \leq c$, define

$$d_j := \begin{cases} \mathbf{1} & \text{if } u_j = a \text{ or } v_j = b, \\ \mathbf{0} & \text{otherwise.} \end{cases} \qquad e_j := \begin{cases} \mathbf{1} & \text{if } u_j = a, \\ \mathbf{0} & \text{otherwise.} \end{cases}$$

Finally, we let $S_{iab} := d_1 \| \cdots \| d_n$ and $T_a := e_1 \| \cdots \| e_n$.

Each of the first $k + 1$ rows and columns corresponds to the truth value (0 or 1) of one or two particular variables in the original formula, as depicted in Fig. 2. The remainder of the rows (c of them) correspond to the clauses of φ.

Here is how this **RCG**$'$ game reflects the original instance of **TQBF** viewed as a game. When Rose plays the ith row (for $1 \leq i \leq k+1$) she is able to choose the truth value of x_{i-1} by placing a 0 or 1 in the corresponding square in Fig. 2 (Rose can play any binary string in the remainder of her row, because $R_i = (\mathbf{0} \cup \mathbf{1})^*$). Then when Colin plays the remainder of the ith column according to C_i, he can similarly choose the truth value of y_{i-1} by placing a 0 or 1 in the corresponding square in Fig. 2. However, because of the S_{iab} component of C_i, Colin is

x_0	?	?	·	?
y_0	x_1	?	⋱	?
?	y_1	x_2	⋱	?
⋮	⋱	⋱	⋱	⋮
?	?	·	y_{k-1}	x_k

Fig. 2. The layout of the variable region. The question marks represent either 0 or 1.

forced to then place a 1 in each of the last c positions corresponding to a clause that is satisfied by the truth settings of these two variables. (The minor exception is the $(k+1)$st column, where there is only the variable x_k to consider.)

Also note that in order for Rose to complete the board, there must be a 1 in at least one of the first $k+1$ positions in every row of the clause region. That is, Rose can win just when the chosen truth values of the variables satisfy all clauses of $\tilde{\varphi}$, making the two games equivalent. Our construction is clearly polynomial time, which finishes the proof.

4.3 Constraining the Players

Theorem 2. RCG$' \leq_p$ RCG.

The rest of this section is a proof of Theorem 2. To reduce from **RCG$'$** to **RCG** we need to provide a method to consolidate the families of regular expressions into one regex per player. Here, we present a generic construction that forces the players to play in order, which can be applied to any **RCG$'$** game—forcing each player to play their families of regexes in index order.

Given an arbitrary instance $G := \langle\langle R_1, \ldots, R_m\rangle, \langle C_1, \ldots, C_n\rangle\rangle$ of **RCG$'$**, we construct an equivalent instance of **RCG**. Our construction requires the alphabet Σ to contain a third symbol "2" that is not part of any string matching any of the R_i or C_i. We currently do not know how to remove this requirement. We can assume that the given **RCG$'$** grid is square, i.e., $m = n$: Suppose this is not the case; for example, suppose $m < n$. Then we can pad the grid with $n - m$ bottom rows by

- concatenating each C_i with $\mathbf{0}^{n-m}$ on the right, and
- defining $R_i := \mathbf{1}^*$ for $m < i \leq n$,

yielding an evidently equivalent $n \times n$ game. We can do something similar if $m > n$. The instance of **RCG** we construct from G will then be a $(2n+1) \times (2n+1)$ game $H := \langle 0^{2n+1}, 0^{2n+1}, R, C\rangle$. We may also assume that $n \geq 2$.

The regular expressions R and C we construct for the respective players are given below, again with explanations afterwards (Fig. 3):

$$R := \mathbf{210^*} \ \cup \tag{2}$$

$$\underbrace{\bigcup_{i=0}^{n-2} \mathbf{0^i 1^3 0^{n-i-2}}}_{\text{I}} \| \underbrace{\mathbf{0^i 1 0^{n-i-1}}}_{\text{II}} \ \cup \tag{3}$$

$$\underbrace{\mathbf{00^{n-2}11}}_{\text{Ir}} \| \underbrace{\mathbf{0^{n-1}1}}_{\text{II}} \ \cup \tag{4}$$

$$\bigcup_{i=1}^{n} \underbrace{\mathbf{0^i 1 0^{n-i}}}_{\text{III}} \| R_i \tag{5}$$

(a) Rose's regular expression. Regex (2) is the 'spine', while regexes (3–4) define the 'calibration' region (I, II). Regex (5) continues calibration in region III while also including the row regexes from G.

$$C := \mathbf{210^*} \ \cup \tag{6}$$

$$\underbrace{\bigcup_{i=0}^{n-2} \mathbf{0^i 1^3 0^{n-i-2}}}_{\text{I}} \| \underbrace{\mathbf{0^i 1 0^{n-i-1}}}_{\text{III}} \ \cup \tag{7}$$

$$\underbrace{\mathbf{00^{n-2}11}}_{\text{Ic}} \| \underbrace{\mathbf{0^{n-1}1}}_{\text{III}} \ \cup \tag{8}$$

$$\bigcup_{i=1}^{n} \underbrace{\mathbf{0^i 1 0^{n-i}}}_{\text{II}} \| C_i \ \cup \tag{9}$$

$$(\mathbf{0} \cup \mathbf{1} \cup \mathbf{100} \cup \mathbf{00^* 10}) \mathbf{2^*} \tag{10}$$

(b) Colin's regular expression. Regex (6) is the 'spine', regexes (7–8) are the calibration region (I and III), regex (9) continues calibration in region II while also including the column regexes from G, and regex (10) is a 'bomb' to punish Rose for cheating.

Fig. 3. The regular expressions wrapping games in **RCG**. Regexes are bracketed with the regions they describe, illustrated in Fig. 4a.

Figure 4a illustrates how H 'wraps' around the game G: players first fill in the spine, then regions I, II, and III before simulating the game G in the lower right square (light grey).

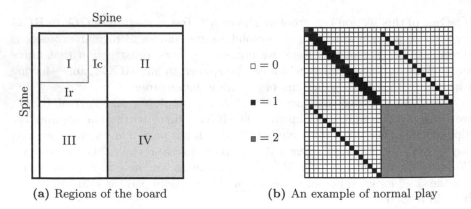

(a) Regions of the board   (b) An example of normal play

Fig. 4. Regions to constrain the players. Each 'block' is a $n \times n$ square.

4.4   Normal Play

By a *round*, we mean a pair of consecutive turns, starting with Rose. We index the rounds starting with round 0. Normal play is in three stages:

Spine: In round 0, both players play the spine, i.e., a string matching 210^*.

Calibration: In round i, where $1 \leq i \leq n$, Rose and Colin each play a 'calibration string,' i.e. either the string matching $0^i 1^3 0^{n-i-2} \| 0^i 10^{n-i-1}$ (if $i < n$) or $00^{n-2} 11 \| 0^{n-1} 1$ (if $i = n$).

Simulation: Rose and Colin now simulate the given $\mathbf{RCG}'$ game: In round $(n + i)$, for $1 \leq i \leq n$, Rose plays a string matching $0^i 10^{n-i-1} \| R_i$ (if she can), and Colin plays a string matching $0^i 10^{n-i-1} \| C_i$ (if he can).

Figure 4b illustrates the state of the grid after round n of normal play (here, $n = 16$). If either player deviates from normal play, we say that the first player to do so is *cheating*. The next lemmas show that Colin cannot cheat, and if Rose cheats, then Colin can force her to lose in a constant number of rounds by playing a *bomb*, i.e., a string matching $(0 \cup 1 \cup 100 \cup 00^* 10) 2^*$, once or twice.

Lemma 6. *In round 0, if Rose does not play the spine, then Colin can win; otherwise, Colin must also play the spine.*

Proof. If Rose does not play 210^*, she has two choices for the first character. If she chooses 0, say, then Colin has a quick kill by playing a bomb (see Fig. 5), with similar results if she cheats with a 1.

In either case, Rose would quickly lose. If Rose does play 210^* on her first turn, Colin must play a string prefixed with 2, his only option matching the regex 210^*.

Lemma 7. *After normal play through round $(i - 1)$ for $1 \leq i \leq n$, Rose prefers regex (3) to regex (5) in round i.*

Proof. If $i = 1$, then Rose must play a string with prefix 1, and so she must play a string matching regex (3). Now suppose $i \geq 2$, and consider the following portion of the board at the start of round i when both players have been playing normally:

1	1	0	0
1	1	1	0
0	1		
0	0		

Rose has a choice of regexes (3) or (5), as each can match a string prefixed by $0^{i-1} 1$. Say Rose chooses regex (5), thus playing a string matching $0^{i-1} 10^{n-i} \| R_{i-1}$. Colin can then respond with a bomb:

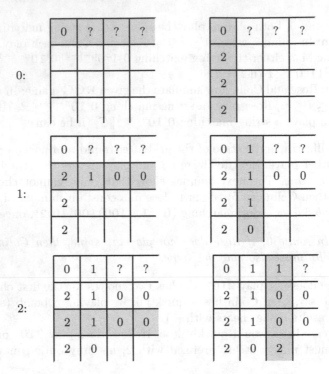

Fig. 5. Each round when Rose cheats with $0\cdots$ in her first move and Colin plays a bomb. Note that Rose has no regex to match the prefix 20. We replace a '?' with a 1 in round 1 to show the worst case, where Colin must survive through round 2 (not required in the 0 case).

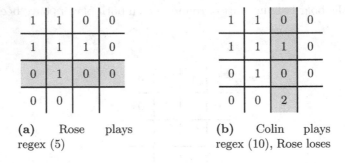

(a) Rose plays regex (5)

(b) Colin plays regex (10), Rose loses

Rose cannot then play any string with prefix $0^i 2$, so she loses in round $(i+1)$.

Lemma 8. *Colin cannot cheat in rounds 1 through n.*

Proof. By Lemma 6, we begin round 1 with the spine having been played by both players. Rose is then forced to play a string prefixed with 1, the only matching regex being regex (3) with $i = 0$: $1110^{n-2}\|10^{n-1}$. From this point on through round n, assuming Rose plays normally, Colin will be faced with prefix $0^{i-1}11$ in round i, and thus must play a string matching regex (7) or (8), i.e., play normally.

The preceding lemmas show that normal play is optimal for both players (even required for Colin) through round n. Thus we can assume normal play through round n, filling regions II and III of the grid with 1's along their diagonals and 0's elsewhere (as with the identity matrix).

Lemma 9. *Assume normal play through round n. For $1 \leq i \leq n$, in round $(n + i)$, Rose must play a string matching $\mathbf{0}^i\mathbf{10}^{n-i}\|R_i$ and Colin must play a string matching $\mathbf{0}^i\mathbf{10}^{n-i}\|C_i$.*

Proof. In round $(n + i)$, Rose and Colin are both faced with prefix 0^i10^{n-i}, and the only regexes that this matches are the respective regexes given above for Rose and Colin.

In rounds $(n + 1)$ through $2n$, the players are essentially playing the game G in region IV, so the winner of H is the winner of G. This completes the proof of Theorem 2.

5   Open Problems

The most immediate question arising from our work is whether **RCG** is **PSPACE**-hard restricted to a binary alphabet. Our proof shows only that it is **PSPACE**-hard for a ternary alphabet. Doing without the third symbol "2" in the alphabet currently seems like a daunting task, despite the fact that under normal play, that symbol appears only once in the upper left-hand corner.

Another question is whether we still get **PSPACE**-hardness if we restrict the regexes R and C to be equal to each other. If one can show **PSPACE**-hardness for **RCG'** restricted so that $R_i = C_i$ for all i, then it may be easy to get $R = C$ for the constructed instance of **RCG**, since these two latter regexes are close to being equal anyway.

Acknowledgments. We would like to thank Thomas Thierauf for several interesting discussions on this topic and to Joshua Cooper for finding for us a particularly challenging and fun regex crossword puzzle. We are also grateful to Klaus-Jörn Lange for suggesting the connection between our work and the theory of picture languages.

References

1. http://regexcrossword.com
2. http://www.regexcrosswords.com
3. MIT Mystery Hunt. http://www.mit.edu/~puzzle
4. Royal dinner. http://regexcrossword.com/challenges/experienced/puzzles/1
5. Slashdot discussion, February 2013. http://games.slashdot.org/story/13/02/13/2346253/can-you-do-the-regular-expression-crossword

6. Berger, R.: The undecidability of the domino problem. No. 66 in memoirs of the American Mathematical Society. American Mathematical Society, Providence, Rhode Island (1966). mR0216954
7. Black, L.: Can you do the regular expression crossword? I programmer, February 2013. http://www.i-programmer.info/news/144-graphics-and-games/5450-can-you-do-the-regular-expression-crossword.html
8. Fenner, S.: The complexity of some regex crossword problems (2014)
9. Giammarresi, D., Restivo, A.: Recognizable picture languages. Int. J. Pattern Recogn. Artif. Intell. 31–46 (1992)
10. Giammarresi, D., Restivo, A.: Two-dimensional languages. In: Rozenberg, G., Salomaa, A. (eds.) Handbook of Formal Languages, vol. 3, pp. 215–267. Springer, Heidelberg (1997). https://doi.org/10.1007/978-3-642-59126-6_4. Chap. 96
11. Latteux, M., Simplot, D.: Recognizable picture languages and domino tiling. Theor. Comput. Sci. **178**(1–2), 275–283 (1997). Note
12. Rosenfeld, A., Rheinboldt, W.: Picture Languages: Formal Models for Picture Recognition. Computer Science and Applied Mathematics. Elsevier Inc., Academic Press Inc., New York (1979)
13. Takahashi, G.: Are regex crosswords NP-hard? cS Stack Exchange question 30143, answered by FrankW, September 2014. http://cs.stackexchange.com/questions/30143/are-regex-crosswords-np-hard
14. Takahashi, G.: Are regex crosswords NP-hard? CS Stack Exchange question 30143, answered by FrankW (2014). http://cs.stackexchange.com/questions/30143/are-regex-crosswords-np-hard

Grammars

Generalized Predictive Shift-Reduce Parsing for Hyperedge Replacement Graph Grammars

Berthold Hoffmann[1] and Mark Minas[2]([envelope])

[1] Universität Bremen, Bremen, Germany
hof@informatik.uni-bremen.de
[2] Universität der Bundeswehr München, Neubiberg, Germany
mark.minas@unibw.de

Abstract. Parsing for graph grammars based on hyperedge replacement (HR) is in general NP-hard, even for a particular grammar. The recently developed predictive shift-reduce (PSR) parsing is efficient, but restricted to a subclass of unambiguous HR grammars. We have implemented a generalized PSR parsing algorithm that applies to all HR grammars, and pursues severals parses in parallel whenever decision conflicts occur. We compare GPSR parsers with the Cocke-Younger-Kasami parser and show that a GPSR parser, despite its exponential worst-case complexity, can be much faster.

Keywords: Hyperedge replacement grammar · Graph parsing

1 Introduction

It is well known that parsing for graph grammars based on hyperedge replacement (HR) is in general NP-hard, even for a particular grammar [8]. In earlier work [6], we have devised predictive shift-reduce parsing (PSR), which lifts Knuth's LR string parsing [9] to graphs, is efficient, but unfortunately restricted to a subclass of unambiguous HR grammars. This makes it unsuitable for applications in natural language processing (NLP) where grammars are often ambiguous. So we extend the PSR algorithm to arbitrary HR grammars in this paper: Just like Tomita's generalized LR string parser [12], the *generalized PSR parser* pursues all possible parses of a graph in parallel whenever ambiguity occurs. We describe the implementation of the generalized PSR parser by Mark Minas,[1] and compare its efficiency with the Cocke-Younger-Kasami parser for arbitrary HR grammars [11].

The remainder of this paper is structured as follows. After recalling HR grammars in Sect. 2 and PSR parsing in Sect. 3, we introduce generalized PSR parsing in Sect. 4, and compare its performance with CYK parsing in Sect. 5.

---

[1] In the graph parser generator *Grappa*, available at www.unibw.de/inf2/grappa.

C. Martín-Vide et al. (Eds.): LATA 2019, LNCS 11417, pp. 233–245, 2019.
https://doi.org/10.1007/978-3-030-13435-8_17

Due to lack of space, our presentation is driven by a small example—a grammar for series-parallel graphs—that exhibits many peculiarities of generalized PSR parsing. In Sect. 6, we conclude by indicating related and future work.

2   Graph Grammars Based on Hyperedge Replacement

Throughout the paper, we assume that X is a global, countably infinite supply of *nodes*, and that Σ is a finite set of *symbols* that comes with an *arity function* $arity \colon \Sigma \to \mathbb{N}$, and is partitioned into disjoint subsets $\mathcal{N}$ of *nonterminals* and $\mathcal{T}$ of *terminals*.

We represent hypergraphs as ordered sequences of edge literals, where each literal represents an edge with its attached nodes. This is convenient as we shall derive (and parse) the edges of a graph in a fixed order.

Definition 1 (Hypergraph). For a symbol $\mathsf{a} \in \Sigma$ and $k = arity(\mathsf{a})$ pairwise distinct nodes $x_1, \ldots, x_k \in X$, $a = \mathsf{a}(x_1, \ldots, x_k)$ represents a *hyperedge* that is labeled with a and attached to $x_1, \ldots, x_k$. $\mathcal{E}_\Sigma$ denotes the set of hyperedges (over Σ).

A *hypergraph* $\langle \gamma, V \rangle$ consists of a sequence $\gamma = e_1 \cdots e_n \in \mathcal{E}_\Sigma^*$ of *hyperedges* and a finite set $V \subseteq X$ of *nodes* that contains all nodes attached to the hyperedges of γ. $\mathcal{G}_\Sigma$ denotes the set of all hypergraphs (over Σ).

In the following, we usually call hypergraphs just *graphs* and hyperedges just *edges*. Moreover, we denote a graph just by its edges γ, and refer to its nodes by V_γ.[2] The "concatenation" of two graphs $\alpha, \beta \in \mathcal{G}_\Sigma$ yields a graph $\gamma = \alpha\beta$ with nodes $V_\gamma = V_\alpha \cup V_\beta$. Two graphs γ and γ' are *equivalent*, written $\gamma \bowtie \gamma'$, if $V_\gamma = V_{\gamma'}$ and γ is a permutation of γ'.

Note that we order the edges of a graph in rules and derivations. However, the relation $\bowtie$ makes graphs with permuted edges equivalent, like in ordinary definitions of graphs. Our parsers will make sure that equivalent graphs are always processed in the same way.

An injective function $\varrho \colon X \to X$ is called a *renaming*, and γ^ϱ denotes the graph obtained by replacing all nodes in γ (and in V_γ) according to ϱ. A *hyperedge replacement rule* $r = (A \to \alpha)$ (*rule* for short) has a nonterminal edge $A \in \mathcal{E}_\mathcal{N}$ as its *left-hand side*, and a graph $\alpha \in \mathcal{G}_\Sigma$ with $V_A \subseteq V_\alpha$ as its *right-hand side*.

Consider a graph $\gamma = \beta \bar{A} \bar{\beta} \in \mathcal{G}_\Sigma$ with a nonterminal edge $\bar{A}$ and a rule $r = (A \to \alpha)$. A renaming $\mu \colon X \to X$ is a *match* (of r in γ) if $A^\mu = \bar{A}$ and if $V_\gamma \cap V_{\alpha^\mu} \subseteq V_{A^\mu}$.[3] A match μ of r *derives* γ to the graph $\gamma' = \beta \alpha^\mu \bar{\beta}$. This is denoted as $\gamma \Rightarrow_{r,\mu} \gamma'$. If $\mathcal{R}$ is a finite set of rules, we write $\gamma \Rightarrow_\mathcal{R} \gamma'$ if $\gamma \Rightarrow_{r,\mu} \gamma'$ for some match μ of some rule $r \in \mathcal{R}$.

Definition 2 (HR Grammar). A *hyperedge replacement grammar* $\Gamma = (\Sigma, \mathcal{T}, \mathcal{R}, Z)$ (*HR grammar* for short) consists of *symbols* Σ with *terminals*

---

[2] V_γ may contain *isolated* nodes that are not attached to any edge in γ.

[3] I.e., a match μ makes sure that the nodes of α^μ that do not occur in $\bar{A} = A^\mu$ do not collide with the other nodes in γ.

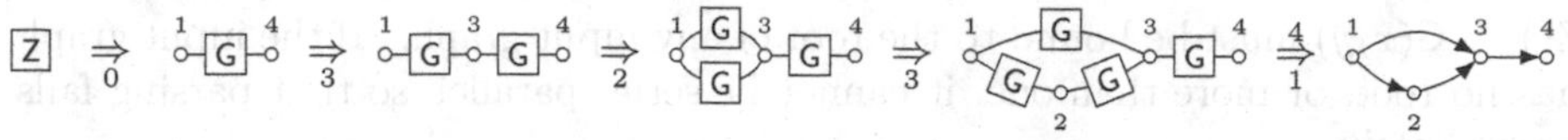

Fig. 1. *A derivation of the graph* $e(1,3)\,e(1,2)\,e(2,3)\,e(3,4)$. Nodes are drawn as circles, nonterminal edges as boxes around their label, with lines to their attached nodes, and terminal edges as arrows from their first to their second attached node; since e is the only terminal label, we omitted it in the terminal graph.

$T \subseteq \Sigma$ as assumed above, a finite set $\mathcal{R}$ of rules, and a *start graph* $Z = \mathsf{Z}()$ with $\mathsf{Z} \in \mathcal{N}$ of arity 0. Γ generates the language

$$\mathcal{L}(\Gamma) = \{g' \in \mathcal{G}_T \mid Z \Rightarrow^*_{\mathcal{R}} g,\, g' \bowtie g\}.$$

In the following, we simply write $\Rightarrow$ and $\Rightarrow^*$ because the rule set $\mathcal{R}$ in question will always be clear from the context.

Example 1 (A HR Grammar for Series-Parallel Graphs). The following rules

$$\mathsf{Z}() \underset{0}{\to} \mathsf{G}(x,y) \qquad\qquad \mathsf{G}(x,y) \underset{1}{\to} e(x,y)$$
$$\mathsf{G}(x,y) \underset{2}{\to} \mathsf{G}(x,y)\,\mathsf{G}(x,y) \qquad\qquad \mathsf{G}(x,y) \underset{3}{\to} \mathsf{G}(x,z)\,\mathsf{G}(z,y)$$

generate series-parallel graphs [8, p. 99]; see Fig. 1 for a derivation with graphs drawn as diagrams.

3   Predictive Shift-Reduce Parsing

The article [6] gives detailed definitions and correctness proofs for PSR parsing. Here we recall the concepts only so far that we can describe its generalization in the next section.

A PSR parser attempts to construct a derivation by reading the edges of a given input graph one after the other.[4] However, the parser must not assume that the edges of the input graph come in the same order as in a derivation. E.g., when constructing the derivation in Fig. 1, it must also accept an input graph $e(2,3)\,e(1,2)\,e(1,3)\,e(3,4)$ where the edges are permuted.

Before parsing starts, a procedure described in [5, Sect. 4] analyzes the grammar for the *unique start node* property, by computing the possible incidences of all nodes created by a grammar. The unique start nodes have to be matched by some nodes in the start rule of the grammar, thus determining where parsing begins. For our example, the procedure detects that every series-parallel graph has a unique root (without ingoing edges), and that the node x in the start rule

---

[4] We silently assume that input graphs do not have isolated nodes. This is no real restriction as one can add special edges to such nodes.

$Z() \rightarrow G(x, y)$ must be bound to the root of any input graph.[5] If the input graph has no root, or more than one, it cannot be series-parallel, so that parsing fails immediately.

A PSR parser is a push-down automaton that is controlled by a *characteristic finite automaton* (CFA). The stack of the PSR parser consists of states of the CFA. The parser makes sure that the sequence of states on its stack always describes a valid walk through its CFA. In order to do so, the parser generator computes a *parsing table* with processing instructions that control the parser.

Table 1 shows the parsing table for our example of series-parallel graphs. It has been generated by the graph parser generator *Grappa* (see footnote 1), using the constructions described in [6]. The rows of the table correspond to states of the CFA. Each of these states has a certain number of parameters. For instance, $Q_2(p, q)$ has two parameters p and q. Parameters remain *abstract* in the CFA and in the parsing table; only the parser will bind them to nodes of the input graph, and store them in *concrete* states on its stack.

When the parser starts, nothing of its input graph has been read yet, and its stack consists of a single concrete state Q_0, where its parameter p is bound to the unique start node, namely the root of the input graph.

The columns of the table correspond to terminal and nonterminal edges as well as the end-of-input marker \$. Column $e(x, y)$ in Table 1 contains all actions that can be taken by the parser if the input graph contains an edge $e(x, y)$ that is still unread. Column \$ contains the actions to be done if the input graph has been read completely. We will come to column $G(x, y)$ later.

The parser looks up its next action in the parsing table by inspecting the top-most state on its stack and all unread edges of the input graph. For illustration, let us assume that the parser has state Q_3 on top of its stack, with its parameters p and q being bound (by an appropriate renaming σ) to the input graph nodes p^σ and q^σ, resp., so that the parser must look into row $Q_3(p, q)$. If the input graph contains an unread e-labeled edge, the entry in column $e(x, y)$ applies. The corresponding table entry contains two possible actions, a shift and a reduce.

The *shift* operation can be selected if the input graph contains an unread e-labeled edge e that connects p^σ, either with q^σ, or with any node that has not yet occurred in the parse, indicated by "$-$" in the condition. If this shift operation is selected, it marks e as read and pushes a new concrete state Q_1 onto the stack, where the parameters p and q of Q_1 are bound to the source and the target node of e.

The *reduce* operation does not require any further condition to be satisfied. It consists of two steps: First, it pops as many states from the top of the stack as the right-hand side of the rule has edges. In our example "reduce 2, $G(p, q)$" refers to rule 2, with two edges in its right-hand side. Second, the reduce operation looks up the row for the new top-most state of the stack, selects the operation for

---

[5] Actually, series-parallel graphs do also have a unique sink (without outgoing edges), which could be used as a second start node bound to y. However, this variation of the grammar would exhibit less peculiarities of the GPSR parser.

Table 1. PSR parsing table for series-parallel graphs.

State	$e(x, y)$	$	$G(x, y)$
$Q_0(p)$	shift $Q_1(x, y)$ if $(x, y) = (p, -)$	error	goto $Q_2(x, y)$ if $(x, y) = (p, -)$
$Q_1(p, q)$	reduce 1, $G(p, q)$		error
$Q_2(p, q)$	shift $Q_1(x, y)$ if $(x, y) = (p, q)$ or $(x, y) = (p, -)$ or $(x, y) = (q, -)$	accept	goto $Q_3(x, y)$ if $(x, y) = (p, q)$ goto $Q_4(x, y, q)$ if $(x, y) = (p, -)$ goto $Q_5(x, y, p)$ if $(x, y) = (q, -)$
$Q_3(p, q)$	shift $Q_1(x, y)$ if $(x, y) = (p, q)$ or $(x, y) = (p, -)$ reduce 2, $G(p, q)$	reduce 2, $G(p,q)$	goto $Q_3(x, y)$ if $(x, y) = (p, q)$ goto $Q_4(x, y, q)$ if $(x, y) = (p, -)$
$Q_4(p, q, u)$	shift $Q_1(x, y)$ if $(x, y) = (p, q)$ or $(x, y) = (q, u)$ or $(x, y) = (p, -)$ or $(x, y) = (q, -)$	error	goto $Q_3(x, y)$ if $(x, y) = (p, q)$ goto $Q_4(x, y, q)$ if $(x, y) = (p, -)$ goto $Q_6(x, y, p)$ if $(x, y) = (q, u)$ goto $Q_7(x, y, u, p)$ if $(x, y) = (q, -)$
$Q_5(p, q, u)$	shift $Q_1(x, y)$ if $(x, y) = (p, q)$ or $(x, y) = (p, -)$ or $(x, y) = (q, -)$ reduce 3, $G(u, q)$	reduce 3, $G(u, q)$	goto $Q_3(x, y)$ if $(x, y) = (p, q)$ goto $Q_4(x, y, q)$ if $(x, y) = (p, -)$ goto $Q_5(x, y, p)$ if $(x, y) = (q, -)$
$Q_6(p, q, u)$	shift $Q_1(x, y)$ if $(x, y) = (p, q)$ or $(x, y) = (p, -)$ reduce 3, $G(u, q)$	reduce 3, $G(u, q)$	goto $Q_3(x, y)$ if $(x, y) = (p, q)$ goto $Q_4(x, y, q)$ if $(x, y) = (p, -)$
$Q_7(p, q, u, v)$	shift $Q_1(x, y)$ if $(x, y) = (p, q)$ or $(x, y) = (q, u)$ or $(x, y) = (p, -)$ or $(x, y) = (q, -)$ reduce 3, $G(v, q)$	reduce 3, $G(v, q)$	goto $Q_3(x, y)$ if $(x, y) = (p, q)$ goto $Q_4(x, y, q)$ if $(x, y) = (p, -)$ goto $Q_6(x, y, p)$ if $(x, y) = (q, u)$ goto $Q_7(x, y, u, p)$ if $(x, y) = (q, -)$

the new nonterminal edge with label G that connects p^σ with q^σ, i.e., in column $G(x, y)$, and pushes the corresponding state onto the stack.

The entries *accept* and *error* in column $ express that, if all edges of the input graph have been read, the parser terminates with success if the top-most state is Q_2, or with failure if it is Q_0 or Q_4.

A PSR parser must always be able to select the correct operation; it must not happen that the parser must choose between two or more operations where one of them leads to a successful parse whereas another one leads to failure. Such a situation is called a *conflict*. It is clear that a PSR parser always selects the correct action if conflicts cannot occur. However, a PSR parser does not know a priori which unread edge must be selected next. Hence, there are not only the shift-reduce or reduce-reduce conflicts (well known from LR parsing [9]). Shift-shift conflicts may also occur if the parser has to choose which input edge should be read next. Moreover, a shift alone may raise a conflict, since the unread input graph may contain more than one edge matching a pattern like $e(p, -)$. Only if the *free edge choice* property holds, the parser knows that any of these edges may be processed, without affecting the result of the parse.[6]

For our example of series-parallel graphs, conflicts arise in all states, except for Q_0 and Q_1. For $Q_3(p^\sigma, q^\sigma)$, e.g., the parser can always select the reduction, but it can also shift any edge connecting p^σ with q^σ or with any other unread node. Apparently, not every choice will lead to a successful parse, even if the input graph is valid.

Thus the parsing table does not always allow to predict the next correct action, and the grammar does not have a PSR parser. This problem can be solved by generalizing PSR parsing as described in the next section.

4   Generalized Predictive Shift-Reduce Parsing

Before we describe GPSR parsing for HR grammars, let us briefly recapitulate LR(k) parsing for context-free string grammars ([9], with $k = 1$ symbols of lookahead) and how this is extended to generalized LR (GLR) parsing [12]. An LR(1) parser is controlled by a parsing table derived from the CFA of the grammar. The parsing table assigns a unique parser action to each state of the CFA and to each terminal symbol: *shift, reduce, accept,* or *error*. In each step, the parser executes the action specified for the current state on top of the stack and the next unread input symbol (the *look-ahead*). However, LR(1) parsing is not possible if the parsing table has conflicts, i.e., if there is a state q and a look-ahead symbol a associated with two actions or more. A parser that reaches q with a look-ahead symbol a has as many choices how it may continue, i.e., the parse stack can be modified in different ways. A search process must then explore which of the resulting parse stacks can be further extended to a successful parse.

A GLR parser organizes this search process as a breadth-first search. It reads the input string from left to right. At any time, it has read a certain prefix α of the input string. It maintains the set of all (parse) stacks which can be obtained by reading α. This set of stacks is in fact processed in rounds as follows: For each stack, the parser determines all possible actions based on the parsing table, the top-most state of the stack, and the look-ahead symbol. The parser has found a successful parse if the action is *accept* and the entire input string has been read.

---

[6] This property can be determined by the parser generator as well. However, it does not hold for the grammar of series-parallel graphs.

(It may proceed if further parses shall be found.) If the action is an *error*, the parser just discards this stack, stops if this has been the last remaining stack, and fails altogether if it has not found a successful parse previously. If the parsing table, however, indicates more than one possible action, the parser duplicates the stack for each of them, and performs each action on one of the copies. If the action is a *shift*, the resulting stack is no longer considered in this round, but only in the next one. This way, at the beginning of the next round, each stack is the result of reading the look-ahead symbol in a *shift* action, and having read the same prefix of the input string.

In fact, a GLR parser does not store complete copies of stacks, but shares their common prefixes and suffixes. The resulting structure is known as a *graph-structured stack* (GSS). An individual stack is represented as a path in the GSS, from some top-most state to the unique initial state.

A GPSR parser generalizes a PSR parser in the same way as a GLR parser generalizes an LR parser. It also maintains a set of parse stacks, which contain concrete states whose parameters are bound to nodes of the input graph. However, a GPSR parser must also deal with the fact that there is no a priori reading sequence of edges of the input graph.

This affects a GPSR parser even more than a PSR parser since a GPSR parser may be forced to pursue different reading sequences in parallel while it performs the search process. This has consequences as follows:

- Each parse stack corresponds to a specific part of the input graph that has been read already. Hence, the parser must store, for each stack separately, which edges of the input graph have been read.
  Sets of stacks are stored as a GSS like in GLR parsers. Each GSS node corresponds to a concrete state. Additionally, each GSS node keeps track of the set of input graph edges that have been read so far. Note that GSS nodes may be shared only if both their concrete states and their sets of read edges coincide.
- GPSR parsers cannot process their sets of stacks in rounds. When a stack is obtained by executing a *shift* action, the parser must not wait until the same edge has been read in all the other stacks; they may read other edges first. As a consequence, a GPSR parser needs other strategies to control the order in which stacks are processed. Strategies are discussed in Sect. 5.

We demonstrate GPSR parsing using the example of series-parallel graphs and the input graph $e(1, 2) e(2, 3) e(1, 3) e(3, 4)$ derived in Fig. 1. We refer to these edges by the letters a, b, c, and d. We write GSS nodes in compact form: e.g., 5_{cd}^{341} refers to the concrete state $Q_5(3, 4, 1)$ and indicates that the edges $c = e(2, 3)$ and $d = e(3, 4)$ have been read already.

The parser determines node 1 as the unique start node, i.e., it starts with concrete state $Q_0(1)$, with all edges of the input graph unread. So the GSS in step 0 consists of $0_\varnothing^1$ (cf. Table 2). The parsing table in Table 1 indicates that the parser can shift both edge $a = e(1, 2)$ and edge $c = e(1, 3)$, resulting in the stacks $0_\varnothing^1 1_a^{12}$ and $0_\varnothing^1 1_c^{13}$. Table 2 shows the corresponding GSS in step 1. (Note that "new" GSS nodes are set in boldface.) Step 1 continues with processing stack

Table 2. Graph-structured stacks and steps of the GPSR parser when parsing the graph consisting of the hyperedges $a = e(1,2), b = e(2,3), c = e(1,3), d = e(3,4)$.

0 $\mathbf{0}^1_\varnothing$ .... shift a, c	**10** $0^1_\varnothing$ with $2^{12}_a \leftarrow 1^{13}_{ac}$, $5^{231}_{ab} \leftarrow 1^{34}_{abd}$ ... reduce 1, $2^{13}_{ab} \leftarrow 1^{13}_{abc}$, $2^{13}_c \leftarrow 1^{12}_{ac}$
1 $0^1_\varnothing \leftarrow 1^{12}_a$ .... reduce 1, $\leftarrow 1^{13}_c$	**11** $0^1_\varnothing$ with $2^{12}_a \leftarrow 1^{13}_{ac}$, $5^{231}_{ab} \leftarrow 5^{342}_{abd}$, $2^{13}_{ab} \leftarrow 5^{341}_{abd}$, $\leftarrow 1^{13}_{abc}$ ... reduce 1, $2^{13}_c \leftarrow 1^{12}_{ac}$
2 $0^1_\varnothing \leftarrow 2^{12}_a$, $\leftarrow 1^{13}_c$ .... reduce 1	**12** $0^1_\varnothing$ with $2^{12}_a \leftarrow 1^{13}_{ac}$, $5^{231}_{ab} \leftarrow 5^{342}_{abd}$ ... reduce 3*, $2^{13}_{ab} \leftarrow 5^{341}_{abd}$, $\leftarrow 3^{13}_{abc}$, $2^{13}_c \leftarrow 1^{12}_{ac}$
3 $0^1_\varnothing \leftarrow 2^{12}_a$, $\leftarrow 2^{13}_c$ ...... shift b, c	**13** $0^1_\varnothing$ with $2^{12}_a \leftarrow 1^{13}_{ac}$, $2^{13}_{ab} \leftarrow 5^{341}_{abd}$ .... reduce 3*, $\leftarrow 3^{13}_{abc}$, $2^{13}_c \leftarrow 1^{12}_{ac}$
4 $0^1_\varnothing$ with $2^{12}_a \leftarrow 1^{13}_{ac}$, $\leftarrow 1^{23}_{ab}$, 2^{13}_c .... shift a, d	**14** $0^1_\varnothing$ with $2^{12}_a \leftarrow 1^{13}_{ac}$, $2^{13}_{ab} \leftarrow 3^{13}_{abc}$ .... reduce 2, $2^{13}_c \leftarrow 1^{12}_{ac}$
5 $0^1_\varnothing$ with $2^{12}_a \leftarrow 1^{13}_{ac}$, $\leftarrow 1^{23}_{ab}$ ...... reduce 1, $2^{13}_c \leftarrow 1^{12}_{ac}$, $\leftarrow 1^{34}_{cd}$	**15** $0^1_\varnothing$ with $2^{12}_a \leftarrow 1^{13}_{ac}$, 2^{13}_{abc} .... shift d, $2^{13}_c \leftarrow 1^{12}_{ac}$
6 $0^1_\varnothing$ with $2^{12}_a \leftarrow 1^{13}_{ac}$, $\leftarrow 5^{231}_{ab}$, $2^{13}_c \leftarrow 1^{12}_{ac}$, $\leftarrow 1^{34}_{cd}$ ...... reduce 1	**16** $0^1_\varnothing$ with $2^{12}_a \leftarrow 1^{13}_{ac}$, $2^{13}_{abc} \leftarrow 1^{34}_{abcd}$ .... reduce 1, $2^{13}_c \leftarrow 1^{12}_{ac}$
7 $0^1_\varnothing$ with $2^{12}_a \leftarrow 1^{13}_{ac}$, $\leftarrow 5^{231}_{ab}$ ... shift d / reduce 3, $2^{13}_c \leftarrow 1^{12}_{ac}$, $\leftarrow 5^{341}_{cd}$	**17** $0^1_\varnothing$ with $2^{12}_a \leftarrow 1^{13}_{ac}$, $2^{13}_{abc} \leftarrow 5^{341}_{abcd}$ .... reduce 3, $2^{13}_c \leftarrow 1^{12}_{ac}$
8 $0^1_\varnothing$ with $2^{12}_a \leftarrow 1^{13}_{ac}$, $\leftarrow 5^{231}_{ab} \leftarrow 1^{34}_{abd}$, 2^{13}_{ab}, $2^{13}_c \leftarrow 1^{12}_{ac}$, $\leftarrow 5^{341}_{cd}$ ... reduce 3*	**18** $0^1_\varnothing$ with $2^{12}_a \leftarrow 1^{13}_{ac}$, 2^{14}_{abcd} .... accept, $2^{13}_c \leftarrow 1^{12}_{ac}$
9 $0^1_\varnothing$ with $2^{12}_a \leftarrow 1^{13}_{ac}$, $\leftarrow 5^{231}_{ab} \leftarrow 1^{34}_{abd}$, 2^{13}_{ab}, $2^{13}_c \leftarrow 1^{12}_{ac}$ .... shift c, d	

$0^1_\varnothing$ 1^{12}_a. (See Sect. 5 for a discussion of strategies.) State $Q_1(1,2)$ just allows a reduction by rule 1, producing a nonterminal edge $G(1,2)$. This pops 1^{12}_a from the stack; processing $G(1,2)$ pushes the concrete state $Q_2(1,2)$, which is represented by 2^{12}_a in step 2.

State 5^{231}_{ab} in step 7 allows both, to shift $d = e(3,4)$, and to reduce by rule 3, with nonterminal edge $G(1,3)$. The resulting GSS nodes are 1^{34}_{abd} and 2^{13}_{ab} in step 8. State 5^{341}_{cd} allows just a reduce action by grammar rule 3. However, this reduce operation with nonterminal edge $G(1,4)$ is invalid. If it were valid, $G(1,4)$ could be derived to the graph consisting of just $c = e(2,3)$ and $d = e(3,4)$, i.e., it would generate node 3. However, this contradicts the fact that the unread edge $b = e(2,3)$ is attached to node 3, which must be generated earlier in the derivation. Therefore, the stack with top-most state 5^{341}_{cd} is discarded in this step (indicated by the asterisk), and analogously in steps 12 and 13.

Note that the shift action in step 9 results in a GSS where 1^{34}_{abd} is the top-most state of two stacks. The reduce operation in step 10, however, removes 1^{34}_{abd} from the GSS and from the corresponding stacks again, and produces the two stacks with top-most states 5^{341}_{abd} and 5^{342}_{abd}.

The GSS in step 18 contains node 2^{14}_{abcd}, i.e., the accept state $Q_2(1,4)$ with the entire input graph being read. The GPSR parser, therefore, has found a successful parse of the input graph.

The current implementation stops when the first successful parse has been found. Another successful parse could have been found if 1^{12}_{ac} had been processed in step 5 or later.

5 Parsing Experiments

We now report on runtime experiments with different parsers applied to series-parallel graphs and to structured flowcharts. The latter are flowcharts that do not allow arbitrary jumps, but represent structured programs with conditional statements and while loops. They consist of rectangles containing instructions, diamonds that indicate conditions, and ovals indicating begin and end of the program. Arrows indicate control flow; see Fig. 2 for an example. Flowcharts are easily represented by graphs as also shown in Fig. 2. Figure 3 defines the rules of an HR grammar generating all graphs representing structured flowcharts. This grammar is not PSR because a state of its CFA has conflicts.

We generated three different parsers for the grammar of series-parallel graphs and for structured flowcharts: a *Cocke-Younger-Kasami* style parser (CYK, [11]) using *DiaGen*[7], and two variants of the GPSR parser using *Grappa* (see footnote 1). The CYK parser was in fact optimized in two ways: the parser creates nonterminal edges by dynamic programming, and each of these edges can be derived to a certain subgraph of the input graph. The optimized parser makes sure that it does not create two or more indistinguishable nonterminals for the same subgraph, even if the nonterminals represent different derivation trees. And it stops as soon as it finds the first derivation of the entire input graph.

---

[7] Homepage: www.unibw.de/inf2/diagen.

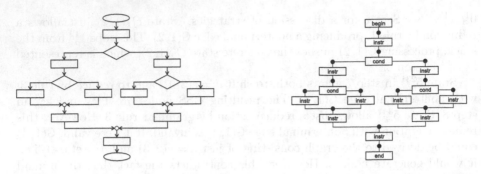

Fig. 2. A structured flowchart (text within the blocks has been omitted) and its graph representation.

$$Z() \rightarrow \text{begin}(x)\, P(x, y)\, \text{end}(y)$$
$$P(x, y) \rightarrow S(x, y) \mid P(x, z)\, S(z, y)$$
$$S(x, y) \rightarrow \text{instr}(x, y) \mid$$
$$\text{cond}(x, u, v)\, P(u, y)\, P(v, y) \mid$$
$$\text{cond}(x, u, y)\, P(u, x)$$

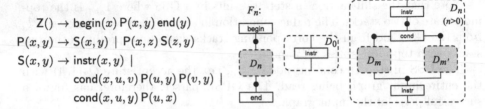

Fig. 3. HR rules for structured flowcharts.

Fig. 4. Definition of flowchart graphs F_n.

The GPSR parsers differ in the strategy that controls which of the currently considered stacks is selected for the next step. GPSR 1 simply maintains a FIFO queue of all such stacks, i.e., new states are enqueued as soon as they are created, and a top-most state is selected for processing as soon as it is next in the queue. GPSR 2, however, applies a more sophisticated strategy. It requires grammar rules to be annotated with either first or second priority. The GPSR 2 parser provides two queues, the first one using FIFO and the second LIFO. New states that result from handling a first priority rule go into the first queue, the others into the second. The parser always tries to select states from the first queue; it selects from the second queue only if the first queue is empty. This way one can control, by annotating grammar rules, which rules should be considered first. This does not affect the correctness of the parser; it can still examine the entire search space. However, it will stop as soon as it finds the first successful parse. By appropriately annotating grammar rules, one can thus speed up the parser if the input graph is valid. However, there is no speed-up for invalid input graphs, since the parser must inspect the entire search space in this case.

The GPSR 2 parser for series-parallel graphs gives rule 3 (series) precedence over rule 2 (parallel); it has been applied to graphs

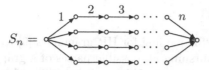

$$S_n =$$

with different values of n. The GPSR 2 parser for structured flowcharts gives sequences priority over conditional statements; it has been applied to flowcharts F_n defined in Fig. 4 and consisting of n conditions and $3n + 1$ instructions. The flowchart in Fig. 2 is in fact F_3. F_n has a subgraph D_n, which, for $n > 0$, contains subgraphs D_m and $D_{m'}$ with $n = m + m' + 1$. Note that the conditions in F_n form a binary tree with n nodes when we ignore instructions. We always choose m and m' such that it is a complete binary tree.

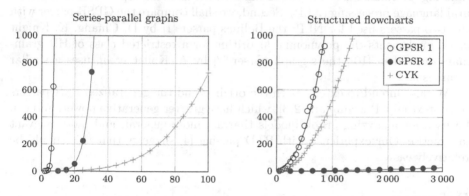

Fig. 5. Runtime (in milliseconds) of different parsers for series-parallel graphs and structured flowcharts.

Figure 5 shows the runtime of the different parsers applied to S_n and F_n with varying value n. Runtime has been measured on an iMac 2017, 4.2 GHz Intel Core i7, Java 1.8.0_181 with standard configuration, and is shown in milliseconds on the y-axis while n is shown on the x-axis.

The experiments first demonstrate that the more sophisticated strategy of GPSR 2 really pays off as GPSR 2 finds a derivation much faster than GPSR 1. For parsing F_{1000}, e.g., GPSR 1 needs 4 013 880 steps, but GPSR 2 just 13 004. The experiments also show that GPSR 1 is in fact much slower than CYK, which demonstrates the need for a sophisticated strategy for the GPSR parser. But for series-parallel graphs, even GPSR 2 is much slower than CYK. Because, the grammar of series-parallel graphs is highly ambiguous. For instance, S_{100} has the ridiculous number of $6.1 \cdot 10^{281}$ derivation trees. The CYK parser has to create 40 422 nonterminal edges for S_{100}, and for S_{40} just 6 582, where most of them represent a high number of different derivation trees. GPSR 2, however, needs 908 122 steps to find a derivation for S_{40}. Apparently, the compactification by the optimized CYK is more effective to cut down the number of choices the parser has to follow.

6   Conclusions

We have generalized PSR parsing for HR grammars [6] to cope with ambiguous graph grammars, by pursuing all possible parses of a graph in parallel until the first derivation has been found. This work is inspired by Tomita's GLR string parsers [12], which extend D.E. Knuth's LR string parsers [9]. For the academic example grammars examined in Sect. 5, in particular the highly ambiguous grammar for series-parallel graphs, comparison of our parser with the CYK parser does not give a clear picture. Moreover, the speed-up obtained by choosing an appropriate strategy only helps when parsing valid graphs, but not when processing invalid graphs. Our experiments shall be extended in two respects: First, we shall study more, and more realistic HR grammars, e.g., the modestly ambiguous (and big) grammars used for processing abstract meaning representations in natural language processing (NLP). Second, we shall compare the GPSR parser with the two parsers used for NLP: the Bolinas parser [1] by D. Chiang, K. Knight *et al.* implements the polynomial algorithm for a restricted class of HR grammars devised in [10]; the s-graph parser [7] by A. Koller *et al.* uses a similar formalism.

We also intend to extend both the original and the generalized PSR parsers to contextual HR grammars [2,3], which have greater generative power, and can be used for analyzing graph models that are more general, and more relevant in practice. Our experience with PTD parsing [4] suggests that this should be relatively easy.

References

1. Chiang, D., Andreas, J., Bauer, D., Hermann, K.M., Jones, B., Knight, K.: Parsing graphs with hyperedge replacement grammars. In: Proceedings of the 51st Annual Meeting Association for Computational Linguistics, vol. 1, pp. 924–932 (2013)
2. Drewes, F., Hoffmann, B.: Contextual hyperedge replacement. Acta Inf. **52**, 497–524 (2015)
3. Drewes, F., Hoffmann, B., Minas, M.: Contextual hyperedge replacement. In: Schürr, A., Varró, D., Varró, G. (eds.) AGTIVE 2011. LNCS, vol. 7233, pp. 182–197. Springer, Heidelberg (2012). https://doi.org/10.1007/978-3-642-34176-2_16
4. Drewes, F., Hoffmann, B., Minas, M.: Predictive top-down parsing for hyperedge replacement grammars. In: Parisi-Presicce, F., Westfechtel, B. (eds.) ICGT 2015. LNCS, vol. 9151, pp. 19–34. Springer, Cham (2015). https://doi.org/10.1007/978-3-319-21145-9_2
5. Drewes, F., Hoffmann, B., Minas, M.: Approximating Parikh images for generating deterministic graph parsers. In: Milazzo, P., Varró, D., Wimmer, M. (eds.) STAF 2016. LNCS, vol. 9946, pp. 112–128. Springer, Cham (2016). https://doi.org/10.1007/978-3-319-50230-4_9
6. Drewes, F., Hoffmann, B., Minas, M.: Formalization and correctness of predictive shift-reduce parsers for graph grammars based on hyperedge replacement. J. Log. Algebr. Meth. Program. (2018). https://doi.org/10.1016/j.jlamp.2018.12.006. https://arxiv.org/abs/1812.11927

7. Groschwitz, J., Koller, A., Teichmann, C.: Graph parsing with s-graph grammars. In: Proceedings of the 53rd Annual Meeting Association for Computational Linguistics, ACL 2015, Volume 1: Long Papers, pp. 1481–1490. The Association for Computer Linguistics (2015)

8. Habel, A.: Hyperedge Replacement: Grammars and Languages. LNCS, vol. 643. Springer, Heidelberg (1992). https://doi.org/10.1007/BFb0013875

9. Knuth, D.E.: On the translation of languages from left to right. Inf. Control 8(6), 607–639 (1965)

10. Lautemann, C.: The complexity of graph languages generated by hyperedge replacement. Acta Inf. 27, 399–421 (1990)

11. Minas, M.: Concepts and realization of a diagram editor generator based on hypergraph transformation. Sci. Comput. Program. 44(2), 157–180 (2002)

12. Tomita, M.: An efficient context-free parsing algorithm for natural languages. In: Proceedings of the 9th International Joint Conference on Artificial Intelligence, pp. 756–764. Morgan Kaufmann (1985)

Transformation of Petri Nets into Context-Dependent Fusion Grammars

Hans-Jörg Kreowski, Sabine Kuske, and Aaron Lye[(⊠)]

University of Bremen, Department of Computer Science and Mathematics,
P.O. Box 33 04 40, 28334 Bremen, Germany
{kreo,kuske,lye}@informatik.uni-bremen.de

Abstract. In this paper, we introduce context-dependent fusion grammars as a new type of hypergraph grammars where the application of fusion rules is restricted by positive and negative context conditions. Our main result is that Petri nets can be transformed into these grammars such that the reachable markings are in one-to-one correspondence to the members of the generated language. As a corollary, we get that the membership problem for context-dependent fusion grammars is at least as hard as the reachability problem of Petri nets.

1 Introduction

In this paper, we introduce context-dependent fusion grammars generalizing fusion grammars that we introduced in [1] as a novel approach to the generation of hypergraph languages. A fusion grammar imitates the basic DNA operations that Adleman employed in his seminal experiment [2] where tubes of molecules are replaced by disjoint unions of connected hypergraphs, the fusion of complementary sticky ends by the fusion of complementary hyperedges, the duplication of DNA molecules based on polymerase chain reaction by a multiplication of connected hypergraphs, and the filtering of molecules of interest by a special extraction of terminal hypergraphs from derived ones. These grammars become context-dependent if the fusion takes only place if certain positive and negative context conditions are satisfied. As the main result, we construct a transformation of Petri nets into context-dependent fusion grammars in such a way that the reachable markings of a Petri net are in a one-to-one relation with the members of the language generated by the corresponding grammar. The key of the transformation is to simulate the firing of a transition by a sequence of applications of fusion rules. The firing of a transition adds as many tokens to the post-places and consumes as many tokens from the pre-places as the weight function requires. While a fusion can do the appropriate addition in a single step, the consumption needs more than one step in general because each fusion consumes exactly two hyperedges. As a corollary, we get that the membership problem for context-dependent fusion grammars is at least EXPSPACE-hard. Recent research results conjecture that reachability problem for Petri nets is non-elementary-hard [3].

© Springer Nature Switzerland AG 2019
C. Martín-Vide et al. (Eds.): LATA 2019, LNCS 11417, pp. 246–258, 2019.
https://doi.org/10.1007/978-3-030-13435-8_18

Petri nets have been related to graph transformation before (cf., e.g., [4–6]). But in these cases, the transformation of Petri nets is easier because the employed graph transformation approaches are universal and hence more flexible than context-dependent fusion grammars. In particular, the simulation of firing a transition can be done by a single rule application.

The paper is organized as follows. In Sect. 2, basic notions and notations of hypergraphs are recalled. Section 3 recalls the notion of Petri nets. In Sect. 4, context-dependent fusion grammars are defined. Section 5 presents the reduction of Petri nets to context-dependent fusion grammars and the main theorem. Section 6 concludes the paper pointing out some open problems.

2   Preliminaries

In this section, the basic notions and notations of hypergraphs are recalled as far as needed.

We consider hypergraphs the hyperedges of which have multiple sources and multiple targets. A *hypergraph* over a given label alphabet Σ is a system $H = (V, E, s, t, lab)$ where V is a finite set of *vertices*, E is a finite set of *hyperedges*, $s, t \colon E \to V^*$ are two functions assigning to each hyperedge a sequence of *sources* and *targets*, respectively, and $lab \colon E \to \Sigma$ is a function, called *labeling*. The components of $H = (V, E, s, t, lab)$ may also be denoted by V_H, E_H, s_H, t_H, and lab_H respectively. The class of all hypergraphs over Σ is denoted by $\mathcal{H}_\Sigma$.

Let $H \in \mathcal{H}_\Sigma$, and let $\equiv$ be an equivalence relation on V_H. Then the *fusion of the vertices in H with respect to* $\equiv$ yields the hypergraph $H/\equiv = (V_H/\equiv, E_H, s_{H/\equiv}, t_{H/\equiv}, lab_H)$ with the set of equivalence classes $V_H/\equiv = \{[v] \mid v \in V_H\}$ and $s_{H/\equiv}(e) = [v_1] \cdots [v_{k_1}]$, $t_{H/\equiv}(e) = [w_1] \cdots [w_{k_2}]$ for each $e \in E_H$ with $s_H(e) = v_1 \cdots v_{k_1}$, $t_H(e) = w_1 \cdots w_{k_2}$.

Given $H, H' \in \mathcal{H}_\Sigma$, a *hypergraph morphism* $g \colon H \to H'$ consists of two mappings $g_V \colon V_H \to V_{H'}$ and $g_E \colon E_H \to E_{H'}$ such that $s_{H'}(g_E(e)) = g_V^*(s_H(e))$, $t_{H'}(g_E(e)) = g_V^*(t_H(e))$ and $lab_{H'}(g_E(e)) = lab_H(e)$ for all $e \in E_H$, where $g_V^* \colon V_H^* \to V_{H'}^*$ is the canonical extension of g_V, given by $g_V^*(v_1 \cdots v_n) = g_V(v_1) \cdots g_V(v_n)$ for all $v_1 \cdots v_n \in V_H^*$.

Given $H, H' \in \mathcal{H}_\Sigma$, H is a *subhypergraph* of H', denoted by $H \subseteq H'$, if $V_H \subseteq V_{H'}$, $E_H \subseteq E_{H'}$, $s_H(e) = s_{H'}(e)$, $t_H(e) = t_{H'}(e)$, and $lab_H(e) = lab_{H'}(e)$ for all $e \in E_H$. $H \subseteq H'$ implies that the two inclusions $V_H \subseteq V_{H'}$ and $E_H \subseteq E_{H'}$ form a hypergraph morphism from $H \to H'$. Given a morphism $g \colon H \to H'$, the image of H in H' under g is a subhypergraph $g(H) \subseteq H'$.

Let $H' \in \mathcal{H}_\Sigma$ as well as $V \subseteq V_{H'}$ and $E \subseteq E_{H'}$. Then the *removal* of (V, E) from H' given by $H = H' - (V, E) = (V_{H'} - V, E_{H'} - E, s_H, t_H, lab_H)$ with $s_H(e) = s_{H'}(e) \; t_H(e) = t_{H'}(e)$ and $lab_H(e) = lab_{H'}(e)$ for all $e \in E_{H'} - E$ defines a subgraph $H \subseteq H'$ if $s_{H'}(e), t_{H'}(e) \in (V_{H'} - V)^*$ for all $e \in E_{H'} - E$.

Let $H \in \mathcal{H}_\Sigma$ and let $att(e)$ be the set of sources and targets for $e \in E_H$. H is *connected* if for each $v, v' \in V_H$, there exists a sequence of triples $(v_1, e_1, w_1) \ldots (v_n, e_n, w_n) \in (V_H \times E_H \times V_H)^*$ such that $v = v_1, v' = w_n$ and $v_i, w_i \in att(e_i)$ for $i = 1, \ldots, n$ and $w_i = v_{i+1}$ for $i = 1, \ldots, n-1$.

A subgraph C of H, denoted by $C \subseteq H$, is a *connected component* of H if it is connected and there is no larger connected subgraph, i.e., $C \subseteq C' \subseteq H$ and C' connected implies $C = C'$. The set of connected components of H is denoted by $\mathcal{C}(H)$.

Given $H, H' \in \mathcal{H}_\Sigma$, the *disjoint union* of H and H' is denoted by $H + H'$. Further, $k \cdot H$ denotes the disjoint union of H with itself k times. We use the *multiplication* of H defined by means of $\mathcal{C}(H)$ as follows. Let $m \colon \mathcal{C}(H) \to \mathbb{N}$ be a mapping, called *multiplicity*, then $m \cdot H = \sum\limits_{C \in \mathcal{C}(H)} m(C) \cdot C$.

3   Petri Nets

In this section, we shortly recall some basic definitions of Petri nets (for more details see, e.g., [7–11] to mention just a few publications among the large number of existing ones.)

A *Petri net* is defined as $PN = (\mathbb{P}, \mathbb{T}, \mathbb{F}, \mathbb{W}, \mathbb{M}_0)$, where $\mathbb{P}$ and $\mathbb{T}$ are disjoint finite sets of *places* and *transitions*, respectively, $\mathbb{F} \subseteq (\mathbb{P} \times \mathbb{T}) \cup (\mathbb{T} \times \mathbb{P})$ is the *flow relation*, $\mathbb{W} \colon \mathbb{F} \to \mathbb{N}_{>0}$ is the *weight function*, and $\mathbb{M}_0 \colon \mathbb{P} \to \mathbb{N}$ is the *initial marking* where a *marking* is a function $\mathbb{M} \colon \mathbb{P} \to \mathbb{N}$ that specifies a number of *tokens* for every place.

If the places are numbered, i.e., $\mathbb{P} = \{p_1, \ldots, p_{|\mathbb{P}|}\}$, where $|X|$ denotes the cardinality of the set X, a marking $\mathbb{M}$ is also denoted by $(\mathbb{M}(p_1), \ldots, \mathbb{M}(p_{|\mathbb{P}|}))$. Figure 1 shows an example of a Petri net with $\mathbb{P} = \{p_1, \ldots, p_5\}$, $\mathbb{T} = \{t_1, \ldots, t_4\}$, $\mathbb{W}(x) = 2$ if $x = (t_4, p_5)$ and $\mathbb{W}(x) = 1$ otherwise, and $\mathbb{M}_0 = (1, 0, 2, 1, 0)$.

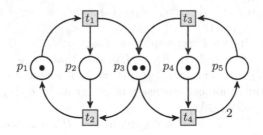

Fig. 1. A sample Petri net

For each $t \in \mathbb{T}$, ${}^\bullet t = \{p \in \mathbb{P} \mid (p, t) \in \mathbb{F}\}$ is the *preset* of t and $t^\bullet = \{p \in \mathbb{P} \mid (t, p) \in \mathbb{F}\}$ is the *postset* of t.

A transition t is *enabled* in some marking $\mathbb{M}$ if $\mathbb{M}(p) \geq \mathbb{W}(p, t)$ for each $p \in {}^\bullet t$. In the Petri net in Fig. 1 only the transitions t_1 and t_4 are enabled. If t is enabled in $\mathbb{M}$, it may *fire* yielding the marking $\mathbb{M}'$ given by $\mathbb{M}'(p) = \mathbb{M}(p) + \mathbb{W}_0(t, p) - \mathbb{W}_0(p, t)$ for each $p \in \mathbb{P}$ where for each $x \in (\mathbb{P} \times \mathbb{T}) \cup (\mathbb{T} \times \mathbb{P})$ $\mathbb{W}_0(x) = \mathbb{W}(x)$ if $x \in \mathbb{F}$ and $\mathbb{W}_0(x) = 0$ otherwise. This means that by firing t in each $p \in {}^\bullet t$ $\mathbb{W}(p, t)$ tokens are consumed and to each $p \in t^\bullet$ $\mathbb{W}(t, p)$ tokens are added. The firing of t from $\mathbb{M}$ to $\mathbb{M}'$ is denoted by $\mathbb{M}[t\rangle\mathbb{M}'$. For example, after firing the transition t_4 in Fig. 1, we get the marking $(1, 0, 1, 0, 2)$.

A marking $\mathbb{M}'$ is *reachable* from a marking $\mathbb{M}$, denoted by $\mathbb{M}[*\rangle\mathbb{M}'$, if $\mathbb{M} = \mathbb{M}'$ or there is a marking $\mathbb{M}''$ and a transition t such that $\mathbb{M}[*\rangle\mathbb{M}''[t\rangle\mathbb{M}'$. For each $PN = (\mathbb{P}, \mathbb{T}, \mathbb{F}, \mathbb{W}, \mathbb{M}_0)$ *Reach*(PN) is the set of markings reachable from $\mathbb{M}_0$.

4   Context-Dependent Fusion Grammars

In this section, we introduce context-dependent fusion grammars as a generalization of fusion grammars introduced in [1].

Fusion grammars generate hypergraph languages from start hypergraphs via successive applications of fusion rules, multiplications of connected components, and a filtering mechanism. The application of a fusion rule consumes two complementary hyperedges and fuses the sources of the one hyperedge with the sources of the other as well as the targets of the one with the targets of the other. Complementarity is defined on a set F of fusion labels that comes together with a complementary label $\overline{A}$ for each $A \in F$. Given a hypergraph, the set of all possible fusions is finite as fusion rules never create anything. To overcome this limitation, arbitrary multiplications of disjoint components within derivations are allowed. The language consists of the terminal part of all resulting connected components that contain no fusion labels and at least one marker label, where marker labels are removed in the end. These marker labels allow us to distinguish between wanted and unwanted terminal components.

Definition 1. 1. $F \subseteq \Sigma$ is a *fusion alphabet* if it is accompanied by a *complementary fusion alphabet* $\overline{F} = \{\overline{A} \mid A \in F\} \subseteq \Sigma$ where $F \cap \overline{F} = \emptyset$ and $\overline{A} \neq \overline{B}$ for $A, B \in F$ with $A \neq B$ and a *type function type*: $F \cup \overline{F} \to (\mathbb{N} \times \mathbb{N})$ with $type(A) = type(\overline{A})$ for each $A \in F$.
2. For each $A \in F$ with $type(A) = (k_1, k_2)$, the *fusion rule* $fr(A)$ is the hypergraph, depicted in Fig. 2, with the following components:
    - $V_{fr(A)} = \{v_i, v_i' \mid i = 1, \ldots, k_1\} \cup \{w_j, w_j' \mid j = 1, \ldots, k_2\}$,
    - $E_{fr(A)} = \{e, \overline{e}\}$,
    - $s_{fr(A)}(e) = v_1 \cdots v_{k_1}$, $s_{fr(A)}(\overline{e}) = v_1' \cdots v_{k_1}'$,
    - $t_{fr(A)}(e) = w_1 \cdots w_{k_2}$, $t_{fr(A)}(\overline{e}) = w_1' \cdots w_{k_2}'$,
    - $lab_{fr(A)}(e) = A$ and $lab_{fr(A)}(\overline{e}) = \overline{A}$.
3. The application of $fr(A)$ to a hypergraph $H \in \mathcal{H}_\Sigma$ proceeds according to the following steps: (1) Choose a *matching morphism* $g \colon fr(A) \to H$. (2) Remove the images of the two hyperedges of $fr(A)$ yielding X. (3) Fuse the sources and targets of the removed edges yielding the hypergraph $H' = X/\equiv$ where $\equiv$ is generated by the relation $\{(v_i, v_i') \mid i = 1, \ldots, k_1\} \cup \{(w_j, w_j') \mid j = 1, \ldots, k_2\}$. The application of $fr(A)$ to H is denoted by $H \underset{fr(A)}{\Longrightarrow} H'$ and called a *direct derivation*.

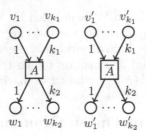

Fig. 2. The fusion rule $fr(A)$ with $type(A) = (k_1, k_2)$

Example 1. Let $F = \{t_4, \circ\}$ with $type(t_4) = (2, 1)$ and $type(\circ) = (0, 1)$. Let H be the hypergraph depicted in Fig. 3a. The rule $fr(t_4)$ can be applied to the hypergraph by fusion of the $\overline{t_4}$-hyperedge with the upper t_4-hyperedge. In this case, one gets the hypergraph in Fig. 3b. Please note that the enumeration of the sources and targets is omitted in drawings if it is clear from the context.

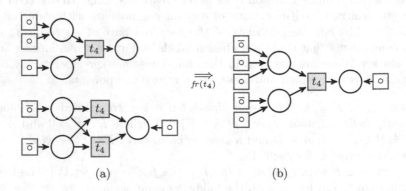

(a)   (b)

Fig. 3. Application of the fusion rule $fr(t_4)$

Definition 2. 1. A *context-dependent fusion rule* is a triple $cdfr = (fr(A), PC, NC)$ for some $A \in F$ where PC and NC are two finite sets of hypergraph morphisms with domain $fr(A)$ defining *positive* and *negative context conditions* respectively.

2. The rule $cdfr$ is applicable to some hypergraph H via a matching morphism $g: fr(A) \to H$ if for each $(c: fr(A) \to C) \in PC$ there exists a hypergraph morphism $h: C \to H$ such that h is injective on the set of hyperedges and $h \circ c = g$, and for no $(c: fr(A) \to C) \in NC$ there exists a hypergraph morphism $h: C \to H$ such that $h \circ c = g$.

3. If $cdfr$ is applicable to H via g, then the direct derivation $H \underset{cdfr}{\Longrightarrow} H'$ is the direct derivation $H \underset{fr(A)}{\Longrightarrow} H'$.

Example 2. Consider the two context-dependent fusion rules $fire(t_4) = (fr(t_4),$ $\{fr(t_4) \rightarrow M_{t_4} + L_{\overline{t_4}}\}, \{fr(t_4) \rightarrow N_{t_4,p_3} + L_{\overline{t_4}}, fr(t_4) \rightarrow N_{t_4,p_4} + L_{\overline{t_4}}\})$ where $M_{t_4}, N_{t_4,p_3}, N_{t_4,p_4}$, and $L_{\overline{t_4}}$ are depicted in Fig. 4. The morphisms are uniquely defined because the matching of the hyperedges is unique. $fire(t_4)$ can be applied to the hypergraph H in Fig. 3a matching the upper t_4-hyperedge and the $\overline{t_4}$-hyperedge because both sources of the t_4-hyperedge are attached to ○-hyperedges and neither the first nor the second source of the t_4-hyperedge is attached to a ⊖-hyperedge. Again $H \underset{fire(t_4)}{\Longrightarrow} H'$ yields the hypergraph in Fig. 3b.

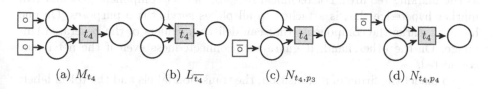

(a) M_{t_4} (b) $L_{\overline{t_4}}$ (c) N_{t_4,p_3} (d) N_{t_4,p_4}

Fig. 4. The hypergraphs of the context-conditions of $fire(t_4)$ in Example 2

Definition 3. 1. A *context-dependent fusion grammar* is a system $cdfg = (Z, F, M, T, P)$ where $Z \in \mathcal{H}_{F \cup \overline{F} \cup T \cup M}$ is a *start hypergraph*, $F \subseteq \Sigma$ is a fusion alphabet, $M \subseteq \Sigma$ with $M \cap (F \cup \overline{F}) = \emptyset$ is a set of *markers*, $T \subseteq \Sigma$ with $T \cap (F \cup \overline{F}) = \emptyset = T \cap M$ is a set of *terminal labels*, and P is a set of context-dependent fusion rules.
2. A *direct derivation* is either a context-dependent fusion rule application $H \underset{cdfr}{\Longrightarrow} H'$ or a multiplication $H \underset{m}{\Longrightarrow} m \cdot H$ for some multiplicity $m \colon \mathcal{C}(H) \rightarrow \mathbb{N}$.
   A *derivation* $H \overset{n}{\Longrightarrow} H'$ of length $n \geq 0$ is a sequence of direct derivations $H_0 \Longrightarrow H_1 \Longrightarrow \ldots \Longrightarrow H_n$ with $H = H_0$ and $H' = H_n$. If the length does not matter, one may write $H \overset{*}{\Longrightarrow} H'$.
3. $L(cdfg) = \{rem_M(Y) \mid Z \overset{*}{\Longrightarrow} H, Y \in \mathcal{C}(H) \cap (\mathcal{H}_{T \cup M} - \mathcal{H}_T)\}$ is the *generated language* where $rem_M(Y)$ is the terminal hypergraph obtained by removing all hyperedges with labels in M from Y.

Remark 1. Graph grammars with context conditions have been studied before (cf., e.g., [12–15]). Our definition is an adaption of the definition in [13].

5 Transformation of Petri Nets into Context-Dependent Fusion Grammars

In this section, Petri nets are transformed into context-dependent fusion grammars in such a way that the set of reachable markings of a Petri net coincides with the language generated by the corresponding context-dependent fusion grammar up to representation. The basic ideas of this transformation, formally constructed in Definition 4, are the following.

Given a Petri net as input, one must choose suitable alphabets, construct the connected components of the start hypergraph and equip fusion rules with appropriate context conditions.

The main connected component, called net component, reflects the Petri net structure and the initial marking. The static structure of the net is a bipartite graph that is represented by a hypergraph where the transitions are considered as hyperedges. Each such transition hyperedge is labeled with the transition itself so that the label identifies the hyperedge uniquely. The initial marking is represented by token-labeled flags, where a flag is a hyperedge without sources and a single target node, attached to places where a place gets as many flags as the marking requires. For technical reasons, the net component gets an extra marker hyperedge that is attached to all places serving two purposes. On one hand, it marks the components that can deliver members of the generated language. On the other hand, it guarantees connectedness even if the net is not connected.

To model the firing of a transition, the transition labels and the token labels are used as fusion labels, and for each transition a connected component is constructed as follows. Copy the corresponding transition hyperedge with its sources, targets and label and accomplish it with a parallel hyperedge that is labeled with the complementary transition. Moreover, the pre- and post-places get flags according to the weight function. While the flags at the post-places are labeled with the token label $\circ$, the flags at the pre-places are labeled with the complementary token label $\bar{\circ}$. If such a connected component is fused with the respective transition hyperedge in the net component, then the transition hyperedge is reconstructed and the pre- and post-places get the additional flags. A flag with the complementary token label can be seen as a reminder that the token has been spent and must be removed from this place. This can be done by a fusion of flags with token and complementary token labels. The context conditions make sure that the firing fusion takes only place if the pre-places carry as many token flags as the enabling requires and that the token fusion happens at a single place. The construction so far guarantees a close relationship between firing and fusion. If the net component has no complementary token flags, then the number of flags at each place defines a marking $\mathbb{M}$. If a transition can be fired yielding the marking $\mathbb{M}'$, then the respective firing fusion can be performed. Moreover, after performing then all possible token fusions the marking represented by the net component is $\mathbb{M}'$.

The rest of the construction concerns the termination of a fusion derivation. All transitions in the net component are removed, all tokens are converted into terminal tokens and the marker is deleted. Each of these termination steps can be done by fusion of hyperedges in the net component with complementary hyperedges in additional components. Finally we get a member of the generated language that represents a reachable marking. This and the converse is formally stated in the theorem after the construction.

Definition 4. Let $PN = (\mathbb{P}, \mathbb{T}, \mathbb{F}, \mathbb{W}, \mathbb{M}_0)$ be a Petri net. Let $\mathbb{P} = \{p_1, \ldots, p_{|\mathbb{P}|}\}$. Let $order(X) = p_{i_1} \ldots p_{i_{|X|}}$ with $i_j < i_{j+1}$ for $j = 1, \ldots, |X| - 1$ for all $X = \{p_{i_1}, \ldots, p_{i_{|X|}}\} \subseteq \mathbb{P}$. Then

$$CDFG(PN) = (Z_{PN}, \{\circ\} \cup \mathbb{T}, \{\mu\}, \{\bullet\}, P_{PN})$$

is the corresponding context-dependent fusion grammar with $type(\circ) = (0, 1)$ and $type(t) = (|{}^\bullet t|, |t^\bullet|)$ for each $t \in \mathbb{T}$ where Z_{PN} and P_{PN} are defined as follows.
$$Z_{PN} = hg(\mathbb{P}, \mathbb{T}, \mathbb{M}_0) + C_\bullet + \sum_{t \in \mathbb{T}} C_t + \sum_{t \in \mathbb{T}} D_t \text{ where}$$

- $hg(\mathbb{P}, \mathbb{T}, \mathbb{M}_0)$ represents the net component and is defined by the hypergraph $(\mathbb{P}, \{pn\} + \mathbb{T} + \{(p, i) \mid p \in \mathbb{P}, i = 1, \ldots, \mathbb{M}_0(p)\}, s_{hg}, t_{hg}, lab_{hg})$ where $+$ denotes the disjoint union of sets and
  - $s_{hg}(pn) = p_1 \ldots p_{|\mathbb{P}|}$, $t_{hg}(pn) = \epsilon$, and $lab_{hg}(pn) = \mu$, where ϵ denotes the empty sequence,
  - $s_{hg}(t) = order({}^\bullet t)$, $t_{hg}(t) = order(t^\bullet)$ and $lab_{hg}(t) = t$, and
  - $s_{hg}((p, i)) = \epsilon$, $t_{hg}((p, i)) = p$ and $lab_{hg}((p, i)) = \circ$.
- $C_\bullet = \boxed{\bullet} \!\!\longrightarrow\!\! \bigcirc \!\!\longleftarrow\!\! \boxed{\overline{\circ}}$ (This component serves for the replacement of $\circ$-flags by terminal $\bullet$-flags.)
- $C_t = ({}^\bullet t \cup t^\bullet, \{e, \overline{e}\} + \{(p, i, pre) \mid p \in {}^\bullet t, i \in \{1, \ldots, \mathbb{W}(p, t)\}\} \cup \{(p, i, post) \mid p \in t^\bullet, i \in \{1, \ldots, \mathbb{W}(t, p)\}\}, s, t, lab)$ with
  - $s(e) = s(\overline{e}) = order({}^\bullet t)$, $t(e) = t(\overline{e}) = order(t^\bullet)$, $lab(e) = t$, $lab(\overline{e}) = \overline{t}$,
  - $s((p, i, pre)) = \epsilon$, $t((p, i, pre)) = p$ and $lab((p, i, pre)) = \overline{\circ}$, and
  - $s((p, i, post)) = \epsilon$, $t((p, i, post)) = p$ and $lab((p, i, post)) = \circ$.
  A sketch of C_t is depicted in Fig. 5a. (The component C_t is used for modeling the firing of t by fusing the $\overline{t}$-hyperedge in C_t with the t-hyperedge in the net component.)
- D_t is obtained from C_t by removing e and all flags, i.e., the t-hyperedge and all hyperedges labeled with $\circ$ or $\overline{\circ}$. A sketch of D_t is depicted in Fig. 5b.

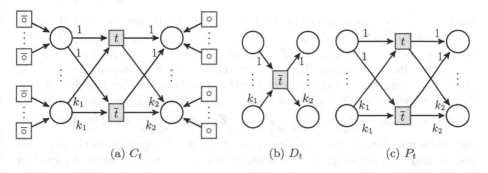

(a) C_t (b) D_t (c) P_t

Fig. 5. A sketch of C_t, D_t and P_t for a transition t

$P_{PN} = \{consume, replace\} \cup \{delete(t), fire(t) \mid t \in \mathbb{T}\}$ where

- $consume = (fr(\circ), \{fr(\circ) \rightarrow \boxed{\circ}\!\!\rightarrow\!\!\bigcirc\!\!\leftarrow\!\!\boxed{\bar{\circ}}\}, \emptyset)$, i.e., the positive context condition requires that only complementary token flags attached to the same vertex are fused.

- $replace = (fr(\circ), \{fr(\circ) \rightarrow \boxed{\circ}\!\!\rightarrow\!\!\bigcirc + \boxed{\bullet}\!\!\rightarrow\!\!\bigcirc\!\!\leftarrow\!\!\boxed{\bar{\circ}}\}, \{fr(\circ) \rightarrow \boxed{\circ}\!\!\rightarrow\!\!\bigcirc\!\!\leftarrow\!\!\boxed{\bar{\circ}} + \boxed{\bar{\circ}}\!\!\rightarrow\!\!\bigcirc, fr(\circ) \rightarrow \boxed{\circ}\!\!\rightarrow\!\!\bigcirc + \boxed{\circ}\!\!\rightarrow\!\!\bigcirc\!\!\leftarrow\!\!\boxed{\bar{\circ}}\})$, where the three hypergraph morphisms map the two connected components of $fr(\circ)$ into different connected components of the codomain each, i.e., the positive context condition requires that the fusion rule can only be applied if a $\bullet$-hyperedge is attached to the same vertex to which the $\bar{\circ}$-hyperedge is attached; the negative context conditions forbids the rule application if $consume$ can be applied and if the matched vertex attached to the $\bar{\circ}$-hyperedge is attached to some $\circ$-hyperedge. As a result only the $\bar{\circ}$-hyperedge attached to $C_\bullet$ is a possible match.

- $delete(t) = (fr(t), \emptyset, \{fr(t) \rightarrow L_t + P_t\})$, where L_t is the t-hyperedge of $fr(t)$ with its sources and targets, P_t is obtained from C_t by removing all flags, and the morphism maps the t-hyperedge into L_t and the $\bar{t}$-hyperedge into P_t, i.e., the negative context condition makes sure that C_t is not used for deletion of transitions. A sketch of P_t is depicted in Fig. 5c.

- $fire(t) = (fr(t), \{fr(t) \rightarrow M_t + L_{\bar{t}}\}, \{fr(t) \rightarrow N_{t,p} + L_{\bar{t}} \mid p \in {}^\bullet t\})$, where
  - $L_{\bar{t}}$ is the $\bar{t}$-hyperedge of $fr(t)$ with its sources and targets.
  - $M_t = ({}^\bullet t \cup t^\bullet, \{e\} \cup \{(p,i) \mid p \in {}^\bullet t, i \in \{1, \ldots, \mathbb{W}(p,t)\}\}, s, t, lab)$
    - $* \; s(e) = order({}^\bullet t), \; t(e) = order(t^\bullet), \; lab(e) = t$
    - $* \; s((p,i)) = \epsilon, t((p,i)) = p$ and $lab((p,i)) = \circ$,
  - $N_{t,p} = ({}^\bullet t \cup t^\bullet, \{e, e_p\}, s, t, lab)$
    - $* \; s(e) = order({}^\bullet t), \; t(e) = order(t^\bullet), \; lab(e) = \bar{t}$
    - $* \; s(e_p) = \epsilon, t(e_p) = p$ and $lab(e_p) = \bar{\circ}$,

i.e., the positive context condition makes sure that each pre-place of t carries at least as many token flags as the weight requires, and the negative context conditions forbid any $\bar{\circ}$-flag on the pre-places.

The mappings are uniquely determined by the labels t and $\bar{t}$.

Example 3. The transformation of the sample Petri net in Sect. 3 yields the context-dependent fusion grammar $(Z_{example}, \{\circ, t_1, t_2, t_3, t_4\}, \{\mu\}, \{\bullet\},$ $\{consume, replace, fire(t_1), \ldots, fire(t_4), delete(t_1), \ldots, delete(t_4)\})$. $Z_{example}$ is depicted in Fig. 6 where flags are not shown, but their labels are depicted inside their targets. The contexts of $fire(t_4)$ are shown in Fig. 4 in Sect. 4. For the other $fire$ rules the contexts are analogue. A sample derivation looks as follows: $Z_{example} \underset{fire(t_4)}{\Longrightarrow} Z_1 \underset{consume}{\overset{2}{\Longrightarrow}} Z_2 \underset{delete}{\overset{4}{\Longrightarrow}} Z_3 \underset{4 \cdot C_\bullet}{\Longrightarrow} Z_4 \underset{replace}{\overset{4}{\Longrightarrow}} Z_5$. The first direct derivation matches the two complementary t_4- and $\bar{t_4}$-hyperedges in the net component and C_{t_4}, respectively. The fusion yields the net component depicted in Fig. 7a. Afterwards, $consume$ is applied twice consuming the two complementary flags yielding the net component depicted in Fig. 7b. Because this net component has no complementary token flags, the numbers of flags at the places define a marking. One may continue with applying a different *fire*-rule followed by consume as long as possible. However, our sample derivation continues with applying *delete* four times deleting all four transition hyperedges in the net component by matching in each step one of the transition hyperedge in the net component and the complementary transition hyperedge in the respective deletion component (out of the $D_{t_1}, \ldots, D_{t_4}$). In order to obtain a terminal hypergraph each ∘-flag is replaced by a •-flag. Because the start hypergraph contains each connected component only once, $C_\bullet$ must be multiplied. Afterwards, *replace* is applied four times. The removal of the μ-hyperedge from the resulting net component yields the terminal hypergraph ● ◯ ● ◯ ●●.

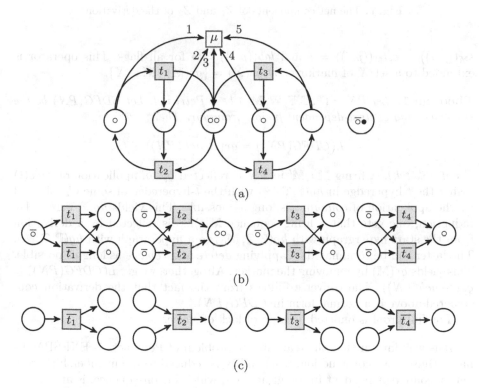

Fig. 6. The start hypergraph $Z_{example}$ where (a) shows the net component and $C_\bullet$; (b) shows $C_{t_1}, \ldots, C_{t_4}$; and (c) shows $D_{t_1}, \ldots, D_{t_4}$.

We can now state the main result of the paper using the following definition of (hyper)graph representation of markings in which the places are the nodes and the number of tokens of each place equals the number of terminal token flags at this place.

Definition 5. Let $M: \mathbb{P} \to \mathbb{N}$ be a marking. Then its (hyper)graph representation is given by $gr(M) = (\mathbb{P}, \{(p, i) \mid p \in \mathbb{P}, i = 1, \ldots, M(p)\}, s_M, t_M, lab_M)$ with

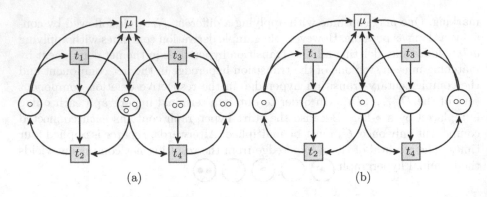

Fig. 7. The net components of Z_1 and Z_2 of the derivation

$s_{\mathbb{M}}((p,i)) = \epsilon, t_{\mathbb{M}}((p,i)) = p$ and $lab((p,i)) = \bullet$ for all flags. This operator is extended to a set X of markings by $gr(X) = \{gr(\mathbb{M}) \mid \mathbb{M} \in X\}$.

Theorem 1. *Let $PN = (\mathbb{P}, \mathbb{T}, \mathbb{F}, \mathbb{W}, \mathbb{M}_0)$ be a Petri net. Let $CDFG(PN)$ be the corresponding context-dependent fusion grammar. Then*

$$L(CDFG(PN)) = gr(Reach(PN)).$$

Proof (Sketch). A firing $\mathbb{M}[t\rangle\mathbb{M}'$ in PN is reflected by an application of $fire(t)$ fusing the t-hyperedge in $hg(\mathbb{P}, \mathbb{T}, \mathbb{M})$ with the $\bar{t}$-hyperedge of some C_t followed by the application of *consume* as long as possible. This yields $hg(\mathbb{P}, \mathbb{T}, \mathbb{M}')$. By induction, one gets that a firing sequence $\mathbb{M}[*\rangle\mathbb{M}'$ is reflected by a derivation from the start hypergraph with $hg(\mathbb{P}, \mathbb{T}, \mathbb{M}_0)$ to a hypergraph with $hg(\mathbb{P}, \mathbb{T}, \mathbb{M})$. The latter can be terminated by applying *delete* and *replace* as long as possible. This yields $gr(\mathbb{M})$ by removing the marker. Altogether, we get $L(CDFG(PN)) \subseteq gr(Reach(PN))$. The converse follows from the fact that the derivation constructed above is a normal form in $CDFG(PN)$.

The full proof is omitted because of lack of space.

It is well-known that the reachability problem of Petri nets is EXPSPACE-hard. Due to our construction, reachability is reduced to the membership problem of context-dependent fusion grammars, which in consequence, is at least as hard as the reachability problem.

Corollary 1. *The membership problem for context-dependent fusion grammars is EXPSPACE-hard.*

In [3], it is claimed that the reachability problem of Petri nets is not elementary. In this case, the mebership problem of context-dependent fusion grammars would be non-elementary-hard.

6 Conclusion

In this paper, we have started the study on context-dependent fusion grammars by transforming Petri nets into this type of hypergraph grammars. To get a better insight, further research is needed including the following issues.

1. Is it true (as we believe) that Petri nets cannot be embedded into fusion grammars without context conditions? Is it even true that only positive or only negative context conditions are not powerful enough to cover Petri nets?
2. Is the membership problem of context-dependent fusion grammars decidable? In this context, it should be noted the even the decidability of the membership problem of fusion grammars is an open problem.
3. In [16], we introduced splicing/fusion grammars enhancing fusion grammars by the inversion of fusions. How does a natural transformation of context-dependent fusion grammars into splicing/fusion grammars look like?
4. A promising application area of fusion grammars is the modeling of bio-chemical reactions. This should be demonstrated by convincing examples.

References

1. Kreowski, H.-J., Kuske, S., Lye, A.: Fusion grammars: a novel approach to the generation of graph languages. In: de Lara, J., Plump, D. (eds.) ICGT 2017. LNCS, vol. 10373, pp. 90–105. Springer, Cham (2017). https://doi.org/10.1007/978-3-319-61470-0_6
2. Adleman, L.M.: Molecular computation of solutions to combinatorial problems. Science **266**, 1021–1024 (1994)
3. Czerwinski, W., Lasota, S., Lazic, R., Leroux, J., Mazowiecki, F.: The reachability problem for Petri nets is not elementary. CoRR. arXiv.org, abs/1809.07115 (2018)
4. Kreowski, H.-J.: A comparison between petri-nets and graph grammars. In: Nolte-meier, H. (ed.) WG 1980. LNCS, vol. 100, pp. 306–317. Springer, Heidelberg (1981). https://doi.org/10.1007/3-540-10291-4_22
5. Corradini, A.: Concurrent computing: from Petri nets to graph grammars. Electron. Notes Theoret. Comput. Sci. **2**, 56–70 (1995)
6. Ehrig, H., Padberg, J.: Graph grammars and Petri net transformations. In: Desel, J., Reisig, W., Rozenberg, G. (eds.) ACPN 2003. LNCS, vol. 3098, pp. 496–536. Springer, Heidelberg (2004). https://doi.org/10.1007/978-3-540-27755-2_14
7. Desel, J., Reisig, W.: Place/transition Petri nets. In: Reisig, W., Rozenberg, G. (eds.) ACPN 1996. LNCS, vol. 1491, pp. 122–173. Springer, Heidelberg (1998). https://doi.org/10.1007/3-540-65306-6_15
8. Girault, C., Valk, R.: Petri Nets for Systems Engineering. Springer, Heidelberg (2003). https://doi.org/10.1007/978-3-662-05324-9
9. Priese, L., Wimmel, H.: Petri-Netze. Springer, Heidelberg (2008). https://doi.org/10.1007/978-3-540-76971-2
10. Best, E., Wimmel, H.: Structure theory of Petri nets. In: Jensen, K., van der Aalst, W.M.P., Balbo, G., Koutny, M., Wolf, K. (eds.) Transactions on Petri Nets and Other Models of Concurrency VII. LNCS, vol. 7480, pp. 162–224. Springer, Heidelberg (2013). https://doi.org/10.1007/978-3-642-38143-0_5

11. Reisig, W.: Understanding Petri Nets - Modeling Techniques, Analysis Methods, Case Studies. Springer, Heidelberg (2013). https://doi.org/10.1007/978-3-642-33278-4
12. Ehrig, H., Habel, A.: Graph grammars with application conditions. In: Rozenberg, G., Salomaa, A. (eds.) The Book of L, pp. 87–100. Springer, Berlin (1986). https://doi.org/10.1007/978-3-642-95486-3_7
13. Habel, A., Heckel, R., Taentzer, G.: Graph grammars with negative application conditions. Fundamenta Inform. **26**(3,4), 287–313 (1996)
14. Dediu, A.-H., Klempien-Hinrichs, R., Kreowski, H.-J., Nagy, B.: Contextual hypergraph grammars – a new approach to the generation of hypergraph languages. In: Ibarra, O.H., Dang, Z. (eds.) DLT 2006. LNCS, vol. 4036, pp. 327–338. Springer, Heidelberg (2006). https://doi.org/10.1007/11779148_30
15. Ehrig, H., Hermann, F., Sartorius, C.: Completeness and correctness of model transformations based on triple graph grammars with negative application conditions. In: Proceedings of the 8th International Workshop on Graph Transformation and Visual Modeling Techniques, vol. 18, pp. 1–18 (2009)
16. Kreowski, H.-J., Kuske, S., Lye, A.: Splicing/fusion grammars and their relation to hypergraph grammars. In: Lambers, L., Weber, J. (eds.) ICGT 2018. LNCS, vol. 10887, pp. 3–19. Springer, Cham (2018). https://doi.org/10.1007/978-3-319-92991-0_1

Generalized Register Context-Free Grammars

Ryoma Senda[1]([⊠]), Yoshiaki Takata[2], and Hiroyuki Seki[1]

[1] Graduate School of Information Science, Nagoya University,
Furo-cho, Chikusa, Nagoya 464-8601, Japan
ryoma.private@sqlab.jp, seki@i.nagoya-u.ac.jp
[2] Graduate School of Engineering, Kochi University of Technology,
Tosayamada, Kami City, Kochi 782-8502, Japan
takata.yoshiaki@kochi-tech.ac.jp

Abstract. Register context-free grammars (RCFG) is an extension of
context-free grammars to handle data values in a restricted way. This
paper first introduces register type as a finite representation of the register contents and shows some properties of RCFG. Next, generalized
RCFG (GRCFG) is defined by permitting an arbitrary relation on data
values in the guard expression of a production rule. We extend register
type to GRCFG and introduce two properties of GRCFG, the simulation
property and the type oracle. We then show that ε-rule removal is possible and the emptiness and membership problems are EXPTIME solvable
for GRCFG that satisfy these two properties.

1 Introduction

This paper focuses on register context-free grammars (abbreviated as RCFG),
which were introduced by Cheng and Kaminsky in 1998 [6]. Recently, register
automata (abbreviated as RA) [10] have been paid attention [11–13] as a core
computational model of query languages for structured documents with data
values such as XPath. For example, XPath can specify both a regular pattern of
node labels (e.g., element names) and a constraint on data values (e.g., attribute
values and PCDATA) in a tree representing an XML document. While RA have
a power sufficient for expressing regular patterns on *paths* of a tree or a graph,
it cannot represent tree patterns (or patterns over branching paths) that can be
represented by some query languages such as XPath. Hence, a computational
model that can represent both local tree patterns and constraints on data values
is expected.

RCFG [6] is defined as an extension of CFG in a similar way to extending
finite automata to RA. In a derivation of a k-RCFG, k data values are associated
with each occurrence of a nonterminal symbol (called a *register assignment*) and
a production rule can be applied only when the guard condition of the rule,
which is a Boolean combination of the equality check between an input data
value and the data value in a register, is satisfied. In [6], properties of RCFG were

© Springer Nature Switzerland AG 2019
C. Martín-Vide et al. (Eds.): LATA 2019, LNCS 11417, pp. 259–271, 2019.
https://doi.org/10.1007/978-3-030-13435-8_19

shown including the decidability of the membership and emptiness problems, and the closure properties. In our previous study [17], the membership problem for RCFG, ε-rule free RCFG and growing RCFG are shown EXPTIME-complete, PSPACE-complete and NP-complete, respectively, and the emptiness problem for these classes are shown EXPTIME-complete.

In this paper, we first show that ε-rules can be removed from a given RCFG without changing the generated language. To prove this property, we introduce a notion called *register type*, which is the quotient of registers by the equivalence classes induced by equality relation among the contents of registers. Next, we move to the main topic of this paper, a generalization of RCFG abbreviated as GRCFG. As we mentioned, what an RCFG (and also an RA) can do when applying a rule is the equality check between the content of a register and an input data value. Then, we come to a natural question that what happens if we allow the check of an arbitrary relation (such as the total order on numbers). Generally, basic problems including membership and emptiness become undecidable. Hence, we want to introduce appropriate conditions for such extensions of RCFG to keep the decidability and complexity of those problems unchanged. For this aim, we extend the above mentioned *register type* for an arbitrary relation and then we introduce two conditions, namely, the simulation and the type oracle. We show that the emptiness and membership are decidable and ε-removal is possible for GRCFG that satisfies these two conditions. As a corollary, we also show that those properties hold for GRCFG with a total order on a dense set.

Related Work. Register automata (RA) was proposed in [10] as finite-memory automata where they show that the membership and emptiness problems are decidable, and the class of languages recognized by RA are closed under union, concatenation and Kleene-star. Later, the computational complexity of the former two problems are analyzed in [7,16]. In [6], register context-free grammars (RCFG) as well as pushdown automata over an infinite alphabet were introduced and the equivalence of the two models were shown. Also, the decidability of membership and emptiness problems and the closure under union, concatenation, Kleene-star were shown in [6]. Extension of RA to a totally ordered set was discussed in [1,9], which is also provided in RCFG in the last section of this paper.

There have been many studies on other extensions of finite models to deal with data values in restricted ways. *Other automata for data words*: As extensions of finite automata other than RA, data automata [5], pebble automata (PA) [14] and nominal automata (NA) [4] are known. Libkin and Vrgoč [13] argue that register automata (RA) is the only model that has efficient data complexity for membership among the above mentioned formalisms. Neven, et al. consider variations of RA and PA, which are either one way or two ways, deterministic, nondeterministic or alternating. They show inclusion and separation relationships among these automata, $FO(\sim, <)$ and $EMSO(\sim, <)$, and give the answer to some open problems including the undecidability of the universality problem for RA [15]. Nominal (G-)automata (NA) is defined by a data set with symmetry and finite supports, and properties of NA are investigated including Myhill-Nerode theorem, closure and determinization in [4]. (Usual) RA with equality and RA with total order can be regarded as NA where the data sets have equality symmetry and total order symmetry, respectively. In [4], nominal CFG is also introduced but decidability of related problems is not discussed. Finiteness of orbits and that of register types in this paper are related, but deeper observation is left as future study. *LTL with freeze quantifier.* Linear temporal logic (LTL) was extended to LTL↓ with freeze quantifier [7,8]. The relationship among subclasses of LTL↓ and RA as well as the decidability and complexity of the satisfiability (nonemptiness) problems are investigated [7]. They especially showed that the emptiness problem for (both nondeterministic and deterministic) RA are PSPACE-complete. *Two-variable logics with data equality.* It is known that two-variable $FO^2(<, +1)$ where $<$ is the ancestor-descendant relation and $+1$ is the parent-child relation is decidable and corresponds to Core XPath. The logic was extended to those with data equality. It was shown in [3] that $FO^2(\sim, <, +1)$ with data equality $\sim$ is decidable on data words. Note that $FO^2(\sim, <, +1)$ is incomparable with LTL↓ of [7]. Also it was shown in [2] that $FO^2(\sim, +1)$ and existential $MSO^2(\sim, +1)$ are decidable on unranked data trees.

2 Register Context-Free Grammars

Let $\mathbb{N} = \{1, 2, \ldots\}$ and $\mathbb{N}_0 = \{0\} \cup \mathbb{N}$. We assume an infinite set D of *data values* as well as a finite alphabet Σ. For a given $k \in \mathbb{N}_0$ specifying the number of *registers*, a mapping $\theta : [k] \to D$ is called an *assignment* (of data values to k registers) where $[k] = \{1, 2, \ldots, k\}$. We assume that a data value $\bot \in D$ is designated as the initial value of a register. Let Θ_k denote the collection of assignments to k registers. For $\theta, \theta' \in \Theta_k$, we write $\theta' = \theta[i \leftarrow d]$ if $\theta'(i) = d$ and $\theta'(j) = \theta(j)$ for $j \neq i$. Let F_k denote the set of *guard expressions* over k registers defined by $\psi := \mathtt{tt} \mid x_i^= \mid \neg\psi \mid \psi \vee \psi$ where $x_i \in \{x_1, \ldots, x_k\}$. Let $\mathtt{ff}, x_i^{\neq}, \psi_1 \wedge \psi_2$ denote $\neg\mathtt{tt}, \neg x_i^=, \neg(\neg\psi_1 \vee \neg\psi_2)$, respectively. The description length of a guard expression ψ, denoted as $\|\psi\|$, is defined as usual where $\|x_i^=\| = 1 + \log k$. For $d \in D$, $\theta \in \Theta_k$ and $\psi \in F_k$, the satisfaction relation $d, \theta \models \psi$ is defined as $d, \theta \models x_i^=$ iff $\theta(i) = d$ and is recursively defined for $\neg$ and $\vee$ in a usual way.

For a finite alphabet Σ and a set D of data values disjoint from Σ, a *data word* over $\Sigma \times D$ is a finite sequence of elements of $\Sigma \times D$ and a *data language* over $\Sigma \times D$ is a subset of $(\Sigma \times D)^*$. $|\beta|$ denotes the cardinality of β if β is a set and the length of β if β is a finite sequence.

For $k \in \mathbb{N}_0$, a *k-register context-free grammar* (k-RCFG) over Σ and D is a triple $G = (V, R, S)$ where

- V is a finite set of nonterminal symbols (abbreviated as nonterminals) where $V \cap (\Sigma \cup D) = \emptyset$,
- R is a finite set of production rules (abbreviated as rules) having either of the following forms: $(A, \psi, i) \to \alpha$ or $(A, \psi) \to \alpha$ where $A \in V$, $\psi \in F_k$, $i \in [k]$ and $\alpha \in (V \cup (\Sigma \times [k]))^*$; we call (A, ψ, i) (or (A, ψ)) the left-hand side and α the right-hand side of the rule, and,

– $S \in V$ is the start symbol.

A rule whose right-hand side is ε is an *ε-rule*. If R contains no ε-rule, G is called *ε-rule free*. A k-RCFG G for some $k \in \mathbb{N}_0$ is just called an RCFG.

In the following, we write $(A, \psi, i)/(A, \psi) \to \alpha \in R$ to represent $(A, \psi, i) \to \alpha \in R$ or $(A, \psi) \to \alpha \in R$. The description length of a k-RCFG $G = (V, R, S)$ is defined as $\|G\| = |V| + |R| \max\{(|\alpha|+1)(\log|V|+\log k)+\|\psi\| \mid (A, \psi, i)/(A, \psi) \to \alpha \in R\}$, where $\|\psi\|$ is the description length of ψ.

We define $\Rightarrow_G$ as the smallest relation containing the instantiations of rules in R and closed under the context as follows. For $A \in V$, $\theta \in \Theta_k$ and $X \in ((V \times \Theta_k) \cup (\Sigma \times D))^*$, we say (A, θ) directly derives X, written as $(A, \theta) \Rightarrow_G X$ if there exist $d \in D$ (regarded as an input data value) and $r = (A, \psi, i) \to c_1 \ldots c_n \in R$ (resp. $r = (A, \psi) \to c_1 \ldots c_n \in R$) such that

$$d, \theta \models \psi, X = c'_1 \ldots c'_n, \theta' = \theta[i \leftarrow d] \text{ (resp. } \theta' = \theta) \text{ where}$$
$$c'_j = \begin{cases} (B, \theta') & \text{if } c_j = B \in V, \\ (b, \theta'(l)) & \text{if } c_j = (b, l) \in \Sigma \times [k]. \end{cases}$$

For $X, Y \in ((V \times \Theta_k) \cup (\Sigma \times D))^*$, we also write $X \Rightarrow_G Y$ if there are $X_1, X_2, X_3 \in ((V \times \Theta_k) \cup (\Sigma \times D))^*$ such that $X = X_1(A, \theta)X_2$, $Y = X_1 X_3 X_2$ and $(A, \theta) \Rightarrow_G X_3$. If we want to emphasize the applied rule r and the input data value d, we write $X \Rightarrow_{G,r}^d Y$.

Let $\overset{*}{\Rightarrow}_G$ and $\overset{+}{\Rightarrow}_G$ be the reflexive transitive closure and the transitive closure of $\Rightarrow_G$, respectively, called the derivation relation of zero or more steps (resp. the derivation relation of one or more steps). We abbreviate $\Rightarrow_G$, $\overset{*}{\Rightarrow}_G$ and $\overset{+}{\Rightarrow}_G$ as $\Rightarrow$, $\overset{*}{\Rightarrow}$ and $\overset{+}{\Rightarrow}$ if G is clear from the context.

We denote by $\bot$ the register assignment that assigns the initial value $\bot$ to every register. We let $L(G) = \{w \mid (S, \bot) \overset{+}{\Rightarrow} w \in (\Sigma \times D)^*\}$. $L(G)$ is called the *data language* generated by G. $(S, \bot) \overset{+}{\Rightarrow} w$ is called a *derivation* of w in G. RCFGs G_1 and G_2 are *equivalent* if $L(G_1) = L(G_2)$.

Example 1. For $\Sigma = \{a, b\}$, let $G = (\{S, A\}, R, S)$ be a 2-RCFG where $R = \{(S, \mathtt{tt}, 1) \to (a, 1)A(a, 1), (A, x_1^{\neq}, 2) \to (b, 2)A(b, 2), (A, x_1^{=}) \to (a, 1)\}$. Then, $L(G) = \{(a, d_0)(b, d_1) \ldots (b, d_n)(a, d_0)(b, d_n) \ldots (b, d_1)(a, d_0) \mid n \geq 0, d_i \neq d_0$ for $i \in [n]\}$.

3   Register Type, Normal Forms and ε-rule Removal

3.1   Register Type

In this subsection, we will define register type, which is useful in expressing equalities among the contents of registers, transforming a given RCFG into a certain normal form and proving some important properties of RCFG. The idea is simple; instead of remembering concrete data values in registers, it suffices to remember the induced equivalence classes of the indices of registers as long as the equalities among data values in the registers are concerned.

Definition 2. *A decomposition of* $[k]$ *into disjoint non-empty subsets is called a* register type *of* k*-RCFG. Let* Γ_k *denote the collection of all register types of* k*-RCFG. For a register type* $\gamma \in \Gamma_k$, *let* $\gamma[i]$ *(*$i \in [k]$*) denote the subset containing* i. $\square$

For example, $\gamma_1 = \{\{1, 2\}, \{3, 5\}, \{4\}\}$ is a register type of 5-RCFG and $\gamma_1[1] = \{1, 2\}$, $\gamma_1[5] = \{3, 5\}$. For a register assignment $\theta \in \Theta_k$ and a register type $\gamma \in \Gamma_k$, we define the typing relation as:

$$\theta \models \gamma :\Longleftrightarrow \forall i, j.(\theta[i] = \theta[j] \Longleftrightarrow \gamma[i] = \gamma[j]).$$

For example, $\theta_1 \in \Theta_5$ such that $\theta_1(1) = \theta_1(2) = 8$, $\theta_1(3) = \theta_1(5) = 10$, $\theta_1(4) = 5$ satisfies $\theta_1 \models \gamma_1$. By definition, for each $\theta \in \Theta_k$, there is exactly one $\gamma \in \Gamma_k$ such that $\theta \models \gamma$. In this case, we say that the type of θ is γ.

3.2   Normal Forms for Guard Expressions

By using register types, we show that a given RCFG can be transformed into an equivalent RCFG G' such that for any rule $r = (A, \psi, i)/(A, \psi) \to \alpha$, r can be applied for any (A, θ), that is, the guard ψ never blocks any (A, θ) and only specifies the equality or inequality among an input data value d and the current contents of the registers. This transformation is the key of the ε-rule removal shown in the next subsection.

First, it is easy to transform a given k-RCFG into an equivalent k-RCFG where the guard expression ψ of every rule has the following form:

$$\psi = (x_{i_1}^{=} \wedge \ldots \wedge x_{i_m}^{=}) \wedge (x_{j_1}^{\neq} \wedge \ldots \wedge x_{j_n}^{\neq}) \tag{1}$$

The above guard can be obtained by the following equivalence transformations:

1. Transform the guard expression of every rule to an equivalent disjunctive normal form.
2. Replace a rule $(A, \psi_1 \vee \psi_2, i) \to \alpha$ into $(A, \psi_1, i) \to \alpha$ and $(A, \psi_2, i) \to \alpha$.

For a guard expression ψ in (1), we let $\psi^{=} = \{i_1, \ldots, i_m\}$ and $\psi^{\neq} = \{j_1, \ldots, j_n\}$. We assume $\psi^{=} \cap \psi^{\neq} = \emptyset$ (the rule with $\psi^{=} \cap \psi^{\neq} \neq \emptyset$ can be removed). For $\gamma \in \Gamma_k$ and $\psi \in F_k$ in the form of (1), we define

$$\gamma \models \psi :\Longleftrightarrow \bigwedge_{i \in \psi^{=}} (\bigwedge_{j \in \psi^{=}} \gamma[i] = \gamma[j] \wedge \bigwedge_{j \in \psi^{\neq}} \gamma[i] \neq \gamma[j]).$$

Note that $\psi^{=} = \emptyset$ implies $\gamma \models \psi$ for any γ. It is easy to see that the following property holds, which means that for an assignment θ that conforms to γ, there is a data value d that satisfies ψ if and only if $\gamma \models \psi$.

$$\theta \models \gamma \;\Rightarrow\; (\gamma \models \psi \Longleftrightarrow \exists d.\ d, \theta \models \psi).$$

Lemma 3. *For an arbitrary* k*-RCFG* G, *we can construct a* k*-RCFG* G' *such that* $L(G') = L(G)$ *and the guard expression of every rule in* G' *is one of the following* $k + 1$ *expressions:* $x_1^{=}, x_2^{=}, \ldots, x_k^{=}, x_1^{\neq} \wedge \cdots \wedge x_k^{\neq}$.

Proof. We assume that the guard expression of every rule of G has the form of (1). For such a guard expression ψ and a register type $\gamma \in \Gamma_k$, let $[\psi, \gamma]$ be the set of guard expressions defined as follows.

$$[\psi, \gamma] = \begin{cases} \{x_i^= \mid i = \min \psi^=\} & \text{if } \psi^= \neq \emptyset, \\ \{x_i^= \mid i \in [k] \setminus \bigcup_{j \in \psi^{\neq}} \gamma[j]\} \cup \{x_1^{\neq} \wedge \ldots \wedge x_k^{\neq}\} & \text{if } \psi^= = \emptyset. \end{cases}$$

Then the following properties hold.

(i) $\psi' \in [\psi, \gamma] \Rightarrow \forall \theta \exists d.\ d, \theta \models \psi'$ (The guard in $[\psi, \gamma]$ is always satisfiable.)

(ii) $\theta \models \gamma \Rightarrow \forall d.\ (d, \theta \models \psi \iff \exists \psi' \in [\psi, \gamma].\ d, \theta \models \psi')$

 (The same input value can be used for ψ and $[\psi, \gamma]$.)

(iii) $\psi' \in [\psi, \gamma] \Rightarrow \forall i \exists \gamma' \forall \theta, d.\ ((\theta \models \gamma \wedge d, \theta \models \psi') \Rightarrow \theta[i \leftarrow d] \models \gamma')$.

 (The register type after the rule application is unique.)

The register type γ' in the above third property is uniquely determined by γ, ψ', and i, and we write the register type as $after(\gamma, \psi', i)$.

We construct k-RCFG $G' = (V', S', R')$ from $G = (V, S, R)$ where $V' = V \times \Gamma_k$, $S' = (S, \{[k]\})$, and R' is the smallest set that satisfies the following inference rule, where $\alpha^{\mathrm{aug}(\gamma')}$ is the sequence obtained from α by replacing every occurrence of every nonterminal A in V with (A, γ'); that is, $(X_1 \ldots X_n)^{\mathrm{aug}(\gamma')} = X_1' \ldots X_n'$ where $X_\ell' = (X_\ell, \gamma')$ if $X_\ell \in V$ and $X_\ell' = X_\ell$ otherwise for every $\ell \in [n]$.

$$\frac{(A, \psi, i) \to \alpha \in R \quad (\text{resp. } (A, \psi) \to \alpha \in R)}{\gamma \models \psi, \ \psi' \in [\psi, \gamma], \ \gamma' = after(\gamma, \psi', i) \quad (\text{resp. } \gamma' = \gamma)}$$
$$\frac{}{((A, \gamma), \psi', i) \to \alpha^{\mathrm{aug}(\gamma')} \in R' \quad (\text{resp. } ((A, \gamma), \psi') \to \alpha^{\mathrm{aug}(\gamma')} \in R')}$$

We can show the following properties, which establish the lemma.

- For a derivation of a data word w in G, if we replace each (A, θ) with $((A, \gamma), \theta)$ where γ is the register type that satisfies $\theta \models \gamma$, then we obtain a derivation of w in G'. Note that by the above property (ii), there must exist $\psi' \in [\psi, \gamma]$ that allows this derivation in G' where ψ is the guard expression of the rule used for the derivation in G.
- For every $((A, \gamma), \theta)$ appearing in a derivation in G', it holds that $\theta \models \gamma$.
- For a derivation of a data word w in G', if we replace each $((A, \gamma), \theta)$ with (A, θ), then we obtain a derivation of w in G.

3.3 ε-rule Removal

Theorem 4. *For an arbitrary k-RCFG, we can construct an equivalent k-RCFG having no ε-rule.*

Proof. By Lemma 3, we can transform any k-RCFG $G = (V, R, S)$ into another k-RCFG $G' = (V', R', S')$ such that $L(G') = L(G)$ and the guard expression of every rule in G' is either $x_1^=, \ldots, x_k^=$, or $x_1^{\neq} \wedge \ldots \wedge x_k^{\neq}$. Because the guard expressions of G' never block the application of each rule, we can compute the set Nu of nullable nonterminals (i.e. the set that consists of every nonterminal A such that $A \Rightarrow^*_{G'} \varepsilon$) in the same way as CFG; that is, we can compute Nu as the smallest set that satisfies the following conditions:

- If $(A, \psi, i)/(A, \psi) \rightarrow \varepsilon \in R'$, then $A \in Nu$.
- If $(A, \psi, i)/(A, \psi) \rightarrow \alpha \in R'$ and α consists of nonterminals in Nu, then $A \in Nu$.

And thus we can remove the ε-rules of G' also in the same way as CFG.

4 Generalized RCFG

4.1 Definitions

We define generalized register context-free grammar by allowing an arbitrary binary relation on the set of data values. Let Σ be a finite alphabet, D be a set of data values such that $\Sigma \cap D = \emptyset$ equipped with a finite set of binary relations $\mathcal{R}$. We call $\mathbb{D} = (D, \mathcal{R})$ a *data structure*. For $k \in \mathbb{N}_0$, a *generalized k-register context-free grammar* (k-GRCFG) is a triple $G = (V, R, S)$ where V, R and S are the same as in k-RCFG except that an atomic formula in a guard expression is $x_i^{\bowtie}$ and $x_i^{\bowtie^{-1}}$ ($i \in [k]$, $\bowtie \in \mathcal{R}$) and its semantics is defined by

$$d, \theta \models x_i^{\bowtie} \text{ iff } \theta(i) \bowtie d \quad \text{and} \quad d, \theta \models x_i^{\bowtie^{-1}} \text{ iff } d \bowtie \theta(i)$$

for any $d \in D$ and $\theta \in \Theta_k$. We sometimes write k-GRCFG($\mathcal{R}$) to emphasize $\mathcal{R}$ and abbreviate it as k-GRCFG($\bowtie$) when $\mathcal{R} = \{\bowtie\}$. Notions and notations for RCFG such as ε-rule, derivation relation $\Rightarrow$, the data language $L(G)$ generated by G are defined in the same way. We also write k-GRCFG(=) to denote a (usual) k-RCFG.

The following properties can be proved in a similar way to the case of k-GRCFG(=) [6].

Theorem 5. *The class of data languages generated by k-GRCFG($\mathcal{R}$) is closed under union, concatenation and Kleene-closure. It is not closed under intersection, complement, homomorphisms or inverse homomorphisms.*

4.2 Simulation Property and Type Oracle

In Sect. 3, we showed that a given RCFG can be transformed to an equivalent RCFG where the guard expression of a production rule never blocks its application by associating a register type with each nonterminal symbol. We can extend register type to GRCFG in a natural way, but the above transformation cannot guarantee the equivalence because the register type no longer has information enough to represent the applicability of a rule in GRCFG.

Example 6. Consider the set of integers with the usual strict total order $\mathbb{Z} = (Z, \{<_Z, >_Z\})$ as a data structure. We might extend register type of GRCFG(=) by introducing $<_Z$ among the equivalence classes of $[k]$. For example, let $\varphi = x_1^< \wedge x_2^>$ be a guard expression of 3-GRCFG($\mathbb{Z}$) and consider register assignments $\theta_1, \theta_2 \in \Theta_3$ such that $\theta_1(1) = \theta_1(3) = 4$, $\theta_1(2) = 7$ and $\theta_2(1) = \theta_2(3) = 5$, $\theta_2(2) = 6$. Also let γ be the register type (informally) defined as $\gamma = \{\{1,3\} <_Z \{2\}\}$. Both $\theta_1 \models \gamma$ and $\theta_2 \models \gamma$ hold. However, there is no $d \in Z$ such that $d, \theta_2 \models \varphi$ while $5, \theta_1 \models \varphi$. □

Similarly, the membership and emptiness lose decidability for GRCFG because a binary relation appearing in a guard expression may be an undecidable relation. To limit the influence of binary relations in a data structure so that GRCFG have mild expressive power, we introduce two properties of a GRCFG, namely, the simulation property and (the existence) of type oracle.

In the rest of this paper, we assume $\mathcal{R}$ is a singleton $\mathcal{R} = \{\bowtie\}$ for simplicity. The properties we show below can be extended in a general case that $\mathcal{R}$ has more than one binary relation. We first extend a register type as a binary relation $\gamma : ([k] \times [k]) \setminus \{(i,i) \mid i \in [k]\} \rightarrow \{\mathtt{tt}, \mathtt{ff}\}[1]$. We say that the type of a register assignment θ is γ (and write $\theta \models \gamma$) iff for all $i, j \in [k]$ $(i \neq j)$,

$$\gamma(i,j) = \mathtt{tt} \text{ iff } \theta(i) \bowtie \theta(j).$$

We write $\theta \sim_\bowtie \theta'$ if the types of register assignments θ and θ' are the same. The collection of all register types of k-GRCFG is denoted by Γ_k as before.

Definition 7 (Simulation). *Let G be a k-GRCFG($\bowtie$) $G = (V, R, S)$. G has the simulation property (with respect to $\bowtie$) if the following condition is met.*

For all $\theta, \theta' \in \Theta_k$, $d \in D$, $r = (A, \varphi, i)/(A, \varphi) \rightarrow \alpha \in R$ such that $\theta \sim_\bowtie \theta'$ and $d, \theta \models \varphi$, there exists $d' \in D$ such that $d', \theta' \models \varphi$, and if the left-hand side of r is (A, φ, i), then $\theta[i \leftarrow d] \sim_\bowtie \theta'[i \leftarrow d']$.

The following diagram illustrates the condition of the simulation property.

$$
\begin{array}{ccc}
(A, \theta) & \Longrightarrow_r^d \ldots (B, \theta[i \leftarrow d]) \ldots \\
\wr & & \wr \\
(A, \theta') & \Longrightarrow_r^{d'} \ldots (B, \theta'[i \leftarrow d']) \ldots
\end{array}
$$

Definition 8 (Type Oracle). *Let $O : \Gamma_k \times F_k \rightarrow \{\mathtt{tt}, \mathtt{ff}\}$ be the predicate defined by: for $\gamma \in \Gamma_k$ and $\psi \in F_k$, $O(\gamma, \psi) = \mathtt{tt}$ iff*

there are $\theta \in \Theta_k$ and $d \in D$ such that $\theta \models \gamma$ and $d, \theta \models \psi$.

We say that D has the type oracle if there is a polynomial time algorithm that answers whether $O(\gamma, \psi) = \mathtt{tt}$ or $\mathtt{ff}$ for given $\gamma \in \Gamma_k$ and $\psi \in F_k$. □

---

[1] We exclude the diagonal elements $\{(i,i) \mid i \in [k]\}$ from the domain of a register type because the applicability of a rule does not depend on whether $\theta(i) \bowtie \theta(i)$.

Finally, we define *data type* as an extension of register type by adding the information on equality between data values in the registers and data values appearing in a given data word w.

Definition 9 (Data Type). *Let w be a data word and D_w be the set of data values appearing in w; i.e. $D_w = \{d_i \mid i \in [n], w = (a_1, d_1) \ldots (a_n, d_n)\}$. Also let $d_{\neq} \notin D_w$ be a newly introduced symbol. We use a function $e : [k] \to D_w \cup \{d_{\neq}\}$, whose codomain is finite, to represent the register assignment by replacing every data value that does not appear in w with $d_{\neq}$. We write $\theta \models e$ iff for all $i \in [k]$,*

$$e(i) = \theta(i) \text{ if } \theta(i) \in D_w \text{ and } e(i) = d_{\neq} \text{ otherwise.}$$

The collection of all such functions $e : [k] \to D_w \cup \{d_{\neq}\}$ is denoted by $E_{w,k}$.

The data type *of a register assignment $\theta \in \Theta_k$ for a data word w is a pair $(\gamma, e) \in \Gamma_k \times E_{w,k}$. We write $\theta \models (\gamma, e)$ iff $\theta \models \gamma$ and $\theta \models e$. We define the simulation property with data type and the data type oracle $O_w(\gamma, e, \varphi)$ of $w \in (\Sigma \times D)^*$ defined for $\gamma \in \Gamma, e \in E_{w,k}, \varphi \in F_k$ in the same way as in the case of register types.*

5   Properties of GRCFG

5.1   ε-rule Removal

Theorem 10. *For an arbitrary GRCFG(⋈) G such that G has the simulation property and D has the type oracle, we can construct an equivalent GRCFG(⋈) G' having no ε-rule.*

Proof. The theorem can be proved in a similar way to Theorem 4 by using the simulation property and the type oracle. Let $G = (V, R, S)$ be a k-GRCFG(⋈). We assume that the guard expression of every rule in R is the conjunction of literals (atomic formulas or their negations). We first construct k-GRCFG(⋈) $G' = (V', R', S')$ from G where

- $V' = V \times \Gamma_k$,
- R' is the smallest set of rules defined as follows. Define the subset of guard expressions Ψ as

$$\Psi = \{\bigwedge_{i \in [k]} \zeta_i \wedge \bigwedge_{i \in [k]} \eta_i \mid \zeta_i \in \{x_i^{\bowtie}, \neg x_i^{\bowtie}\}, \eta_i \in \{x_i^{\bowtie^{-1}}, \neg x_i^{\bowtie^{-1}}\}\}.$$

Let $r = (A, \varphi, i) \to \alpha \in R$. (A rule $(A, \varphi) \to \alpha$ can be processed in a similar way.) Also let $\gamma \in \Gamma_k$ and $\psi \in \Psi$. If $O(\gamma, \varphi \wedge \psi) = \mathtt{tt}$,

$$((A, \gamma), \varphi \wedge \psi, i) \to \alpha^{\mathrm{aug}(\gamma')} \in R'$$

where $\gamma' \in \Gamma_k$ is a register type that satisfies $\theta[i \leftarrow d] \models \gamma'$ for any θ and d such that $\theta \models \gamma$ and $d, \theta \models \varphi \wedge \psi$ (see the proof of Lemma 3 for the definition of $\alpha^{\mathrm{aug}(\gamma')}$). Note that γ' must exist and γ' is uniquely determined by γ, $\varphi \wedge \psi$ and i because γ specifies whether $\theta(i) \bowtie \theta(j)$ holds or not for each pair $i, j \in [k]$ ($i \neq j$) and also ψ specifies whether $\theta(i) \bowtie d$ and $d \bowtie \theta(i)$ hold or not for each $i \in [k]$ and an input data value d.

– $S' = (S, \gamma_0)$ where $\perp^k \models \gamma_0$.

See an example of the construction in Example 11. We can show $L(G) = L(G')$ by induction on the length of derivations in G and G', using the simulation property (to show $L(G') \subseteq L(G)$) and the type oracle (to show both inclusions).

The rest of the proof is similar to the one in Theorem 4.

Example 11. Let $k = 2$ and consider a rule $r = (A, \varphi, 1) \to \alpha$ where $\varphi = x_1^{\bowtie} \wedge x_2^{\bowtie^{-1}}$. The possible register types are $\gamma_1 = (\delta_{12} \wedge \delta_{21})$, $\gamma_2 = (\delta_{12} \wedge \neg\delta_{21})$, $\gamma_3 = (\neg\delta_{12} \wedge \delta_{21})$ and $\gamma_4 = (\neg\delta_{12} \wedge \neg\delta_{21})$ where $\delta_{12} = (\theta(1) \bowtie \theta(2))$ and $\delta_{21} = (\theta(2) \bowtie \theta(1))$.[2] After the elimination of the unsatisfiable ones and Boolean simplification, we can assume that $\Psi = \{\psi_1, \psi_2, \psi_3, \psi_4\}$ where $\psi_1 = x_1^{\bowtie^{-1}} \wedge x_2^{\bowtie}$, $\psi_2 = x_1^{\bowtie^{-1}} \wedge \neg x_2^{\bowtie}$, $\psi_3 = \neg x_1^{\bowtie^{-1}} \wedge x_2^{\bowtie}$ and $\psi_4 = \neg x_1^{\bowtie^{-1}} \wedge \neg x_2^{\bowtie}$. If $O(\gamma_i, \varphi \wedge \psi_j) = \mathtt{tt}$, the register type γ' after the rule application is $\gamma_1, \gamma_2, \gamma_1, \gamma_2$ for $\psi_1, \psi_2, \psi_3, \psi_4$, respectively. In this example, the type γ' is determined depending only on ψ_j and independent of γ_i because $k = 2$ and an input data value is loaded to the first register when r is applied.

5.2 Emptiness and Membership

Theorem 12. *The emptiness problem for GRCFG($\bowtie$) such that G has the simulation property and D has the type oracle, is EXPTIME-complete.*

Proof. Let $G = (V, R, S)$ be such a k-GRCFG($\bowtie$) and $G' = (V', R', S')$ be the k-GRCFG($\bowtie$) constructed from G in the proof of Theorem 10. As shown in that proof, $L(G') = L(G)$. We construct CFG $G'' = (V', R'', S')$ from G' where

$$R'' = \{(A, \gamma) \to X_1 \ldots X_n \mid ((A, \gamma), \varphi, i)/((A, \gamma), \varphi) \to X_1' \ldots X_n' \in R' \text{ for}$$
some φ and i, and $X_j = X_j'$ if $X_j' \in V'$ and $X_j = a$ if $X_j \notin V'$ for each $j \in [n]\}$.

We can easily show $L(G') = \emptyset \Leftrightarrow L(G'') = \emptyset$ because a rule application is never blocked in G'.

Because the size of the CFG constructed in this way is exponential to k and the emptiness problem for CFG is decidable in linear time, the emptiness problem for GRCFG is decidable in deterministic time exponential to k.

The lower bound can be obtained from EXPTIME-completeness of the emptiness problem for k-GRCFG($=$) [17].

Theorem 13. *The membership problem for GRCFG($\bowtie$) such that G has the simulation property with data type and D has the data type oracle, is EXPTIME-complete.*

(This theorem can be proved in a similar way to Theorem 12.)

---

[2] For readability, we denote a register type as a Boolean formula on a register assignment θ. For example, $\gamma_2(1, 2) = \mathtt{tt}$ and $\gamma_2(2, 1) = \mathtt{ff}$ if we follow the notation defined in Sect. 4.2.

5.3 GRCFG with a Total Order on a Dense Set

Lemma 14. *Every GRCFG($<_Q$) has the simulation property and Q has the type oracle where $<_Q$ is the strict total order on the set Q of all rational numbers. Similarly, it has the simulation property with data type and Q has the data type oracle.*

Proof. We abbreviate $<_Q$ as $<$. Let $G = (V, R, S)$ be a k-GRCFG($<$), $\theta \in \Theta_k$, $\gamma \in \Gamma_k$ and $r = (A, \varphi, i) \rightarrow \alpha \in R$ where φ is the conjunction of literals (of the form $x_i^<$ or $\neg x_j^<$). (The case $r = (A, \varphi) \rightarrow \alpha \in R$ can be treated in a similar way.) Assume that $\theta \models \gamma$. The rule r can be applied to (A, θ) iff there is $d \in Q$ such that $d, \theta \models \varphi$. The condition $d, \theta \models \varphi$ as well as the assumption $\theta \models \gamma$ can be represented as a set of inequations on $d, \theta(1), \ldots, \theta(k)$. Whether this set of inequations has a contradiction does not depend on the concrete values $\theta(1), \ldots, \theta(k)$, and if it does not have a contradiction, then there must exist $d \in Q$ that satisfies it because Q is dense. Moreover, whether $\theta[i \leftarrow d] \models \gamma'$ holds for a given γ', which can also be represented as the consistency of a set of inequations on $d, \theta(1), \ldots, \theta(k)$, does not depend on θ. Hence, if $\theta \models \gamma$, $\theta' \models \gamma$, $d, \theta \models \varphi$ and $\theta[i \leftarrow d] \models \gamma'$, there is $d' \in Q$ satisfying $d', \theta \models \varphi$ and $\theta'[i \leftarrow d'] \models \gamma'$ and the simulation property holds.

Similarly, for deciding $O(\gamma, \varphi) = \mathtt{tt}$, it suffices to represent the condition

$$d, \theta \models \varphi \wedge \theta \models \gamma$$

as a set of inequations on $d, \theta(1), \ldots, \theta(k)$ as above and solve it.

We can show the simulation property with data type and the existence of the data type oracle in a similar way.

Example 15. Consider a 2-GRCFG($<$)$= (V, R, S)$ and a rule $(A, \varphi, 1) \rightarrow B \in R$ where $\varphi = x_1^< \wedge \neg x_2^<$. We see that $d, \theta \models \varphi \Leftrightarrow \theta(1) < d \leq \theta(2)$. Because $k = 2$ and $<$ is a total order on Q, there are three possible register types $\gamma_1 = (\theta(1) < \theta(2))$, $\gamma_2 = (\theta(2) < \theta(1))$ and $\gamma_3 = (\theta(1) = \theta(2))$. As easily known, (i) there is $d \in Q$ such that $d, \theta \models \varphi$ and $\theta \models \gamma$ if and only if $\gamma = \gamma_1$, and (ii) if $\gamma = \gamma_1$ then such $d \in Q$ satisfies either (ii-a) $d < \theta(2)$, $\theta[1 \leftarrow d] \models \gamma_1$ or (ii-b) $d = \theta(2)$, $\theta[1 \leftarrow d] \models \gamma_3$.

Corollary 16. *For a given GRCFG($<_Q$), we can construct an equivalent GRCFG($<_Q$) having no ε-rule. The emptiness and membership problems are both EXPTIME-complete for GRCFG($<_Q$).*

Proof. By Lemma 14 and Theorems 10, 12 and 13.

6 Conclusion

We have introduced register type to RCFG and shown an equivalence transformation to RCFG that never blocks a rule application by associating a register type with each nonterminal symbol. Then we have defined generalized RCFG

(GRCFG) that can use an arbitrary relation in the guard expression. Using the technique of register type and making two reasonable assumptions, the simulation property and the existence of type oracle, the decidability of emptiness and membership for GRCFG and a transformation to an ε-free GRCFG have been provided.

Nominal CFG [4] with equality symmetry, total order symmetry and integer symmetry correspond to GRCFG($=$), GRCFG($<_Q$) (Sect. 5.3) and GRCFG($<_Z$) (Example 6), respectively. Investigating the relation between nominal CFG and GRCFG in depth is future work.

References

1. Benedikt, M., Ley, C., Puppis, G.: What you must remember when processing data words. In: 4th Alberto Mendelzon International Workshop on Foundations of Data Management (2010)
2. Bojańczyk, M., Muscholl, A., Schwentick, T., Segoufin, L.: Two-variable logic on data trees and XML reasoning. J. ACM **56**(3), 13:1–13:48 (2009). https://doi.org/10.1145/1516512.1516515
3. Bojańczyk, M., David, C., Muscholl, A., Schwentick, T., Segoufin, L.: Two-variable logic on data words. ACM Trans. Comput. Log. **12**(4), 27:1–27:26 (2011). https://doi.org/10.1145/1970398.1970403
4. Bojańczyk, M., Klin, B., Lasota, S.: Automata theory in nominal sets. Log. Methods Comput. Sci. **10**(3) (2014). https://doi.org/10.2168/LMCS-10(3:4)2014
5. Bouyer, P.: A logical characterization of data languages. Inf. Process. Lett. **84**(2), 75–85 (2002). https://doi.org/10.1016/S0020-0190(02)00229-6
6. Cheng, E.Y., Kaminski, M.: Context-free languages over infinite alphabets. Acta Inf. **35**(3), 245–267 (1998). https://doi.org/10.1007/s002360050120
7. Demri, S., Lazić, R.: LTL with the freeze quantifier and register automata. ACM Trans. Comput. Log. **10**(3), 16:1–16:30 (2009). https://doi.org/10.1145/1507244.1507246
8. Demri, S., Lazić, R., Nowak, D.: On the freeze quantifier in constraint LTL: decidability and complexity. Inf. Comput. **205**(1), 2–24 (2007). https://doi.org/10.1016/j.ic.2006.08.003
9. Figueira, D., Hofman, P., Lasota, S.: Relating timed and register automata. Math. Struct. Comput. Sci. **26**(6), 993–1021 (2016). https://doi.org/10.1017/S0960129514000322
10. Kaminski, M., Francez, N.: Finite-memory automata. Theor. Comput. Sci. **134**(2), 329–363 (1994). https://doi.org/10.1016/0304-3975(94)90242-9
11. Libkin, L., Martens, W., Vrgoč, D.: Querying graphs with data. J. ACM **63**(2), 14:1–14:53 (2016). https://doi.org/10.1145/2850413
12. Libkin, L., Tan, T., Vrgoč, D.: Regular expressions for data words. J. Comput. Syst. Sci. **81**(7), 1278–1297 (2015). https://doi.org/10.1016/j.jcss.2015.03.005
13. Libkin, L., Vrgoč, D.: Regular path queries on graphs with data. In: 15th International Conference on Database Theory (ICDT 2012), pp. 74–85 (2012). https://doi.org/10.1145/2274576.2274585
14. Milo, T., Suciu, D., Vianu, V.: Typechecking for XML transformers. In: 19th ACM Symposium on Principles of Database Systems (PODS 2000), pp. 11–22 (2000). https://doi.org/10.1145/335168.335171

15. Neven, F., Schwentick, T., Vianu, V.: Finite state machines for strings over infinite alphabets. ACM Trans. Comput. Log. **5**(3), 403–435 (2004). https://doi.org/10.1145/1013560.1013562

16. Sakamoto, H., Ikeda, D.: Intractability of decision problems for finite-memory automata. Theor. Comput. Sci. **231**(2), 297–308 (2000). https://doi.org/10.1016/S0304-3975(99)00105-X

17. Senda, R., Takata, Y., Seki, H.: Complexity results on register context-free grammars and register tree automata. In: Fischer, B., Uustalu, T. (eds.) ICTAC 2018. LNCS, vol. 11187, pp. 415–434. Springer, Cham (2018). https://doi.org/10.1007/978-3-030-02508-3_22

Languages

Logic and Rational Languages
of Scattered and Countable
Series-Parallel Posets

Amazigh Amrane and Nicolas Bedon[(✉)]

LITIS (EA 4108), Université de Rouen, Rouen, France
Amazigh.Amrane@etu.univ-rouen.fr, Nicolas.Bedon@univ-rouen.fr

Abstract. We show that an extension of MSO with Presburger arithmetic, named P-MSO, is as expressive as branching automata over scattered and countable N-free posets. As a consequence of the effectiveness of the constructions from one formalism to the other, the P-MSO theory of the scattered and countable N-free posets is decidable.

Keywords: Automata and logic · Transfinite N-free posets ·
Series-parallel posets · Series-parallel rational languages ·
Branching automata · Monadic second-order logic ·
Presburger arithmetic

1 Introduction

Since their introduction in computer science by Kleene [12], finite automata on words have been extended in many directions, because of the variety of their uses. One of the early extensions of automata are from Büchi [6]. First, Büchi, and independently Elgot [10] and Trakhtenbrot [21], showed that finite automata and monadic second-order logic (MSO) are expressively equivalent for languages of finite words, with effective constructions from one formalism to the other. A decision procedure for the MSO theory of finite words immediately follows. This early connection between automata on words and logic has been quickly developed in many ways. Büchi extended automata over finite words to infinite words and continued to study their connections with MSO. With automata over ω-words [7] he gave in particular a decision procedure for the first-order logic theory of $(\mathbb{N}, +)$, retrieving a result of Presburger. With automata over ordinals, he proved that the MSO theory of all countable ordinals is decidable [8]. All those decision procedures are relative to the theory of one successor, and Büchi asked if automata over words could be extended to obtain decision procedures for logics of many successors. This question has been answered positively by Rabin [18] with automata on infinite trees. Encoding linear orderings into trees, Rabin deduced the decidability of MSO over countable linearly ordered sets. Automata over linear orderings were introduced more recently [5] and used to

© Springer Nature Switzerland AG 2019
C. Martín-Vide et al. (Eds.): LATA 2019, LNCS 11417, pp. 275–287, 2019.
https://doi.org/10.1007/978-3-030-13435-8_20

give another proof of the decidability of MSO over countable and scattered linear orderings [3].

Finite automata on words are natural models for finite sequential processes. Many extensions have been proposed in order to model concurrency. Rabin automata on trees are one of them. We focus in this work on *branching automata* introduced by Lodaya and Weil [14–17]. They recognize languages of finite series-parallel partially ordered sets (posets), or equivalently [19,22] finite N-free posets. Lodaya and Weil developed rational expressions equivalent to branching automata, and their algebraic approach. Branching automata have been shown effectively equivalent [1] to an extension of MSO with Presburger arithmetic named P-MSO, providing a decision procedure for the P-MSO theory of finite N-free posets. Branching automata have been extended to ω-N-free posets by Kuske [13], with a connection with MSO in the particular case of languages of N-free posets with bounded-size antichains.

In this paper we focus on the class $SP^\diamond(A)$ of N-free labeled posets with finite antichains and countable and scattered chains. By extension of branching automata of Lodaya and Weil, a model of branching automata for languages of posets of $SP^\diamond(A)$ have been introduced in [4] as well as equivalent rational expressions. The main result presented in this paper is that $L \subseteq SP^\diamond(A)$ is rational if and only if L is definable in P-MSO, with effective constructions from one formalism to the other. As a consequence, the P-MSO theory of $SP^\diamond(A)$ is decidable. Well-known techniques can be easily adapted for the construction of a rational expression from a P-MSO formula. Since this is not true for the converse, we particularly focus on this part.

2   Notation and Basic Definitions

We let $|E|$ denote the cardinality of a set E, 2^E its powerset, $[n]$ the set $\{1, \ldots, n\}$ (for any non-negative integer $n \in \mathbb{N}$), and $\pi_i(c)$ the ith component of a tuple c.

We start by some basic definitions on linear orderings (see [20] for a detailed presentation). Let J be a set equipped with an order $<$. The ordering J is *linear* if either $j < k$ or $k < j$ for any distinct $j, k \in J$. We denote by $-J$ the backward linear ordering obtained from the set J with the reverse ordering. A linear ordering J is *dense* if for any $j, k \in J$ such that $j < k$, there exists an element i of J such that $j < i < k$. It is *scattered* if it contains no dense sub-ordering. The ordering ω of natural integers is scattered. Ordinals are also scattered orderings. We let $\mathcal{O}$ and $\mathcal{S}$ denote respectively the class of countable ordinals and the class of countable scattered linear orderings. We also let 0 denote the empty linear ordering. Let $J \in \mathcal{S}$. An *interval* K of J is a subset $K \subseteq J$ such that for all $k_1, k_2 \in K$ and $j \in J$, if $k_1 < j < k_2$ then $j \in K$. A *cut* (K, L) of J consists of a pair of two disjoint intervals K and L of J such that $K \cup L = J$ and $k < l$ for all $(k, l) \in K \times L$. The set $\hat{J}$ of all cuts of J is naturally equipped with the ordering $(K_1, L_1) < (K_2, L_2)$ if and only if $K_1 \subsetneq K_2$. This linear ordering can be extended to $J \cup \hat{J}$ by keeping the orderings on the elements of J and of $\hat{J}$, and, for any $j \in J$ and $c = (K, L) \in \hat{J}$, by setting $j < c$ (resp. $c < j$) whenever $j \in K$ (resp. $j \in L$). We let $\hat{J}^*$ denote $\hat{J} \setminus \{(\emptyset, J), (J, \emptyset)\}$.

A *poset* $(P, <)$ is a set P partially ordered by $<$. For short we often denote the poset $(P, <)$ by P. The *width* of P is $\text{wd}(P) = \sup\{|E| : E \text{ is an antichain of } P\}$ where sup denotes the least upper bound of the set. In this paper, we restrict to posets with finite antichains and countable and scattered chains. We let ϵ denote the empty poset. Let $(P, <_P)$ and $(Q, <_Q)$ be two disjoint posets. The *union* (or *parallel composition*) $P \cup Q$ of $(P, <_P)$ and $(Q, <_Q)$ is the poset $(P \cup Q, <_P \cup <_Q)$. The *sum* (or *sequential composition*) $P + Q$ of P and Q is the poset $(P \cup Q, <_P \cup <_Q \cup P \times Q)$. The sum of two posets can be generalized to any linearly ordered sequence $((P_j, <_j))_{j \in J}$ of pairwise disjoint posets by $\sum_{j \in J} P_j = (\bigcup_{j \in J} P_j, (\bigcup_{j \in J} <_j) \cup (\bigcup_{j,j' \in J, \ j<j'} P_j \times P_{j'}))$. The sequence $((P_j, <_j))_{j \in J}$ is called a *J-factorization*, or (sequential) *factorization* for short, of the poset $\sum_{j \in J} P_j$. A poset P is *sequential* if it admits a J-factorization where J contains at least two elements $j \neq j'$ with $P_j, P_{j'} \neq \epsilon$, or P is a singleton. It is *parallel* when $P = P_1 \parallel P_2$ for some $P_1, P_2 \neq \epsilon$. A sequential factorization is *irreducible* when all the P_j are either singletons or parallel posets. The notion of irreducible parallel factorization is defined similarly. The class $SP^\diamond$ of *series-parallel* scattered and countable posets is the smallest class of posets containing ϵ, the singleton and being closed under finite parallel composition and sum indexed by countable scattered linear orderings. It has a nice characterization in terms of graph properties: $SP^\diamond$ coincides with the class of scattered and countable N-free posets without infinite antichains [4]. Recall that $(P, <)$ is *N-free* if there is no $X = \{x_1, x_2, x_3, x_4\} \subseteq P$ such that $< \cap X^2 = \{(x_1, x_2), (x_3, x_2), (x_3, x_4)\}$. We let $SP^{\diamond+}$ denote $SP^\diamond \setminus \{\epsilon\}$. When $P \in SP^\diamond$ and $P = R + P' + S$ or $P = P' \cup R$ for some $R, S, P' \in SP^\diamond$ then P' is a *factor* of P; the factors of P', R and S are also factors of P.

An *alphabet* A is a non-empty finite set whose elements are called *letters*. A poset P is *labeled* by A when it is equipped with a *labeling* total map $l: P \to A$. The notion of a labeled poset corresponds to *pomset* in the literature. Also, the finite labeled posets of width at most 1 correspond to the usual notion of words. For short, the singleton poset labeled by $\{a\}$ is denoted by a, and we often make no distinction between a poset and a labeled poset, except for operations. The *sequential product* (or *concatenation*, denoted by $P \cdot P'$ or PP' for short) and the *parallel product* $P \parallel P'$ of labeled posets are respectively obtained by the sequential and parallel compositions of the corresponding (unlabeled) posets. The sequential product of a linearly ordered sequence of labeled posets is denoted by $\prod$. The class of posets of $SP^\diamond$ labeled by A (or over A) is denoted by $SP^\diamond(A)$, and $SP^\diamond(A) \setminus \{\epsilon\}$ by $SP^{\diamond+}(A)$. Observe that the elements of $A^\diamond = \{P \in SP^\diamond(A) : \text{wd}(P) \leq 1\}$ are the words on scattered and countable linear orderings, as defined in [5]. A *language* of a set S is a subset of S. The sequential product is extended from posets to languages of posets by $L \cdot L' = \{P \cdot P' : P \in L, P' \in L'\}$. A similar extension holds for the parallel product. Let A and B be two alphabets and $P \in SP^\diamond(A)$, $L \subseteq SP^\diamond(B)$ and $\xi \in A$. The language consisting of the labeled poset P in which each element labeled by the letter ξ is non-uniformly replaced by a labeled poset of L is denoted by $L \circ_\xi P$. By *non-uniformly* we mean that the elements labeled by ξ may be replaced by different elements of L. This substitution $L\circ_\xi$ is the homomorphism from $(SP^\diamond(A), \|, \prod)$ into the powerset algebra $(2^{SP^\diamond(A\cup B)}, \|, \prod)$ with $a \mapsto a$ and $\xi \mapsto L$.

3   Rational Languages

Let A be an alphabet, $\xi \in A$, L and L' be languages of $SP^\diamond(A)$. Set

$$L \circ_\xi L' = \bigcup_{P \in L'} L \circ_\xi P \qquad\qquad L^* = \{\prod_{j \in [n]} P_j : n \in \mathbb{N}, P_j \in L\}$$

$$L^{*\xi} = \bigcup_{i \in \mathbb{N}} L^{i\xi} \text{ with } L^{0\xi} = \{\xi\} \text{ and } L^{(i+1)\xi} = (\bigcup_{j \le i} L^{j\xi}) \circ_\xi L$$

$$L^\omega = \{\prod_{j \in \omega} P_j : P_j \in L\} \qquad\qquad L^{-\omega} = \{\prod_{j \in -\omega} P_j : P_j \in L\}$$

$$L^\natural = \{\prod_{j \in \alpha} P_j : \alpha \in \mathcal{O}, P_j \in L\} \qquad L^{-\natural} = \{\prod_{j \in -\alpha} P_j : \alpha \in \mathcal{O}, P_j \in L\}$$

$$L \diamond L' = \{\prod_{j \in J \cup \hat{J}^*} P_j : J \in \mathcal{S} \setminus \{0\} \text{ and } P_j \in L \text{ if } j \in J \text{ and } P_j \in L' \text{ if } j \in \hat{J}^*\}$$

Set $op = \{\|, \circ_\xi, {}^{*\xi}, \cup, \cdot, *, \diamond, \omega, -\omega, \natural, -\natural\}$. The class of *rational languages* [4] of $SP^\diamond(A)$ is the smallest class containing $\emptyset$, $\{\epsilon\}$, $\{a\}$ for all $a \in A$, and being closed under the operations of op, provided $\epsilon \notin L$ in $L \circ_\xi L'$, and with conditions on ${}^{*\xi}$: $L^{*\xi}$ is rational if L is rational, $\epsilon \notin L$ and for every $P \in L$, if ξ is the label of some element x of P, there exists $y \in P$ such that x and y are distinct and incomparable. This latter condition excludes from the rational languages those of the form $(a\xi b)^{*\xi} = \{a^n \xi b^n : n \in \mathbb{N}\}$, for example, which are known to be not Kleene rational. Observe that the usual Kleene rational languages [12] of A^* are a particular case of the rational languages defined above, in which the operators $\|, \circ_\xi, {}^{*\xi}, \omega, -\omega, \natural, -\natural$ and $\diamond$ are not allowed. Note also that the rational languages of $SP^\diamond(A)$ are precisely those of Bruyère and Carton [5] of $A^\diamond$ when $\|, \circ_\xi$ and ${}^{*\xi}$ are not allowed, and are also precisely those of Lodaya and Weil [14–17] of finite N-free posets when $\omega, -\omega, \natural, -\natural$ and $\diamond$ are not allowed. A *rational expression* e is a term of the free algebra over $\{\emptyset\} \cup A$ using the operations of op as functions, where the union is denoted as usual by $+$ instead of $\cup$. Its language $L(e)$ is defined inductively using the definitions of the operations of op.

Example 1. Let $A = \{a, b, c\}$ and $L = c \circ_\xi (a \| (b\xi))^{*\xi}$. Then L is the smallest language containing c and such that if $x \in L$, then $a \| (bx) \in L$. Thus $L = \{c, a \| (bc), a \| (b(a \| (bc))), \dots\}$.

Let L be a language where the letter ξ is not used. Define $L^\circledast = \{\epsilon\} \circ_\xi (L \| \xi)^{*\xi} = \{\|_{i<n} P_i : n \in \mathbb{N}, P_i \in L\}$. When A is an alphabet, $A^\circledast$ is the class of all finite antichains over A, or equivalently, the class of all finite commutative words over A. Recall that a language of a commutative monoid is rational (as defined by Kleene for finite words) if and only if it is semi-linear (see e.g. [9]). In order to avoid confusion between rational languages of a commutative monoid and

rational languages of transfinite words or posets we call *semi-linear* a rational language of a commutative monoid.

4   P-MSO

Presburger arithmetic and MSO are two classical logics in computer science.

Recall that Presburger arithmetic is the first-order logic of $(\mathbb{N}, +)$. The Presburger set $L(\rho)$ of a Presburger formula $\rho(x_1, \ldots, x_n)$ whose free variables are $x_1, \ldots, x_n$ consists of all interpretations of $(x_1, \ldots, x_n)$ satisfying ρ. A language $L \subseteq \mathbb{N}^n$ is a *Presburger set* of $\mathbb{N}^n$ if it is the Presburger set of some Presburger formula. We let $\mathcal{P}_n$ denote the class of all Presburger formulæ with n free variables and we set $\mathcal{P} = \bigcup_{i \in \mathbb{N}} \mathcal{P}_i$. Presburger logic provides tools to manipulate semi-linear sets of commutative monoids with formulæ. Indeed, let $A = \{a_1, \ldots, a_n\}$ be an alphabet totally ordered by the indexes of the a_is. As $u \in A^{\circledast}$ can be thought of as a n-tuple $(|u|_{a_1}, \ldots, |u|_{a_n}) \in \mathbb{N}^n$, where $|u|_a$ denotes the number of occurrences of letter a in u, then $A^{\circledast}$ is isomorphic to $(\mathbb{N}^n, +)$. It is known from [11] that a language L of $A^{\circledast}$ is semi-linear if and only if it is the Presburger set $L(\rho)$ of some Presburger formula $\rho(x_1, \ldots, x_n)$, i.e. $(|u|_{a_1}, \ldots, |u|_{a_n}) \in L(\rho)$ if and only if $u \in L$. Observe that the ordering of the free variables $x_1, \ldots, x_n$ of ρ is related to the ordering of A. Note that $\{\epsilon\}$ is the Presburger set of any closed tautology.

Let $\rho(x_1, \ldots, x_k)$ and $\rho'(x_1', \ldots, x_{k'}')$ be Presburger formulæ and let $A = \{a_1, \ldots, a_k\}$ and $B = \{b_1, \ldots, b_{k'}\}$ be two totally ordered alphabets such that $A \cap B = \emptyset$ or $A = B$. Consider the Presburger sets of ρ and ρ' as semi-linear languages L and L' of respectively $A^{\circledast}$ and $B^{\circledast}$. For all $i \in [k']$, $L \circ_{b_i} L'$ (resp. L'^{*b_i}) is a semi-linear language of $X^{\circledast}$ where X is the totally ordered alphabet $A \circ_{b_i} B$ when $A \cap B = \emptyset$ or A when $A = B$ (resp. B). It is also the Presburger set of some formula that we denote by $\rho \circ_{x_i'} \rho'$ (resp. $\rho'^{*x_i'}$).

From the point of view of syntax, formulæ of P-MSO obey the following grammar:

$$\psi ::= a(x) \mid x \in X \mid x < y \mid \psi_1 \vee \psi_2 \mid \psi_1 \wedge \psi_2 \mid \neg\psi$$
$$\mid \exists x\psi \mid \exists X\psi \mid \forall x\psi \mid \forall X\psi \mid \mathcal{Q}(Z, \psi_1, \ldots, \psi_n, \rho(x_1, \ldots, x_n))$$

The 10 first items of the grammar syntactically define MSO. The last one extends MSO to P-MSO. We interpret P-MSO over posets of $SP^{\diamond}(A)$. Here lowercase variables x, y are first-order variables interpreted over elements of posets, $x_1, \ldots, x_n$ first-order variables interpreted over non-negative integers, and uppercase variables X, Y, Z are second-order variables interpreted over sets of elements of posets. Atomic formulæ $x < y$ and $x \in X$ are self-explanatory. For each letter $a \in A$, the atomic formula $a(x)$ tells if the element x is labeled by a. In the last item of the grammar, Z is the name of a free second-order variable, each ψ_i a P-MSO formula and ρ a Presburger formula with n free variables $x_1, \ldots, x_n$.

Semantics of P-MSO formulæ is defined by extension of semantics of Presburger and MSO logics. The notions of a language and definability naturally extend from MSO to P-MSO. Let us turn to the semantics of $\phi(Z) \equiv$

$\mathcal{Q}(Z, \psi_1, \dots, \psi_n, \rho(x_1, \dots, x_n))$. Let $P \in SP^{\diamond+}(A)$ and $Z \subseteq P$. Then Z satisfies ϕ if it is a non-empty factor of P, and there exist $(v_1, \dots, v_n) \in L(\rho)$ and sequential posets $Z_{1,1}, \dots, Z_{1,v_1}, \dots, Z_{n,1}, \dots, Z_{n,v_n} \in SP^{\diamond+}(A)$ such that $Z = \|_{i \in [n]} \|_{j \in [v_i]} Z_{i,j}$ and $Z_{i,j}$ satisfies ψ_i for all $i \in [n]$ and $j \in [v_i]$.

Example 2. Let $A = \{a_1, a_2\}$, and let $\psi_i \equiv \exists x \ a_i(x)$, $i \in [2]$. Let $\rho(x_1, x_2) \equiv \exists k_1, k_2 \ x_1 = 2k_1 \wedge x_2 = 2k_2 + 1 \in \mathcal{P}$. Let $P_1 = a_1(a_1 \| a_1)a_1$, $P_2 = a_2 a_2$ and $P_3 = (a_1 \| a_2)a_1$. Then $P = P_1 \| P_2 \| P_3$ satisfies $\mathcal{Q}(P, \psi_1, \psi_2, \rho(x_1, x_2))$ since $\{1, 2, 3\}$ can be partitioned into $(K_1, K_2) = (\{1, 3\}, \{2\})$ with $(|K_1|, |K_2|) \in L(\rho)$, and for all $i \in K_j$, $j \in [2]$, P_i satisfying ψ_j.

The main result of this paper is the following:

Theorem 3. *Let A be an alphabet. A language L of $SP^{\diamond}(A)$ is rational if and only if it is P-MSO definable. Furthermore the constructions from one formalism to the other are effective.*

5   From Rational Expressions to P-MSO

Let A be an alphabet. In this section we build by induction on a rational expression e a P-MSO formula that checks if a poset $P \in SP^{\diamond}(A)$ has the structure induced by e. In this construction, we need a slightly modified (but equivalent) notion of a rational expression, that we named >1-expression. The construction is divided in two steps. First, we build by induction on the >1-expression e an intermediary structure called the D-graph D_e of e. Then, we inductively parse D_e in order to construct a P-MSO formula equivalent to e.

>1-*expressions* are built using sequential operations that compose at least two non-empty posets. They use new sequential operators $op^{>1}$ when $op \in \{\cdot, *, \diamond, \omega, -\omega, \natural, -\natural\}$ instead of op. When $L, L' \subseteq SP^{\diamond}(A)$ let

$$L^{\cdot>1}L' = (L \setminus \{\epsilon\}) \cdot (L' \setminus \{\epsilon\}) \qquad L^{*>1} = \{\prod_{i \in [n]} P_i : n > 1, P_i \in L \setminus \{\epsilon\}\}$$

$$L^{\omega>1} = \{\prod_{i \in \omega} P_i : P_i \in L \text{ for all } i \in \omega \text{ and } P_i, P_j \neq \epsilon \text{ for some } i, j \text{ with } i \neq j\}$$

$$L\diamond^{>1}L' = \{\prod_{j \in J \cup \hat{J}^*} P_j : J \in \mathcal{S} \setminus \{0\}, P_j \in L \text{ if } j \in J, P_j \in L' \text{ if } j \in \hat{J}^*$$

$$\text{and } P_i, P_j \neq \epsilon \text{ for some } i, j \in J \cup \hat{J}^*, i \neq j\}$$

We let $L +_c L'$ denote $L + L'$ when condition c is verified, L otherwise. Then $L \cdot L' = L^{\cdot>1}L' +_{\epsilon \in L} L' +_{\epsilon \in L'} L$, $L^* = L^{*>1} + L + \epsilon$, $L^\omega = L^{\omega>1} +_{\epsilon \in L} L^*$ and $L \diamond L' = L\diamond^{>1}L' + L +_{\epsilon \in L} L'$. Similar definitions and equalities hold for $-\omega$, $\natural$ and $-\natural$. Every rational expression can be transformed into an >1-expression. Considering the equalities above as rewriting rules this transformation is unique.

A *D-graph* D is a rooted, directed and ordered finite graph whose vertices are labeled and edges are of two disjoint kinds: *normal* and *special*. Here, *ordered* graph means that the edges outgoing from a node n are totally ordered. The root $r(D)$ of D is unique. The set of all normal and special edges of D are respectively denoted by $E_N(D)$ and $E_S(D)$. We call *leaf* a node without *normal* outgoing edges. We say that a node n is *edged* by $\text{out}(n) = e_1 \ldots e_k$ to express that the ordered sequence of edges that outgoes from n is $e_1 \ldots e_k$. The label of a leaf n of D is either a letter of the alphabet, and in this case $\text{out}(n)$ is empty, or from the class $\mathcal{P}$ of Presburger formulæ, and in this case $\text{out}(n)$ is composed exclusively of special edges. Labels of other nodes are from $\mathcal{P} \cup \{\cdot^{>1}, *^{>1}, \diamond^{>1}, \omega^{>1}, -\omega^{>1}, \natural^{>1}, -\natural^{>1}\}$. The length of $\text{out}(n)$ is consistent with the label $\gamma(n)$ of a node n: 0 for letters, 2 for $\cdot^{>1}$ and $\diamond^{>1}$, k for a label in $\mathcal{P}_k$, 1 otherwise. We let $n \to m$ denote an edge from n to m; n is a *direct ascendant* of m, and m a *direct descendant* of n. We often see sequences as words. For example, we let $e \circ_{e'} s$ denote the sequence of edges obtained by replacing in the sequence of edges s each occurrence of the edge e' by the edge e. We let $n' \circ_n^{\text{src}} s$ denote the sequence of edges obtained from s by replacing every occurrence of n in sources of edges by n'.

The existence of a path labeled by $P \in SP^\diamond(A)$ from a node n is defined by induction on P as follows. When $\gamma(n) = a \in A$ there is a path from n labeled by a. When $\gamma(n)$ is some sequential operation $op^{>1}$ then the existence of a path from n and its label are defined consistently with the definition of $op^{>1}$ and the existing paths from the direct descendants of n. When $\gamma(n)$ is some Presburger formula ρ and the sequence of its direct descendants is some $n_1, \ldots, n_k$ then there is a path α from n and labeled by P when $P = \|_{i \in [k]} \|_{j \in [x_i]} P_{i,j}$ for some $(x_1, \ldots, x_k) \in L(\rho)$ and $P_{i,j}$ label of some path $\alpha_{i,j}$ from n_i, $i \in [k]$, $j \in [x_i]$. When $n \to n_i \in E_S(D)$ we say that the factors $P_{i,j}$, $j \in [x_i]$, of P are *marked* by $n \to n_i$ in α. The $\alpha_{i,j}$s are *sub-paths* of α. Sub-paths and marking are hereditary notions of paths: sub-paths of sub-paths of α are sub-paths of α, and every factor of P marked by a special edge e in a sub-path of α is also considered marked by e in α. The language $L(n)$ of n consists of all labels of paths from n, and the language $L(D)$ of D is $L(D) = L(r(D))$.

We build D_e from e such that $L(D_e) = L(e)$ and D_e fulfills the properties:

PP: there is no edge $n \to m$ such that both n and m are labeled in $\mathcal{P}$;

SS: there is no special edge $n \to m$ such that m is labeled in $\mathcal{P}$;

DAG: D_e *without its special edges* has a structure of rooted directed acyclic graph.

Property PP is used in particular in order to compute, during the construction of D_e from e, the Presburger formulæ that will appear later in the P-MSO formula built from D_e. Property SS ensures that $L(n)$ do not contain parallel posets when n is the destination of a special edge. In a D-graph with those properties, the above definition of the existence of a path is well-founded.

5.1 From >1-expressions to D-graphs

The D-graph D_e is built by induction on the >1-expression e. Except when the contrary is specified, new edges added during the constructions of this section are normal. Before starting the construction we need to introduce a new notion. A D-graph D is ξ-*normalized* if any node n labeled by some Presburger formula has at most one direct descendant m such that $\xi \in L(m)$. As every D-graph D with Property PP can easily be transformed into a ξ-normalized D-graph D' with $L(D') = L(D)$, we assume further that D-graphs are ξ-normalized.

Let us start the construction of D_e. When $e = \epsilon$ (resp. $e = a \in A$), D_e is just a node labeled by any closed tautology (resp. labeled by a), without edges.

Assume e has the form $e = e_1 \ op \ e_2$ (resp. $e = e'^{op}$) with $op \in \{\cdot^{>1}, \diamond^{>1}\}$ (resp. $op \in \{*^{>1}, \omega^{>1}, -\omega^{>1}, \natural^{>1}, -\natural^{>1}\}$). Then D_e is built from the union of D_{e_1} and D_{e_2} (resp. from $D_{e'}$), with one more node n as a root, labeled by op, and edged by $n \to r(D_{e_1}), n \to r(D_{e_2})$ (resp. $n \to r(D_{e'})$).

Let us turn now to the more tricky case $e = e_1 \circ_\xi e_2$. When $e_2 = \xi$ then D_e is identical to D_{e_1}. Otherwise, D_e is built from the union of D_{e_1} and D_{e_2} as follows. For any node n of D_{e_2} with label $\gamma(n) = \xi$, we change $\gamma(n)$ by $\gamma(r(D_{e_1}))$ and set $out(n)$ to be $n \circ_{r(D_{e_1})}^{src} out(r(D_{e_1}))$. We now need to ensure Property PP. If $\gamma(r(D_{e_1})) \notin \mathcal{P}$ this is done. Otherwise, it is some $\rho(x_1, \ldots, x_k) \in \mathcal{P}$. For each direct ascendant p of n that is labeled by some $\rho'(x_1', \ldots, x_{k'}')$ and edged by some $out(p) = f_1 \ldots f_{k'}$ with n the destination of some f_i, $i \in [k']$, change $\gamma(p)$ by $\rho \circ_{x_i'} \rho'$ and $out(p)$ by $(p \circ_n^{src} out(n)) \circ_{f_i} out(p)$. If the inner degree of n becomes 0 after this modification of its direct ascendants, then remove n from the D-graph. Remove also $r(D_{e_1})$. The root of the new D-graph is $r(D_{e_2})$.

The construction of $D_{e*\xi}$ from D_e relies on the same principle: each node of D_e labeled by ξ should be replaced by a copy of the root. Again, this may be not so simple in some cases because of Property PP. The construction is as follows, starting from D_e. We proceed in two steps. The first step consists in transforming the root. If $\gamma(r(D_e))$ is some $\rho(x_1, \ldots, x_n)$ there are two cases. Assume that $out(r(D_e)) = e_1 \ldots e_n$ for some $e_1 \ldots e_n$. The first case is when there exists $e_i \colon r(D_e) \to n_i$, $i \in [n]$ such that $\gamma(n_i) = \xi$. Since D_e is ξ-normalized i is unique. Replace $\gamma(r(D_e))$ by ρ^{*x_i}. Otherwise, add a new node x labeled by ξ, transform $\gamma(r(D_e))$ into $(\rho(x_1, \ldots, x_n) \wedge x_{n+1} = 0) \vee (\wedge_{i \in [n]} x_i = 0 \wedge x_{n+1} = 1)$ and $out(r(D_e))$ into $out(r(D_e))(r(D_e) \to x)$ where $r(D_e) \to x$ is a new normal edge. If $\gamma(r(D_e)) \notin \mathcal{P}$ then consider $D_{e+\xi}$ instead of D_e for the remainder of the construction. This ends the first step of the construction. After this first step, the root r of the D-graph is labeled by some $\rho(x_1, \ldots, x_k)$. Let X be the set of nodes labeled by ξ which are not direct descendants of r. The second step consists in replacing, for each node $n \in X$, its label by $\gamma(r)$ and its edging by $n \circ_r^{src} out(r)$, considering that those new edges are special. Some additional transformations are necessary in order to ensure Property PP. For each $n \in X$, and for each of its direct ascendants p that is labeled by some $\rho'(x_1', \ldots, x_{k'}')$ and edged by some $out(p) = f_1 \ldots f_{k'}$ with $f_i \colon p \to n$ for some $i \in [k']$, change $\gamma(p)$ by $\rho \circ_{x_i'} \rho'$ and the edging of p by $(p \circ_n^{src} out(n)) \circ_{f_i} out(p)$. The new edges introduced here are special. Then remove n from the D-graph if its inner degree is 0. This is the only case where special edges are added. Note that after the first step, $\gamma(r) \in \mathcal{P}$ and the D-graph fulfills Property PP. In particular, none of the direct descendants $n_1, \ldots, n_k$ of r has its label in $\mathcal{P}$. Since the new special edges have their destinations in $n_1, \ldots, n_k$ then the construction preserves Property SS.

Now assume e has the form $e = e_1 + e_2$ (resp. $e = e_1 \parallel e_2$). If $\gamma(r(D_{e_1}))$ and $\gamma(r(D_{e_2}))$ are not in $\mathcal{P}$, then D_e is built from the union of D_{e_1} and D_{e_2}, with a new node n labeled by $\rho(x_1, x_2) \equiv \sum_{i \in [2]} x_i = 1$ (resp. $x_1 = x_2 = 1$), edged by $n \to r(D_{e_1}), n \to r(D_{e_2})$, which is the root of D_e. If $\gamma(r(D_{e_i}))$ is some $\rho_i(x_{i,1}, \ldots, x_{i,n_i})$ for all $i \in [2]$, then D_e is built from the union of D_{e_1} and D_{e_2}, with a new node n as a root labeled by $\rho_1 \circ_{x_1} \rho_2 \circ_{x_2} \rho(x_1, x_2)$ and edged by $(n \circ_{r(D_{e_1})}^{\mathrm{src}} \mathrm{out}(r(D_{e_1})))(n \circ_{r(D_{e_2})}^{\mathrm{src}} \mathrm{out}(r(D_{e_2})))$, and with $r(D_{e_1})$ and $r(D_{e_2})$ deleted. The construction is similar in the other cases.

In a D-graph, two edges $s_1 \to d_1$ and $s_2 \to d_2$ are *consecutive* if $d_1 = s_2$. Roughly speaking, the following proposition is a consequence of the fact that by definition of $L^{*\xi}$ in rational languages, in any poset of L, any element labeled by ξ must be in parallel with some other element. In the remainder of the paper we let D_e denote the D-graph of e when e is a >1-expression, or of the >1-expression of e when e is a rational expression.

Proposition 4. *Let D_f be the D-graph of some rational expression f. For any sequence $\alpha = e_1 \ldots e_l$ of consecutive edges of D_f with $e_1, e_l \in E_S(D_f)$, $l > 1$, there exists a node n source of some e_i, $i \in [l]$, such that n is labeled by some Presburger formula $\rho(x_1, \ldots, x_m)$, $\mathrm{out}(n)$ is some $g_1, \ldots, g_m$, $e_i = g_r$ for some $r \in [m]$ and for all $(y_1, \ldots, y_m) \in L(\rho)$, if $y_r > 0$ then $\sum_{i \in [m]} y_i > 1$.*

As a consequence, when there is a path from $r(D_f)$ labeled by some P, if it contains two sub-paths labeled by F_1 and F_2 both marked by some $e \in E_S(D_f)$, then F_1 and F_2 are necessarily sequential posets (Property SS), and either

1. $F_1 \cap F_2 = \emptyset$. Possibly, $F_1 F_2$ is a sequential factor of P;
2. one is strictly included into the other, wlog. $F_1 \subsetneq F_2$. In this case, there is some $x \in F_2 \setminus F_1$ such that x is incomparable to all the elements of F_1.

5.2 From D-graphs to P-MSO

Let D_e be the D-graph of (the >1-expression of) some rational expression e. We are now going to recursively parse D_e in order to compute a P-MSO sentence ϕ_{D_e} such that $P \in SP^\diamond(A)$ is a model for ϕ_{D_e} if and only if $P \in L(D_e)$. For each node n of D_e, we define a P-MSO formula $\phi_n(X)$ which depends on a second-order parameter X. We want a factor X of P to satisfy ϕ_n if and only if $X \in L(n)$. When $\gamma(n) = \cdot^{>1}$, n has two direct descendants n_1 and n_2, and $\phi_n(X)$ expresses that there exists a partition of X into X_1, X_2 such that $X_1 < X_2$ and X_i, $i \in [2]$, satisfies ϕ_{n_i}. Here we let $X_1 < X_2$ denote that $x_1 < x_2$ for all $x_1 \in X_1, x_2 \in X_2$. This construction for $\cdot^{>1}$ is P-MSO definable. The cases of other labels are mere adaptations of the case of linear orderings [3], except for Presburger formulæ. Indeed, nodes labeled in $\mathcal{P}$ may be sources of special edges, which may cause circular dependencies between the ϕ_ns. We use a technique named *s-coloring* in order to avoid such circular dependencies.

Let C be a finite set whose elements are named *colors* and $P \in SP^\diamond$. Let us denote by $F_s(P)$ the class of all sequential factors of P, and set $\mathbb{B} = \{\text{false}, \text{true}\}$. A *s-coloring* $\mathfrak{c}$ of P is a partial map $\mathfrak{c} \colon F_s(P) \to C$. While in general maps cannot be expressed with MSO, Proposition 5 states that s-colorings can be encoded by means of MSO, under assumptions on the sequential factors F of P on which $\mathfrak{c}(F)$ is defined. This encoding s_C^X, that we do not give here, uses a bunch of second-order variables that we also denote by s_C^X for convenience. We let $s_C^X(F) = c$ denote that it associates the color c to $F \in F_s(P)$. We use s-coloring in order to associate a special edge to some $F \in F_s(P)$ as follows.

Proposition 5. *Let D_e be the D-graph of a rational expression e, $C = \mathbb{B} \times E_S(D_e)$, n a node of D_e. Assume there is a path γ from n labeled by some P. There exist a s-coloring of P with C and its encoding s_C^X with MSO such that $\pi_2(s_C^X(F)) = c \in E_S(D_e)$ if and only if F is marked by c in γ, for any $F \in F_s(P)$.*

The encoding for s-colorings does not allow $s_C^X(F) = s_C^X(F')$ when $F, F', FF' \in F_s(P)$. Alternation of the booleans in C is used when we need a s-coloring to associate the same special edge to F and F'.

Set C as in Proposition 5. When the label of a node n is some $\rho(x_1, \ldots, x_k)$ and $out(n) = n \to n_1, \ldots, n \to n_k$, set

$$\phi_n(X) \equiv \mathcal{Q}(X, \chi_1, \ldots, \chi_k, \rho(x_1, \ldots, x_k))$$

where $\chi_i \equiv \forall Y (\forall y \ y \in Y) \to \phi_{n_i}(Y)$ when $n \to n_i \in E_N(D_e)$, $\chi_i \equiv \forall Y (\forall y \ y \in Y) \to \vee_{b \in \mathbb{B}} s_C^X(Y) = (b, n \to n_i)$ when $n \to n_i \in E_S(D_e)$. The sentence ϕ_{D_e} consists in claiming that there exists an encoding of a s-coloring of $P \in SP^\diamond(A)$ on which it is interpreted, such that

- for all $n \to m \in E_S(D_e)$ and any $F \in F_s(P)$, if $\pi_2(s_C^X(F)) = n \to m$ then F satisfies ϕ_m;
- P satisfies $\phi_{r(D_e)}$.

6   From P-MSO to Rational Expressions

Provided a P-MSO formula, the first step is to construct an equivalent branching automaton (see [4]) over scattered and countable N-free posets. This is done merely by melting several techniques; those of the translation of a MSO$^\diamond$ formula to an automaton over countable and scattered linear orderings [3], and those of the translation of a P-MSO formula to a branching automaton over finite N-free posets [1]. The crucial argument here is the effective closure of rational languages of $SP^\diamond(A)$ under boolean operations [2]. The final step is the construction of a rational expression from a branching automaton over $SP^\diamond(A)$ [4]. As those constructions are effective, and because the emptiness of rational languages is decidable, we conclude:

Theorem 6. *Let A be an alphabet. The P-MSO theory of $SP^\diamond(A)$ is decidable.*

7   An Example

When $L \subseteq SP^\diamond(A)$ set $L^+ = L^{*^{>1}} + L$. In this section we detail the construction of a P-MSO formula from the extended rational expression $(a \parallel b) \circ_\xi ((\xi \parallel \xi)^+)^{*\xi}$. The corresponding >1-expression is $e = (a \parallel b) \circ_\xi ((\xi \parallel \xi)^{*^{>1}} + (\xi \parallel \xi))^{*\xi}$. The different steps of the construction of D_e are shown in Fig. 1. We have $E_S(D_e) = \{n_3 \to n_2\}$, $C = \{(\text{false}, n_3 \to n_2), (\text{true}, n_3 \to n_2)\}$.

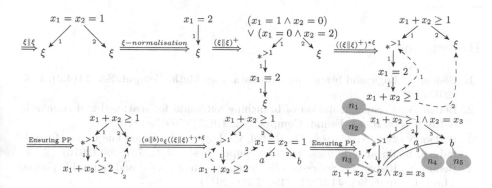

Fig. 1. The step-by-step construction of the D-graph of $(a \parallel b) \circ_\xi ((\xi \parallel \xi)^{*^{>1}} + (\xi \parallel \xi))^{*\xi}$

Let us turn to the construction of ϕ_{D_e}. The notion of an interval of a linear ordering naturally extends to partial orderings. We use MSO-definable short-cuts [3] $X \subseteq_{\max} Y$, $\text{Finite}(X)$, $\text{Partition}(X, X_1, X_2)$ and $\text{Trace}(Y, T)$ that are satisfied respectively when X is an interval of Y maximal with respect to inclusion, when X is a finite linear ordering, when X_1 and X_2 form a partition of X and when T consists of exactly one element of each $Z \subseteq_{\max} Y$. Then

$$\phi_{n_1}(X) \equiv \mathcal{Q}(X, \overline{\phi_{n_2}(Y)}, \overline{\phi_{n_4}(Y)}, \overline{\phi_{n_5}(Y)}, x_1 + x_2 \geq 1 \wedge x_2 = x_3)$$

$$\phi_{n_2}(X) \equiv \exists X_1, X_2, T_1, T_2 \ \text{Partition}(X, X_1, X_2) \wedge \text{Trace}(X_1, T_1) \wedge \text{Trace}(X_2, T_2)$$
$$\wedge \text{Finite}(T_1 \cup T_2) \wedge \forall Z \ ((Z \subseteq_{\max} X_1) \vee (Z \subseteq_{\max} X_2)) \to \phi_{n_3}(Z)$$

$$\phi_{n_3}(X) \equiv \mathcal{Q}(X, \overline{\bigvee_{b \in B} s_C^X(Y) = (b, n_3 \to n_2)}, \overline{\phi_{n_4}(Y)}, \overline{\phi_{n_5}(Y)}, x_1 + x_2 \geq 2 \wedge x_2 = x_3)$$

$$\phi_{n_4}(X) \equiv |X| = 1 \wedge \forall x \ (x \in X \to a(x)) \qquad \phi_{n_5}(X) \equiv |X| = 1 \wedge \forall x \ (x \in X \to b(x))$$

$$\phi_{D_e} \equiv \exists R \exists s_C^X \ (\forall x \ x \in R) \wedge \text{s-Coloring}(R, s_C^X) \wedge \phi_{n_1}(R)$$
$$\wedge \ (\forall F \ (F_s(F, R) \wedge \bigvee_{b \in \mathbb{B}} s_C^X(F) = (b, n_3 \to n_2)) \to \phi_{n_2}(F))$$

where $\overline{\psi(Y)} \equiv \forall Y \ (\forall y \ y \in Y) \to \psi(Y)$, $\text{s-Coloring}(R, s_C^X)$ checks if s_C^X encodes a s-coloring of R and $F_s(F, R)$ checks if F belongs to $F_s(R)$.

Let $F_1' = F_2' = (\xi \parallel \xi) \cdot (\xi \parallel \xi)$, $F_3' = (F_2' \parallel \xi) \cdot (\xi \parallel \xi)$. Let $F_i = \{(a \parallel b)\} \circ_\xi F_i'$ for all $i \in [3]$. Let $P_1 = (F_1 \parallel F_3) \cdot ((a \parallel b) \parallel (a \parallel b))$, $P_2 = a$ and $P_3 = b$. Note that $P = P_1 \parallel P_2 \parallel P_3 \in L(e)$. Regarding D_e, $F_1, F_2, F_3, P_1 \in L(n_2)$,

$F_1 \parallel F_3 \in L(n_3)$, $P_2 \in L(n_4)$ and $P_3 \in L(n_5)$. Thus $P \in L(n_1)$. There is a path α from n_1 labeled by P such that $F \in \{F_1, F_2, F_3\}$ if and only if F is marked by $n_3 \rightarrow n_2$ in α. Hence, each F_i, $i \in [3]$, must be s-colored by $(b_i, n_3 \rightarrow n_2)$ for some $b_i \in \mathbb{B}$. Under this s-coloring observe that F_1, F_2 and F_3 satisfy $\overline{\vee_{b \in \mathcal{B}} s_G^X(Y)} = (b, n_3 \rightarrow n_2)$ as well as ϕ_{n_2}, $F_1 \parallel F_3$, P_1, P_2 and P_3 satisfy respectively ϕ_{n_3}, ϕ_{n_2}, ϕ_{n_4} and ϕ_{n_5}. Thus P satisfies ϕ_{n_1} and is a model for ϕ_{D_e}.

Acknowledgments. The authors would like to thank all the referees for their helpful comments.

References

1. Bedon, N.: Logic and branching automata. Log. Meth. Comput. Sci. **11**(4:2), 1–38 (2015)
2. Bedon, N.: Complementation of branching automata for scattered and countable N-free posets. Int. J. Found. Comput. Sci. **19**(25), 769–799 (2018)
3. Bedon, N., Bès, A., Carton, O., Rispal, C.: Logic and rational languages of words indexed by linear orderings. Theory Comput. Syst. **46**(4), 737–760 (2010)
4. Bedon, N., Rispal, C.: Series-parallel languages on scattered and countable posets. Theor. Comput. Sci. **412**(22), 2356–2369 (2011)
5. Bruyère, V., Carton, O.: Automata on linear orderings. J. Comput. Syst. Sci. **73**(1), 1–24 (2007)
6. Büchi, J.R.: Weak second-order arithmetic and finite automata. Zeit. Math. Logik. Grund. Math. **6**, 66–92 (1960)
7. Büchi, J.R.: On a decision method in the restricted second-order arithmetic. In: 1960 Proceedings of the International Congress on Logic, Methodology and Philosophy of Science, Berkeley, pp. 1–11. Stanford University Press (1962)
8. Büchi, J.R.: Transfinite automata recursions and weak second order theory of ordinals. In: 1964 Proceedings of the International Congress Logic, Methodology, and Philosophy of Science, pp. 2–23. North Holland Publishing Company (1965)
9. Eilenberg, S., Schützenberger, M.P.: Rational sets in commutative monoids. J. Algebra **13**(2), 173–191 (1969)
10. Elgot, C.C.: Decision problems of finite automata design and related arithmetics. Trans. Am. Math. Soc. **98**, 21–51 (1961)
11. Ginsburg, S., Spanier, E.H.: Semigroups, Presburger formulas, and languages. Pac. J. Math. **16**(2), 285–296 (1966)
12. Kleene, S.C.: Representation of events in nerve nets and finite automata. In: Automata Studies, pp. 3–41. Princeton University Press (1956)
13. Kuske, D.: Towards a language theory for infinite N-free pomsets. Theor. Comput. Sci. **299**, 347–386 (2003)
14. Lodaya, K., Weil, P.: A Kleene iteration for parallelism. In: Arvind, V., Ramanujam, S. (eds.) FSTTCS 1998. LNCS, vol. 1530, pp. 355–366. Springer, Heidelberg (1998). https://doi.org/10.1007/978-3-540-49382-2_33
15. Lodaya, K., Weil, P.: Series-parallel posets: algebra, automata and languages. In: Morvan, M., Meinel, C., Krob, D. (eds.) STACS 1998. LNCS, vol. 1373, pp. 555–565. Springer, Heidelberg (1998). https://doi.org/10.1007/BFb0028590
16. Lodaya, K., Weil, P.: Series-parallel languages and the bounded-width property. Theor. Comput. Sci. **237**(1–2), 347–380 (2000)

17. Lodaya, K., Weil, P.: Rationality in algebras with a series operation. Inf. Comput. **171**, 269–293 (2001)
18. Rabin, M.O.: Decidability of second-order theories and automata on infinite trees. Trans. Am. Math. Soc. **141**, 1–5 (1969)
19. Rival, I.: Optimal linear extension by interchanging chains. Proc. AMS **89**(3), 387–394 (1983)
20. Rosenstein, J.G.: Linear Orderings. Academic Press, Cambridge (1982)
21. Trakhtenbrot, B.A.: Finite automata and monadic second order logic. Siberian Math. **3**, 101–131 (1962). (Russian). Translation AMS Transl. 59 23–55 (1966)
22. Valdes, J., Tarjan, R.E., Lawler, E.L.: The recognition of series parallel digraphs. SIAM J. Comput. **11**, 298–313 (1982)

Toroidal Codes and Conjugate Pictures

Marcella Anselmo[1], Maria Madonia[2(⊠)], and Carla Selmi[3]

[1] Dipartimento di Informatica, Università di Salerno,
Via Giovanni Paolo II, 132, 84084 Fisciano, (SA), Italy
manselmo@unisa.it
[2] Dipartimento di Matematica e Informatica, Università di Catania,
Viale Andrea Doria 6/a, 95125 Catania, Italy
madonia@dmi.unict.it
[3] LITIS, Université de Rouen Normandie,
76830 Saint Etienne du Rouvray, Rouen, France
carla.selmi@univ-rouen.fr

Abstract. Toroidal codes of pictures are introduced as the generalization of circular codes of strings in two dimensions. They are characterized by a property of very pureness on a generated language. The class of such codes is compared with other close classes of codes of pictures. In analogy to the string case, toroidal codes are investigated in relation to the conjugate pictures. Conjugacy between pictures is here defined and many properties and characterizations are shown.

Keywords: Two-dimensional languages · Circular codes · Conjugacy

1 Introduction

Circular strings are different from linear strings in that the last symbol is considered to precede the first symbol. A circular string is sometimes referred to as a *necklace*. Circular strings have played, and still play, an important role in many areas of computer science and related fields, notably bioinformatics. As an example, the properties of circular strings are of interest in the circular string matching problem, in the alignment of circular strings problem [10,18] and in the investigation of circular splicing systems which are inspired by a recombinant behaviour of circular DNA [13,14].

Coding with circular strings is a classical topic in the theory of (variable-length) codes (see [11,19,22]). A *circular code* is a set of strings such that any circular string has at most one decomposition with strings in the set. The investigation on circular codes of strings relies on the notion of *conjugacy* of strings. Two strings (of same length) are conjugate if one can be obtained from the other one by a cyclic permutation. In other words, two conjugate strings can be read on the same necklace.

Partially supported by INdAM-GNCS Project 2018, FARB Project ORSA175982 of University of Salerno and CREAMS Project of University of Catania.

© Springer Nature Switzerland AG 2019
C. Martín-Vide et al. (Eds.): LATA 2019, LNCS 11417, pp. 288–301, 2019.
https://doi.org/10.1007/978-3-030-13435-8_21

The aim of this paper is to extend this theory to the two-dimensional world.

The generalization of the classical notion of string to the two dimensions leads to the definition of polyomino, in its different declinations - labeled polyominoes, directed polyominoes, as well as rectangular labeled polyominoes, that we will refer to as *pictures*. In the literature, one can find different attempts to generalize the notion of code to 2D objects [1,12,21]. In this paper, we consider the definition of code of pictures introduced in [3,4]. A set X is a code if any picture is tileable, without holes or overlapping, in at most one way with pictures in X. In this framework, one can find the definition of prefix and strong prefix code of pictures, as well as that of code of pictures with finite deciphering delay and comma-free code of pictures [2,4,7,8]. Moreover, recently, a generalization of the circular codes of strings has been proposed in [9], where the *cylindric codes* of pictures are defined. A language X is a cylindric code if the pictures of X cannot tile the lateral surface of any cylinder (for any height and radius) in two different ways. In this paper, we take a further step forward and consider a *torus* instead of a cylinder; we define the *toroidal codes* of pictures.

In order to study the torus, in an easier and handier way, we cut it along a vertical line and then along a horizontal line, in such a way to obtain a picture (to come back to the torus, just let the top and the bottom sides of the picture coincide, as well as the left and the right ones). Subsequently, instead of investigating the tiling of a torus with pictures in a given language X, we will consider the *toroidal decomposition* over X of the associated picture. Observe that when cutting a labeled torus in all the possible ways, we obtain several pictures of the same size which are characterized by a nice relation; they are conjugate. We introduce the *conjugacy* relation on pictures and show several properties and characterizations of conjugate pictures. Note that a (partial) notion of horizontal/vertical conjugacy was already introduced in [20] in order to capture the 2D horizontal/vertical periodicity of a picture, and to yield a succinct and efficient algorithm for 2D dictionary matching. Similar periodicity properties are obtained, in a wider framework, considering the overlapping of a picture with itself [5,6].

A particular attention is devoted to the self-conjugate pictures, i.e., pictures that are proper conjugate of themselves. *Non-self-conjugate* pictures play in 2D the role of primitive strings in 1D. Observe that the notion of non-self-conjugacy in 1D coincides with that of primitiveness; this does not hold in 2D. The counting of the number of conjugacy classes of strings is classically based on the counting of primitive classes [11,19]. We give an upper bound on the number of non-self-conjugate conjugacy classes of pictures which yields an upper bound on the cardinality of a toroidal code. As a main result, the toroidal codes are characterized by a property of very pureness on a generated language. At the end, we compare all the families of codes of pictures considered in this paper and show examples of languages that separate them.

2 Preliminaries

We recall some definitions about two-dimensional languages (see [16]). A *picture* over a finite alphabet Σ is a two-dimensional rectangular array of elements of Σ. Given a picture p, $|p|_{row}$ and $|p|_{col}$ denote the number of rows and columns, respectively, while $size(p) = (|p|_{row}, |p|_{col})$ denotes the picture *size*. We also consider all the empty pictures, referred to as $\lambda_{m,0}$ and $\lambda_{0,n}$, for all $m, n \geq 0$; they correspond to the pictures of size $(m, 0)$ or $(0, n)$, respectively. The set of all pictures over Σ of fixed size (m, n) is denoted by $\Sigma^{m,n}$, while Σ^{m*} and Σ^{*n} denote the set of all pictures over Σ with a fixed number m of rows and n of columns, respectively. The set of all pictures over Σ is denoted by Σ^{**}, while Σ^{++} refers to the set of all non-empty pictures on Σ. A *two-dimensional language* (or *picture language*) over Σ is a subset of Σ^{**}.

In order to locate a position in a non-empty picture, it is necessary to put the picture in a reference system. The set of coordinates $dom(p) = \{1, 2, \ldots, |p|_{row}\} \times \{1, 2, \ldots, |p|_{col}\}$ is referred to as the *domain* of a picture p. We let $p(i, j)$ denote the symbol in p at coordinates (i, j). Moreover, to easily detect the border positions of pictures, we use the initials of the words "top", "bottom", "left" and "right"; for example the *tl-corner* of p refers to position $(1, 1)$ the *tl-corner* of p.

A *subdomain* of $dom(p)$ is a set d of the form $\{i, i + 1, \ldots, i'\} \times \{j, j + 1, \ldots, j'\}$, where $1 \leq i \leq i' \leq |p|_{row}$, $1 \leq j \leq j' \leq |p|_{col}$, also specified by $[(i, j), (i', j')]$. The portion of p corresponding to the subdomain $[(i, j), (i', j')]$ is denoted $p[(i, j), (i', j')]$. Then, a non-empty picture x is *subpicture of p* if $x = p[(i, j), (i', j')]$, for some $1 \leq i \leq i' \leq m$, $1 \leq j \leq j' \leq n$; it will be referred to as the subpicture *associated* with $[(i, j), (i', j')]$ and we will say that x *occurs* at position (i, j) (its tl-corner).

Dealing with pictures, two "classical" concatenation products are defined. Let $p, q \in \Sigma^{**}$ be pictures of size (m, n) and (m', n'), respectively. The *column concatenation* of p and q (denoted by $p \oslash q$) and the *row concatenation* of p and q (denoted by $p \ominus q$) are partial operations, defined only if $m = m'$ and if $n = n'$, respectively, as:

$$p \oslash q = \boxed{\begin{array}{|c|c|} p & q \end{array}} \qquad\qquad p \ominus q = \boxed{\begin{array}{|c|} p \\ \hline q \end{array}}.$$

These definitions can be extended to define row- and column- concatenations, and *row-* and *column- stars* of two-dimensional languages. We also consider another star operation for picture languages, the tiling star. The idea is to compose pictures in a way to cover a rectangular area as, for example, in the figure below.

Let $X \subseteq \Sigma^{**}$. The set X^{++} is the set of all the non-empty pictures p over Σ whose domain can be partitioned into disjoint subdomains $\{d_1, d_2, \ldots, d_k\}$ such that any subpicture p_h of p associated with the subdomain d_h belongs to X, for all $h = 1, ..., k$. Then, the *tiling star* of X, denoted by X^{**}, is the union of the set X^{++} with all the empty pictures. Language X^{**} is called the set of all tilings by X in [23]. In the sequel, if $p \in X^{++}$, the corresponding partition $\{d_1, d_2, \ldots, d_k\}$ of $dom(p)$, is called a *tiling decomposition* of p over X.

3   Two-Dimensional Codes and Cylindric Decompositions

In this section, we recall the definitions of code of pictures, given in [4], comma-free code and cylindric code of pictures, given in [9], together with some examples. They are strictly related to the toroidal codes which are a central topic of this paper.

Let Σ be a finite alphabet. A language $X \subseteq \Sigma^{++}$ is a *code* if any $p \in \Sigma^{++}$ has at most one tiling decomposition over X. For example, let $\Sigma = \{a, b\}$. It is easy to see that $X_1 = \left\{ \boxed{a\ b}, \begin{array}{c}\boxed{\begin{array}{c}a\\b\end{array}}\end{array}, \boxed{\begin{array}{cc}a&a\\a&a\end{array}} \right\}$ is a code. On the other hand, the language $X_2 = \left\{ \boxed{a\ b}, \boxed{b\ a}, \boxed{\begin{array}{c}a\\a\end{array}} \right\}$ is not a code. Indeed, picture $\boxed{\begin{array}{ccc}a&b&a\\a&b&a\end{array}}$ has two different tiling decompositions over X_2, $t_1 = \boxed{\begin{array}{c|c}a\ b&a\\\hline a\ b&a\end{array}}$ and $t_2 = \boxed{\begin{array}{c|c}a&b\ a\\\hline a&b\ a\end{array}}$.

The comma-free codes of pictures are a generalization of the comma-free, or self-synchronizing, codes of strings. It is worthy to mention that their definition does not need a privileged decoding direction, as it will be for the toroidal codes. A language $X \subseteq \Sigma^{++}$ is *comma-free* if no picture $p \in X$ is covered by pictures in X. Informally, a picture p is *covered* by pictures in a set X, if p can be tiled (without holes and overlapping) with pictures in X which possibly exceed the borders (cf. [3,8]). It is immediate to observe that any comma-free set is a code, that we will call a comma-free code.

Comma-free codes of strings are studied inside the class of circular codes. In the literature, the circular codes are usually defined as follows. A language $X \subseteq \Sigma^+$ is a circular code of strings if for all $m, n \geq 1$ and $x_1, x_2, \ldots, x_n \in X$, $y_1, y_2, \ldots, y_m \in X$, $t \in \Sigma^*$ and $s \in \Sigma^+$, the equalities $sx_2 \ldots x_n t = y_1 y_2 \ldots y_m$, $x_1 = ts$ imply that t is the empty string, $m = n$ and $x_i = y_i$ for $1 \leq i \leq n$. In view of the generalization of such definition to two dimensions, let us introduce the following definition. A *circular decomposition* of a string $w \in \Sigma^*$ is a sequence s, $x_2, \ldots, x_n, t$ such that $w = sx_2 \ldots x_n t$ and $ts, x_2, \ldots, x_n \in X$. When $s = 1$ then $w \in X^*$.

The translation of the definition of circular code of strings into the world of pictures leads to some new situations. Following [9], the role of a circle in the plane can be played, in the space, by a cylinder, either horizontally or vertically placed. Then, a set X of pictures is a (horizontal or vertical) cylindric code if the pictures of X cannot tile the lateral surface of any cylinder (for any height and radius) in two different ways. In order to avoid the difficulty of handling objects in the space, we are going to cut the surface of the cylinder and investigate the rectangular picture we obtain. Let us introduce the following notations.

Given a picture p of size (m,n) a *horizontal across subdomain* of $dom(p)$ is a set d of the form $\{i, i+1, \ldots, i'\} \times \{j, j+1, \ldots, n, 1, 2, \ldots, j'\}$, where $1 \le i \le i' \le m$, $1 \le j' < j \le n$; d will be denoted by the pair $[(i,j), (i',j')]$.

In an analogous way, a *vertical across subdomain* of $dom(p)$ is a set d of the form $\{i, i+1, \ldots, m, 1, 2, \ldots, i'\} \times \{j, j+1, \ldots, j'\}$, where $1 \le i' < i \le m$, $1 \le j \le j' \le n$; d will be denoted by the pair $[(i,j), (i',j')]$.

In order to stress the difference between a horizontal or vertical across subdomain and a subdomain, as defined in Sect. 2, the latter will be sometimes called an *internal subdomain*. The portion of p corresponding to the positions in an across subdomain $[(i,j), (i',j')]$ is denoted by $p[(i,j), (i',j')]$. It is a union of two (internal) subdomains.

A *horizontal (vertical, resp.) cylindric tiling decomposition* of a picture $p \in \Sigma^{++}$ over X is a partition of $dom(p)$ into disjoint internal and/or horizontal (vertical, resp.) across subdomains $\{d_1, d_2, \ldots, d_k\}$ such that, for all $h = 1, ..., k$, the subpicture p_h of p associated with the subdomain d_h belongs to X. A horizontal cylindric tiling decomposition is simply called a cylindric tiling decomposition in [9].

Finally, a language $X \subseteq \Sigma^{++}$ is a *horizontal (vertical, resp.) cylindric code* if any picture in Σ^{++} has at most one horizontal (vertical, resp.) cylindric tiling decomposition over X. A horizontal cylindric code is simply called a cylindric code in [9]. As an example, the language $X =$

$$\left\{ \boxed{a\ b}, \quad \boxed{a\ a\ b\ a}, \quad \boxed{\begin{array}{ccc} a & a & a \\ b & a & b \end{array}}, \quad \boxed{\begin{array}{ccc} a & a & a \\ b & b & a \end{array}}, \quad \boxed{\begin{array}{cccc} a & a & a & b \\ b & a & b & a \end{array}} \right\}$$ is not a horizontal cylindric

code. One can easily show that the picture $p = \boxed{\begin{array}{ccccccc} a & a & a & a & a & a & b \\ b & b & a & b & a & b & a \end{array}}$ has two different

horizontal cylindric tiling decompositions over X.

4   Primitive Pictures and Conjugacy

In this section, we are going to introduce the notion of conjugacy in pictures and to explore some combinatorial properties of primitive and conjugate pictures. They are involved in the investigation on the toroidal codes introduced in the next section.

In 1D, two strings x, y are called conjugate if there exist strings u, v such that $x = uv$ and $y = vu$. We frequently say that y is a conjugate of x. The conjugacy is an equivalence relation and a class of conjugacy is often called a necklace. The necklaces are classically investigated in connection with circular codes (see [11,22]). Let us introduce the following definitions and notations for the pictures.

Given $p \in \Sigma^{++}$ and $m, n > 0$, $p^{m\ominus}$ denotes the picture $p^{m\ominus} = p \ominus p \ominus \cdots \ominus p$ obtained by row concatenation of m copies of p, while $p^{n\oplus}$ denotes the picture $p^{n\oplus} = p \oplus p \oplus \cdots \oplus p$ obtained by column concatenation of n copies of p. Furthermore, $p^{m,n}$ is defined as $p^{m,n} = (p^{m\ominus})^{n\oplus} = (p^{n\oplus})^{m\ominus}$.

Given $p_1, p_2, p_3, p_4 \in \Sigma^{**}$, with $p_1 \in \Sigma^{h,k}$, $p_2 \in \Sigma^{h,j}$, $p_3 \in \Sigma^{i,k}$, $p_4 \in \Sigma^{i,j}$, let us denote by $\oplus(p_1, p_2, p_3, p_4)$ the picture $p = (p_1 \oplus p_2) \ominus (p_3 \oplus p_4)$.

A picture $p \in \Sigma^{++}$ is called *primitive* if it is not a power of another picture. Thus, p is primitive iff $p = q^{m,n}$ with $m, n > 0$ implies $p = q$. Each non-empty picture p is a power of a unique primitive picture (see [15]). The unique primitive picture r such that $p = r^{m,n}$, for some integer $m > 0$ or $n > 0$, is called the *primitive root* of p; the pair (m, n) is the *exponent* of p (see [15,17]).

Definition 1. *Let $p, p' \in \Sigma^{m,n}$. The pictures p and p' are conjugate, and we write $p \sim p'$, if there exist $p_1, p_2, p_3, p_4 \in \Sigma^{**}$, where p_1 is not empty, such that $p = \oplus(p_1, p_2, p_3, p_4)$ and $p' = \oplus(p_4, p_3, p_2, p_1)$.*

We will also say that p' is a conjugate of p. In the case that $h = m$ and $k = n$ then $p = p_1$ and we have that p is trivially conjugate of itself. If $h < m$ and $k < n$ then we will say that p' is the conjugate of p in position $(h + 1, k + 1)$. Note that $(h + 1, k + 1)$ is the position of the tl-corner of p_4 inside p. If $h = m$ and $k < n$, we will say that p' is the conjugate of p in position $(1, k + 1)$ (the position of the tl-corner of p_2). If $h < m$ and $k = n$, we will say that p' is the conjugate of p in position $(h + 1, 1)$ (the position of the tl-corner of p_3).

The following results characterize the conjugates of a picture. The first one generalizes the result stating that two conjugate strings are obtained from each other by a cyclic permutation. The second one can be obtained as a corollary. Let us introduce the following notations (see [20]). Two pictures $p, q \in \Sigma^{m,n}$ are *horizontal conjugate*, denoted $p \sim_h q$, if $p = p_L \oslash p_R$ and $q = p_R \oslash p_L$, for some $p_L, p_R \in \Sigma^{m,*}$. Similarly, $p, q \in \Sigma^{m,n}$ are *vertical conjugate*, denoted $p \sim_v q$, if $p = p_U \ominus p_D$ and $q = p_D \ominus p_U$, for some $p_U, p_D \in \Sigma^{*,n}$. It can be proved that the relations $\sim_h$ and $\sim_v$ are equivalence relations.

Proposition 2. *Two pictures $p, q \in \Sigma^{m,n}$ are conjugate if and only if there exists $r \in \Sigma^{m,n}$ such that $q \sim_v r$ and $r \sim_h p$.*

Proof. Let $p, q \in \Sigma^{m,n}$ be two conjugate pictures. By definition, we have that $p = \oplus(p_1, p_2, p_3, p_4)$ and $q = \oplus(p_4, p_3, p_2, p_1)$, for some pictures p_1, p_2, p_3, p_4. Let $r = \oplus(p_2, p_1, p_4, p_3)$. Then $q \sim_v r$ and $r \sim_h p$.

Vice versa, let $r \in \Sigma^{m,n}$ be a picture such that $q \sim_v r$ and $r \sim_h p$. By definition, we have $q = q_U \ominus q_D$, $r = q_D \ominus q_U$ and $r = r_L \oslash r_R$, $p = r_R \oslash r_L$. Then, there exist $r_1, r_2, r_3, r_4 \in \Sigma^{**}$, as in Definition 1, such that $r = \oplus(r_1, r_2, r_3, r_4)$, with $q_D = r_1 \oslash r_2$, $q_U = r_3 \oslash r_4$, $r_L = r_1 \ominus r_3$ and $r_R = r_2 \ominus r_4$. Then $p = r_R \oslash r_L = (r_2 \ominus r_4) \oslash (r_1 \ominus r_3) = \oplus(r_2, r_1, r_4, r_3)$ and $q = q_U \ominus q_D = \oplus(r_3, r_4, r_1, r_2)$. We obtain that p and q are conjugate. $\square$

Corollary 3. *Two pictures $p, q \in \Sigma^{m,n}$ are conjugate if and only if q is a subpicture of $p^{2,2}$.*

Proof. Let $p, q \in \Sigma^{m,n}$. Applying Proposition 2, p and q are conjugate pictures iff there exists $r \in \Sigma^{m,n}$ such that $q \sim_v r$ and $r \sim_h p$. Let p, q, r be decomposed as in the proof of Proposition 2. Then, $p^{2,2} = (r_R \ominus r_R) \oslash (q_D \ominus q_U \ominus q_D \ominus q_U) \oslash (r_L \ominus r_L)$. Hence, p and q are conjugate iff q is a subpicture of $p^{2,2}$. $\square$

Proposition 4. *The conjugacy relation is an equivalence relation.*

Proof. It follows by definition that conjugacy is a reflexive and symmetric relation. We prove that it is a transitive relation. Let $p, q, r \in \Sigma^{n,m}$ be such that $p \sim q$ and $q \sim r$. We will show that $p \sim r$. By Corollary 3, we have that q is a subpicture of $p^{2,2}$. From the fact that p and q have the same size follows that $q^{2,2}$ is a subpicture of $p^{3,3}$. Applying Corollary 3, we obtain that r is a subpicture of $q^{2,2}$ and then r is a subpicture of $p^{3,3}$. Since p, q, r have the same size, r is also a subpicture of $p^{2,2}$. So, again by Corollary 3, we obtain that r and p are conjugate pictures. $\qquad \square$

A conjugacy class is a class of this equivalence relation. The class of all the conjugates of p is denoted $p^\sim$. Among all the conjugacy classes, it is possible to point out some special ones that will be important in the counting of conjugacy classes.

We observed that any picture is trivially a conjugate of itself. On the other hand, we distinguish the case when a picture is a "strict" conjugate of itself, as stated in the following definition. Corollary 7 and Proposition 8 justify the definition of primitive and of non-self-conjugate conjugacy classes.

Definition 5. *Let $p \in \Sigma^{++}$. The picture p is* self-conjugate *if there exist p_1, p_2, p_3, $p_4 \in \Sigma^{**}$, $p_1 \neq p$, such that $p = \oplus(p_1, p_2, p_3, p_4)$ and $p = \oplus(p_4, p_3, p_2, p_1)$.*

Proposition 6. *Two conjugate pictures have the same exponent and their primitive roots are conjugate.*

Proof. Let $p, q \in \Sigma^{m,n}$ be two conjugate pictures. By definition, we have that $p = \oplus(p_1, p_2, p_3, p_4)$ and $q = \oplus(p_4, p_3, p_2, p_1)$, with p_1, p_2, p_3, p_4 as in Definition 1. Suppose that $p = r^{m,n}$, for $r \in \Sigma^{++}$. It can be proved that there exist $r_i \in \Sigma^{**}$, $1 \leq i \leq 4$, such that $r = \oplus(r_1, r_2, r_3, r_4)$ and $q = \bar{r}^{m,n}$ with $\bar{r} = \oplus(r_4, r_3, r_2, r_1)$ (r_i's are obtained cutting r along the vertical and horizontal lines giving q from p). $\qquad \square$

As a consequence of Proposition 6, we obtain the following corollary.

Corollary 7. *Let $p \in \Sigma^{++}$. If p is primitive then all the conjugates of p are primitive.*

Proposition 8. *Let $p \in \Sigma^{m,n}$. If p is self-conjugate then all the conjugates of p are self-conjugate.*

Proof. Let p be a self-conjugate picture and let q be a conjugate of p. By Corollary 3, p and q are subpictures of $p^{2,2}$. Let (i, j) be the position of the topmost and leftmost non-trivial occurrence of p inside $p^{2,2}$ and let (s, t) be the position of the topmost and leftmost occurrence of q inside $p^{2,2}$. One can prove that $1 \leq i, s \leq m$ and $1 \leq j, t \leq n$. Since $p^{2,2}$ is a subpicture of $p^{3,3}$, then $q^{2,2}$ is the subpicture of $p^{3,3}$ with tl-corner in (s, t). Now, suppose $s = i + h$ and $t = j + k$, for some $h, k \geq 0$ (the other cases can be proved similarly). Then $(m + h, n + k)$ is the position of a non-trivial occurrence of q inside $p^{3,3}$. Hence, q is a subpicture of $q^{2,2}$. $\qquad \square$

In 1D, the counting of the conjugacy classes is based on the counting of the primitive conjugacy classes (see for example [11,22]). Note that a string is primitive if and only if it is non-self-conjugate. On the other hand, one can easily show by contradiction that the following result holds in 2D.

Proposition 9. *Let $p \in \Sigma^{++}$. If p is non-self-conjugate then p is a primitive picture.*

The converse of Proposition 9 does not hold. Indeed, there exist primitive and self-conjugate pictures, as, for example, $\begin{array}{|cc|} \hline a & b \\ b & a \\ \hline \end{array}$. In 2D, the counting of the conjugacy classes of pictures will focus on non-self-conjugate conjugacy classes.

Proposition 10. *Let $p \in \Sigma^{m,n}$. If p is non-self-conjugate then the cardinality of its conjugacy class is $|p^{\sim}| = mn$.*

Proof. Let p be a non-self-conjugate picture. We claim that the conjugates of p in each of its positions are all different. By the contrary, let p' and p'' be the conjugates of p in two different positions, and suppose $p' = p''$. By the transitive property of the conjugacy, p' and p'' are conjugate, and they are conjugate in a non-trivial way, that is p' is self-conjugate against Proposition 8. □

Let Σ be an alphabet with $|\Sigma| = k$, $\psi_k(m,n)$ ($\nu_k(m,n)$, resp.) be the number of primitive (non-self-conjugate, resp.) pictures in $\Sigma^{m,n}$ and $\phi_k(m,n)$ be the number of non-self-conjugate conjugacy classes in $\Sigma^{m,n}$. In the sequel, μ is the Möbius function.

Proposition 11. *Let Σ be an alphabet with $|\Sigma| = k$. Then,*
$$\phi_k(m,n) \leq \left(\sum_{d_1|m} \sum_{d_2|n} \mu(d_1)\mu(d_2) k^{mn/(d_1 d_2)} \right)/mn.$$

Proof. It follows by Proposition 10 that $\phi_k(m,n) \leq \nu_k(m,n)/mn$. From Proposition 9, $\nu_k(m,n) \leq \psi_k(m,n)$. Hence, $\phi_k(m,n) \leq \nu_k(m,n)/mn \leq \psi_k(m,n)/mn$. Finally, in [15] it is proved that $\psi_k(m,n)$ equals the sum in the formula. □

5   Toroidal Codes

The translation of the definition of circular code of strings into the picture world leads to some new situations. The role of a circle on the plane can be played in the 3D space, not only by a cylinder, either horizontally or vertically placed (as discussed in Sect. 3), but also by a torus. This more general approach has the advantage to be independent from a decoding direction or border constraint.

A set X of pictures is a toroidal code if the pictures of X cannot tile any torus (of any dimension) in two different ways. In order to avoid the difficulty of handling objects in a 3D space, we are going to cut the surface of the labeled torus and consider the rectangular picture p we obtain. The picture p will be tiled by some pictures in X which may exceed a border and, consequently, let uncovered the corresponding positions of p in the opposite side. This special kind of tiling of p will be called a toroidal tiling decomposition of p.

In order to formalize this notion, let us introduce another type of subdomain of the domain of a picture, besides the internal, the horizontal and the vertical across subdomains introduced in Sect. 3. A *corner across subdomain* of $dom(p)$ is a set d of the form

$$\{i, i+1, \ldots, m, 1, 2, \ldots, i'\} \times \{j, j+1, \ldots, n, 1, 2, \ldots, j'\}$$

where $1 \leq i' < i \leq |p|_{row}$, $1 \leq j' < j \leq |p|_{col}$; d will be denoted by the pair $[(i,j),(i',j')]$. The portion of p corresponding to the positions in a corner across subdomain $[(i,j),(i',j')]$ is denoted by $p[(i,j),(i',j')]$. It is a union of four (internal) subdomains, each one containing a different corner of p.

Let us introduce the definition of toroidal tiling decomposition of a picture as the generalization of the notion of circular decomposition of a string (see Sect. 3).

Definition 12. *A* toroidal tiling decomposition *of a picture* $p \in \Sigma^{++}$ *over* X *is a partition of* $dom(p)$ *into disjoint internal and/or across subdomains* $\{d_1, d_2, \ldots, d_k\}$ *such that, for all* $h = 1, \ldots, k$, *the subpicture* p_h *of* p *associated with the subdomain* d_h *belongs to* X.

Remark 13. Observe that a toroidal tiling decomposition of a picture p provides a tiling of a torus; it suffices to let the top border of p coincide with its bottom border, and the left border with its right one. The converse is true, too. Note that a toroidal tiling decomposition of p *induces* a toroidal tiling decomposition for any conjugate of p. In fact, each conjugate p' of p can be obtained by cutting the surface of the torus, corresponding to p, along some vertical and some horizontal lines. So, we will talk about the toroidal tiling decomposition of a class of conjugacy, i.e., a labeled torus (just as in 1D for necklaces). Indeed, the toroidal tiling decompositions of a conjugacy class have to be considered just as one single decomposition (and not many different ones).

Example 14. Let $X = \{x_1, x_2, x_3\}$ where $x_1 = \begin{array}{|cc|} \hline b\ a\ b \\ a\ b\ a \\ \hline \end{array}$, $x_2 = \boxed{b\ a\ b}$, $x_3 = \begin{array}{|c|} \hline a \\ a \\ a \\ \hline \end{array}$ and let $p = \begin{array}{|ccccc|} \hline a\ a\ a\ a\ b \\ b\ a\ a\ b\ a \\ b\ a\ a\ b\ a \\ \hline \end{array}$. A toroidal tiling decomposition of p over X is $\{d_1, d_2, d_3, d_4\}$, where $d_1 = [(3,4),(1,1)]$ is a corner across subdomain, $d_2 = [(1,2),(3,2)]$ is an internal subdomain, $d_3 = [(2,3),(1,3)]$ is a vertical across subdomain and $d_4 = [(2,4),(2,1)]$ is a horizontal across subdomain (see also Fig. 1). The subpictures associated to the subdomains d_1, d_2, d_3 and d_4 are $p_1 = x_1$, $p_2 = p_3 = x_3$ and $p_4 = x_2$.

We are now ready to introduce the definition of toroidal code.

Definition 15. *A language* $X \subseteq \Sigma^{++}$ *is a* toroidal code *if any picture in* Σ^{++} *has at most one toroidal tiling decomposition over* X.

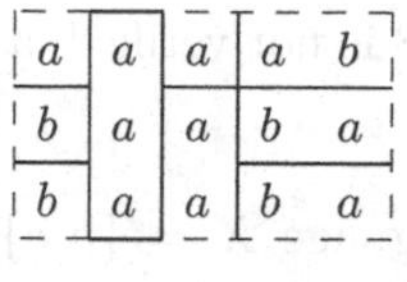

a	a	a	a	b
b	a	a	b	a
b	a	a	b	a

Fig. 1. The toroidal tiling decompositions of p given in Example 14.

Remark 16. Let us sketch a method to construct toroidal codes. Choose two disjoint languages P and Q such that $\forall p \in P$, p cannot be covered by pictures in X and $\forall q \in Q$, q can be covered by pictures in X, but each covering of q makes use of at least one picture of P. Then, X is a toroidal code. Suppose by contradiction that there exists a picture $t \in \Sigma^{++}$ which admits two different toroidal tiling decompositions c and d over X. Note that any subpicture of t associated with a subdomain in c is covered by pictures of X associated with some subdomains of d. It follows by definition of P that, if a subpicture $t_h \in P$ of t is associated with a subdomain belonging to c then t_h can be covered only by itself in d. But, since the two toroidal decompositions of t over X are different, it must exist at least a subpicture $t_k \in Q$ of t, associated with a subdomain of c and, for what has been said before, t_k must be covered only by pictures of Q, associated with some subdomains of d. This contradicts the definition of Q. Moreover, X is not a comma-free code since the pictures of Q are covered by pictures in X.

Example 17. Let $X = \{x_1, x_2, x_3, x_4, x_5, x_6, x_7\}$ where $x_1 = \begin{array}{|cc|} a & b \\ b & b \end{array}$, $x_2 = \begin{array}{|cc|} a & a \\ b & c \end{array}$,

$x_3 = \boxed{c\,b\,a\,c\,d\,d}$, $x_4 = \begin{array}{|ccc|} a & d & d \\ b & d & d \end{array}$, $x_5 = \begin{array}{|ccc|} a & d & b \\ c & c & b \\ b & a & c \end{array}$, $x_6 = \begin{array}{|cc|} a & c \\ c & c \\ b & d \end{array}$, $x_7 = \boxed{d\,b\,b\,b\,d}$. The

language X is a toroidal code. In fact, $X = P \cup Q$, with $P = \{x_1, x_2, x_3, x_4\}$ and $Q = \{x_5, x_6, x_7\}$, which satisfy the requirements in Remark 16.

Proposition 18. *Let $X \subseteq \Sigma^{++}$ be a toroidal code. Then, the following properties hold.*

(1) X cannot contain two different conjugate pictures.
(2) X cannot contain a self-conjugate picture.

Proof. (1) Suppose by contradiction that there exist $p, q \in X$ with $p \sim q$ i.e. there exist $p_1, p_2, p_3, p_4 \in \Sigma^{**}$, such that $p = \oplus(p_1, p_2, p_3, p_4)$ and $q = \oplus(p_1, p_2, p_3, p_4)$. Consider the case with p_i not empty, for $i = 1, 2, 3, 4$ (the other cases can be similarly handled) and suppose that q is the conjugate of p in position (i, j). The picture p has two different toroidal tiling decompositions over X, say t and t', with $t = \{d\} = \{[(1, 1), (|p|_{row}, |p|_{col})]\}$ and $t' = \{d'\} = \{[(i, j), (i - 1, j - 1)]\}$. This is a contradiction to X toroidal code.

(2) The proof is the analogous of the one for item 1). □

The inverse of Proposition 18 is not verified, as shown in the following example.

Example 19. Consider the language $X = \{p, q\}$ with $p = \begin{array}{|cc|} a & a \\ a & b \end{array}$ and $q = \begin{array}{|cc|} b & b \\ b & a \end{array}$.
The pictures p and q are non-self-conjugate and they are not conjugate each
other, but X is not a toroidal code. Actually, the picture $r = \begin{array}{|cccc|} a & a & b & b \\ a & b & b & a \\ b & b & a & a \\ b & a & a & b \end{array}$ has
two different toroidal decompositions. The first one has only internal across
subdomains. It is $\{d_1, d_2, d_3, d_4\}$, with $d_1 = [(1,1), (2,2)]$, $d_2 = [(1,3), (2,4)]$,
$d_3 = [(3,1), (4,2)]$, $d_4 = [(3,3), (4,4)]$, where the pictures of X associated with
are p, q, q, p, respectively. The second decomposition is $\{s_1, s_2, s_3, s_4\}$, with $s_1 = [(2,2), (3,3)]$, $s_2 = [(2,4), (3,1)]$, $s_3 = [(4,2), (1,3)]$, $s_4 = [(4,4), (1,1)]$, where
the pictures of X associated with are q, p, p, q, respectively.

Applying Propositions 11 and 18, we obtain an upper bound on the cardinality of a toroidal code in the uniform case. In the general case of $X \subseteq \Sigma^{++}$, the
bound applies to each $X \cap \Sigma^{m,n}$.

Proposition 20. *Let $X \subseteq \Sigma^{m,n}$ be a toroidal code and $|\Sigma| = k$. Then,*
$$|X| \leq \left(\sum_{d_1 | m} \sum_{d_2 | n} \mu(d_1)\mu(d_2) k^{mn/(d_1 d_2)} \right) / mn.$$

In 1D, the definition of circular code is related to the notion of pureness of
the monoid X^* generated by X. In 2D, the situation is more involved. There
is no proper definition of a generated monoid. We are going to associate to X
the language X^{tor}, which will play in 2D the role of X^*, and then introduce a
proper definition of pureness.

Definition 21. *The* toroidal tiling star *of a language $X \subseteq \Sigma^{++}$, denoted by
X^{tor}, is the set of pictures p whose domain can be partitioned into disjoint internal and/or across subdomains $\{d_1, d_2, \ldots, d_k\}$ such that any subpicture p_h of p
associated with the subdomain d_h belongs to X, for all $h = 1, \ldots, k$, and the
subdomain containing $(1,1)$ is an internal domain.*

Definition 22. *Let $X \subseteq \Sigma^{++}$. The language X^{tor} is* very pure *if for any
$p, p' \in X^{tor}$, $p \sim p'$, with p_1, p_2, p_3, p_4 as in Definition 1, and any toroidal
decomposition d of p and d' of p' over X, we have that, for any $i = 1, 2, 3, 4$, the
restriction of d to $dom(p_i)$ is equal to the restriction of d' to $dom(p_i)$.*

Let us state the main result of this section.

Proposition 23. *The language X is a toroidal code if and only if X^{tor} is very
pure.*

Proof. Let X be a toroidal code and suppose, by contradiction, that X^{tor} is
not very pure. Then, there exist $p, p' \in X^{tor}$, $p \sim p'$, with $p = \oplus(p_1, p_2, p_3, p_4)$
and $p' = \oplus(p_4, p_3, p_2, p_1)$, a toroidal decomposition d of p over X and a toroidal

decomposition d' of p' over X such that, for some $i \in \{1, 2, 3, 4\}$, the restriction of d to $dom(p_i)$ is different from the restriction of d' to $dom(p_i)$. Note that the toroidal decomposition d' of p' over X, induces a toroidal decomposition d'' of p over X, different from d and this is a contradiction to X toroidal code.

Now, let $X \subseteq \Sigma^{++}$ such that X^{tor} is very pure and suppose, by contradiction, that X is not a toroidal code. Then, there exists $q \in \Sigma^{++}$ with two different toroidal tiling decompositions over X, say $t = \{d_1, d_2, \ldots, d_s\}$ and $\bar{t} = \{\bar{d}_1, \bar{d}_2, \ldots, \bar{d}_r\}$. Note that there exist two different positions $(i, j), (\bar{i}, \bar{j}) \in dom(q)$ such that, the position (i, j) is covered in t by the tl-corner of a picture of X and the position $(\bar{i}, \bar{j})$ is covered in $\bar{t}$ by the tl-corner of a picture of X.

Then, set $p_1 = q[(1,1), (i-1, j-1)]$, $p_2 = q[(1,j), (i-1, n)]$, $p_3 = q[(i,1), (m, j-1)]$, $p_4 = q[(i,j), (m, n)]$ and $p = \oplus(p_4, p_3, p_2, p_1)$. Clearly, p is the conjugate of q in position (i, j) and $p \in X^{tor}$. In an analogous way, set $p'_1 = q[(1,1), (\bar{i}-1, \bar{j}-1)]$, $p'_2 = q[(1, \bar{j}), (\bar{i}-1, n)]$, $p'_3 = q[(\bar{i}, 1), (m, \bar{j}-1)]$, $p'_4 = q[(\bar{i}, \bar{j}), (m, n)]$ and $p' = \oplus(p'_4, p'_3, p'_2, p'_1)$. Clearly, p' is the conjugate of q in position $(\bar{i}, \bar{j})$ and $p' \in X^{tor}$.

Note that, since conjugacy is a transitive relation, we have that p and p' are conjugate. Moreover, the toroidal tiling decomposition t of q induces a toroidal tiling decomposition d of p over X and the toroidal tiling decomposition $\bar{t}$ of q induces a toroidal tiling decomposition d' of p' over X. Since t and $\bar{t}$ are different, this implies that, for some $i \in \{1, 2, 3, 4\}$, the restriction of t to $dom(p_i)$ is different from the restriction of $\bar{t}$ to $dom(p_i)$. This, in turn, implies that, for some $i \in \{1, 2, 3, 4\}$, the restriction of d to $dom(p_i)$ is different from the restriction of d' to $dom(p_i)$. This contradicts our assumption X^{tor} very pure.   □

We now compare all the classes of two-dimensional codes considered. We denote them as follows. Let C be the class of (two-dimensional) codes, $CYLV$ be the class of vertical cylindric codes, $CYLH$ be the class of horizontal cylindric codes, TOR be the class of toroidal codes, CF be the class of comma-free codes.

Using the definitions, one can prove the following result. The strictness of the inclusions is shown in Example 25.

Proposition 24. $CF \subsetneq TOR \subsetneq CYLV \cap CYLH$.
   Moreover, $CYLV \cap CYLH \subsetneq CYLH \subsetneq C$ and $CYLV \cap CYLH \subsetneq CYLV \subsetneq C$.

Example 25. The language $X_1 = \left\{ \boxed{\begin{smallmatrix} a & a \\ a & a \end{smallmatrix}} \right\}$ is a code, but it is neither vertical nor horizontal cylindric code. The language $X_2 = \left\{ \boxed{\begin{smallmatrix} a & b \\ a & b \end{smallmatrix}} \right\}$ is a horizontal cylindric code which is not a vertical cylindric one. The language $X_3 = \left\{ \boxed{\begin{smallmatrix} a & b \\ b & a \end{smallmatrix}} \right\}$ is both a horizontal and a vertical cylindric code, but it is not a toroidal code. Actually, $\boxed{\begin{smallmatrix} a & b \\ b & a \end{smallmatrix}}$ has two different toroidal tiling decompositions, $\{d_1\}$ and $\{d_2\}$, where $d_1 = [(1,1), (2,2)]$ and $d_2 = [(1,2), (1,1)]$.

An example of toroidal code that is not a comma-free code is given in Example 17.

References

1. Aigrain, P., Beauquier, D.: Polyomino tilings, cellular automata and codicity. Theor. Comput. Sci. **147**, 165–180 (1995)
2. Anselmo, M., Giammarresi, D., Madonia, M.: Strong prefix codes of pictures. In: Muntean, T., Poulakis, D., Rolland, R. (eds.) CAI 2013. LNCS, vol. 8080, pp. 47–59. Springer, Heidelberg (2013). https://doi.org/10.1007/978-3-642-40663-8_6
3. Anselmo, M., Giammarresi, D., Madonia, M.: Two dimensional prefix codes of pictures. In: Béal, M.-P., Carton, O. (eds.) DLT 2013. LNCS, vol. 7907, pp. 46–57. Springer, Heidelberg (2013). https://doi.org/10.1007/978-3-642-38771-5_6
4. Anselmo, M., Giammarresi, D., Madonia, M.: Prefix picture codes: a decidable class of two-dimensional codes. Int. J. Found. Comput. Sci. **25**(8), 1017–1032 (2014)
5. Anselmo, M., Giammarresi, D., Madonia, M.: Unbordered pictures: properties and construction. In: Maletti, A. (ed.) CAI 2015. LNCS, vol. 9270, pp. 45–57. Springer, Cham (2015). https://doi.org/10.1007/978-3-319-23021-4_5
6. Anselmo, M., Giammarresi, D., Madonia, M.: Avoiding overlaps in pictures. In: Pighizzini, G., Câmpeanu, C. (eds.) DCFS 2017. LNCS, vol. 10316, pp. 16–32. Springer, Cham (2017). https://doi.org/10.1007/978-3-319-60252-3_2
7. Anselmo, M., Giammarresi, D., Madonia, M.: Picture codes and deciphering delay. Inf. Comput. **253**, 358–370 (2017)
8. Anselmo, M., Giammarresi, D., Madonia, M.: Structure and properties of strong prefix codes of pictures. Math. Struct. Comput. Sci. **27**(2), 123–142 (2017)
9. Anselmo, M., Madonia, M.: Two-dimensional comma-free and cylindric codes. Theor. Comput. Sci. **658**, 4–17 (2017)
10. Barton, C., Iliopoulos, C.S., Pissis, S.P.: Fast algorithms for approximate circular string matching. Algorithms Mol. Biol. **9**, 9 (2014)
11. Berstel, J., Perrin, D., Reutenauer, C.: Codes and Automata. Cambridge University Press, Cambridge (2009)
12. Bozapalidis, S., Grammatikopoulou, A.: Picture codes. ITA **40**(4), 537–550 (2006)
13. De Felice, C., Zaccagnino, R., Zizza, R.: Unavoidable sets and regularity of languages generated by (1, 3)-circular splicing systems. In: Dediu, A.-H., Lozano, M., Martín-Vide, C. (eds.) TPNC 2014. LNCS, vol. 8890, pp. 169–180. Springer, Cham (2014). https://doi.org/10.1007/978-3-319-13749-0_15
14. De Felice, C., Zaccagnino, R., Zizza, R.: Unavoidable sets and circular splicing languages. Theor. Comput. Sci. **658**, 148–158 (2017)
15. Gamard, G., Richomme, G., Shallit, J., Smith, T.J.: Periodicity in rectangular arrays. Inf. Process. Lett. **118**, 58–63 (2017)
16. Giammarresi, D., Restivo, A.: Two-dimensional languages. In: Rozenberg, G., Salomaa, A. (eds.) Handbook of Formal Languages, vol. III, pp. 215–267. Springer, Heidelberg (1997). https://doi.org/10.1007/978-3-642-59126-6_4
17. Kulkarni, M.S., Mahalingam, K.: Two-dimensional palindromes and their properties. In: Drewes, F., Martín-Vide, C., Truthe, B. (eds.) LATA 2017. LNCS, vol. 10168, pp. 155–167. Springer, Cham (2017). https://doi.org/10.1007/978-3-319-53733-7_11
18. Lee, T., Na, J.C., Park, H., Park, K., Sim, J.S.: Finding consensus and optimal alignment of circular strings. Theor. Comput. Sci. **468**, 92–101 (2013)

19. Lothaire, M.: Combinatorics on Words. Cambridge University Press, Cambridge (1997)
20. Marcus, S., Sokol, D.: 2D Lyndon words and applications. Algorithmica **77**(1), 116–133 (2017)
21. Moczurad, M., Moczurad, W.: Some open problems in decidability of brick (labelled polyomino) codes. In: Chwa, K.-Y., Munro, J.I.J. (eds.) COCOON 2004. LNCS, vol. 3106, pp. 72–81. Springer, Heidelberg (2004). https://doi.org/10.1007/978-3-540-27798-9_10
22. Perrin, D., Restivo, A.: Enumerative combinatorics on words. In: Bona, M. (ed.) Handbook of Enumerative Combinatorics. CRC Press (2015)
23. Simplot, D.: A characterization of recognizable picture languages by tilings by finite sets. Theor. Comput. Sci. **218**(2), 297–323 (1991)

Geometrical Closure of Binary $V_{3/2}$ Languages

Jean-Philippe Dubernard[1], Giovanna Guaiana[1], and Ludovic Mignot[2(✉)]

[1] LITIS EA 4108, Université de Rouen Normandie, Avenue de l'Université,
76801 Saint-Étienne-du-Rouvray, France
{jean-philippe.dubernard,giovanna.guaiana}@univ-rouen.fr
[2] GR²IF, Université de Rouen Normandie, Avenue de l'Université,
76801 Saint-Étienne-du-Rouvray, France
ludovic.mignot@univ-rouen.fr

Abstract. We define the geometrical closure of a language over a j-ary alphabet, and we prove that in the case of dimension 2 the family $V_{3/2}$ in the Straubing-Thérien hierarchy of languages is closed under this operation. In other words, the geometrical closure of a $V_{3/2}$ binary language is still a $V_{3/2}$ language. This is achieved by carrying out some transformations over a regular expression representing the $V_{3/2}$ language, which leads to a $V_{3/2}$ regular expression for the geometrical closure.

Keywords: Regular language · Geometrical language ·
Regular expression · Straubing-Thérien hierarchy

1 Introduction

A connex figure in a j-dimensional space is a set of points of $\mathbb{N}^j$ such that, for any point in the figure, there exists a path from the origin to this point. A path is intended to proceed by an incremental step in one of the j dimensions at a time. For a language L over a j-ary alphabet, the figure of L is the set of the points in $\mathbb{N}^j$ corresponding to the Parikh vectors of the prefixes of its words. Conversely, the language of a connex figure is the set of words covering all the possible paths from the origin in the figure. A language L is geometrical if the set of its prefixes is equal to the language of the figure of L.

Geometrical languages have been introduced in [4]. Initially, their study was motivated by their application to the modeling of real-time task systems [2], via regular languages [7] or discrete geometry [7,9]. This led to the study of their behaviour in language theory. While geometrical languages occur in any level of the Chomsky hierarchy, it is the geometrical regular languages that have been mainly studied. There exist two polynomial algorithms [3,6] to check if a binary regular language is geometrical, and an exponential one [3] for regular languages.

In this paper we investigate the following property: a family of languages is geometrically closed if, for any of its languages L, the geometrical closure of L,

© Springer Nature Switzerland AG 2019
C. Martín-Vide et al. (Eds.): LATA 2019, LNCS 11417, pp. 302–314, 2019.
https://doi.org/10.1007/978-3-030-13435-8_22

which we define as the language of its figure, belongs to the family. We remark that the class of regular languages is not geometrically closed. Indeed let L be the language $(aa^*b)^*$. The figure is composed of all the points below and on the diagonal $\{(n,n) \mid n \geq 0\}$. Then the language of the figure of L is the set of words whose prefixes contain a number of b less than or equal to the number of a. We take into account the Straubing-Thérien hierarchy of regular languages over an alphabet A, first considered implicitly in [16] and explicitly in [15], and we focus on the family of languages corresponding to the level 3/2. This hierarchy is defined starting by the empty set and the free monoid A^*, and alternating polynomial closures and Boolean closures. For recent references on the Straubing-Thérien hierarchy, the reader is referred to [11,13,14]. The level $V_{3/2}$ has been proved to be decidable [1,12]. Moreover, it is shown in [1] that the languages of $V_{3/2}$ are finite unions of languages of the form $A_0^* a_1 A_1^* a_2 \cdots a_k A_k^*$, with $a_i \in A$ and $A_i \subseteq A$. We rely on this combinatorial characterization to yield our main result. We also recall that $V_{3/2}$ is one of the few interesting families of regular languages which are known to be closed under partial commutations [5,8].

The main result of this paper is that the family $V_{3/2}$ over a two-letters alphabet is also geometrically closed, and we effectively provide a $V_{3/2}$ expression for the language of the figure of a $V_{3/2}$ language. We first show that any binary $V_{3/2}$ expression can be converted, preserving the figure, into a particular kind of regular expression, a sum of components, made of planes, wires and strips, which have at most one starred sub-alphabet each. Then we explain how to normalize and reduce these sums in order to get independent parts whose geometrical closure can be easily computed.

The paper is organized as follows. In the second section we recall the definitions concerning geometrical languages. In Sect. 3 we show the first transformation of a regular expression denoting a $V_{3/2}$ language (in sum of components). In Sect. 4 we present the next transformations, normalization and reduction. Finally, in Sect. 5 we show how to compute a regular expression denoting the geometrical closure of the starting $V_{3/2}$ language. In Sect. 6 we present a web application allowing to display in the 2-dimensional space the transformations we operate, together with the corresponding regular expressions.

2   Preliminaries

In the following, we consider the alphabets $A_j = \{a_1, \ldots, a_j\}$ for any integer $j > 0$. The *vector* Vector(w) of a word w in A_j^* is its Parikh vector, i.e. the j-tuple $(|w|_{a_1}, \ldots, |w|_{a_j})$, where $|w|_{a_k}$ is the number of occurrences of a_k in w, for any integer $1 \leq k \leq j$. The *set of the prefixes* of a language L is denoted by Pref(L). The *figure* of a language $L \subseteq A_j^*$ is the set $\mathcal{F}(L)$ of j-tuples $\{\text{Vector}(w) \mid w \in \text{Pref}(L)\}$. More generally, a *figure* of *dimension* j is a subset of $\mathbb{N}^j$. Two j-tuples $c = (c_1, \ldots, c_j)$ and $c' = (c_1', \ldots, c_j')$ are *consecutive* if there exists an integer $i \leq j$ such that $|c_i - c_i'| = 1$ and $c_k = c_k'$ for any integer $k \leq j$ distinct from i. In this case, the integer i is said to be the *shift* of (c, c'), denoted by Shift(c, c'). Moreover, we say that c *precedes* c' if $c_i' - c_i = 1$. A figure F is *connex* if for any tuple c in $F \setminus \{0^j\}$, F contains a tuple c' preceding c. A *path* is a non-empty finite sequence of consecutive j-tuples $[p_1, \ldots, p_n]$ such that p_k precedes p_{k+1} for any integer $1 \leq k \leq n - 1$. The set $\mathrm{Paths}(F)$ is the set of the paths of a connex figure F that start from the origin. The word $\mathrm{Word}(p)$ associated with a path $p = [p_1, p_2, \ldots]$ is inductively defined by $\mathrm{Word}([p_1]) = \varepsilon$, $\mathrm{Word}([p_1, \ldots, p_k]) = \mathrm{Word}([p_1, \ldots, p_{k-1}]) \cdot a_{\mathrm{Shift}(p_{k-1}, p_k)}$. The language of a connex figure F is $\mathcal{L}(F) = \{\mathrm{Word}(p) \mid p \in \mathrm{Paths}(F)\}$. A language L is *geometrical* if $\mathrm{Pref}(L) = \mathcal{L}(\mathcal{F}(L))$.

Example 1. Let us consider the two figures F_1 and F_2 over A_2

$$F_1 = \{(0,0), (1,1), (1,2), (2,1), (2,2), (3,3), (3,4), (4,4)\}, \quad F_2 = F_1 \cup \{(0,1), (3,2)\}$$

respectively represented in Figs. 1 and 2. By definition, F_1 is not connex, whereas F_2 is. Moreover, if we consider the language L defined by

$$L = \{a_2 a_1 a_1, a_2 a_1 a_2 a_1 a_1 a_2 a_2 a_1\},$$

it can be checked that $F_2 = \mathcal{F}(L)$. However, L is not geometrical, since the word $a_2 a_1 a_1 a_2$ is in $\mathcal{L}(F_2) \setminus \mathrm{Pref}(L)$.

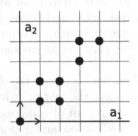

Fig. 1. A non connex figure.

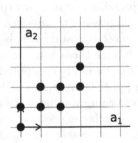

Fig. 2. A connex figure.

In the sequel, we only consider connex figures.

A *regular expression* E over A_j is inductively defined by

$$E = a_i, \qquad\qquad E = \varepsilon, \qquad\qquad E = \emptyset,$$
$$E = E_1 + E_2, \qquad E = E_1 \cdot E_2, \qquad E = E_1^*,$$

where $1 \leq i \leq j$, and E_1 and E_2 are any two regular expressions over A_j. The *language denoted by* a regular expression E over A_j is the language inductively defined by:

$$L(a_i) = \{a_i\}, \qquad\qquad L(\varepsilon) = \{\varepsilon\}, \qquad\qquad L(\emptyset) = \emptyset,$$
$$L(E_1 + E_2) = L(E_1) \cup L(E_2), \quad L(E_1 \cdot E_2) = L(E_1) \cdot L(E_2), \quad L(E_1^*) = (L(E_1))^*,$$

where $1 \leq i \leq j$ and E_1 and E_2 are any two regular expressions over A_j. By extension the *figure* $\mathcal{F}(E)$ of a regular expression E is the set of j-tuples $\mathcal{F}(L(E))$. A regular expression E is:

– a *block* if it equals $u(a_{i_1} + \cdots + a_{i_k})^*$, with u in A_j^* and $\{a_{i_1}, \ldots, a_{i_k}\} \subseteq A_j$,
– *elementary* if it is a concatenation of blocks.

An expression is $V_{3/2}$ if it is a finite sum of elementary regular expressions. A language is $V_{3/2}$ if it is denoted by a $V_{3/2}$ expression. The family of $V_{3/2}$ languages corresponds to the level $3/2$ of the Straubing-Thérien hierarchy [1].

3   Geometrical Equivalence and Sums of Components

There are several different languages, geometrical or not, that have the same figure. This leads to the definition of an equivalence of languages.

Definition 2. *Two languages L and L' over A_j are geometrically equivalent, denoted by $L \sim L'$, if $\mathcal{F}(L) = \mathcal{F}(L')$. Two regular expressions E and E' are geometrically equivalent if $\mathcal{F}(E) = \mathcal{F}(E')$.*

Example 3. The languages $L = \{a_2 a_1 a_1, a_2 a_1 a_2 a_1 a_1 a_2 a_2 a_1\}$ of Example 1 and the language $L' = \{a_2 a_1 a_2, a_2 a_1 a_1 a_2 a_1 a_2 a_2 a_1\}$ are geometrically equivalent.

Definition 4. *The geometrical closure $\overline{L}$ of a language L over A_j is the language $\mathcal{L}(\mathcal{F}(L))$.*

If a language is geometrical, then its geometrical closure is just its prefix-closure. Remark that two geometrically equivalent languages have the same geometrical closure.

In the sequel we apply these definitions to the family of $V_{3/2}$ languages. A $V_{3/2}$ language is not necessarily geometrical. Take for example $L = L(a_1^* a_2)$. The words in $L(a_2 a_1^+)$ belong to $\mathcal{L}(\mathcal{F}(L))$, but they are not prefixes of L. Besides, a $V_{3/2}$ language is not in general closed by prefixes, but its closure by prefixes is still a $V_{3/2}$ language. Our aim is to prove that the geometrical closure of a $V_{3/2}$ language is still a $V_{3/2}$ language, and we achieve our purpose here in the case of a 2-ary alphabet.

N.B. From now on, we implicitly consider binary languages. Moreover, in order to simplify the notations, we set $a = a_1$ and $b = a_2$.

Example 5. The geometrical closure $\mathcal{L}(\mathcal{F}(L))$ of L in Example 1 is

$\{\varepsilon, b, ba, baa, bab, baab, baba, baaba, babaa, baabab, babaab,$

$$baababb, babaabb, baababba, babaabba\}.$$

This example also shows that the geometrical closure of a $V_{3/2}$ expression is not necessarily equal to the sum of the closures of its elementary expressions. Hence, we make several transformations on a $V_{3/2}$ expression, in order to get an expression which is geometrically equivalent to the starting expression, and that we are able to close easily. In this section we make a first transformation, preserving the figure: a $V_{3/2}$ expression will be converted into a sum of special expressions, called components, like wires, strips and planes. Throughout the paper a sum of components will always be intended as finite.

A regular expression E is:

- a *plane* if it equals $u(a + b)^*$,
- a *vertical (resp. horizontal) strip* of width n if it equals ub^*a^{n-1} (resp. ua^*b^{n-1}),
- a *wire* if it equals u,

for some word u in A_2^*. In all these cases, E is said to be a *component*. If E is a strip or a plane, then u is said to be the *affix* of E. In the case of a vertical (resp. horizontal) strip, the word ua^{n-1} (resp. ub^{n-1}) is the *end* of the strip.

Example 6. Let us consider the expression E defined by

$$E = baa + aa + bb^*a + bbaa(a + b)^* + ba + a^*.$$

It is the sum of three wires (baa, aa and ba), a plane ($bbaa(a + b)^*$), a vertical strip of width two (bb^*a) and a horizontal strip of width one (a^*). The figure of E is represented in Fig. 3.

To improve readability, we will use in the sequel the graphic representation defined as follows: the planes and the strips are represented by colored areas (red for the planes, green for the horizontal strips, blue for the vertical strips), whereas the wires, the affixes and the ends are represented as black lines. As an example, the figure of the previous example is represented in Fig. 4. Notice that the wire aa is covered by the green strip a^*.

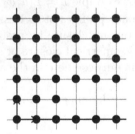

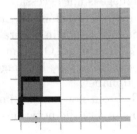

Fig. 3. The figure of E.

Fig. 4. The geometrical representation of $\mathcal{F}(E)$. (Color figure online)

We immediately get the following result.

Proposition 7. *A sum of components denotes a $V_{3/2}$ language.*

Moreover, let us show that any $V_{3/2}$ language is geometrically equivalent to a language denoted by a sum of components. For this purpose, we focus on some geometrical equivalences.

Lemma 8. *Let L be a language over A_2, u and v be two words in A_2^*, and* $\mathrm{Vector}(v) = (x, y)$. *Then:*

$$\{u\} \cdot A_2^* \cdot L \sim \{u\} \cdot A_2^*, \tag{1}$$

$$\{u\} \cdot \{a\}^* \cdot \{v\} \sim \{ua^x\} \cdot \{a\}^* \cdot \{b^y\} \cup \{ua^i v \mid 0 \leq i \leq x\}, \tag{2}$$

$$\{u\} \cdot \{b\}^* \cdot \{v\} \sim \{ub^y\} \cdot \{b\}^* \cdot \{a^x\} \cup \{ub^i v \mid 0 \leq i \leq y\}, \tag{3}$$

$$\{u\} \cdot \{\alpha\}^* \cdot \{v\} \cdot \{\alpha\}^* \cdot L \sim \{u\} \cdot \{\alpha\}^* \cdot \{v\} \cdot L, \quad \alpha \in A_2, \tag{4}$$

$$\{u\} \cdot \{a\}^* \cdot \{v\} \cdot \{b\}^* \cdot L \sim \{ua^x\} \cdot A_2^* \cup \{ua^i v \mid 0 \leq i \leq x\}, \tag{5}$$

$$\{u\} \cdot \{b\}^* \cdot \{v\} \cdot \{a\}^* \cdot L \sim \{ub^y\} \cdot A_2^* \cup \{ub^i v \mid 0 \leq i \leq y\}, \tag{6}$$

$$\{u\} \cdot \{a\}^* \cdot \{v\} \cdot A_2^* \cdot L \sim \{ua^x\} \cdot A_2^* \cup \{ua^i v \mid 0 \leq i \leq x\}, \tag{7}$$

$$\{u\} \cdot \{b\}^* \cdot \{v\} \cdot A_2^* \cdot L \sim \{ub^y\} \cdot A_2^* \cup \{ub^i v \mid 0 \leq i \leq y\}. \tag{8}$$

Hence,

- according to Eq. (1), if the first starred expression is a plane, then all the following blocks can be dropped,
  according to Eqs. (2) and (3), a $V_{3/2}$ expression with a unique starred symbol can be replaced by a strip and a sum of wires,
- according to Eq. (4), if the two first starred expressions are two identical starred symbols separated by a word, then the second one can be dropped,
- according to Eqs. (5), (6), (7) and (8), if the two first starred expressions are two distinct starred nonempty alphabets separated by a word, then the expression can be transformed into a plane and a sum of wires.

Consequently,

Proposition 9. *Any $V_{3/2}$ language is geometrically equivalent to a language denoted by a sum of components.*

4   Reduction of a Sum of Components

Let us now show how to transform a sum of components, preserving the figure, in order to obtain an easy-to-close expression. We proceed in two steps. The *normalization* is an operation where some wires are added and some planes and strips are shifted, w.r.t. to the frontiers. The next step, the *reduction*, deals with planes and strips by getting rid of unnecessary components, and merging together some others.

4.1   Normalization

We show how to *normalize* a sum of components, that allows us to determine what is the part of the language we need to close. This process will allow **(1)** to highlight a rectangle in the figure, containing only wires and all the wires, and determined by the frontiers, and **(2)** to shift the strips and the planes beyond this rectangle by adding some wires. More formally, the frontiers are two integers l and h such that

– any point (x, y) of the figure satisfying $x < l$ and $y < h$ belongs to at least one wire and only to wires;
– any point (x, y) of the figure satisfying $x = l$ or $y = h$ belongs to at least one wire, and can appear in a strip or in a plane;
– any point (x, y) of the figure satisfying $x > l$ or $y > h$ only belongs to planes or strips.

From now on, we use the notation $\text{Vector}(u) = (x_u, y_u)$ for any $u \in A_2^*$.

Definition 10. *A regular expression E is* normalized *if it is a sum of planes, strips and wires such that there exist two integers l and h, satisfying:*

1. for any of its wires u,

$$x_u \leq l, \qquad\qquad y_u \leq h, \qquad\qquad (9)$$

2. for any of its vertical (resp. horizontal) strips of affix u, of width n and of end e, the following conditions hold:

$$(y_e = h) \wedge (x_e \leq l) \qquad (resp. \ (x_e = l) \wedge (y_e \leq h)), \qquad (10)$$

$$E \ contains \ the \ wire \ ua^{n-1} \ (resp. \ ub^{n-1}), \qquad (11)$$

3. for any of its planes of affix u, the following conditions hold:

$$(y_u = h) \wedge (x_u \leq l) \vee (x_u = l) \wedge (y_u \leq h), \qquad (12)$$

$$E \ contains \ the \ wires \ ua^{l-x_u}, \ ub^{h-y_u}. \qquad (13)$$

The two integers l and h are the frontiers *of E.*

Let us then show how to normalize a sum of components. Let E be a non-empty sum of components. We denote by X the set

$$X = \{u \in A_2^* \mid u \text{ is a wire or an affix or an end of a component in } E\},$$

and we consider the two integers l and h defined by

$$l = \max\{x_w \mid w \in X\}, \qquad\qquad h = \max\{y_w \mid w \in X\}.$$

Notice that E satisfies Eq. (9). Let us show how to transform E into a geometrically equivalent normalized expression.

Consider a plane p of affix u. Then $x_u \leq l$ and $y_u \leq h$. We set $l' = l - x_u$ and $h' = h - y_u$. The figure of p is equal to the figure of the sum of the planes $p' = ua^{l'}(a+b)^*$ and $p'' = ub^{h'}(a+b)^*$ and the expression $E' = ua^{l'} + ub^{h'} + u(a^{l'} \sqcup\!\sqcup b^{h'})$, where $\sqcup\!\sqcup$ is the classical shuffle product. We recall that for any two words u and v over an alphabet A, $u \sqcup\!\sqcup v$ is the sum of the words $u_1 v_1 \cdots u_n v_n$ with $u_1 \cdots u_n = u$, $v_1 \cdots v_n = v$ and $u_i, v_i \in A^*$ for

$1 \le i \le n$. Since E' denotes a finite language, it is equivalent to a sum E'' of wires, which satisfies Eq. (9). Moreover, the wires $ua^{l'}$ and $ua^{l'}b^{h'}$ needed by p' to satisfy Eq. (13), and the wires $ub^{h'}$ and $ub^{h'}a^{l'}$ needed by p'' to satisfy Eq. (13), are all included in E''. Finally, p' and p'' satisfy Eq. (12), and therefore $E'' + p' + p''$ is a normalized sum of components geometrically equivalent to p.

Consider now a vertical strip s of affix u and of width n. If $y_u < h$, let $h' = h - y_u$. The figure of s is equal to the figure of the sum of the vertical strip $s' = ub^{h'}b^*a^{n-1}$ and the expression $E' = u(a^{n-1} \sqcup b^{h'})$. Since E' denotes a finite language, it is equivalent to a sum E'' of wires, which satisfies Eq. (9). Finally s' satisfies Eq. (10), and the wire $ub^{h'}a^{n-1}$ needed for Eq. (11) is included in E''. Therefore $E'' + s'$ is a normalized sum of components geometrically equivalent to s. If a horizontal strip is not normalized, then the normalization can be performed similarly.

As a direct consequence of the previous conclusions, since each plane and each strip can be treated independently, we get the following results.

Proposition 11. *Any sum of components can be normalized into a geometrically equivalent sum of components.*

Example 12. Let us consider the expression E of Example 6. The normalization of E produces the expression E' defined by

$$E' = baa + aa + bbb^*a + bba + bab + bbaa(a + b)^* + bbaa + ba + aaa^*.$$

Indeed,

- the frontiers are $l = 2$ and $h = 2$,
- the strip bb^*a is removed and produces the strip bbb^*a and the wires bba and bab (according to the normalization),
- the plane $bbaa(a + b)^*$ produces the wire $bbaa$ (Eq. (13)),
- the strip a^* is transformed into the strip aaa^* (Eq. (10)), satisfying Eq. (11) since aa is already in E.

Fig. 5. The figure of E'.

The associated figure is represented in Fig. 5. The frontiers l and h are marked by a yellow point.

4.2   Reduction of a Normalized Expression

Let us now explain how to reduce a normalized expression preserving the figure. As an example, if a component is included in a plane, then it can be removed. As another example, if two vertical strips overlap, or if the end of a strip and the affix of the other have the same vector or correspond to two consecutive points, then they can be merged.

Definition 13. *A regular expression E with frontiers l and h is reduced if it is normalized and satisfies the following three conditions:*

1. *E contains at most two planes of affix u and u' with $x_u = l$ and $y_{u'} = h$, such that either $x_{u'} < l$ and $y_u < h$, or $u = u'$,*
2. *for any plane of affix u and for any vertical (resp. horizontal) strip of end e, $x_e < x_u - 1$ (resp. $y_e < y_u - 1$),*
3. *for any two vertical (resp. horizontal) strips of affixes u and u' and of ends e and e', $x_e < x_{u'} - 1$ or $x_{e'} < x_u - 1$ (resp. $y_e < y_{u'} - 1$ or $y_{e'} < y_u - 1$).*

Let us show how to reduce a normalized expression E with frontiers l and h.

Suppose that E contains two planes p and p' of affixes u and u', with $x_u \leq x_{u'}$ and $y_u = y_{u'}$. Then removing p' from E produces a geometrically equivalent expression. Notice that the case when $y_u \leq y_{u'}$ and $x_u = x_{u'}$ can be treated similarly. Consequently, by recurrence on the number of planes, it can be shown that E is geometrically equivalent to a normalized expression with

- at most two planes,
- at most one plane $u(a + b)^*$ such that $x_u \leq l$,
- at most one plane $u(a + b)^*$ such that $y_u \leq h$.

Consider now two components of E, a vertical strip s of affix u and of end e, and a plane p of affix u' (resp. a vertical strip s' of affix u' and of end e'), such that $x_e \geq x_{u'} - 1$ (resp. $x_e \geq x_{u'} - 1$ and $x_{e'} \geq x_e$). If $x_u \geq x_{u'}$ (and necessarily $y_u = y_{u'}$), then removing s from E produces a geometrically equivalent expression. Otherwise, s can be replaced by the plane $p' = u(a + b)^*$ and by the wire $w = u$ (resp. s and s' can be merged into an equivalent vertical strip $p' = ub^*a^{x_{e'} - x_u}$ and they are equivalent to $p' + w$, where w is the wire $ua^{x_{e'} - x_u}$). Therefore, replacing s (resp. s and s') by $p' + w$ in E produces a geometrically equivalent expression. In the case of a new plane creation, the reduction step can be performed one more time if needed to get rid of an unnecessary plane (see Example 14). Notice that this creation is needed when p is a plane with $x_{u'} = l = x_e$ and $y_{u'} \neq h$. The case of horizontal strips can be performed similarly.

The expression obtained by the reduction step is still normalized, with the same frontiers l and h of the starting expression.

Example 14. Let us consider the normalized expressions E_1 and E_2 defined by

$$E_1 = b(a + b)^* + b + aa + aaa^* + baa, \quad E_2 = baa(a + b)^* + baa + aa + aaa^*.$$

In both of these cases, the strip aaa^* is replaced by $aa(a + b)^*$ during the reduction step. However, in E_2, the other plane $baa(a + b)^*$ has to be removed to obtain a reduced expression, while $b(a + b)^*$ is not removed in E_1.

From the previous construction, we directly get the following result.

Proposition 15. *Any normalized expression E can be reduced to a geometrically equivalent expression, which is normalized with the same frontiers as E.*

As a direct consequence of Propositions 9, 11 and 15,

Proposition 16. *Any $V_{3/2}$ language is geometrically equivalent to a language denoted by a reduced sum of components.*

Example 17. Let us consider the normalized expression E' of Example 12. The reduction of E' produces the expression E'' defined by

$$E'' = baa + aa + bba + bab + bb(a + b)^*$$
$$+ bb + bbaa + ba + aaa^*.$$

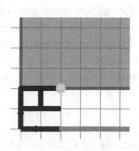

Indeed, the strip bbb^*a and the plane $bbaa(a + b)^*$ are merged into the plane $bb(a + b)^*$. The wire bb is added in order to satisfy Eq. (13). The associated figure is represented in Fig. 6.

Fig. 6. The figure of $L(E'')$.

5 Geometrical Closure of a Reduced Expression

In this section we show how to compute a regular expression denoting the geometrical closure of a reduced expression.

For a reduced expression E of frontiers l and h, the points (x, y) of $\mathcal{F}(E)$ satisfying $x = l \wedge y \leq h$ or $y = h \wedge x \leq l$ will also be called *frontiers*.

In order to compute the closure, notice that since E is reduced, the closure of a plane or a strip can be performed independently of every other component. Indeed, the closure of a language L will add some words when there exist two consecutive points p_1 and p_2 in $\mathcal{F}(L)$ and a word w in $\mathrm{Pref}(L)$ such that $\mathrm{Vector}(w) = p_1$ and $wa_{\mathrm{Shift}(p_1, p_2)} \notin \mathrm{Pref}(L)$. Now, two consecutive points cannot belong to two distinct strips, or to a plane and a strip (if two consecutive points belong to two distinct planes, then they both belong to the same plane too). Moreover, if p is a point included in the part of the figure beyond the frontiers, then a path in the figure, from the origin to p, necessarily goes through a point in the frontiers, and this point is included in a wire. Therefore, it is sufficient to compute **(1)** the geometrical closure of the set of wires (by enumerating the finite set of paths of its associated finite figure) and **(2)** all the paths from the frontiers to any point in a component (plane or strip), and **(3)** to combine these two parts appropriately.

First, in a plane, the set of paths from the frontier to any point of the plane is denoted by $(a + b)^*$.

Let s be a vertical strip of affix u and of width n, and consider a point $(x_u + k, y_u)$ on the frontier of s (with $k \leq n - 1$). Then the set of the paths inside the strip, starting from $(x_u + k, y_u)$, is denoted by

$$\sum_{0 \leq j \leq n - 1 - k} b^*(ab^*)^j.$$

Likewise, the set of the paths from a point $(x_u, y_u + k)$ on the frontier of a horizontal strip of affix u and of width n inside the strip is denoted by

$$\sum_{0 \leq j \leq n-1-k} a^*(ba^*)^j.$$

Let us now show how to compute the geometrical closure of a reduced expression. Let E be a reduced expression and X be the set of the wires of E.

First, the geometrical closure $\overline{X}$ of X is a finite language that can be computed directly from the figure. Moreover, for any couple (x, y) of integers, we denote by $\overline{X}_{(x,y)}$ the set defined by $\{w \in \overline{X} \mid \text{Vector}(w) = (x, y)\}$. An expression F is a *geometrical representation* of a component c of E if

$$L(F) = \{u \in \overline{L(E)} \mid \text{Vector}(u) \in \mathcal{F}(c)\}.$$

Denoting by C the set of strips and planes of E and by F_c a geometrical representation of c, we get

$$\overline{L(E)} = \overline{X} \cup L(\sum_{c \in C} F_c).$$

Let us now compute geometrical representations of strips and planes. If $c = ub^*a^{n-1}$ is a vertical strip of E, then we set

$$\hat{c} = \sum_{\substack{0 \leq k \leq n-1, \\ w \in \overline{X}_{\text{Vector}(ua^k)}, \\ 0 \leq j \leq n-1-k}} wb^*(ab^*)^j.$$

Remark that $\hat{c}$ geometrically represents c and that $L(\hat{c})$ is a $V_{3/2}$ language. If $c = ua^*b^{n-1}$ is a horizontal strip of E, then we set

$$\hat{c} = \sum_{\substack{0 \leq k \leq n-1, \\ w \in \overline{X}_{\text{Vector}(ub^k)}, \\ 0 \leq j \leq n-1-k}} wa^*(ba^*)^j.$$

Similarly to the previous case, $\hat{c}$ geometrically represents c and $L(\hat{c})$ is a $V_{3/2}$ language. Finally, if $c = u(a + b)^*$ is a plane of E, and l and h are the frontiers of E, then

– if $x_u = l$,

$$\hat{c} = \sum_{\substack{0 \leq i \leq h-y_u, \\ w \in \overline{X}_{\text{Vector}(ub^i)}}} w(a + b)^*;$$

– if $y_u = h$,

$$\hat{c} = \sum_{\substack{0 \leq i \leq l-x_u, \\ w \in \overline{X}_{\text{Vector}(ua^i)}}} w(a + b)^*.$$

Similarly to the previous cases, $\hat{c}$ geometrically represents c and $L(\hat{c})$ is a $V_{3/2}$ language. As a conclusion,

Theorem 18. *The family of $V_{3/2}$ languages is geometrically closed.*

Example 19. Let us consider the reduced expression E'' of Example 17. Hence,

$$X = \{baa, aa, bba, bab, bb, bbaa, ba\}, \qquad \overline{X} = \bigcup_{0 \leq i,j \leq 2} (a^i \sqcup\!\!\!\sqcup b^j),$$

$$\overline{X}_{((x,y)|x \leq 2, y \leq 2)} = (a^x \sqcup\!\!\!\sqcup b^y), \qquad\qquad F_{aaa^*} = aaa^*,$$

$$F_{bb(a+b)^*} = bb(a+b)^* + abb(a+b)^* + bab(a+b)^* + bba(a+b)^*$$
$$+ aabb(a+b)^* + abab(a+b)^* + abba(a+b)^* + baab(a+b)^*$$
$$+ baba(a+b)^* + bbaa(a+b)^*.$$

6 Web Application

The computation of a component sum, its normalization, its reduction and its closure have been implemented in Haskell (made in Haskell, compiled in Javascript using the REFLEX PLATFORM) in order to help the reader to manipulate the notions. This web application can be found here [10]. As an example, the expression $baa + aa + bb^*a + bbaa(a+b)^* + ba + a^*$ of Example 6 can be defined from the literal input b.a.a+a.a+b.b*.a+b.b.a.a.(a+b)*+b.a+a*.

7 Perspectives

We plan to investigate our constructions in higher dimensional spaces to determine whether the whole $V_{3/2}$ family is geometrically closed. The adjunction of a third dimension implies the consideration of more complex objects, but we hope that our two steps (normalization and reduction) can still be applied.

References

1. Arfi, M.: Opérations polynomiales et hiérarchies de concaténation. Theoret. Comput. Sci. **91**(1), 71–84 (1991)
2. Baruah, S.K., Rosier, L.E., Howell, R.R.: Algorithms and complexity concerning the preemptive scheduling of periodic, real-time tasks on one processor. Real-Time Syst. **2**(4), 301–324 (1990)
3. Béal, M.-P., Champarnaud, J.-M., Dubernard, J.-P., Jeanne, H., Lombardy, S.: Decidability of geometricity of regular languages. In: Yen, H.-C., Ibarra, O.H. (eds.) DLT 2012. LNCS, vol. 7410, pp. 62–72. Springer, Heidelberg (2012). https://doi.org/10.1007/978-3-642-31653-1_7
4. Blanpain, B., Champarnaud, J.M., Dubernard, J.P.: Geometrical languages. In: LATA, Report 35/07, pp. 127–138. Research Group on Mathematical Linguistics, Universitat Rovira i Virgili, Tarragona (2007)

5. Bouajjani, A., Muscholl, A., Touili, T.: Permutation rewriting and algorithmic verification. Inf. Comput. **205**(2), 199–224 (2007)
6. Champarnaud, J.-M., Dubernard, J.-P., Jeanne, H.: Geometricity of binary regular languages. In: Dediu, A.-H., Fernau, H., Martín-Vide, C. (eds.) LATA 2010. LNCS, vol. 6031, pp. 178–189. Springer, Heidelberg (2010). https://doi.org/10.1007/978-3-642-13089-2_15
7. Geniet, D., Largeteau, G.: WCET free time analysis of hard real-time systems on multiprocessors: a regular language-based model. Theoret. Comput. Sci. **388**(1–3), 26–52 (2007)
8. Guaiana, G., Restivo, A., Salemi, S.: On the trace product and some families of languages closed under partial commutations. J. Automata Lang. Comb. **9**(1), 61–79 (2004)
9. Largeteau, G., Geniet, D., Andrès, É.: Discrete geometry applied in hard real-time systems validation. In: Andres, E., Damiand, G., Lienhardt, P. (eds.) DGCI 2005. LNCS, vol. 3429, pp. 23–33. Springer, Heidelberg (2005). https://doi.org/10.1007/978-3-540-31965-8_3
10. Mignot, L.: Application: geometrical closure of $V_{3/2}$ binary expressions. http://ludovicmignot.free.fr/programmes/geomClosure/index.html. Accessed 27 Oct 2018
11. Pin, J.É.: The dot-depth hierarchy, 45 years later. In: The Role of Theory in Computer Science, pp. 177–202. World Scientific (2017)
12. Pin, J.É., Weil, P.: Ponynominal closure and unambiguous product. Theory Comput. Syst. **30**(4), 383–422 (1997)
13. Place, T., Zeitoun, M.: Concatenation hierarchies: new bottle, old wine. In: Weil, P. (ed.) CSR 2017. LNCS, vol. 10304, pp. 25–37. Springer, Cham (2017). https://doi.org/10.1007/978-3-319-58747-9_5
14. Place, T., Zeitoun, M.: Generic results for concatenation hierarchies. Theory Comput. Syst. 1–53 (2017). https://doi.org/10.1007/s00224-018-9867-0
15. Straubing, H.: Finite semigroup varieties of the form $v * d$. J. Pure Appl. Algebra **36**, 53–94 (1985)
16. Thérien, D.: Classification of finite monoids: the language approach. Theoret. Comput. Sci. **14**, 195–208 (1981)

Deterministic Biautomata and Subclasses of Deterministic Linear Languages

Galina Jirásková[1] and Ondřej Klíma[2]([✉])

[1] Mathematical Institute, Slovak Academy of Sciences,
Grešákova 6, 040 01 Košice, Slovak Republic
jiraskov@saske.sk
[2] Department of Mathematics and Statistics, Masaryk University,
Kotlářská 2, 611 37 Brno, Czech Republic
klima@math.muni.cz

Abstract. We propose the notion of a deterministic biautomaton, a machine reading an input word from both ends. We focus on various subclasses of deterministic linear languages and give their characterizations by certain subclasses of deterministic biautomata. We use these characterizations to establish closure properties of the studied subclasses of languages and to get basic decidability results concerning them.

1 Introduction

Formal languages can be generated by grammars or, alternatively, they can be recognized by various devices. The second approach is useful when we want to decide whether a given word belongs to a given language. The devices may be nondeterministic or deterministic, and the latter ones are usually more efficient in the task. Nevertheless, this advantage is often compensated by a weaker expressive power of that machines.

The machines considered in this paper are based on the notion of nondeterministic biautomata introduced by Holzer and Jacobi [5]. A nondeterministic biautomaton is a device consisting of a finite control which reads symbols from a read-only input tape using a pair of input heads. The left head reads the symbols from left to right, and the right one from right to left. In one step of a computation, the finite control nondeterministically chooses one of the heads, reads the next symbol by it, and moves into a new state. The choice of the new state is again made nondeterministically. A computation ends when the heads finish the reading of the input word and meet somewhere inside the tape. And as usually, the input word is accepted if there is a computation which ends in a final state.

G. Jirásková—Research supported by VEGA grant 2/0132/19 and grant APVV-15-0091.
O. Klíma—Research supported by Institute for Theoretical Computer Science (ITI), project No. P202/12/G061 of the Grant Agency of the Czech Republic.

C. Martín-Vide et al. (Eds.): LATA 2019, LNCS 11417, pp. 315–327, 2019.
https://doi.org/10.1007/978-3-030-13435-8_23

Nondeterministic biautomata recognize the class of linear context-free languages that are generated by context-free grammars in which every production has at most one nonterminal in its right hand side. The conversion from a biautomaton to a linear grammar, or vice versa, is straightforward [5].

In this paper we consider deterministic biautomata for which at most one computation is possible for every input word. Our motivation comes from the papers by Nasu, Honda [15], Ibarra, Jiang, Ravikumar [9], de la Higuera, Oncina [4], and Holzer, Lange [6] that studied subclasses of linear languages obtained by applying a kind of determinism to linear grammars. Our aim is to provide characterizations of the above mentioned subclasses by deterministic biautomata.

We propose the notion of a deterministic biautomaton in the next section. In Sect. 3, we describe certain subclasses of deterministic biautomata that recognize exactly the classes of languages generated by DL, $LinLL(1)$, and NH-DL grammars considered in [4]. We use these characterizations in Sect. 4, to study closure properties of the considered subclasses. Using these properties we are able to answer basic decidability question in Sect. 5.

To conclude this introduction, let us mention that several other devices recognizing linear languages have been introduced in the literature. Nondeterministic linear automata have been studied by Loukanova [14] and their deterministic counterparts have been considered by Bedregal in [3], where some partial observations concerning deterministic linear languages were also given. Another example of devices recognizing linear languages has been provided by Rosenberg [16] who used some kind of 2-tape automata.

2   Linear Languages and Nondeterministic Biautomata

For a finite non-empty alphabet Σ, let Σ^* denote the set of all words over Σ including the empty word λ. Let $\mathcal{P}(X)$ denote the set of all subsets of a set X.

A *linear grammar* is a tuple (N, Σ, P, S) where N and Σ are disjoint sets of *nonterminals* and *terminals*, respectively, P is a set of production rules of the form $A \to w$ where $A \in N$ and $w \in \Sigma^* N \Sigma^* \cup \Sigma^*$, and S is the initial nonterminal. We use the standard notation for grammars like $\Rightarrow$ for one derivation step and $\overset{*}{\Rightarrow}$ for the transitive-reflexive closure of the relation $\Rightarrow$. The *language generated by a grammar* (N, Σ, P, S) is the set $\{w \in \Sigma^* \mid S \overset{*}{\Rightarrow} w\}$. A language is *linear* if it is generated by a linear grammar.

A *nondeterministic biautomaton* (NB) is a sextuple $\mathcal{B} = (Q, \Sigma, \cdot, \circ, I, F)$ where Q is a finite set of *states*, Σ is an *input alphabet*, $\cdot : Q \times \Sigma \to \mathcal{P}(Q)$ is a *left action*, $\circ : Q \times \Sigma \to \mathcal{P}(Q)$ is a *right action*, $I \subseteq Q$ is the set of *initial* states and $F \subseteq Q$ is the set of *final* states.

A *configuration* is a pair (q, w), where q is a state and w is a word which remains to be read. The relation $\vdash$ on the set of all configurations is defined as follows: For $q, q' \in Q$, $w \in \Sigma^*$ and $a \in \Sigma$, we have $(q, aw) \vdash (q', w)$ if $q' \in q \cdot a$ and $(q, wa) \vdash (q', w)$ if $q' \in q \circ a$. Furthermore, the reflexive and transitive closure of $\vdash$ is denoted by $\vdash^*$. Let $L_\mathcal{B}(q) = \{w \in \Sigma^* \mid (q, w) \vdash^* (f, \lambda) \text{ for some } f \in F\}$ be the set of words accepted from the state q. Then the *language recognized* by $\mathcal{B}$ is $L_{\mathcal{B}} = \bigcup_{i \in I} L_{\mathcal{B}}(i)$. The class of all languages recognized by NBs is denoted by **NB**. It is known that it coincides with the class of linear languages [5].

Remark 1. For each NB, one can use the power-set construction [10] and obtain a biautomaton in which both $\cdot$ and $\circ$ are mappings from $Q \times \Sigma$ into Q. The resulting biautomaton is deterministic in the terminology used in [10]. However, it is far from being a deterministic biautomaton in our sense since either the left or the right reading head can still be nondeterministically chosen.

3 Determinism for Biautomata and Linear Languages

In this section we introduce a formal definition of a deterministic biautomaton. Furthermore, we define certain subclasses of deterministic biautomata which correspond to the classes of deterministic linear languages studied in the literature.

We are interested in biautomata which admit at most one computation for every input word. In particular, they have just one initial state. Furthermore, each state must determine which head will be used in the next step of the computation. And of course, the actions $\cdot$ and $\circ$ must be deterministic.

Definition 2. *A biautomaton* $\mathcal{B} = (Q, \Sigma, \cdot, \circ, I, F)$ *is deterministic (DB) if*

(1) $|I| = 1$;
(2) for each $q \in Q$ *and* $a \in \Sigma$, *we have* $|q \cdot a| \le 1$ *and* $|q \circ a| \le 1$,
(3) if $q \cdot a \ne \emptyset$ *for some* $a \in \Sigma$, *then* $q \circ b = \emptyset$ *for each* $b \in \Sigma$,
(4) if $q \circ a \ne \emptyset$ *for some* $a \in \Sigma$, *then* $q \cdot b = \emptyset$ *for each* $b \in \Sigma$.
   The class of all languages which are recognized by DBs is denoted by **DB**.

It follows from the definition that there are states of two different kinds. The first ones, called *left states*, have the property that $q \circ a = \emptyset$ for each $a \in \Sigma$, and the second ones, called *right states*, have the property that $q \circ a \ne \emptyset$ for some $a \in \Sigma$ and consequently $q \cdot b = \emptyset$ for every $b \in \Sigma$. We denote the set of all left states by Q_L and the set of all right states by Q_R. Therefore Q is a disjoint union of Q_L and Q_R. Now a deterministic biautomaton can be written as $\mathcal{B} = (Q_L, Q_R, \Sigma, \cdot, \circ, i, F)$ where $i \in Q_L \cup Q_R$ is the initial state, $\cdot$ is a partial function from Q_L to $Q_L \cup Q_R$ and $\circ$ is a partial function from Q_R to $Q_L \cup Q_R$. Note that every state q satisfying $q \cdot a = q \circ a = \emptyset$ for each $a \in \Sigma$ belongs to Q_L. This definition is almost identical with the definition of deterministic linear automaton from [3]; the main difference is that we have a unique initial state while in [3] more initial states are allowed.

Notice that every DB can be modified, by adding an additional state, to an equivalent *complete* DB which satisfies $|q \cdot a| = 1$ for each $a \in \Sigma$ and $q \in Q_L$, and $|q \circ a| = 1$ for each $a \in \Sigma$ and $q \in Q_R$.

Remark 3. Consider a complete DB $\mathcal{B} = (Q_L, Q_R, \Sigma, \cdot, \circ, i, F)$ recognizing a language L. Then, for every $w \in \Sigma^*$, there is a uniquely determined state $q \in Q$ such that $(i, w) \vdash^* (q, \lambda)$. Therefore, DB $\mathcal{B}^c = (Q_L, Q_R, \Sigma, \cdot, \circ, i, Q \backslash F)$ recognizes the complement of L. Consequently, the class **DB** is closed under complementation. On the other hand, the class **NB** is not closed under complementation since it is closed under union, but it is not closed under intersection as shown by a folklore example of two linear languages the intersection of which is not context-free: $\{a^n b^m c^m \mid m,n \geq 0\} \cap \{a^n b^n c^m \mid m,n \geq 0\}$. Thus **DB** $\subsetneq$ **NB**.

3.1   Characterization of Deterministic Linear Languages

The following definition of deterministic linear grammars is taken from [4]. The first property of these grammars is that they do not contain production rules with the right hand side starting with a non-terminal. Moreover, for a fixed nonterminal A, the first terminal on the right hand side of a rule having A on the left hand side uniquely determines the rule and it is followed by a nonterminal.

Definition 4 [4]. *A deterministic linear (DL) grammar $G = (N, \Sigma, S, P)$ is a grammar where all productions are of the form $A \to aBu$ or $A \to \lambda$ and which satisfies the following condition: for each $A, B, C \in N, a \in \Sigma, u, v \in \Sigma^*$, if the rules $A \to aBu$ and $A \to aCv$ are in P, then $B = C$ and $u = v$.*

Theorem 6 in [3] shows a construction of a DB recognizing a language given by a deterministic linear grammar. Our goal is to specify the corresponding class of DBs more precisely. Since left states of the constructed DB are obtained from nonterminals, the following definition assumes that the initial state and all final states are left states, and every right state has exactly one out-going transition.

Definition 5. *A DB $\mathcal{B} = (Q_L, Q_R, \Sigma, \cdot, \circ, i, F)$ is weak from the right (DBW) if $\{i\} \cup F \subseteq Q_L$ and for each $q \in Q_R$, there is a unique $a \in \Sigma$ such that $q \circ a$ is defined. The class of all languages recognized by DBWs is denoted by **DBW**.*

Theorem 6. *A language is generated by a deterministic linear grammar if and only if it is recognized by a deterministic biautomaton weak from the right.*

Proof (Sketch). Let $G = (N, \Sigma, S, P)$ be a DL grammar generating a language L. We construct a DB $\mathcal{B} = (N, Q_R, \Sigma, \cdot, \circ, S, F)$ in which for each production in P of the form $A \to aBb_1 \cdots b_k$, where $k \geq 1$ and $a, b_1, \ldots, b_k \in \Sigma$, we put the state $(B, b_1 \cdots b_\ell)$ into Q_R for each ℓ with $1 \leq \ell \leq k$, and we set $A \cdot a = (B, b_1 \cdots b_k)$. Furthermore, for each ℓ with $2 \leq \ell \leq k$, we set $(B, b_1 \cdots b_\ell) \circ b_\ell = (B, b_1 \cdots b_{\ell-1})$, and finally, we set $(B, b_1) \circ b_1 = B$. For each production of the form $A \to aB$, we set $A \cdot a = B$. For each production of the form $A \to \lambda$, we just put A into F. The resulting biautomaton is a DBW and it recognizes the language L.

Let $\mathcal{B} = (Q_L, Q_R, \Sigma, \cdot, \circ, i, F)$ be a DBW for L. We may assume that $\mathcal{B}$ is trim, that is, each its state is reachable, and some final state can be reached from each of its states. We construct the grammar (Q_L, Σ, i, P) where rules are of the form $A \to aBu$, where B is the uniquely determined first possible left state accessible from $A \cdot a$ and u is the word which is read by the right head during the move from $A \cdot a$ into B. Finally, we add rules $A \to \lambda$ for $A \in F$. $\square$

3.2   Characterization of Linear $LL(1)$ Languages

Inside context-free languages, the hierarchy of LL languages is defined; see [1]. Since we are interested only in linear languages and since we deal with deterministic grammars, we concentrate on linear $LL(1)$ grammars. The next definition is from [6] and it uses the following technical notation. Let $G = (N, \Sigma, S, P)$ be a grammar and $\alpha \in (N \cup \Sigma)^*$. Then $FIRST_G(\alpha)$ is defined as the set $\{a \in \Sigma \mid \alpha \overset{*}{\Rightarrow} a\beta$ for some $\beta \in (N \cup \Sigma)^*\}$ extended by λ whenever $\alpha \overset{*}{\Rightarrow} \lambda$.

Definition 7 [6]. *A linear grammar $G = (N, \Sigma, S, P)$ is a linear $LL(1)$ grammar if the following condition is satisfied for every $A \in N$: for every sentential form $S \overset{*}{\Rightarrow} uAv$, with $u, v \in \Sigma^*$, and every pair of distinct productions $A \to \alpha_1$ and $A \to \alpha_2$ in P, the sets $FIRST_G(\alpha_1 v)$ and $FIRST_G(\alpha_2 v)$ are disjoint.*

Although linear $LL(1)$ grammar need not be deterministic, it is known that we can construct an equivalent linear $LL(1)$ grammar which is deterministic: First, each linear $LL(1)$ grammar has an equivalent linear $LL(1)$ grammar with productions of the forms $A \to a\alpha$ or $A \to \lambda$ [13, Theorem 4]. Then, every production of the form $A \to a\alpha$ can be replaced by a sequence of productions which satisfy the conditions in Definition 4; cf. [4, Theorem 2].

Our aim is to define a subclass of DBWs which recognize exactly linear $LL(1)$ languages. The problematic situation occurs with pairs of rules $A \to \lambda$ and $A \to a\alpha$ with $S \overset{*}{\Rightarrow} wAaw'$. In the proof of Theorem 6, the rule $A \to \lambda$ implies that $A \in F$. Therefore, we cannot have a final state A in which the transition on a symbol a is defined, and at the same time a right state p and symbols $a_1, a_2 \ldots, a_\ell$ with $(p \circ a) \cdot a_1 \cdot a_2 \cdots a_\ell = A$. Therefore, the pattern in Fig. 1 is forbidden. In all our figures, left states are in circles and right states are in squares. Triangles mean that we do not know whether the state is left or right. Moreover, we use solid lines for the action $\cdot$ and dashed lines for $\circ$. This combines notations from both [3] and [12].

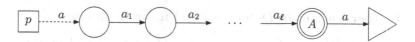

Fig. 1. A forbidden pattern in a DBWR.

Definition 8. *We say that a DBW $\mathcal{B} = (Q_L, Q_R, \Sigma, \cdot, \circ, i, F)$ has restricted final states (DBWR) if for every final state f, such that there is a state p in Q_R, and letters $a, a_1, \ldots, a_\ell$ satisfying $(p \circ a) \cdot a_1 \cdots a_\ell = f$, the transition $f \cdot a$ is undefined. The class of languages recognized by DBWRs is denoted by* **DBWR***.*

Theorem 9. *A language is generated by a linear $LL(1)$ grammar if and only if it is recognized by a DBWR.*

Proof (Sketch). Let L be generated by a linear $LL(1)$ grammar $G = (N, \Sigma, S, P)$. Then L is generated by a deterministic linear grammar [13, Theorem 4]. Therefore L is recognized by a DBW $\mathcal{B} = (Q_L, Q_R, \Sigma, \cdot, \circ, i, F)$ by Theorem 6. We can show that $\mathcal{B}$ is a DBWR if and only if G is a linear $LL(1)$ grammar.   □

Since **DBWR** $\subseteq$ **DBW**, there is a natural question whether the inclusion is strict. The following example of a language L in **DBW\DBWR** is taken from [4]. However, there is no formal proof of this fact in [4], because there is a fault in the argument − it is overlooked that all regular languages are linear $LL(1)$ languages. The next proof can be seen as a first application of Theorem 9.

Lemma 10. *Let* $L = \{a^k b^\ell \mid 0 \le k \le \ell\}$. *Then* $L \in$ **DBW\DBWR**.

Proof (Sketch). The language L is recognized by the DBW shown in Fig. 2.

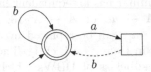

Fig. 2. A DBW recognizing the language $\{a^k b^\ell \mid 0 \le k \le \ell\}$.

Suppose, to get a contradiction, that the language L is recognized by a trim DBWR $\mathcal{B} = (Q_L, Q_R, \Sigma, \cdot, \circ, i, F)$. Let $n = |Q_L| + |Q_R|$. Now the idea is that $\mathcal{B}$ must enter a cycle while reading the word $w = a^n b^n$, and it may leave this cycle only in a left state by reading b. Then depending on the number of a's and b's read while working in the cycle, we can either find a word in L which is not accepted by $\mathcal{B}$, or a word which is accepted by $\mathcal{B}$ but is not in L.   □

3.3 Another Variant of Deterministic Linear Grammars

Nasu and Honda [15] introduced another class of deterministic linear languages called NH-deterministic languages by Higuera and Oncina [4] who also proved that these languages form a subclass of $LinLL(1)$. We present a modification of DBWs to get a characterization of NH-deterministic languages.

Definition 11 (Definition 3 in [4]). *A NH-DL grammar* $G = (N, \Sigma, S, P)$ *is a linear grammar where all productions are of the form* $A \to aBu$ *or* $A \to a$ *and which satisfies the following condition: for all* $A \in N, a \in \Sigma, \alpha, \beta \in N\Sigma^* \cup \{\lambda\}$, *if* $A \to a\alpha, A \to a\beta \in P$, *then* $\alpha = \beta$.

Definition 12. *A DBW* $\mathcal{B} = (Q_L, Q_R, \Sigma, \cdot, \circ, i, \{f\})$ *is called* weak from the right with a passive final state *(DBWP) if* $i \ne f$ *and* $f \cdot a = \emptyset$ *for every* $a \in \Sigma$. *The class of all languages recognized by DBWPs is denoted by* **DBWP**.

Theorem 13. *A language L is generated by an NH-DL grammar if and only if $L \in$ DBWP.*   □

Note that Higuera and Oncina [4] considered also the class of languages generated by IJR-DL grammars [9]. However, the difference between IJR-DL grammars and NH-DL grammars is very subtle: there is exactly one language which is generated by a IJR-DL grammar but it is not generated by any NH-DL grammar, namely the language $\{\lambda\}$ as mentioned in [6, Theorem 3]. Thus we could obtain a characterization of languages generated by IJR-DL grammars if we remove the condition $i \neq f$ in the definition of DBWPs. In the case of $i = f$, the considered modification of DBWP recognizes exactly the language $\{\lambda\}$.

4   Closure Properties of Deterministic Linear Languages

First of all, we state that **DBWP $\subsetneq$ DBWR $\subsetneq$ DBW $\subsetneq$ DB $\subsetneq$ NB**. The first two strict inclusions were stated in [4] via corresponding grammars; however, with some incompleteness of arguments as mentioned before Lemma 10. Since **DB** is closed under complementation by Remark 3 and, as we show later, **DBW** is not, we obtain **DBW $\subsetneq$ DB**. Finally, **DB $\subsetneq$ NB** was discussed in Remark 3.

Another interesting question is the relation with the class of all regular languages **Reg**. Since every finite deterministic automaton can be viewed as a DBW with the empty set of right states, we have **Reg $\subseteq$ DBWR**. On the other hand, the class **DBWP** is incomparable with **Reg** since over a unary alphabet, the languages in **DBWP** are exactly languages consisting of a single word. On the other hand, the DBWP in Fig. 3 recognizes a non-regular language.

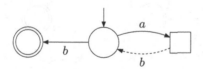

Fig. 3. A DBWP recognizing the language $\{a^n b^{n+1} \mid n \geq 1\}$.

The main aim of this section is a better understanding of closure properties of the considered subclasses of linear languages. We use these properties in the next section, to answer some decidability questions related to the class of linear languages. Table 1 summarizes the results of this section.

The class **NB** forms a full trio, that is, it is closed under homomorphisms, inverse homomorphisms and intersection with regular languages [8, Exercise 11.1]. Every full trio is closed under quotient with a regular language [8, Theorem 11.3] and under regular substitutions [8, Theorem 11.4]. It is an easy exercise to show that **NB** is closed under concatenation with regular languages. On the other hand, as mentioned in Remark 3, the class **NB** is not closed under intersection and complementation. It is also not closed under concatenation since the

Table 1. Closure properties of the studied subclasses of linear languages.

class of languages	NB	DB	DBW	DBWR	DBWP
grammars	linear		DL	$LinLL(1)$	NH-DL
intersection	No	No			
intersection with regular	Yes	Yes			
concatenation	No	No			
concatenation with regular	Yes	No			
complementation	No	Yes	No		
union	Yes	No			
reversal	No	Yes	No		
right quotient by a letter	Yes	Yes			No
homomorphic images	Yes	No			

language $\{a^n b^n \mid n \geq 0\}\{a^n b^n \mid n \geq 0\}$ is not linear (see e.g. [1, Section 6.1]). This gives the second column in Table 1. The following example shows that all the remaining subclasses are not closed under intersection.

Example 14. We slightly modify the example from Remark 3 to get a pair of languages in class **DBWP**. Consider $K = \{a^n b^m c^{n+1} \mid m \geq 1, n \geq 0\}$ and $L = \{a^m b^n c^{n+1} \mid m \geq 0, n \geq 1\}$. The DBWPs recognizing K and L are shown in Fig. 4, while $K \cap L = \{a^n b^n c^{n+1} \mid n \geq 1\}$ is not a context-free language.

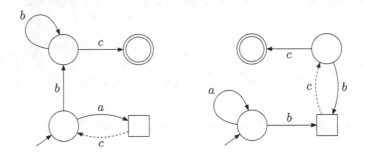

Fig. 4. DBWPs for $\{a^n b^m c^{n+1} \mid m \geq 1, n \geq 0\}$ and $\{a^m b^n c^{n+1} \mid m \geq 0, n \geq 1\}$.

Remark 15. Since all the classes in Table 1 are not closed under intersection, they cannot be closed under union and complementation at the same time.

In the rest of this section we demonstrate the techniques that use deterministic biautomata to get the results in Table 1.

Theorem 16. *The classes of languages* **DB, DBW, DBWR, DBWP** *are closed under intersection with regular languages.*

Proof. For a given regular language L, the syntactic relation $\equiv_L$ of the language L is an equivalence relation on Σ^* and it is defined in the following way. For $u, v \in \Sigma^*$ we have $u \equiv_L v$ if and only if $(\forall s, t \in \Sigma^*)(sut \in L \iff svt \in L)$. It is known that the relation $\equiv_L$ is a congruence on the monoid Σ^* and it has a finite index. Then the finite quotient $\Sigma^* /_{\equiv_L}$ is called the *syntactic monoid* of L and we denote it as M. The *syntactic morphism* is the mapping $\eta : \Sigma^* \to M$ given by $\eta(u) = [u]$, where $[u]$ is $\equiv_L$-class containing u. The neutral element of M is $[\lambda]$ and it is denoted by 1. If we denote $N = \eta(L)$, then $L = \{u \in \Sigma^* \mid [u] \in N\}$.

Assume that $(Q_L, Q_R, \Sigma, \cdot, \circ, i, F)$ is a DB for K. We construct the following DB $(Q'_L, Q'_R, \Sigma, \cdot', \circ', i', F')$, where $Q'_L = Q_L \times M \times M$, $Q'_R = Q_R \times M \times M$, $i' = (i, 1, 1)$, $F' = \{(q, m, n) \mid q \in F, m \cdot n \in N\}$, and $(q, m, n) \cdot' a = (q \cdot a, m \cdot [a], n)$ if $(q, m, n) \in Q'_L$, and $(q, m, n) \circ' a = (q \circ a, m, [a] \cdot n)$ if $(q, m, n) \in Q'_R$. The resulting deterministic biautomaton recognizes exactly the language $K \cap L$.

In the proof above, if the original DB is a DBW, DBWR, or DBWP, then the resulting DB is a DBW, DBWR, or DBWP, respectively, as well.   □

Theorem 17. *The classes* **DB**, **DBW**, **DBWR**, *and* **DBWP** *are not closed under concatenation with regular languages.*

Proof (Sketch). Let $L = \{a^n b a^n \mid n \geq 0\}$ be the language recognized by the DBWP $(\{i, f\}, \{q\}, \{a, b\}, \cdot, \circ, i, \{f\})$ with $i \cdot a = q$, $i \cdot b = f$, $q \circ a = i$. Let us show that the languages $\Sigma^* L$ and $L\Sigma^*$ are not in **DB**.

Assume that $\Sigma^* L$ is recognized by a DB $\mathcal{B} = (Q_L, Q_R, \Sigma, \cdot, \circ, i, F)$. Assume that $\mathcal{B}$ has n states and consider a word $u = a^n b a^n$ in $\Sigma^* L$. In the first n steps of the computation, the heads read only symbols a from the beginning or from the end of the input word u, and moreover, the computation must visit some state twice. Thus, there is a state q and two distinct pairs (k_1, ℓ_1) and (k_2, ℓ_2) of numbers in the set $\{0, 1, \ldots, n\}$ such that, for both $i \in \{1, 2\}$, after $k_i + \ell_i$ steps when k_i copies of a are read by the left head and ℓ_i copies of a are read by the right head, the computation finishes in the state q. Now depending on whether $\ell_1 \neq \ell_2$ or $k_1 \neq k_2$ we can describe a word w such that exactly one of $a^{k_1} w a^{\ell_1}$ and $a^{k_2} w a^{\ell_2}$ is in $\Sigma^* L$, which gives a contradiction.

Now, consider the language $L\Sigma^*$. If it would be in **DB**, then its reversal $\Sigma^* L^R$ would be in **DB** since it is enough to exchange the role of left and right states. However, since $L = L^R$, we have $\Sigma^* L^R = \Sigma^* L$, and as shown above, the language $\Sigma^* L$ is not recognized by any **DB**.   □

Theorem 18. *The classes* **DB**, **DBW**, *and* **DBWR** *are closed under right quotient by a letter.*

Proof. Let L be recognized by a DB $\mathcal{B} = (Q_L, Q_R, \Sigma, \cdot, \circ, i, F)$ and $a \in \Sigma$. First, assume that $i \in Q_R$. If $i \circ a$ is not defined, then $La^{-1} = \emptyset$, so $La^{-1} \in$ **DB**. If $i \circ a$ is defined, then the DB $\mathcal{B}' = (Q_L, Q_R, \Sigma, \cdot, \circ, i \circ a, F)$ recognizes La^{-1}. So, in the rest of the proof we can assume that $i \in Q_L$.

Consider a disjoint copy of Q_L denoted by $\overline{Q_L} = \{\overline{q} \mid q \in Q_L\}$, and define a new DB $\mathcal{B}' = (\overline{Q_L} \cup Q_L, Q_R, \Sigma, \cdot', \circ, \overline{i}, F')$ where $F' = \{\overline{q} \mid q \in Q_L, q \cdot a \in F\} \cup F$. Then, for each $q \in Q_L$ and $b \in \Sigma$, we set $q \cdot' b = q \cdot b$. Next, we set $\overline{q} \cdot' b = \overline{q \cdot b}$

if $q \cdot b \in Q_L$ and $\overline{q} \cdot' b = (q \cdot b) \circ a$ if $q \cdot b \in Q_R$; thus, in the first case we have $\overline{q} \cdot' b \in \overline{Q_L}$, and in the second case we have $\overline{q} \cdot' b \in Q_L \cup Q_R$ under the assumption that the state $(q \cdot b) \circ a$ exists. Let us show that $\mathcal{B}'$ recognizes La^{-1}. The part $\overline{Q_L}$ of $\mathcal{B}'$ reads the input word w from left to right and accepts w if and only if wa is accepted by $\mathcal{B}$ without using the right head. Now, if w is accepted by $\mathcal{B}'$ in a final state in F, then we need to use some move from $\overline{Q_L}$ to $Q_L \cup Q_R$, which exists if and only if the corresponding computation for wa exists in $\mathcal{B}$. If the original DB is a DBW or a DBWR, then $F \subseteq Q_L$, so $F' \subseteq \overline{Q_L} \cup Q_L$. Since the right action in $\mathcal{B}'$ is the same as in $\mathcal{B}$, there is nothing else to check.   □

Example 19. Consider the language $L = b^*a$ which is recognized by the DBWP $(\{i, f\}, \emptyset, \{a, b\}, \cdot, \circ, i, \{f\})$ with $i \cdot b = i$ and $i \cdot a = f$. We have $La^{-1} = b^*$. The language b^* is not in **DBWP** since every unary language in **DBWP** consists of a single word.

5   Basic Decidability Questions

To get some undecidability results, the following Greibach's theorem can be used. The theorem requires some closure properties and it is stated as follows.

Theorem 20 (Greibach's Theorem, [8, Theorem 8.14]). *If $\mathcal{C}$ is a class of languages that is effectively closed under concatenation with regular sets and union, and for which "$= \Sigma^*$" is undecidable for some sufficiently large Σ, then for every nontrivial subset P of $\mathcal{C}$ that contains all regular sets and is closed under right quotient by a letter, it is undecidable whether a language in $\mathcal{C}$ is in P.*

Both **CF** and **NB** are effectively closed under concatenation with regular sets and union, and in both of them the universality is undecidable [2]. Therefore, we can set $\mathcal{C} = $ **CF** and $P = $ **NB** in Greibach's theorem, and get that it is undecidable whether a language in **CF** is in **NB** [7]. By setting $\mathcal{C} = $ **NB** and $P = $ **Reg**, we get that is undecidable whether a language in **NB** is regular [5, Theorem 9]. As shown in the previous section, the classes **DB**, **DBW**, and **DBWR** contain regular languages and are closed under right quotient by a letter, so we get the next result.

Theorem 21. *Let $P \in \{$**DB**, **DBW**, **DBWR**$\}$. It is undecidable whether a language in **NB** is in P.*

As shown in Sect. 4, the class **DBWP** is incomparable to **Reg** and it is not closed under right quotient by a letter. Thus Greibach's theorem cannot be used in this case, and decidability of **DBWP** in **NB** remains open.

We next modify the proof from [2] which shows that $K \cap L = \emptyset$ is undecidable in **NB** to get undecidability of this problem for the class **DBWP**.

Theorem 22. *The emptiness of intersection is undecidable in **DBWP**.*

Proof. The proof is a reduction from PCP. Let (u_i, v_i) for $i = 1, 2 \ldots, k$, where u_i and v_i are words over Σ, be an instance of PCP. Our aim is to construct two DBWPs $\mathcal{A}$ and $\mathcal{B}$ such that $L_{\mathcal{A}} \cap L_{\mathcal{B}} = \emptyset$ if and only if PCP does not have a solution for this instance.

Let the input alphabet of $\mathcal{A}$ and $\mathcal{B}$ be $\{1, 2, \ldots, k\} \cup \Sigma \cup \{\#\}$. Let the initial state of $\mathcal{A}$ be a left state p_0 which has the transitions on $1, 2, \ldots, k$ to pairwise distinct right states $p_1, p_2, \ldots, p_k$. In each p_i, automaton reads the symbols of the word u_i^R by its right head, and after reading its last symbol it reaches the initial state p_0. In p_0 it reads $\#$ and reaches the unique left final state in which no transitions are defined. The automaton $\mathcal{B}$ is constructed in a similar way using words v_i. Notice that $\mathcal{A}$ and $\mathcal{B}$ accept words in a form $i_1 i_2 \ldots i_\ell \#(u_{i_1} u_{i_2} \ldots u_{i_\ell})^R$ and $j_1 j_2 \ldots j_t \#(v_{j_1} v_{j_2} \ldots v_{j_t})^R$, respectively. Therefore, $L_{\mathcal{A}} \cap L_{\mathcal{B}} \neq \emptyset$ if and only if PCP has a solution for the instance $\{(u_i, v_i) \mid i = 1, 2 \ldots, k\}$. $\square$

To test the emptiness of a language in **DB**, it is enough to test the reachability of a final state. This can be done in NL, and it is NL-hard for **DFA**s, and even for partial DFAs with a unique final state which has no out-going transitions [11]. Next, **DB** is closed under complementation, and therefore universality is in NL, it is NL-hard for DFAs, while no language in **DBWP** is universal. Moreover, **DB** is closed under intersection with regular sets. Thus testing the equality $L = R$ for $L \in$ **DB** and a given regular set R is equivalent to testing the emptiness of languages $L \cap \overline{R}$ and $\overline{L} \cap R$ which can be done in NL, while $L = \emptyset$ is NL-hard for **DBWP**. The next theorem summarizes these observations. Table 2 displays all our decidability results and compares them to the known results for **NB**.

Theorem 23. *The emptiness, universality, and equality to a given regular set are decidable in* **DB**, *and they are* NL-*complete for* **DB**, **DBW**, *and* **DBWR**.

Table 2. Decidability properties in the subclasses of linear languages.

class of languages	NB	DB	DBW	DBWR	DBWP
grammars	linear		DL	$LinLL(1)$	NH-DL
Does L in **NB** belong to:	-	undecidable			?
emptiness of intersection	undecidable	undecidable			
emptiness	NL-complete	NL-complete			
equality to a given regular set	undecidable	NL-complete			
universality	undecidable	NL-complete			trivial

6   Conclusions

We proposed a notion of deterministic biautomata and considered their certain subclasses to get characterizations of subclasses of linear context-free languages resulting from applying a kind of determinism to linear grammars. In particular, we were able to characterize the classes of languages generated by DL, $LinLL(1)$ and NH-DL grammars by deterministic biautomata weak from the right, weak from the right with restricted final states, and weak from the right with a passive final state. Using these characterizations, we studied closure properties in the considered classes. We proved, for example, that all these classes are closed under intersection with regular languages, and all, except for the last one, are closed under right quotient by a letter.

We used the closure properties to answer basic decidability questions. We showed that the question whether a given linear language is in a considered subclass is undecidable, except for the smallest class. We also proved that the emptiness of intersection is undecidable in all considered subclasses, while emptiness and equality to a given regular set are NL-complete in all of them.

Some questions remain open. For example, we do not know whether the considered classes are closed under left quotient under a letter. We do not know either whether the problem of language equality is decidable in the considered subclasses; here we cannot use the equality to Σ^* since, contrarily to linear languages, universality is decidable in all considered subclasses. The full version of this paper can be found at: http://im.saske.sk/~jiraskov/DB.pdf.

Acknowledgments. We would like to thank Professor Erkki Mäkkinen who proposed the topic of deterministic linear languages to us. We are also grateful to Libor Polák for useful discussions in the beginning of this research.

References

1. Autebert, J.-M., Berstel, J., Boasson, L.: Context-free languages and pushdown automata. In: Rozenberg, G., Salomaa, A. (eds.) Handbook of Formal Languages, vol. 1, pp. 111–174. Springer, Heidelberg (1997). https://doi.org/10.1007/978-3-642-59136-5_3
2. Baker, B.S., Book, R.V.: Reversal-bounded multipushdown machines. J. Comput. Syst. Sci. **8**(3), 315–332 (1974)
3. Bedregal, B.R.C.: Some subclasses of linear languages based on nondeterministic linear automata. Preprint (2016). http://arxiv.org/abs/1611.10276
4. de la Higuera, C., Oncina, J.: Inferring deterministic linear languages. In: Kivinen, J., Sloan, R.H. (eds.) COLT 2002. LNCS (LNAI), vol. 2375, pp. 185–200. Springer, Heidelberg (2002). https://doi.org/10.1007/3-540-45435-7_13
5. Holzer, M., Jakobi, S.: Minimization and characterizations for biautomata. Fundam. Informaticae **136**(1–2), 113–137 (2015)
6. Holzer, M., Lange, K.-J.: On the complexities of linear LL(1) and LR(1) grammars. In: Ésik, Z. (ed.) FCT 1993. LNCS, vol. 710, pp. 299–308. Springer, Heidelberg (1993). https://doi.org/10.1007/3-540-57163-9_25
7. Hoogeboom, H.J.: Undecidable problems for context-free grammars. Unpublished (2015). https://liacs.leidenuniv.nl/~hoogeboomhj/second/codingcomputations.pdf
8. Hopcroft, J.E., Ullman, J.D.: Introduction to Automata Theory. Languages and Computation. Addison-Wesley, Boston (1979)

9. Ibarra, O.H., Jiang, T., Ravikumar, B.: Some subclasses of context-free languages in NC1. Inf. Process. Lett. **29**(3), 111–117 (1988)
10. Jakobi, S.: Modern Aspects of Classical Automata Theory: Finite Automata, Biautomata, and Lossy Compression. Logos Verlag, Berlin (2015)
11. Jones, N.D.: Space-bounded reducibility among combinatorial problems. J. Comput. Syst. Sci. **11**(1), 68–85 (1975)
12. Klíma, O., Polák, L.: On biautomata. RAIRO - Theor. Inf. Appl. **46**(4), 573–592 (2012)
13. Kurki-Suonio, R.: On top-to-bottom recognition and left recursion. Commun. ACM **9**(7), 527–528 (1966)
14. Loukanova, R.: Linear context free languages. In: Jones, C.B., Liu, Z., Woodcock, J. (eds.) ICTAC 2007. LNCS, vol. 4711, pp. 351–365. Springer, Heidelberg (2007). https://doi.org/10.1007/978-3-540-75292-9_24
15. Nasu, M., Honda, N.: Mappings induced by pgsm-mappings and some recursively unsolvable problems of finite probabilistic automata. Inf. Control **15**(3), 250–273 (1969)
16. Rosenberg, A.L.: A machine realization of the linear context-free languages. Inf. Control **10**(2), 175–188 (1967)

Learning Unions of k-Testable Languages

Alexis Linard[1(✉)], Colin de la Higuera[2], and Frits Vaandrager[1]

[1] Institute for Computing and Information Science, Radboud University,
Nijmegen, The Netherlands
{a.linard,f.vaandrager}@cs.ru.nl
[2] Laboratoire des Sciences du Numérique de Nantes, Université de Nantes,
Nantes, France
cdlh@univ-nantes.fr

Abstract. A classical problem in grammatical inference is to identify a language from a set of examples. In this paper, we address the problem of identifying a union of languages from examples that belong to several *different* unknown languages. Indeed, decomposing a language into smaller pieces that are easier to represent should make learning easier than aiming for a too generalized language. In particular, we consider k-testable languages in the strict sense (k-TSS). These are defined by a set of allowed prefixes, infixes (sub-strings) and suffixes that words in the language may contain. We establish a Galois connection between the lattice of all languages over alphabet Σ, and the lattice of k-TSS languages over Σ. We also define a simple metric on k-TSS languages. The Galois connection and the metric allow us to derive an efficient algorithm to learn the union of k-TSS languages. We evaluate our algorithm on an industrial dataset and thus demonstrate the relevance of our approach.

Keywords: Grammatical inference · k-Testable languages ·
Union of languages · Galois connection

1 Introduction

A common problem in grammatical inference is to find, i.e. *learn*, a regular language from a set of examples of that language. When this set is divided into positive examples (belonging to the language) and negative examples (not belonging to the language), the problem is typically solved by searching for the smallest deterministic finite automaton (DFA) that accepts the positive examples, and rejects the negative ones. Moreover there exist algorithms which *identify in the limit* a DFA, that is, they eventually learn correctly any language/automaton from such examples [6].

We consider in this work a setting where one can observe positive examples from multiple different languages, but they are given together and it is not clear

This research is supported by the Dutch Technology Foundation (STW) under the Robust CPS program (project 12693).

© Springer Nature Switzerland AG 2019
C. Martín-Vide et al. (Eds.): LATA 2019, LNCS 11417, pp. 328–339, 2019.
https://doi.org/10.1007/978-3-030-13435-8_24

to which language each example belongs to. For example, given the following set of strings $S = \{aa, aaa, aaaa, abab, ababab, abba, abbba, abbbba\}$, learning a single automaton will be less informative than learning several DFAs encoding respectively the languages a^*, $(ab)^*$ and ab^*a. There is a trade-off between the number of languages and how specific each language should be. That is, covering all words through a single language may not be the desired result, but having a language for each word may also not be desired. The problem at hand is therefore double: to cluster the examples and learn the corresponding languages.

In this paper, we focus on k-testable languages in the strict sense (k-TSS) [10]. A k-TSS language is determined by a finite set of substrings of length at most k that are allowed to appear in the strings of the language. It has been proved that, unlike for regular languages, algorithms can learn k-TSS languages in the limit from text [16]. Practically, this learning guarantee has been used in a wide range of applications [2,3,12,13]. However, all these applications consider learning of a sole k-TSS language [2], or the training of several k-TSS languages in a context of supervised learning [13]. Learning unions of k-TSS languages has been suggested in [14].

A first contribution of this paper is a Galois connection between the lattice of all languages over alphabet Σ and the lattice of k-TSS languages over Σ. This result provides a unifying and abstract perspective on known properties of k-TSS languages, but also leads to several new insights. The Galois connection allows to give an alternative proof of the learnability in the limit of k-TSS languages, and suggests an algorithm for learning unions of k-TSS languages. A second contribution is the definition of a simple metric on k-TSS languages. Based on this metric, we define a clustering algorithm that allows us to efficiently learn unions of k-TSS languages.

Our research was initially motivated by a case study of print jobs that are submitted to large industrial printers. These print jobs can be represented by strings of symbols, where each symbol denotes a different media type, such as a book cover or a newspaper page. Together, this set of print jobs makes for a fairly complicated 'language'. Nevertheless, we observed that each print job can be classified as belonging to one of a fixed set of categories, such as 'book' or 'newspaper'. Two print jobs that belong to the same category are typically similar, to the extent that they only differ in terms of prefixes, infixes and suffixes. Therefore, the languages stand for the different families of print jobs. Our goal is to uncover these k-TSS languages.

This paper is organized as follows. In Sect. 2 we recall preliminary definitions on k-TSS languages and define a Galois connection that characterizes these languages. We then present in Sect. 3 our algorithm for learning unions of k-TSS languages. Finally, we report on the results we achieved for the industrial case study in Sect. 4. We refer to the full version of our paper for all the proofs.[1]

---

[1] For missing proofs, see http://arxiv.org/abs/1812.08269.

2 k-Testable Languages

The class of k-testable languages in the strict sense (k-TSS) has been introduced
by McNaughton and Papert [10]. Informally, a k-TSS language is determined by
a finite set of substrings of length at most k that are allowed to appear in
the strings of the language. This makes it possible to use as a parser a sliding
window of size k, which rejects the strings that at some point do not comply
with the conditions. Concepts related to k-TSS languages have been widely used
e.g. in information theory, pattern recognition and DNA sequence analysis [4,16].
Several definitions of k-TSS languages occur in the literature, but the differences
are technical. In this section, we present a slight variation of the definition of
k-TSS languages from [7], which in turn is a variation of the definition occurring
in [4,5]. We establish a Galois connection that characterizes k-TSS languages,
and show how this Galois connection may be used to infer a learning algorithm.

We write $\mathbb{N}$ to denote the set of natural numbers, and let i, j, k, m, and n
range over $\mathbb{N}$.

2.1   Strings

Throughout this paper, we fix a finite set Σ of *symbols*. A *string* $x = a_1 \ldots a_n$ is
a finite sequence of symbols. The *length* of a string x, denoted $\mid x \mid$ is the number
of symbols occurring in it. The empty string is denoted λ. We denote by Σ^* the
set of all strings over Σ, and by Σ^+ the set of all nonempty strings over Σ (i.e.
$\Sigma^* = \Sigma^+ \cup \{\lambda\}$). Similarly, we denote by $\Sigma^{<i}$, Σ^i and $\Sigma^{>i}$ the sets of strings
over Σ of length less than i, equal to i, and greater than i, respectively.

Given two strings u and v, we will denote by $u \cdot v$ the concatenation of u and
v. When the context allows it, $u \cdot v$ shall be simply written uv. We say that u is
a *prefix* of v iff there exists a string w such that $uw = v$. Similarly, u is a *suffix*
of v iff there exists a string w such that $wu = v$. We denote by $x[: k]$ the prefix
of length k of x and $x[-k :]$ the suffix of length k of x.

A *language* is any set of strings, so therefore a subset of Σ^*. Concatenation
is lifted to languages by defining $L \cdot L' = \{u \cdot v \mid u \in L$ and $v \in L'\}$. Again, we
will write LL' instead of $L \cdot L'$ when the context allows it.

2.2 k-Testable Languages

A k-TSS language is determined by finite sets of strings of length $k - 1$ or k that
are allowed as prefixes, suffixes and substrings, respectively, together with all the
short strings (with length at most $k - 1$) contained in the language. The finite
sets of allowed strings are listed in what McNaughton and Papert [10] called
a k-*test vector*. The following definition is taken from [7], except that we have
omitted the fixed alphabet Σ as an element in the tuple, and added a technical
condition ($I \cap F = C \cap \Sigma^{k-1}$) that we need to prove Theorem 7.

Definition 1. *Let $k > 0$. A k-test vector is a 4-tuple $Z = \langle I, F, T, C \rangle$ where*

  – *$I \subseteq \Sigma^{k-1}$ is a set of allowed* prefixes,

- $F \subseteq \Sigma^{k-1}$ is a set of allowed suffixes,
- $T \subseteq \Sigma^k$ is a set of allowed segments, and
- $C \subseteq \Sigma^{<k}$ is a set of allowed short strings satisfying $I \cap F = C \cap \Sigma^{k-1}$.

We write $\mathcal{T}_k$ for the set of k-test vectors.

Note that the set $\mathcal{T}_k$ of k-test vectors is finite. We equip set $\mathcal{T}_k$ with a partial order structure as follows.

Definition 2. Let $k > 0$. The relation $\sqsubseteq$ on $\mathcal{T}_k$ is given by

$$\langle I, F, T, C \rangle \sqsubseteq \langle I', F', T', C' \rangle \Leftrightarrow I \subseteq I' \text{ and } F \subseteq F' \text{ and } T \subseteq T' \text{ and } C \subseteq C'.$$

With respect to this ordering, $\mathcal{T}_k$ has a least element $\bot = \langle \emptyset, \emptyset, \emptyset, \emptyset \rangle$ and a greatest element $\top = \langle \Sigma^{k-1}, \Sigma^{k-1}, \Sigma^k, \Sigma^{<k} \rangle$. The union, intersection and symmetric difference of two k-test vectors $Z = \langle I, F, T, C \rangle$ and $Z' = \langle I', F', T', C' \rangle$ are given by, respectively,

$$Z \sqcup Z' = \langle I \cup I', F \cup F', T \cup T', C \cup C' \cup (I \cap F') \cup (I' \cap F) \rangle,$$
$$Z \sqcap Z' = \langle I \cap I', F \cap F', T \cap T', C \cap C' \rangle,$$
$$Z \triangle Z' = \langle I \triangle I', F \triangle F', T \triangle T', C \triangle C' \triangle (I' \cap F) \triangle (I \cap F') \rangle.$$

The reader may check that $Z \sqcup Z'$, $Z \sqcap Z'$ and $Z \triangle Z'$ are k-test vectors indeed, preserving the property $I \cap F = C \cap \Sigma^{k-1}$. The reader may also check that $(\mathcal{T}_k, \sqsubseteq)$ is a lattice with $Z \sqcup Z'$ the least upper bound of Z and Z', and $Z \sqcap Z'$ the greatest lower bound of Z and Z'. The symmetric difference operation $\triangle$ will be used further on to define a metric on k-test vectors.

We can associate a k-test vector $\alpha_k(L)$ to each language L by taking all prefixes of length $k - 1$ of the strings in L, all suffixes of length $k - 1$ of the strings in L, and all substrings of length k of the strings in L. Any string which is both an allowed prefix and an allowed suffix is also a short string, as well as any string in L with length less than $k - 1$.

Definition 3. Let $L \subseteq \Sigma^*$ be a language and $k \in \mathbb{N}$. Then $\alpha_k(L)$ is the k-test vector $\langle I_k(L), F_k(L), T_k(L), C_k(L) \rangle$ where

- $I_k(L) = \{u \in \Sigma^{k-1} \mid \exists v \in \Sigma^* : uv \in L\}$,
- $F_k(L) = \{w \in \Sigma^{k-1} \mid \exists v \in \Sigma^* : vw \in L\}$,
- $T_k(L) = \{v \in \Sigma^k \mid \exists u, w \in \Sigma^* : uvw \in L\}$, and
- $C_k(L) = (L \cap \Sigma^{<k-1}) \cup (I_k(L) \cap F_k(L))$.

It is easy to see that operation $\alpha_k : 2^{\Sigma^*} \to \mathcal{T}_k$ is monotone.

Proposition 4. For all languages L, L' and for all $k > 0$,

$$L \subseteq L' \Rightarrow \alpha_k(L) \sqsubseteq \alpha_k(L').$$

Conversely, we associate a language $\gamma_k(Z)$ to each k-test vector $Z = \langle I, F, T, C \rangle$, consisting of all the short strings from C together with all strings of length at least $k - 1$ whose prefix of length $k - 1$ is in I, whose suffix of length $k - 1$ is in F, and where all substrings of length k belong to T.

Definition 5. *Let $Z = \langle I, F, T, C \rangle$ be a k-test vector, for some $k > 0$. Then*

$$\gamma_k(Z) = C \cup ((I\Sigma^* \cap \Sigma^* F) \setminus (\Sigma^*(\Sigma^k \setminus T)\Sigma^*)).$$

We say that a language L is k-testable in the strict sense (k-TSS) if there exists a k-test vector Z such that $L = \gamma_k(Z)$. Note that all k-TSS languages are regular.

Again, it is easy to see that operation $\gamma_k : \mathcal{T}_k \to 2^{\Sigma^*}$ is monotone.

Proposition 6. *For all $k > 0$ and for all k-test vectors Z and Z',*

$$Z \sqsubseteq Z' \Rightarrow \gamma_k(Z) \subseteq \gamma_k(Z').$$

The next theorem, which is our main result about k-testable languages, asserts that α_k and γ_k form a (monotone) Galois connection [11] between lattices $(\mathcal{T}_k, \sqsubseteq)$ and $(2^{\Sigma^*}, \subseteq)$.

Theorem 7 (Galois connection). *Let $k > 0$, let $L \subseteq \Sigma^*$ be a language, and let Z be a k-test vector. Then $\alpha_k(L) \sqsubseteq Z \Leftrightarrow L \subseteq \gamma_k(Z)$.*

The above theorem generalizes results on strictly k-testable languages from [4,16]. Composition $\gamma_k \circ \alpha_k$ is commonly called the *associated closure operator*, and composition $\alpha_k \circ \gamma_k$ is known as the *associated kernel operator*. The fact that we have a Galois connection has some well-known consequences for these associated operators.

Corollary 8. *For all $k > 0$, $\gamma_k \circ \alpha_k$ and $\alpha_k \circ \gamma_k$ are monotone and idempotent.*

Monotony of $\gamma_k \circ \alpha_k$ was established previously as Theorem 3.2 in [4] and as Lemma 3.3 in [16].

Corollary 9. *For all $k > 0$, $L \subseteq \Sigma^*$ and $Z \in \mathcal{T}_k$,*

$$\alpha_k \circ \gamma_k(Z) \sqsubseteq Z \tag{1}$$

$$L \subseteq \gamma_k \circ \alpha_k(L) \tag{2}$$

Inequality (1) asserts that the associated kernel operator $\alpha_k \circ \gamma_k$ is *deflationary*, while inequality (2) says that the associated closure operator $\gamma_k \circ \alpha_k$ is *inflationary* (or *extensive*). Inequality (2) was established previously as Lemma 3.1 in [4] and (also) as Lemma 3.1 in [16].

Another immediate corollary of the Galois connection is that in fact $\gamma_k \circ \alpha_k(L)$ is the smallest k-TSS language that contains L. This has been established previously as Theorem 3.1 in [4].

Corollary 10. *For all $k > 0$, $L \subseteq \Sigma^*$, and $Z \in \mathcal{T}_k$,*

$$L \subseteq \gamma_k(Z) \Rightarrow \gamma_k \circ \alpha_k(L) \subseteq \gamma_k(Z).$$

As a final corollary, we mention that $\alpha_k \circ \gamma_k(Z)$ is the smallest k-test vector that denotes the same language as Z. This is essentially Lemma 1 of [16].

Corollary 11. *For all $k > 0$ and $Z \in \mathcal{T}_k$, $\gamma_k \circ \alpha_k \circ \gamma_k(Z) = \gamma_k(Z)$. Moreover, for any $Z' \in \mathcal{T}_k$,*

$$\gamma_k(Z) = \gamma_k(Z') \Rightarrow \alpha_k \circ \gamma_k(Z) \sqsubseteq Z'.$$

We can provide a simple characterization of $\alpha_k \circ \gamma_k(Z)$ as the *canonical k-test vector* obtained by removing all the allowed prefixes, suffixes and segments that do not occur in the k-testable language generated by Z.

Definition 12. *Let $Z = \langle I, F, T, C \rangle$ be a k-test vector, for some $k > 0$. We say that $u \in I$ is a* junk prefix *of Z if u does not occur as a prefix of any string in $\gamma_k(Z)$. Similarly, we say that $u \in F$ is a* junk suffix *of Z if u does not occur as a suffix of any string in $\gamma_k(Z)$, and we say that $u \in T$ is a* junk segment *of Z if u does not occur as a substring of any string in $\gamma_k(Z)$. We call Z* canonical *if it does not contain any junk prefixes, junk suffixes, or junk segments.*

Proposition 13. *Let Z be a k-test vector, for some $k > 0$, and let Z' be the canonical k-test vector obtained from Z by deleting all junk prefixes, junk suffixes, and junk segments. Then $\alpha_k \circ \gamma_k(Z) = Z'$.*

Proposition 13 implies that if we restrict the lattice $(\mathcal{T}_k, \sqsubseteq)$ to the canonical k-test vectors, our Galois connection becomes a Galois insertion.

2.3 Learning k-TSS Languages

It is well-known that any k-TSS language can be identified in the limit from positive examples [4,5]. Below we recall the basic argument; we refer to [4,5,16] for efficient algorithms.

Theorem 14. *Any k-TSS language can be identified in the limit from positive examples.*

Proof. Let L be a k-TSS language and let $w_1, w_2, w_3, \ldots$ be an enumeration of L. Let $L_0 = \emptyset$ and $L_i = L_{i-1} \cup \{w_i\}$, for $i > 0$. We then have

$$L_1 \subseteq L_2 \subseteq L_3 \subseteq \cdots$$

By monotonicity of α_k (Proposition 4) we obtain

$$\alpha_k(L_1) \sqsubseteq \alpha_k(L_2) \sqsubseteq \alpha_k(L_3) \sqsubseteq \cdots \tag{3}$$

and by monotonicity of γ_k (Proposition 6)

$$\gamma_k \circ \alpha_k(L_1) \subseteq \gamma_k \circ \alpha_k(L_2) \subseteq \gamma_k \circ \alpha_k(L_3) \subseteq \cdots \tag{4}$$

Since $\gamma_k \circ \alpha_k$ is inflationary (Corollary 9), L is a k-TSS language and, for each i, $\gamma_k \circ \alpha_k(L_i)$ is the smallest k-TSS language that contains L_i (Corollary 10), we have

$$L_i \subseteq \gamma_k \circ \alpha_k(L_i) \subseteq L \tag{5}$$

Because $(\mathcal{T}_k, \sqsubseteq)$ is a finite partial order it does not have an infinite ascending chain. This means that sequence (3) converges. But then sequence (4) also converges, that is, there exists an n such that, for all $m \geq n$, $\gamma_k \circ \alpha_k(L_m) = \gamma_k \circ \alpha_k(L_n)$. By Eqs. (4) and (5) we obtain, for all i,

$$L_i \subseteq \gamma_k \circ \alpha_k(L_i) \subseteq \gamma_k \circ \alpha_k(L_n) \subseteq L$$

This implies $L = \gamma_k \circ \alpha_k(L_n)$, meaning that the sequence (4) of k-TSS languages converges to L.

3   Learning Unions of k-TSS Languages

In this section, we present guarantees concerning learnability in the limit of unions of k-TSS languages. Then, we present an algorithm merging closest and compatible k-TSS languages.

3.1   Generalities

It is well-known that the class of k-testable languages in the strict sense is not closed under union. Take for instance the two 3-testable languages, represented by their DFA's in Fig. 1a, that are generated by the following 3-test vectors:

$$Z = \langle \{aa\}, \{aa\}, \{aaa\}, \{aa\} \rangle$$
$$Z' = \langle \{ba, bb\}, \{ab, bb\}, \{baa, bab, aaa, aab\}, \{bb\} \rangle$$

with $\Sigma = \{a, b\}$. The union $\gamma_3(Z) \cup \gamma_3(Z')$ of these languages, represented by its DFA in Fig. 1a, is not a 3-testable language. Indeed, it is not a k-testable language for any value of $k > 0$. For $k = 1$, the only k-testable language that extends $\gamma_3(Z) \cup \gamma_3(Z')$ is Σ^*. For $k \geq 2$, the problem is that since a^{k-1} is an allowed prefix, $a^{k-1}b$ is an allowed segment, and $a^{k-2}b$ is an allowed suffix, $a^{k-1}b$ has to be in the language, even though it is not an element of $\gamma_3(Z) \cup \gamma_3(Z')$.

It turns out that we can generalize Theorem 14 to unions of k-TSS languages.

Theorem 15. *Any language that is a union of k-TSS languages can be identified in the limit from positive examples.*

Proof. Let $L = L_1 \cup \cdots \cup L_l$, where all the L_p are k-TSS languages, and let $w_1, w_2, w_3, \ldots$ be an enumeration of L. Define, for $i > 0$,

$$K_i = \bigcup_{j=1}^{i} \gamma_k \circ \alpha_k(\{w_j\}).$$

Since each w_j is included in a k-TSS language contained in L, and $\gamma_k \circ \alpha_k(\{w_j\})$ is the smallest k-TSS language that contains w_j, we conclude that, for all j, $\gamma_k \circ \alpha_k(\{w_j\}) \subseteq L$, which in turn implies $K_i \subseteq L$. Since there are only finitely many k-test vectors and finitely many k-TSS languages, the sequence

$$K_1 \subseteq K_2 \subseteq K_3 \subseteq \cdots \tag{6}$$

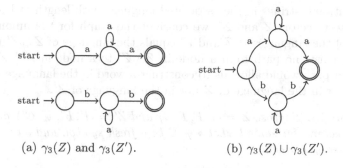

(a) $\gamma_3(Z)$ and $\gamma_3(Z')$.   (b) $\gamma_3(Z) \cup \gamma_3(Z')$.

Fig. 1. k-testable languages are not closed under union.

converges, that is there exists an n such that, for all $m \geq n$, $K_m = K_n$. This implies that all w_j are included in K_n, that is $L \subseteq K_n$. In combination with the above observation that all K_i are contained in L, this proves that sequence (6) converges to L.

The proof of Theorem 15 provides us with a simple first algorithm to learn unions of k-TSS languages: for each example word that we see, we compute the k-test vector and then we take the union of the languages denoted by all those k-test vectors. The problem with this algorithm is that potentially we end up with a huge number of different k-test vectors. Thus we would like to cluster as many k-test vectors in the union as we can, without changing the overall language. Before we can introduce our clustering algorithm, we first need to define a metric on k-test vectors.

Definition 16. *The* cardinality *of a k-test vector $Z = \langle I, F, T, C \rangle$ is defined by:*

$$|Z| = |I| + |F| + |T| + |C \cap \Sigma^{<k-1}|.$$

Intuitively, the distance between two k-test vectors is the number of prefixes, suffixes, substrings and short words that must be added/removed to transform one k-test vector into the other. For examples, see Fig. 2b.

Definition 17. *The function $d: \mathcal{T}_k \times \mathcal{T}_k \mapsto \mathbb{R}^+$, which defines the* distance *between a pair of k-test vectors, is given by: $d(Z, Z') = |Z \triangle Z'|$.*

The next proposition provides a necessary and sufficient condition for when the γ_k operator preserves least upper bounds, that is, when the union of the languages of two k-test vectors equals the language of the union of these vectors. The basic idea is that, for each k-test vector, we may construct a directed graph in which the segments are the nodes. The graph contains an edge from segment u to segment v if, when the content of the sliding window is u at some point, it may become v when the sliding window advances one step. There exists a 1-to-1 correspondence between paths in this graph from an initial segment to a final segment, and strings in the associated language with length at least $k - 1$. Given two test vectors Z and Z', we consider the graph for the union $Z \sqcup Z'$. The union of the languages of Z and Z' equals the language of $Z \sqcup Z'$ iff in this graph there exists no path from a node in $Z \setminus Z'$ to a node in $Z' \setminus Z$, or vice versa. Such a path would allow us to construct a word in the language of $Z \sqcup Z'$ that is neither in the language of Z nor in the language of Z'.

Proposition 18. *Suppose $Z = \langle I, F, T, C \rangle$ and $Z' = \langle I', F', T', C' \rangle$ are canonical k-test vectors, for some k. Let $\bullet \notin \Sigma$ be a fresh symbol, and let $G = (V, E)$ be the directed graph with*

$$V = \{\bullet u \mid u \in I \cup I'\} \cup T \cup T' \cup \{u\bullet \mid u \in F \cup F'\},$$
$$E = \{(au, ub) \in V \times V \mid a, b \in \Sigma \cup \{\bullet\}, u \in \Sigma^{k-1}\}.$$

Suppose each vertex in V is colored either red, blue or white. Vertices in $T \setminus T'$ are red, vertices in $T' \setminus T$ are blue, and vertices in $T \cap T'$ are white. A vertex $\bullet u$ is red if $u \in I \setminus I'$, blue if $u \in I' \setminus I$, and white if $u \in I \cap I'$. A vertex $u\bullet$ is red if $u \in F \setminus F'$, blue if $u \in F' \setminus F$, and white if $u \in F \cap F'$. Then $\gamma_k(Z \sqcup Z') = \gamma_k(Z) \cup \gamma_k(Z')$ iff there exists no path in G from a red vertex to a blue vertex, nor from a blue vertex to a red vertex.

Suppose alphabet Σ contains n elements. Then the size of graph G from Proposition 18 is in $O(n \cdot |Z \cup Z'|)$, and we can construct G from Z and Z' in time $O((n+k) \cdot |Z \cup Z'|)$. Since the reachability property in Proposition 18 can be decided in a time that is linear in the size of G, we obtain an $O((n+k) \cdot |Z \cup Z'|)$-time algorithm for deciding $\gamma_k(Z \sqcup Z') = \gamma_k(Z) \cup \gamma_k(Z')$.

3.2   Efficient Algorithm

Our algorithm to learn unions of k-testable languages is based on hierarchical clustering. Given a set $\mathcal{S}$ of n words, we compute its related set of k-test vectors $S = \{\alpha_k(\{x\}) \mid x \in \mathcal{S}\}$. Note that the k-test vectors are canonical. Then, an $n \times n$ distance matrix is computed. To that end, the distance used is the pairwise distance between k-test vectors defined in Definition 17. Next, the algorithm finds the closest pair of compatible k-test vectors Z and Z', such that $\gamma_k(Z \sqcup Z') = \gamma_k(Z) \cup \gamma_k(Z')$ and computes their union. An efficient implementation for finding closest k-test vectors is the nearest-neighbor chain algorithm [1], which finds pairs of k-test vectors such that these two closest k-test vectors are the nearest neighbors of each other. The distance between the merged k-test vectors and the remaining k-test vectors in S is updated. These two operations are repeated until all initial k-test vectors have been merged into one, or that no allowed union of two k-test vectors such that $\gamma_k(Z \sqcup Z') = \gamma_k(Z) \cup \gamma_k(Z')$ is possible. We gather at the end of the process a linkage between k-test vectors, which can lead to the computation of a dendrogram. When the number of k-test vectors to learn is known, one can use this expected number of languages to find the threshold that would, given the hierarchical clustering, return the desired unions of k-test vectors.

Example 19. Let $k = 3$. Given the sample of strings S in Fig. 2a, compute the associate sample of 3-test vectors $S = \{Z_1, Z_2, \ldots, Z_8\}$. Then, compute its distance matrix (Fig. 2b) using the metric defined in Definition 17. Using classical linkage algorithms (for instance nearest-neighbor chain algorithm), compute the related linkage matrix depicted in Fig. 2c. We gather the dendrogram shown in Fig. 2d, where the 3 remaining 3-test vectors $Z_1 \sqcup Z_8$, $Z_2 \sqcup Z_5 \sqcup Z_7$ and $Z_3 \sqcup Z_4 \sqcup Z_6$ cannot be merged. Indeed:

- $\gamma_k(Z_1 \sqcup Z_8 \sqcup Z_2 \sqcup Z_5 \sqcup Z_7) \neq \gamma_k(Z_1 \sqcup Z_8) \cup \gamma_k(Z_2 \sqcup Z_5 \sqcup Z_7)$.
- $\gamma_k(Z_1 \sqcup Z_8 \sqcup Z_3 \sqcup Z_4 \sqcup Z_6) \neq \gamma_k(Z_1 \sqcup Z_8) \cup \gamma_k(Z_3 \sqcup Z_4 \sqcup Z_6)$.
- $\gamma_k(Z_2 \sqcup Z_5 \sqcup Z_7 \sqcup Z_3 \sqcup Z_4 \sqcup Z_6) \neq \gamma_k(Z_2 \sqcup Z_5 \sqcup Z_7) \cup \gamma_k(Z_3 \sqcup Z_4 \sqcup Z_6)$.

With a desired number of 3-TSS languages to learn of 3, the returned languages are $\gamma_k(Z_1 \sqcup Z_8)$ and $\gamma_k(Z_2 \sqcup Z_5 \sqcup Z_7)$ and $\gamma_k(Z_3 \sqcup Z_4 \sqcup Z_6)$. With a desired number of 3-TSS languages to learn of 4, the returned languages would be $\gamma_k(Z_1 \sqcup Z_8)$ and $\gamma_k(Z_2 \sqcup Z_5 \sqcup Z_7)$ and $\gamma_k(Z_3)$ and $\gamma_k(Z_4 \sqcup Z_6)$ instead.

S	S
baba	$Z_1 = \langle\{ba\}, \{ba\}, \{bab, aba\}, \{\}\rangle$
abba	$Z_2 = \langle\{ab\}, \{ba\}, \{abb, bba\}, \{\}\rangle$
abcabc	$Z_3 = \langle\{ab\}, \{bc\}, \{abc, bca, cab\}, \{\}\rangle$
cbacba	$Z_4 = \langle\{cb\}, \{ba\}, \{cba, bac, acb\}, \{\}\rangle$
abbbba	$Z_5 = \langle\{ab\}, \{ab\}, \{abb, bbb, bba\}, \{\}\rangle$
cbacbacba	$Z_6 = \langle\{cb\}, \{ba\}, \{cba, bac, acb\}, \{\}\rangle$
abbba	$Z_7 = \langle\{ab\}, \{ba\}, \{abb, bbb, bba\}, \{\}\rangle$
babababc	$Z_8 = \langle\{ba\}, \{bc\}, \{bab, aba, abc\}, \{\}\rangle$

(a) Dataset and corresponding 3-test vectors.

	Z_1	Z_2	Z_3	Z_4	Z_5	Z_6	Z_7	Z_8
Z_1	0	6	9	7	7	7	7	3
Z_2	6	0	7	7	1	7	1	9
Z_3	9	7	0	10	8	10	8	6
Z_4	7	7	10	0	8	0	8	10
Z_5	7	1	8	8	0	8	0	10
Z_6	7	7	10	0	8	0	8	10
Z_7	7	1	8	8	0	8	0	10
Z_8	3	9	6	10	10	10	10	0

(b) Distance matrix.

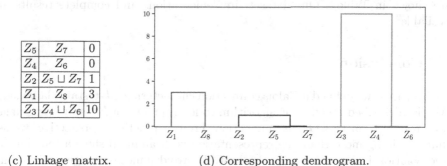

Z_5	Z_7	0
Z_4	Z_6	0
Z_2	$Z_5 \sqcup Z_7$	1
Z_1	Z_8	3
Z_3	$Z_4 \sqcup Z_6$	10

(c) Linkage matrix.

(d) Corresponding dendrogram.

Fig. 2. Learning unions of k-test vectors.

We can see here that the lower bound on the number of returned languages is the number of unions of k-test vectors satisfying the compatibility constraint $\gamma_k(Z \sqcup Z') = \gamma_k(Z) \cup \gamma_k(Z')$. However, in case this constraint is relaxed, it is possible to obtain a clustering into less parts, up to a single cluster standing for $\gamma_k(\bigsqcup_{Z \in S})$.

4   Case Study

Job Dataset. Our case study has been inspired by an industrial problem related to the domain of cyber-physical systems. Recent work [15] focused on the impact of design parameters of a flexible manufacturing system on its productivity. It appeared in the aforementioned study that the productivity depends on the jobs being rendered. To that end, the prior identification of the different job patterns is crucial to enabling engineers to optimize parameters related to the flexible manufacturing system.

Table 1. Sample of identified job patterns.

Job	Pattern	3-test vector	Type of job
aaaaa	a^+	$Z = \langle \{aa\}, \{aa\}, \{aaa\}, \{aa\} \rangle$	Homogeneous
aaaaaaaaaa			
aaaaa . . . aaa			
abababab	$(ab)^+$	$Z = \langle \{ab\}, \{ab\}, \{aba, bab\}, \{ab\} \rangle$	Heterogeneous
ababababab			
abcabcabc	$(abc)^+$	$Z = \langle \{ab\}, \{bc\}, \{abc, bca, cab\}, \emptyset \rangle$	
abcabcabcabc			
abcbcbcbca	$a(bc)^+a$	$Z = \langle \{ab\}, \{ca\}, \{abc, bcb, cbc, cba\}, \emptyset \rangle$	Miscellaneous

We consider a dataset containing strings, each representing a job. Our job patterns are also represented by 3-testable languages, the 3-test vectors of which are shown in Table 1. Our dataset, implementations and complete results are available[2].

5   Conclusion

In this paper, we defined a Galois connection characterizing k-testable languages. We also described an efficient algorithm to learn unions of k-testable languages that results from this Galois connection. From a practical perspective, we see that obtaining more than one representation is meaningful since a too generalized solution is not necessarily the best. To avoid unnecessary generalizations, the union of two k-testable languages that would not be a k-testable language is not allowed. Note also that depending on the applications, expert knowledge can provide an indication on the number of languages the returned union should contain. In further work, we would like to extend the learning of unions of languages to regular languages. An attempt to learn pairwise disjoint regular languages has been made in [8,9]. However, no learnability guarantee has been provided so far.

---

[2] See https://gitlab.science.ru.nl/alinard/learning-union-ktss.

References

1. Benzécri, J.P.: Construction d'une classification ascendante hiérarchique par la recherche en chaîne des voisins réciproques. Les cahiers de l'analyse des données **7**(2), 209–218 (1982)
2. Bex, G.J., Neven, F., Schwentick, T., Tuyls, K.: Inference of concise DTDs from XML data. In: Proceedings of the 32nd International Conference on Very Large Data Bases, pp. 115–126 (2006)
3. Coste, F.: Learning the language of biological sequences. In: Heinz, J., Sempere, J.M. (eds.) Topics in Grammatical Inference, pp. 215–247. Springer, Heidelberg (2016). https://doi.org/10.1007/978-3-662-48395-4_8
4. García, P., Vidal, E.: Inference of k-testable languages in the strict sense and application to syntactic pattern recognition. IEEE Trans. Pattern Anal. Mach. Intell. **12**(9), 920–925 (1990)
5. Garcia, P., Vidal, E., Oncina, J.: Learning locally testable languages in the strict sense. In: First International Workshop Algorithmic Learning Theory (ALT), pp. 325–338 (1990)
6. Gold, M.: Language identification in the limit. Inf. Control **10**(5), 447–474 (1967)
7. de la Higuera, C.: Grammatical Inference: Learning Automata and Grammars. Cambridge University Press, Cambridge (2010)
8. Linard, A.: Learning several languages from labeled strings: state merging and evolutionary approaches. arXiv preprint arXiv:1806.01630 (2018)
9. Linard, A., Smetsers, R., Vaandrager, F., Waqas, U., van Pinxten, J., Verwer, S.: Learning pairwise disjoint simple languages from positive examples. arXiv preprint arXiv:1706.01663 (2017)
10. McNaughton, R., Papert, S.A.: Counter-Free Automata (M.I.T. Research Monograph No. 65). The MIT Press (1971)
11. Nielson, F., Nielson, H., Hankin, C.: Principles of Program Analysis. Springer, Heidelberg (1999). https://doi.org/10.1007/978-3-662-03811-6
12. Rogers, J., Pullum, G.K.: Aural pattern recognition experiments and the subregular hierarchy. J. Log. Lang. Inf. **20**(3), 329–342 (2011)
13. Tantini, F., Terlutte, A., Torre, F.: Sequences classification by least general generalisations. In: Sempere, J.M., García, P. (eds.) ICGI 2010. LNCS (LNAI), vol. 6339, pp. 189–202. Springer, Heidelberg (2010). https://doi.org/10.1007/978-3-642-15488-1_16
14. Torres, I., Varona, A.: k-TSS language models in speech recognition systems. Comput. Speech Lang. **15**(2), 127–148 (2001)
15. Umar, W., et al.: A fast estimator of performance with respect to the design parameters of self re-entrant flowshops. In: Euromicro Conference on Digital System Design, pp. 215–221 (2016)
16. Yokomori, T., Kobayashi, S.: Learning local languages and their application to dna sequence analysis. IEEE Trans. Pattern Anal. Mach. Intell. **20**(10), 1067–1079 (1998)

Graphs, Trees and Rewriting

Regular Matching and Inclusion on Compressed Tree Patterns with Context Variables

Iovka Boneva[1], Joachim Niehren[2], and Momar Sakho[2(✉)]

[1] Université de Lille, Lille, France
[2] Inria, Lille, France
momar.sakho@inria.fr

Abstract. We study the complexity of regular matching and inclusion for compressed tree patterns extended by context variables. The addition of context variables to tree patterns permits us to properly capture compressed string patterns but also compressed patterns for unranked trees with tree and hedge variables. Regular inclusion for the latter is relevant to certain query answering on XML streams with references.

Keywords: Computational complexity · Patterns · Trees ·
Tree languages and tree automata

1 Introduction

A pattern is a term with variables describing a string, a tree, or some other algebraic value. The following generic problems for patterns were widely studied:

Pattern matching: Is a given algebraic value an instance of a given pattern?
Pattern unification: Do two given patterns have some common instance?
Regular pattern matching: Does some instance of a given pattern belong to a given regular language?
Regular pattern inclusion: Do all instances of a given pattern belong to a given regular language?

As inputs, these problems receive descriptors of patterns, values, and regular languages. Most typically, a string pattern may be described in a compressed manner by using a singleton context-free grammar (also called straight-line program), and a regular string language may be represented by a nondeterministic finite automaton (NFA) or by a deterministic finite automaton (DFA). The problem of string pattern matching is well-known to be NP-complete for NFAs [1] but in P for DFAs, with and without compression [4]. The more general problem of string unification is known to be PSPACE-complete [8].

Compressed patterns were called hyperstreams in [6]. Regular inclusion on compressed patterns is the problem of certain query answering on hyperstreams for queries defined by automata. This application motivated the study of regular

© Springer Nature Switzerland AG 2019
C. Martín-Vide et al. (Eds.): LATA 2019, LNCS 11417, pp. 343–355, 2019.
https://doi.org/10.1007/978-3-030-13435-8_25

	DFAs	NFAs
Regular Matching	PSPACE-c	PSPACE-c
Regular Inclusion	PSPACE-c	PSPACE-c

Fig. 1. (Compressed) string patterns.

	DFAs	NFAs
Regular Matching	P	P
Regular Inclusion	P	PSPACE-c

Fig. 2. Linear restriction.

	DTAs	NTAs
Regular Matching	NP-c	EXP-c
Regular Inclusion	coNP-c	EXP-c

Fig. 3. (Compressed) tree patterns.

	DTAs	NTAs
Regular Matching	P	P
Regular Inclusion	P	EXP-c

Fig. 4. Linear restriction.

	DTAs	NTAs
Regular Matching	EXP-c	EXP-c
Regular Inclusion	EXP-c	EXP-c

Fig. 5. Adding context variables.

	DTAs	NTAs
Regular Matching	P	P
Regular Inclusion	P	EXP-c

Fig. 6. Linear restriction.

inclusion and matching in [2]. For string patterns, both problems were shown to be PSPACE-complete, for DFAs and NFAs, with and without compression. See Fig. 1 for an overview. When restricted to linear string patterns, the complexity goes down to polynomial time in 3 of the 4 cases, as summarized in Fig. 2. The problem which remains PSPACE-complete is regular inclusion on linear string patterns for NFAs.

The complexity landscapes of regular matching and inclusion for tree patterns look quite different to the case of string patterns, see Figs. 3 and 4. Here, regular languages are defined by tree automata, which may either be nondeterministic (NTAs) or (bottom-up) deterministic (DTAs), while compressed descriptions of tree patterns can be obtained by singleton tree grammars. Regular matching for tree patterns against NTAs is EXP-complete with and without compression. For DTAs, however, regular matching is NP-complete and regular inclusion coNP-complete. For linear tree patterns, three of the four problems are in P except for the case of regular inclusion against NTAs. In [3], regular matching for tree patterns with neither compression nor context variables is studied as the *ground instance intersection problem*. Recently [9] studied the problem of matching compressed terms represented as singleton tree grammars, which is incomparable with regular matching that we study here.

The prime reason for the asymmetry of the complexity landscapes in the case of strings and trees is that string patterns cannot be encoded as tree patterns with a monadic signature without adding context variables. For instance, the string pattern $aZZbY$ corresponds to the tree pattern $a(Z(Z(b(Y))))$ with context variable Z and tree variable Y. The interest of adding context variables to tree patterns was already noticed when generalizing string pattern matching to context pattern matching [4], which are both NP-complete, with or without compression. The same was noticed when generalizing string unification to context unification, that are both in PSPACE [5]. Since we are interested in a proper generalization of regular matching and inclusion from string to tree patterns, we propose to study these problems for tree patterns with context variables.

The main contributions of the present paper are the complexity classes of regular matching and inclusion for compressed tree patterns with context variables, which are summarized in Figs. 5 and 6. The results are fully symmetric to those for compressed string patterns, except that PSPACE-completeness is replaced by EXP-completeness. The main reason for this change is that the central problem of context inhabitation is EXP-complete for tree automata, while the inhabitation problem is PSPACE-complete for word automata.

Finally, we show that regular pattern matching and inclusion have the same complexity for (compressed) patterns on unranked trees with tree and hedge variables, mainly since such patterns can be encoded into (compressed ranked) tree patterns with context variables. Compressed patterns for unranked trees capture XML streams with references [7]. They permit to generalize the notion of hyperstreams in [2] from strings to unranked trees.

Outline. We introduce tree patterns with context variables in Sect. 2. The inhabitation problem for Σ-algebras is defined in Sect. 3. The complexity of context inhabitation for tree automata is discussed in Sect. 4. Compressed tree patterns with context variables are introduced in Sect. 5 and then studied for regular matching and inclusion in Sect. 6. Due to space limitation, the discussion of the special cases of linear patterns and patterns with context variables, as well as the missing proofs, are not included in this extended abstract.

2   Tree Patterns with Context Variables

We consider the set of types $\mathbf{T} = \{e, c\}$ with the type e for trees and the type $c = e \multimap e$ for contexts. The latter linear type is inspired by linear logic and is different from the usual (nonlinear) function type $e \to e$. We assume sets $\mathcal{V}^e$ of tree variables and $\mathcal{V}^c$ of context variables. The tree variables are ranged over by x, y, z and the context variables by X, Y. The set of all variables is $\mathcal{V} = \mathcal{V}^e \cup \mathcal{V}^c$.

We fix a finite ranked signature $\Sigma = \uplus_{n \geq 0} \Sigma^{(n)}$ of function symbols $f \in \Sigma^{(n)}$ of arity n. We assume that Σ contains at least one constant and one symbol of arity at least 2. The set of trees $\mathcal{T}_\Sigma$ is the least set that contains all elements $f(t_1, \ldots, t_n)$ where $f \in \Sigma^{(n)}$ for some $n \geq 0$ and $t_1, \ldots, t_n \in \mathcal{T}_\Sigma$. Atomic trees $a() \in \mathcal{T}_\Sigma$ are deliberately identified with $a \in \Sigma^{(0)}$. The set of contexts $C \in \mathcal{C}_\Sigma$ is the set of all terms $\lambda x.p$ such that $p \in \mathcal{T}_{\Sigma \uplus \{x\}}$ for some tree variable $x \in \mathcal{V}^e$ that occurs exactly once in p. The set of all values of both types is $Val_\Sigma = \mathcal{T}_\Sigma \cup \mathcal{C}_\Sigma$.

The sets of all *tree patterns* $\mathcal{P}_\Sigma^e$ and of all *context patterns* $\mathcal{P}_\Sigma^c$ are defined in Fig. 7. Note that both types of patterns may contain context variables. The set of all *patterns* is $\mathcal{P}_\Sigma = \mathcal{P}_\Sigma^e \cup \mathcal{P}_\Sigma^c$. For a (tree or context) pattern π, its sets of free variables $fv(\pi)$ and of bound variables $bv(\pi)$ can be defined as usual. The set $\mathcal{G}_\Sigma^\tau$ of ground patterns of type $\tau \in \mathbf{T}$ is the subset of patterns in $\mathcal{P}_\Sigma^\tau$ without free variables. The set of all ground patterns is denoted by $\mathcal{G}_\Sigma = \mathcal{G}_\Sigma^e \cup \mathcal{G}_\Sigma^c$. Clearly, any *tree* $t \in \mathcal{T}_\Sigma$ is a ground pattern of type e and any *context* $C \in \mathcal{C}_\Sigma$ is a

ground pattern of type c. A pattern is called *linear* if each of its free variables
has at most one free occurrence.

Tree patterns $p, p_1, \ldots, p_n \in \mathcal{P}_\Sigma^e ::= x \mid f(p_1, \ldots, p_n) \mid P@p$
Context patterns $P \in \mathcal{P}_\Sigma^c ::= X \mid \lambda x.p$ where x occurs exactly once in p

Fig. 7. Tree patterns and context patterns, with $x \in \mathcal{V}^e$, $X \in \mathcal{V}^c$, $n \geq 0$, $f \in \Sigma^{(n)}$.

We can apply β-reduction to both kinds of patterns. Each β-reduction step
replaces some redex of the form $(\lambda x.p)@p'$ in a bigger pattern by $p[x/p']$ if
$x \notin bv(p)$ and otherwise renames x apart before. Any ground tree pattern $p \in \mathcal{G}_\Sigma^e$
can be β-reduced in a linear number of steps to some tree in polynomial time,
since all λ-binders are assumed to be linear. The semantics $\llbracket p \rrbracket$ of a ground
pattern p is the tree obtained from p by exhaustive β-reduction. Similarly, any
ground context pattern $P \in \mathcal{G}_\Sigma^c$ can be β-reduced in a linear number of steps to
a unique context $\lambda x.p \in \mathcal{C}_\Sigma$. The semantics $\llbracket P \rrbracket$ is the function $\llbracket P \rrbracket : \mathcal{T}_\Sigma \to \mathcal{T}_\Sigma$
such that $\llbracket P \rrbracket(t) = p[x/t]$ for any tree t. Note that $\llbracket \mathcal{G}_\Sigma^e \rrbracket = \llbracket \mathcal{T}_\Sigma \rrbracket$ is equal to $\mathcal{T}_\Sigma$
while $\llbracket \mathcal{G}_\Sigma^c \rrbracket = \llbracket \mathcal{C}_\Sigma \rrbracket$ is a proper subset of the set of functions of type $\mathcal{T}_\Sigma \to \mathcal{T}_\Sigma$.

A substitution $\sigma : V \to \mathcal{G}_\Sigma$ where $V \subseteq \mathcal{V}$ is called well-typed if it maps
tree variables to $\mathcal{G}_\Sigma^e$ and context variables to $\mathcal{G}_\Sigma^c$. For any pattern $p \in \mathcal{P}_\Sigma^e$, the
grounding $\sigma(p) \in \mathcal{G}_\Sigma^e$ is obtained by applying σ to the free variables in p. The
set of all instances of p is obtained by β-normalizing all groundings:

$$Inst(p) = \{ \llbracket \sigma(p) \rrbracket \mid \sigma : fv(p) \to \mathcal{G}_\Sigma \text{ well-typed} \}.$$

Clearly, $Inst(p) \subseteq \mathcal{T}_\Sigma$. For example, consider the tree pattern $p = X@(X@a)$
and the substitution σ where $\sigma(X) = \lambda x.f(b, x)$ and $\sigma(x) = a$. Then the β-
normalization of the grounding $\sigma(p) = \sigma(X)@(\sigma(X)@\sigma(x))$ is the tree $t =
f(b, f(b, a))$, i.e. $t \in Inst(p)$. Similarly, for any $P \in \mathcal{P}_\Sigma^c$, we can define the
grounding $\sigma(P) \in \mathcal{G}_\Sigma^c$. The set of instances $Inst(P)$ contains the semantics of all
groundings of P. Clearly $Inst(P) \subseteq \llbracket \mathcal{C}_\Sigma \rrbracket$.

3   Inhabitation for Σ-Algebras

We recall the notion of inhabitation by trees and contexts in Val_Σ for Σ-algebras,
and then relate it to the notion of pattern evaluation in Σ-algebras.

A Σ-algebra $\Delta = (dom^\Delta, .^\Delta)$ consists of a set $D = dom^\Delta$ called the
domain, and a mapping $.^\Delta$ that interprets symbols $f \in \Sigma^{(n)}$ as functions
$f^\Delta : D^n \to D$. In particular, the set of trees $\mathcal{T}_\Sigma$ yields a Σ-algebra, known
as the *term algebra*, whose domain is $\mathcal{T}_\Sigma$ and whose interpretation satisfies
$f^{\mathcal{T}_\Sigma}(t_1, \ldots, t_n) = f(t_1, \ldots, t_n)$. Depending on their type, we can interpret values
in Val_Σ as elements of dom^Δ or as functions on dom^Δ. The *interpretation of a
tree* $t = f(t_1, \ldots, t_n) \in \mathcal{T}_\Sigma$ is the domain element $\llbracket t \rrbracket^\Delta = f^\Delta(\llbracket t_1 \rrbracket^\Delta, \ldots, \llbracket t_n \rrbracket^\Delta)$,
while the *interpretation of a context* $C = \lambda x.p \in \mathcal{C}_\Sigma$ is the function $\llbracket C \rrbracket^\Delta : D \to
D$ with $\llbracket C \rrbracket^\Delta(d) = \llbracket p[x/d] \rrbracket^\Delta$ for all $d \in D$.

Definition 1. *Let Δ be a Σ-algebra. An element $d \in dom^\Delta$ is called Δ-inhabited, if there exists a tree $t \in \mathcal{T}_\Sigma$ such that $d = [\![t]\!]^\Delta$. A function $S : dom^\Delta \to dom^\Delta$ is called Δ-inhabited if there exists a context $C \in \mathcal{C}_\Sigma$ such that $S = [\![C]\!]^\Delta$.*

The subset of all Δ-inhabited elements and functions is $[\![Val_\Sigma]\!]^\Delta = [\![\mathcal{T}_\Sigma]\!]^\Delta \cup [\![\mathcal{C}_\Sigma]\!]^\Delta$. We next lift algebra interpretation on values to algebra evaluation on patterns. We call a variable assignment $\sigma : V \to [\![Val_\Sigma]\!]^\Delta$ with $V \subseteq \mathcal{V}$ well-typed, if σ maps tree variables to $[\![\mathcal{T}_\Sigma]\!]^\Delta$ and context variables to $[\![\mathcal{C}_\Sigma]\!]^\Delta$. In Fig. 8, we define for any tree pattern p and any well-typed variable assignment $\sigma : V \to [\![Val_\Sigma]\!]^\Delta$ with $fv(p) \subseteq V$ the evaluation $[\![p]\!]^{\Delta,\sigma} \in [\![\mathcal{T}_\Sigma]\!]^\Delta$, and similarly $[\![P]\!]^{\Delta,\sigma} \in [\![\mathcal{C}_\Sigma]\!]^\Delta$ for all context patterns P with $fv(P) \subseteq V$. The evaluation of a ground pattern $\pi \in \mathcal{G}_\Sigma$ in Δ does not depend on the variable assignment σ. Therefore we can write $[\![\pi]\!]^\Delta$ instead of $[\![\pi]\!]^{\Delta,\sigma}$. Clearly, algebra evaluation restricted to values is equal to algebra interpretation. Furthermore, note that $[\![Val_\Sigma]\!]^\Delta = [\![\mathcal{G}_\Sigma]\!]^\Delta$ since any ground pattern in $\mathcal{G}_\Sigma$ can be β-reduced to some value in Val_Σ which has the same interpretation. Note also that the notion of Δ-inhabitation does not change when based on ground patterns instead of values.

$$[\![x]\!]^{\Delta,\sigma} = \sigma(x), \qquad [\![f(p_1,\dots,p_n)]\!]^{\Delta,\sigma} = f^\Delta([\![p_1]\!]^{\Delta,\sigma},\dots,[\![p_n]\!]^{\Delta,\sigma}),$$
$$[\![X]\!]^{\Delta,\sigma} = \sigma(X) \qquad [\![\lambda x.p]\!]^{\Delta,\sigma}(d) = [\![p[x/d]]\!]^{\Delta,\sigma}, \qquad [\![P@p]\!]^{\Delta,\sigma} = \sigma(P)([\![p]\!]^{\Delta,\sigma}).$$

Fig. 8. Algebra evaluation of patterns.

Consider a well-typed variable assignment $\sigma : V \to \mathcal{G}_\Sigma$. Then $[\![.]\!]^\Delta \circ \sigma$ is a well-typed variable assignment into $[\![\mathcal{G}_\Sigma]\!]^\Delta = [\![Val_\Sigma]\!]^\Delta$, such that $[\![\sigma(p)]\!]^\Delta = [\![p]\!]^{\Delta,[\![.]\!]^\Delta \circ \sigma}$ for all tree patterns p with $fv(p) \subseteq V$. As a consequence for the term algebra, the set of instances $Inst(p)$ of a tree pattern p is equal to $\{[\![p]\!]^{\mathcal{T}_\Sigma,[\![.]\!]^{\mathcal{T}_\Sigma} \circ \sigma} \mid \sigma : fv(p) \to \mathcal{G}_\Sigma$ well-typed$\}$, and similarly for context patterns P.

4   Inhabitation for Tree Automata

We recall the notion of tree automata for recognizing *regular languages of trees* and discuss tree and context inhabitation problems for tree automata. As we will see in the following section, these inhabitation problems are closely related to regular matching and inclusion for patterns with tree and context variables.

Definition 2. *A (nondeterministic) tree automaton (NTA) over Σ is a tuple $A = (Q, \Sigma, F, \Delta)$ where Q is a finite set of states, $F \subseteq Q$ is the set of final states, and $\Delta \subseteq \cup_{n \geq 0} \Sigma^{(n)} \times Q^{n+1}$ is the transition relation.*

A rule $(f, q_1, \dots, q_n, q) \in \Delta$ is written as $f(q_1, \dots, q_n) \to q$. The transition Σ-algebra of the NTA A – that we equally denote by Δ – has as its domain 2^Q

and interprets the function symbols $f \in \Sigma^{(n)}$ where $n \geq 0$ as the n-ary functions f^Δ such that for all subsets of states $Q_1 \ldots, Q_n \subseteq Q$:

$$f^\Delta(Q_1, \ldots, Q_n) = \{q \mid \exists q_1 \in Q_1 \ldots \exists q_n \in Q_n.\ f(q_1, \ldots, q_n) \to q \text{ in } \Delta\}.$$

The *regular language* $L(A)$ recognized by A is defined as the set of all trees in $\mathcal{T}_\Sigma$ whose evaluation in the Σ-algebra Δ yields some final state in F:

$$L(A) = \{t \in \mathcal{T}_\Sigma \mid [\![t]\!]^\Delta \cap F \neq \emptyset\}.$$

An NTA is *(bottom-up) deterministic* or equivalently a DTA if no two distinct rules of Δ have the same left-hand side, i.e., if Δ is a partial function from $\cup_{n\geq0}\Sigma^{(n)} \times Q^n$ to Q. The *determinization* of an NTA A is the tree automaton $\det(A) = (2^Q, \Sigma, \det(\Delta), \det(F))$ where $\det(\Delta) = \{(f, Q_1, \ldots, Q_n, f^\Delta(Q_1, \ldots, Q_n)) \mid f \in \Sigma^{(n)}, Q_1, \ldots, Q_n \subseteq Q\}$, and $\det(F) = \{Q' \subseteq Q \mid Q' \cap F \neq \emptyset\}$. It is well-known that $\det(A)$ is a DTA with $L(A) = L(\det(A))$. Furthermore, for any tree $t \in \mathcal{T}_\Sigma$ it holds that $[\![t]\!]^{\det(\Delta)} = \{[\![t]\!]^\Delta\}$.

Tree Inhabitation. Let NTA_Σ be the set of all NTAs with signature Σ, and similarly DTA_Σ. We call NTA and DTA automata classes. For any automaton class $\mathcal{A}$ and any signature Σ, tree inhabitation is the following problem:

Inhab$_\Sigma^e(\mathcal{A})$. *Input:* A tree automaton $A = (Q, \Sigma, F, \Delta) \in \mathcal{A}_\Sigma$, $Q' \subseteq Q$.
Output: The truth value of whether Q' is Δ-inhabited.

Theorem 1 (Folklore). *Tree inhabitation* $\text{INHAB}_\Sigma^e(\text{NTA})$ *is* EXP-*complete, while its restriction* $\text{INHAB}_\Sigma^e(\text{DTA})$ *to deterministic tree automata is in* P.

Proof. Let $A = (Q, \Sigma, F, \Delta)$ be an NTA and $Q' \subseteq Q$. By definition, Q' is Δ-inhabited iff there exists a tree $p \in \mathcal{T}_\Sigma$ such that $[\![p]\!]^\Delta = Q'$, which is equivalent to that $[\![p]\!]^{\det(\Delta)} = \{Q'\}$. Thus Q' is Δ-inhabited iff Q' is accessible in the tree automaton $\det(A)$. This can be tested in polynomial time from $\det(A)$ which is computed in exponential time. Thus $\text{INHAB}_\Sigma^e(\text{NTA})$ is in EXP. If A is a DTA, then there is no need to determinize it and Q' is a singleton. It is thus sufficient to test whether Q' is accessible in A. Hence $\text{INHAB}_\Sigma^e(\text{DTA})$ is in polynomial time.

We now have to show that $\text{INHAB}_\Sigma^e(\text{NTA})$ is EXP-hard. This is achieved by reduction from the problem of non-emptiness of the intersection of a sequence of DTAs, which is well known to be EXP-complete [10]. Let $A_1, \ldots, A_n$ be a sequence of DTAs with alphabet Σ. Suppose that $A_i = (Q^i, \Sigma, \Delta^i, F^i)$. Without loss of generality, we can assume that each of them has a single final state $F^i = \{q_f^i\}$. Let A be the disjoint union of all A_i, that is $A = (Q, \Sigma, F, \Delta)$ where $Q = \uplus_{i=1}^n Q^i$, $\Delta = \uplus_{i=1}^n \Delta^i$ and $F = \{q_f^1, \ldots, q_f^n\}$. Since all A_i are deterministic, we can then show that $t \in \cap_{i=1}^n L(A_i)$ iff F is Δ-inhabited by t. $\qquad\square$

Context Inhabitation. Contexts evaluate to very particular functions in transition algebras of tree automata, since they use their bound variable once.

Definition 3. *A* union homomorphism *on* 2^Q *is a function* $S : 2^Q \to 2^Q$ *such that* $S(\emptyset) = \emptyset$ *and for all* $Q', Q'' \subseteq Q$, $S(Q' \cup Q'') = S(Q') \cup S(Q'')$.

Lemma 1 (Folklore). *For any context $C \in \mathcal{C}_\Sigma$ and NTA $A = (Q, \Sigma, F, \Delta)$ the semantics $[\![C]\!]^\Delta$ is a union homomorphism on 2^Q.*

The main reason to restrict ourselves to contexts is that Lemma 1 would fail for nonlinear λ-terms such as $N = \lambda x.f(x,x)$. In order to see this, consider the signature $\Sigma = \{a, f\}$ where a is a constant and f a symbol of arity 2, and the NTA $A = (Q, \Sigma, F, \Delta)$ with $Q = \{q_1, q_2, q_{ok}\}$, $F = \{q_{ok}\}$ and $\Delta = \{a \to q_1, a \to q_2, f(q_1, q_2) \to q_{ok}\}$. We have $[\![N]\!]^\Delta(\{q_1\}) = [\![N]\!]^\Delta(\{q_2\}) = \emptyset$, while $[\![N]\!]^\Delta(\{q_1, q_2\}) = \{q_{ok}\}$. Hence, $[\![N]\!]^\Delta(\{q_1, q_2\}) \neq [\![N]\!]^\Delta(\{q_1\}) \cup [\![N]\!]^\Delta(\{q_2\})$, so that $[\![N]\!]^\Delta$ is not a union homomorphism and cannot be represented by a function $s : Q :\to 2^Q$ as stated in Lemma 2. Since union homomorphisms are determined by their images on singletons, they can be represented by functions $s : Q \to 2^Q$. Conversely, every such function defines the union homomorphism $\hat{s} : 2^Q \to 2^Q$ such that for any $Q' \subseteq Q$: $\hat{s}(Q') = \cup_{q \in Q'} s(q)$.

Lemma 2. *If $S : 2^Q \to 2^Q$ is a union homomorphism then $S = \hat{s}$ where $s : Q \to 2^Q$ is the function with $s(q) = S(\{q\})$ for all $q \in Q$.*

We next consider the problem of context-inhabitation for tree automata. Here, the input is a succinct descriptor of a union homomorphism:

Inhab$_\Sigma^c(\mathcal{A})$. *Input:* An automaton $A = (Q, \Sigma, F, \Delta) \in \mathcal{A}_\Sigma$, $s : Q \to 2^Q$.
Output: The truth value of whether $\hat{s}$ is Δ-inhabited.

Context inhabitation is a restriction of the more general λ-*definability* problem, which is undecidable [11,12]. However, λ-definability for orders up to 3 is decidable [13], and context-inhabitation is a special case of second-order λ-definability. Its precise complexity, however, has not been studied so far to the best of our knowledge.

Proposition 1. *Let $A = (Q, \Sigma, F, \Delta)$ be an NTA and $s : Q \to 2^Q$. Then $\hat{s}$ is Δ-inhabited iff there exists $C \in \mathcal{C}_\Sigma$ such that for all $q \in Q$, $s(q) = [\![C]\!]^\Delta(\{q\})$.*

Proof. The forward implication is straightforward. For the backwards direction, let $C \in \mathcal{C}_\Sigma$ be a context with $s(q) = [\![C]\!]^\Delta(\{q\})$ for all $q \in Q$. Since $\hat{s}$ is a union homomorphism, we have for all $Q' \subseteq Q$ that $\hat{s}(Q') = \cup_{q \in Q'} s(q) = \cup_{q \in Q'} [\![C]\!]^\Delta(\{q\}) = [\![C]\!]^\Delta(Q')$ since $[\![C]\!]^\Delta$ is a union-homomorphism by Lemma 1. Thus $\hat{s}$ is Δ-inhabited. $\square$

Theorem 2. *For both classes of tree automata $\mathcal{A} \in \{\text{NTA}, \text{DTA}\}$ the context-inhabitation problem $\text{INHAB}_\Sigma^c(\mathcal{A})$ is EXP-complete.*

Proof. Since $\text{INHAB}_\Sigma^c(\text{DTA})$ is EXP-complete, a naive exponential time reduction from $\text{INHAB}_\Sigma^c(\text{NTA})$ to $\text{INHAB}_\Sigma^c(\text{DTA})$ would lead to a doubly exponential time algorithm. Nevertheless, we will present a single exponential time algorithm for $\text{INHAB}_\Sigma^c(\text{NTA})$ based on determinization. Let $A = (Q, \Sigma, F, \Delta)$ be an NTA where $Q = \{q_1, \ldots, q_n\}$ and $s : Q \to 2^Q$. We fix a variable $x \in \mathcal{V}^e$ arbitrarily. For each $i \in \{1, \ldots, n\}$, let $A_i = (Q, \Sigma \uplus \{x\}, \Delta \cup \{x \to q_i\}, F)$. For any context $C = \lambda x.p$,

$[\![p]\!]^{A_i}$ is the set of states to which C can be evaluated when starting at the hole marker x with state q_i. Let $\tilde{A}$ be the product DTA $\tilde{A} = \det(A_1) \times \ldots \times \det(A_n)$. Note that the number of states of $\tilde{A}$ is at most $(2^n)^n = 2^{n^2}$, which is exponential. Furthermore, the tuple $(s(q_1), \ldots, s(q_n))$ is an accessible state of $\tilde{A}$ if and only if there is a context $\lambda x.p \in \mathcal{C}_\Sigma$ such that for all $1 \le i \le n$, $[\![\lambda x.p]\!]^\Delta(\{q_i\}) = s(q_i)$. By Proposition 1 this is equivalent to that $\hat{s}$ is Δ-inhabited. Testing whether $(s(q_1), \ldots, s(q_n))$ is accessible in $\tilde{A}$ is in polynomial time in the size of $\tilde{A}$ and thus in exponential time too. The EXP-hardness of $\textsc{Inhab}_\Sigma^c(\textsc{Dta})$ can be shown by reduction from the intersection problem of DTAs. The idea of the proof is similar to that of the PSPACE-hardness proof of DFA-inhabitation (see [2]), so we omit the details.   □

5 Compressed Tree Patterns

We now recall compressed tree patterns with context variables that are defined by singleton tree grammars.

Definition 4. *A compressed pattern (with context variables) of type $\tau \in \mathbf{T}$ is an acyclic context-free tree grammar $G = (N, \Sigma, R, S)$ where $N \subseteq \mathcal{V}$ is a finite set of nonterminals, $S \in N \cap \mathcal{V}^\tau$ is the start symbol, R is a partial well-typed function from N to patterns in $\mathcal{P}_\Sigma$ with free variables in N. The set of all compressed tree patterns of type τ is denoted by $c\mathcal{P}_\Sigma^\tau$.*

For instance, consider the compressed tree pattern $G \in c\mathcal{P}_\Sigma^e$ with the nonterminals $N = \{x, X, Y, Z, y\}$, with $S = x$ and with two rules $R(x) = X@a(X@b, Y@c)$, and $R(X) = \lambda x.Z@a(x, y)$. This grammar is acyclic, in that no variable on the left hand side of some rule can appear in any subsequent rule. It should be noticed that the tree language of the grammar G is $\emptyset$. What interests us instead is the tree pattern $pat(G) = (\lambda x.Z@a(x,y))@a((\lambda x.Z@a(x,y))@b, Y@c)$ that G represents in a compressed manner. By exhaustive β-reduction of $pat(G)$ we obtain the tree pattern with context variables $[\![pat(G)]\!] = Z@a(a(Z@a(b,y), Y@c), y)$.

We define the free variables of a compressed tree pattern G as the free variables of $pat(G)$, and the bound variables of G as the nonterminals in $dom(R)$ and the bound variables on the right-hand sides of these rules.

In what follows we will identify any tree pattern $p \in \mathcal{P}_\Sigma^e$ with the compressed tree pattern in $c\mathcal{P}_\Sigma^e$ that has a single rule mapping a new start symbol to p. This compressed tree pattern has no compression. In this sense, $\mathcal{P}_\Sigma^e \subseteq c\mathcal{P}_\Sigma^e$. A compressed tree pattern G is called *linear* if its tree pattern $pat(G)$ is linear.

Let $A = (Q, \Sigma, F, \Delta)$ be an NTA, $V \subseteq \mathcal{V}$ a finite subset of variables, and σ a function with domain V that maps any tree variable $x \in V$ to $\sigma(x) \subseteq Q$ and any context variable $X \in V$ to a function $\sigma(X) : Q \to 2^Q$. Note that $\sigma(X)$ represents the union homomorphism $\widehat{\sigma(X)} : 2^Q \to 2^Q$. Let $\hat{\sigma}$ be such that $\hat{\sigma}(x) = \sigma(x)$ for all $x \in V$ and $\hat{\sigma}(X) = \widehat{\sigma(X)}$ for all $X \in V$.

Lemma 3. *For any $G = (N, \Sigma, R, S) \in c\mathcal{P}_\Sigma^e$ with $fv(G) \subseteq V$ we can compute $[\![pat(G)]\!]^{\Delta,\hat{\sigma}}$ in polynomial time from A, G, and σ.*

Proof. The algorithm evaluates the pattern inductively along the partial order on the nonterminals of G; the latter exists because G is acyclic. For any $v \in V$, let G_v be the compressed tree pattern equal to G except that the start symbol is changed to v. Then we can show for all $v \in V$ that $[\![pat(G_v)]\!]^{\Delta,\hat{\sigma}}$ can be computed in polynomial time from A, G, and σ. In particular this holds for $[\![pat(G)]\!]^{\Delta,\hat{\sigma}} = [\![pat(G_S)]\!]^{\Delta,\hat{\sigma}}$. □

6 Regular Matching and Inclusion

We now study the complexity of regular matching and inclusion for classes $\mathcal{H}$ of compressed tree patterns with context variables such as $\mathcal{P}^e$ and $c\mathcal{P}^e$.

Definition 5. *For any class $\mathcal{H}$ of compressed tree patterns, any class $\mathcal{A}$ of NTAs, and for any ranked alphabet Σ we define two decision problems:*

Regular pattern inclusion $\mathrm{INCL}_\Sigma(\mathcal{H}, \mathcal{A})$. *Input: A compressed tree pattern $G \in \mathcal{H}_\Sigma$ and a tree automaton $A \in \mathcal{A}_\Sigma$.*
   Output: The truth value of whether $Inst(pat(G)) \subseteq L(A)$.
Regular pattern matching $\mathrm{MATCH}_\Sigma(\mathcal{H}, \mathcal{A})$. *Input: A compressed tree pattern $G \in \mathcal{H}_\Sigma$ and a tree automaton $A \in \mathcal{A}_\Sigma$.*
   Output: The truth value of whether $Inst(pat(G)) \cap L(A) \neq \emptyset$.

The following characterization of regular matching induces a decision procedure by reduction to context inhabitation, and is useful in the hardness proof.

Lemma 4. *Let $A = (Q, \Sigma, F, \Delta)$ be an NTA, $p \in \mathcal{P}_\Sigma^e$ be a tree pattern. Then $Inst(p) \cap L(A) \neq \emptyset$ if and only if there exists a well-typed assignment into Δ-inhabited subset of states and union-homomorphisms $\sigma : fv(p) \to [\![Val_\Sigma]\!]^\Delta$ such that $[\![p]\!]^{\Delta,\sigma} \cap F \neq \emptyset$.*

Proposition 2 (Lower Bound Matching). $\mathrm{MATCH}_\Sigma(\mathcal{P}^e, \mathrm{DTA})$ *is* EXP-*hard.*

Proof. We reduce $\mathrm{INHAB}_\Sigma^c(\mathrm{DTA})$ to $\mathrm{MATCH}_\Sigma(\mathcal{P}^e, \mathrm{DTA})$ in polynomial time, then the result follows from Theorem 2. Let $A = (Q, \Sigma, F, \Delta)$ be a DTA and $s : Q \to 2^Q$ be a function. We set $Q = \{q_1, \ldots, q_n\}$ and consider a new symbol $\# \notin \Sigma$ of arity n and a new state $q_\#$. From this we build a new DTA $\tilde{A} = (\tilde{Q}, \tilde{\Sigma}, \tilde{F}, \tilde{\Delta})$ where $\tilde{Q} = Q \cup \{q_\#\} \cup \{s(q_i) \mid 1 \leq i \leq n\}$, $\tilde{\Sigma} = \Sigma \cup \{\#\} \cup Q$, $\tilde{F} = \{q_\#\}$ and $\tilde{\Delta} = \Delta \cup \{\#(s(q_1), \ldots, s(q_n)) \to q_\#\}$. Let $X \in \mathcal{V}^c$ and $p = \#(X@q_1, \ldots, X@q_n) \in \mathcal{P}_\Sigma^e$. The reduction is induced by the following claim, whose technical proof is based on Lemma 4 without any special tricks.

Claim. The function $\hat{s}$ is Δ-inhabited if and only if $Inst(p) \cap L(\tilde{A}) \neq \emptyset$. □

Lemma 5 (Complementation). *Regular inclusion and matching are complementary problems for deterministic automata: For any class of compressed tree patterns $\mathcal{H}$, $\mathrm{INCL}_\Sigma(\mathcal{H}, \mathrm{DTA})$ and $\mathrm{coMATCH}_\Sigma(\mathcal{H}, \mathrm{DTA})$ are equivalent modulo P.*

Proof. For a compressed tree pattern G and an NTA A, $Inst(p) \subseteq L(A)$ iff $Inst(p) \cap \overline{L(A)} = \emptyset$ iff $= Inst(p) \cap L(\overline{A}) = \emptyset$, and the complementation operation is polynomial for DTAs and exponential for NTAs– since it requires determinization.

Proposition 3 (Lower Bound Inclusion). $\mathrm{INCL}_\Sigma(\mathcal{P}^e, \mathrm{DTA})$ *is* EXP-*hard.*

Proof. Lemma 5 states that $\mathrm{INCL}_\Sigma(\mathcal{P}^e, \mathrm{DTA}) = \mathrm{coMATCH}_\Sigma(\mathcal{P}^e, \mathrm{DTA})$ modulo P. By Proposition 2, $\mathrm{MATCH}_\Sigma(\mathcal{P}^e, \mathrm{DTA})$ is EXP-hard and since EXP is closed by complement, $\mathrm{coMATCH}_\Sigma(\mathcal{P}^e, \mathrm{DTA})$ is EXP-hard too. It then holds that $\mathrm{INCL}_\Sigma(\mathcal{P}^e, \mathrm{DTA})$ is EXP-hard.   □

We next reduce the problems of regular matching and inclusion to context inhabitation for tree automata in order to obtain upper complexity bounds.

Proposition 4 (Upper Bounds). $\mathrm{MATCH}_\Sigma(c\mathcal{P}^e, \mathrm{NTA})$ *and* $\mathrm{INCL}_\Sigma(c\mathcal{P}^e, \mathrm{NTA})$ *are in* EXP.

Proof. Let $G \in c\mathcal{P}_\Sigma^e$ be a compressed tree pattern with start symbol $S \in \mathcal{V}^e$ and set of nonterminals N, and $A = (Q, \Sigma, F, \Delta)$ be an NTA. According to Lemma 4, to decide whether $pat(G)$ matches $L(A)$ it is sufficient to find a well-typed assignment σ with domain $fv(G)$ such that $\sigma(x) \subseteq Q$ for all $x \in fv(G)$ and $\sigma(X) : Q \to 2^Q$ for all $X \in fv(G)$. Furthermore, $\widehat{\sigma} : fv(G) \to [\![Val_\Sigma]\!]$ must map to Δ-inhabited subsets of Q and Δ-inhabited union-homomorphisms of type $2^Q \to 2^Q$ such that $\widehat{\sigma}(pat(G)) \cap F \neq \emptyset$. Thus the algorithm iterates over all such σ, tests the inhabitation of $\widehat{\sigma}(v)$ for all $v \in N$, and checks that $\widehat{\sigma}(pat(G)) \cap F \neq \emptyset$. It is successful if the test succeeds for some σ. The number of iterations is at most $2^{|Q|^2 \cdot |fv(G)|}$. Moreover inhabitation can be tested in time $O(2^{|Q|^2})$ by Theorems 2 and 1 while $\widehat{\sigma}(pat(G))$ is computed in polynomial time from A, G, and σ by Lemma 3. Thus the algorithm is in EXP. For $\mathrm{INCL}_\Sigma(c\mathcal{P}^e, \mathrm{NTA})$, the algorithm is similar except that the condition $\widehat{\sigma}(pat(G)) \cap F \neq \emptyset$ must hold for all $\widehat{\sigma}$ mapping $fv(G)$ to Δ-inhabited sets of states and functions.   □

7   Encoding Patterns for Unranked Trees

The original motivation of the present work was to understand the problems of regular matching and inclusion for hedge patterns. We next show that these problems can be solved using reductions to the corresponding problems of (ranked) tree patterns with context variables.

Unlike ranked trees, unranked trees are constructed from symbols without fixed arities. We fix a finite set Γ of such symbols. The set of hedges $\mathcal{H}_\Gamma$ is the least set that contains all words of hedges in $\mathcal{H}_\Gamma{}^*$ and all pairs $a(H)$ where $a \in \Gamma$ and $H \in \mathcal{H}_\Gamma$ is a hedge. The set of unranked trees $\mathcal{U}_\Gamma$ is the subset of hedges of the form $a(H)$.

We assume a set of variables for unranked trees $Y \in \mathcal{V}^u$ and a set of hedge variables $Z \in \mathcal{V}^h$. The set of hedge patterns $H \in \mathcal{P}_\Gamma^h$ with these two types of variables is then defined by the abstract syntax in Fig. 9. The set $\mathcal{P}_\Gamma^u$ of *patterns for unranked trees* is the subset of hedge patterns of the forms $a(H)$ or $Y \in \mathcal{V}^u$. The set of free variables $fv(H)$ is defined as usual. A well-typed variable assignment $\sigma : V \to \mathcal{H}_\Gamma$ where $V \subseteq \mathcal{V}^u \uplus \mathcal{V}^h$ is a function that maps variables from $\mathcal{V}^u$ to unranked trees in $\mathcal{U}_\Gamma$ and variables from $\mathcal{V}^h$ to hedges in $\mathcal{H}_\Gamma$. The application $\sigma(H)$ is the hedge obtained from H by replacing all variables Y by the unranked tree $\sigma(Y)$ and all variables Z by the hedge $\sigma(Z)$. The instance set of H is denoted $Inst(H) = \{\sigma(H) \mid \sigma : fv(H) \to \mathcal{H}_\Gamma$ well-typed$\}$. Note that $Inst(H) \subseteq \mathcal{U}_\Gamma$ for any unranked tree pattern $H \in \mathcal{P}_\Gamma^u$.

Hedge patterns	$H, H' \in \mathcal{P}_\Gamma^h ::= Y \mid a(H) \mid \varepsilon \mid Z \mid HH'$
Encoding	$\langle Y \rangle^c = Y, \quad \langle a(H) \rangle^c = \lambda y.a(\langle H \rangle^c @\#, y), \quad \langle \varepsilon \rangle^c = \lambda y.y,$
	$\langle Z \rangle^c = Z, \quad \langle HH' \rangle^c = \lambda y.(\langle H \rangle^c @(\langle H' \rangle^c @y)), \quad \langle H \rangle^e = \langle H \rangle^c @\#.$

Fig. 9. Encoding of a hedge pattern $H \in \mathcal{P}_\Gamma^h$ into a context pattern $\langle H \rangle^c \in \mathcal{P}_\Sigma^c$, where $Y \in \mathcal{V}^u$, $Z \in \mathcal{V}^h$, $a \in \Gamma$, and ε is the empty word.

We next show in Fig. 9 how to encode hedge patterns into (ranked) context patterns over the signature $\Sigma = \Sigma^{(2)} \uplus \Sigma^{(0)}$ where $\Sigma^{(2)} = \Gamma$ and $\Sigma^{(0)} = \{\#\}$ for $\#$ is a fresh symbol not in Γ. For instance, the hedge pattern $H_0 = a(ZbcY)$ is encoded into the context pattern $\langle H_0 \rangle^c = \lambda y.a(Z@(b(\#, c(\#, Y@\#))), y)$. The concatenation operation on hedges is simulated by the application operation of contexts. The set of context variables used in the encoding is $\mathcal{V}^c = \mathcal{V}^u \uplus \mathcal{V}^h$, while the set $\mathcal{V}^e$ of tree variables is left arbitrary. Finally, we define for any unranked tree $H \in \mathcal{P}_\Gamma^u$ its encoding as a tree pattern $\langle H \rangle^e \in \mathcal{P}_\Sigma^e$ by $\langle H \rangle^e = \langle H \rangle^c @\#$.

In order to show the soundness of this encoding (Lemma 6 below), we need to restrict the instantiation operation. Intuitively, we cannot allow arbitrary substitutions to be applied to $\langle H \rangle^e$ because then the resulting tree pattern might not be a correct encoding of an unranked tree. A variable assignment $\sigma : V \to Val_\Sigma$ is called *unranked* if it maps unranked tree variables to $\langle \mathcal{U}_\Gamma \rangle^c$ and hedge variables to $\langle \mathcal{H}_\Gamma \rangle^c$. The *unranked-restricted instance set* of a tree pattern p is defined by $Inst^{unr}(p) = \{[\![\sigma(p)]\!] \mid \sigma : fv(p) \to Val_\Sigma$ well-typed and unranked$\}$ and similarly for $Inst^{unr}(P)$.

Lemma 6. $[\![\langle Inst(H) \rangle^e]\!] = Inst^{unr}(\langle H \rangle^e)$ *for any* $H \in \mathcal{P}_\Gamma^u$.

Proof idea. We can prove for any $H \in \mathcal{P}_\Gamma^u$ that $[\![\langle Inst(H) \rangle^c]\!] = Inst^{unr}(\langle H \rangle^c)$ by induction of the structure of H. This claim implies the lemma.

Let $c\mathcal{P}_\Gamma^u$ be the set of compressed unranked trees over Γ, defined in an analogous way as compressed tree patterns. For a class of automata $\mathcal{A} \in \{\text{DTA}, \text{NTA}\}$, the problem $\text{MATCH}_\Gamma(c\mathcal{P}^u, \mathcal{A})$ of regular matching of compressed unranked tree patterns takes as input an unranked tree pattern $H \in c\mathcal{P}^u$ and an automaton

A in class $\mathcal{A}$, and outputs the truth value of whether $Inst^{\mathrm{unr}}(\langle H \rangle^e) \cap L(A) \neq \emptyset$. The problem INCL_Γ ($c\mathcal{P}^u$, $\mathcal{A}$) of regular inclusion for compressed patterns of unranked trees is defined in an analogous way. Note that using tree automata in the above definitions is not a restriction, as it is well known [3] that for any unranked tree language L recognizable by a hedge automaton, there exists an NTA that recognizes the encodings as ranked trees of the elements of L.

Proposition 5. *For any $\mathcal{A} \in \{\text{DTA}, \text{NTA}\}$ there exist polynomial time reductions from $\text{MATCH}_\Gamma(c\mathcal{P}^u, \mathcal{A})$ to $\text{MATCH}_{\Sigma'}(c\mathcal{P}^e, \mathcal{A})$ and from $\text{INCL}_\Gamma(c\mathcal{P}^u, \mathcal{A})$ to $Incl_{\Sigma'}(c\mathcal{P}^e, \mathcal{A})$ for some signature Σ' derived from Σ.*

Proof idea. The basic idea is to use Lemma 6, but we also need to constrain the variable assignments for the encoded patterns to be unranked. We illustrate how this works on an example for the case of regular matching. Consider the unranked tree pattern $H = a(Z) \in \mathcal{P}_\Gamma^u$, a language $\mathcal{L} \subseteq \mathcal{U}_\Gamma$ and an NTA A over Σ with $L(A) = \langle \mathcal{L} \rangle^e$. From $\langle H \rangle^e = a(Z@\#, \#)$, we build the tree pattern $p_H = a(root_Z(Z@hole_Z(\#)), \#)$, where $root_Z$ and $hole_Z$ are new unary symbols. We also construct a NTA A' from A so that $Inst^{\mathrm{unr}}(\langle H \rangle^e) \cap L(A) \neq \emptyset$ iff $Inst(p_H) \cap L(A') \neq \emptyset$. Basically in p_H any variable Z is "enclosed" between the $root_Z$ and $hole_Z$ symbols, and A' tests that any context between $root_Z$ and $hole_Z$ is a correct encoding of an unranked hedge.

Theorem 3. *For any class of tree automata $\mathcal{A} \in \{\text{DTA}, \text{NTA}\}$ the problems $\text{MATCH}_\Gamma(c\mathcal{P}^u, \mathcal{A})$ and $\text{INCL}_\Gamma(c\mathcal{P}^u, \mathcal{A})$ are EXP-complete.*

Proof. The upper bounds for $\text{MATCH}_\Sigma(c\mathcal{P}^u, \text{NTA})$ and $\text{INCL}_\Sigma(c\mathcal{P}^u, \text{NTA})$ follow via the polynomial time reduction from Proposition 5 from the upper bounds for $\text{MATCH}_\Sigma(c\mathcal{P}^e, \text{NTA})$ and $\text{INCL}_\Sigma(c\mathcal{P}^e, \text{NTA})$ in Proposition 4. The EXP-hardness of $\text{MATCH}_\Gamma(c\mathcal{P}^u, \text{DTA})$ and thus of the other 3 problems can be shown in analogy to the proof of Proposition 2. $\qquad \square$

8   Conclusion

We have shown that regular matching and inclusion for ranked tree patterns with context variables is EXP-complete with and without compression. The complexity goes down to P for linear compressed tree patterns in 3 of 4 cases. The same result holds for unranked tree patterns with hedge variables, which is relevant to certain query answering on hyperstreams. Previous approaches were limited to hyperstreams containing words (compressed string patterns), while the present approach can deal with hyperstreams containing unranked data trees (compressed unranked tree patterns).

Acknowledgments. We are grateful to Sylvain Salvati for pointing out and helping to solve difficulties. This work was partially supported by a grant from CPER Nord-Pas de Calais/FEDER DATA Advanced data science and technologies 2015–2020.

References

1. Angluin, D.: Finding patterns common to a set of strings. JCSS **21**, 46–62 (1980)
2. Boneva, I., Niehren, J., Sakho, M.: Certain query answering on compressed string patterns: from streams to hyperstreams. In: Potapov, I., Reynier, P.-A. (eds.) RP 2018. LNCS, vol. 11123, pp. 117–132. Springer, Cham (2018). https://doi.org/10.1007/978-3-030-00250-3_9
3. Comon, H., et al.: TATA, October 2007. http://tata.gforge.inria.fr
4. Gascón, A., Godoy, G., Schmidt-Schauß, M.: Context matching for compressed terms. In: LICS 2008, USA, pp. 93–102. IEEE CS (2008)
5. Jeż, A.: Context unification is in PSPACE. In: Esparza, J., Fraigniaud, P., Husfeldt, T., Koutsoupias, E. (eds.) ICALP 2014. LNCS, vol. 8573, pp. 244–255. Springer, Heidelberg (2014). https://doi.org/10.1007/978-3-662-43951-7_21
6. Labath, P., Niehren, J.: A functional language for hyperstreaming XSLT. Technical report, INRIA Lille (2013)
7. Maneth, S., Ordóñez, A., Seidl, H.: Transforming XML streams with references. In: Iliopoulos, C., Puglisi, S., Yilmaz, E. (eds.) SPIRE 2015. LNCS, vol. 9309, pp. 33–45. Springer, Cham (2015). https://doi.org/10.1007/978-3-319-23826-5_4
8. Plandowski, W.: Satisfiability of word equations with constants is in PSPACE. J. ACM **51**(3), 483–496 (2004)
9. Schmidt-Schauß, M.: Linear pattern matching of compressed terms and polynomial rewriting. Math. Struct. Comput. Sci. **28**(8), 1415–1450 (2018)
10. Seidl, H.: Deciding equivalence of finite tree automata. SIAM J. Comput. **19**(3), 424–437 (1990)
11. Loader, R.: The undecidability of λ-definability. In: Anderson, C.A., Zelëny, M. (eds.) Logic, Meaning and Computation. SYLI, vol. 305, pp. 331–342. Springer, Dordrecht (2001). https://doi.org/10.1007/978-94-010-0526-5_15
12. Joly, T.: Encoding of the halting problem into the monster type and applications. In: Hofmann, M. (ed.) TLCA 2003. LNCS, vol. 2701, pp. 153–166. Springer, Heidelberg (2003). https://doi.org/10.1007/3-540-44904-3_11
13. Zaionc, M.: Probabilistic approach to the lambda definability for fourth order types. Electr. Notes Theoret. Comput. Sci. **140**, 41–54 (2005)

Rule-Based Unification in Combined Theories and the Finite Variant Property

Ajay K. Eeralla[1], Serdar Erbatur[2], Andrew M. Marshall[3],
and Christophe Ringeissen[4(✉)]

[1] University of Missouri, Columbia, USA
[2] Ludwig-Maximilians-Universität, München, Germany
[3] University of Mary Washington, Fredericksburg, USA
[4] Université de Lorraine, CNRS, Inria, LORIA, Nancy, France
`Christophe.Ringeissen@loria.fr`

Abstract. We investigate the unification problem in theories defined by rewrite systems which are both convergent and forward-closed. These theories are also known in the context of protocol analysis as theories with the finite variant property and admit a variant-based unification algorithm. In this paper, we present a new rule-based unification algorithm which can be seen as an alternative to the variant-based approach. In addition, we define forward-closed combination to capture the union of a forward-closed convergent rewrite system with another theory, such as the Associativity-Commutativity, whose function symbols may occur in right-hand sides of the rewrite system. Finally, we present a combination algorithm for this particular class of non-disjoint unions of theories.

Keywords: Term rewriting · Unification · Combination · Forward-closure

1 Introduction

Unification plays a central role in logic systems based on the resolution principle, to perform the computation in declarative programming, and to deduce new facts in automated reasoning. Syntactic unification is particularly well-known for its use in logic programming. Being decidable and unitary are remarkable properties of syntactic unification. More generally, we may consider equational unification, where the problem is defined modulo an equational theory E, like for instance the Associativity-Commutativity. Equational unification, say E-unification, is undecidable in general. However, specialized techniques have been developed to solve the problem for particular classes of equational theories, many of high practical interest. It is not uncommon to have such equational theories include

C. Ringeissen—This work has received funding from the European Research Council (ERC) under the H2020 research and innovation program (grant agreement No. 645865-SPOOC).

ⓒ Springer Nature Switzerland AG 2019
C. Martín-Vide et al. (Eds.): LATA 2019, LNCS 11417, pp. 356–367, 2019.
https://doi.org/10.1007/978-3-030-13435-8_26

Associativity-Commutativity, which is useful to represent arithmetic operators. Nowadays, security protocols are successfully analyzed using dedicated reasoning tools [4,5,13,18] in which protocols are usually represented by clauses in first-order logic with equality. In these protocol analyzers, equational theories are used to specify the capabilities of an intruder [1]. To support the reasoning in these equational theories E, one needs to use E-unification procedures. When the equational theory E has the Finite Variant Property (FVP) [8], there exists a reduction from E-unification to syntactic unification via the computation of finitely many variants of the unification problem. When this reduction is used, we talk about variant-based unification. The class of equational theories with the FVP has attracted a considerable interest since it contains theories that are crucial in protocol analysis [6,7,9,14,19]. The concept of narrowing is another possible unification technique when E is given by a convergent term rewrite system (TRS). Narrowing is a generalization of rewriting which is widely used in declarative programming. It is complete for E-unification, but it terminates only in some very particular cases. A particular narrowing strategy, called folding variant narrowing, has been shown complete and terminating for any equational theory with the FVP [14]. When E has the property of being *syntactic* [16,20], it is possible to apply a rule-based unification procedure in the same vein as the one known for syntactic unification, which is called a mutation-based unification procedure. Unfortunately, being syntactic is not a sufficient condition to insure the termination of this unification procedure. Finally, another important scenario is given by an equational theory E defined as a union of component theories. To solve this case, it is quite natural to proceed in a modular way by reusing the unification algorithms available in the component theories. There are terminating and complete combination procedures for signature-disjoint unions of theories [3, 21], but the non-disjoint case remains a challenging problem [12].

In this paper, we investigate the impact of considering an equational theory with the FVP in order to get a terminating mutation-based unification procedure and a terminating combination procedure for some non-disjoint unions of theories. Instead of directly talking about the FVP, we study the equivalent class of theories defined by forward-closed convergent TRSs [6]. Actually, a forward-closed convergent TRS is a syntactic theory admitting a terminating mutation-based unification procedure. Here, we consider the unification problem in the class of *forward-closed combinations* defined as unions of a forward-closed convergent TRS plus an equational theory over function symbols that may only occur in the right-hand sides of the TRS. To solve this problem we need a mutation procedure for the forward-closed component of the combination. Rather than reusing the mutation procedure given in [17] we develop a new mutation procedure which is more conducive to combination. By adding some standard combination rules, we show how to extend this new mutation procedure in order to solve the unification problem in forward-closed combinations.

The rest of the paper is organized as follows. Section 2 recalls the standard notions and Sect. 3 introduces the class of forward-closed theories. In Sect. 4, we present a terminating mutation-based unification procedure for forward-closed theories. In Sect. 5, we introduce forward-closed combinations. The related combination method is given in Sect. 6, by proving its termination and correctness. Finally, Sect. 8 discusses some limitations and possible extensions of this work. Omitted proofs and additional unification procedures can be found in [10].

2 Preliminaries

We use the standard notation of equational unification and term rewriting systems [2]. Given a first-order signature Σ and a (countable) set of variables V, the set of Σ-terms over variables V is denoted by $T(\Sigma, V)$. The set of variables in a term t is denoted by $Var(t)$. A term t is *ground* if $Var(t) = \emptyset$. A term is *linear* if all its variables occur only once. For any position p in a term t (including the root position ϵ), $t(p)$ is the symbol at position p, $t|_p$ is the subterm of t at position p, and $t[u]_p$ is the term t in which $t|_p$ is replaced by u. A substitution is an endomorphism of $T(\Sigma, V)$ with only finitely many variables not mapped to themselves. A substitution is denoted by $\sigma = \{x_1 \mapsto t_1, \ldots, x_m \mapsto t_m\}$, where the domain of σ is $Dom(\sigma) = \{x_1, \ldots, x_m\}$. Application of a substitution σ to t is written $t\sigma$. Given a set E of Σ-axioms (i.e., pairs of Σ-terms, denoted by $l = r$), the *equational theory* $=_E$ is the congruence closure of E under the law of substitutivity (by a slight abuse of terminology, E is often called an equational theory). Equivalently, $=_E$ can be defined as the reflexive transitive closure $\leftrightarrow_E^*$ of an equational step $\leftrightarrow_E$ defined as follows: $s \leftrightarrow_E t$ if there exist a position p of s, $l = r$ (or $r = l$) in E, and substitution σ such that $s|_p = l\sigma$ and $t = s[r\sigma]_p$. An axiom $l = r$ is *regular* if $Var(l) = Var(r)$. An axiom $l = r$ is *linear* (resp., *collapse-free*) if l and r are linear (resp. non-variable terms). An equational theory is *regular* (resp., linear/collapse-free) if all its axioms are regular (resp., linear/collapse-free). A theory E is *syntactic* if it has finite *resolvent presentation* S, defined as a finite set of axioms S such that each equality $t =_E u$ has an equational proof $t \leftrightarrow_S^* u$ with at most one equational step $\leftrightarrow_S$ applied at the root position. A Σ-equation is a pair of Σ-terms denoted by $s =^? t$ or simply $s = t$ when it is clear from the context that we do not refer to an axiom. An E-unification problem is a set of Σ-equations, $G = \{s_1 =^? t_1, \ldots, s_n =^? t_n\}$, or equivalently a conjunction of Σ-equations. The set of variables in G is denoted by $Var(G)$. A solution to G, called an *E-unifier*, is a substitution σ such that $s_i\sigma =_E t_i\sigma$ for all $1 \leq i \leq n$, written $E \models G\sigma$. A substitution σ is *more general modulo E* than θ on a set of variables V, denoted as $\sigma \leq_E^V \theta$, if there is a substitution τ such that $x\sigma\tau =_E x\theta$ for all $x \in V$. An E-unification algorithm computes a (finite) *Complete Set of E-Unifiers* of G, denoted by $CSU_E(G)$, which is a set of substitutions such that each $\sigma \in CSU_E(G)$ is an E-unifier of G, and for each E-unifier θ of G, there exists $\sigma \in CSU_E(G)$ such that $\sigma \leq_E^{Var(G)} \theta$. Given a unifiable equation $s =^? t$, a syntactic unification algorithm computes a unique most general unifier denoted by $mgu(s, t)$. A set of equations $G = \{x_1 =^? t_1, \ldots, x_n =^? t_n\}$ is said to be in *tree solved form* if each x_i is a variable occurring once in G. Given an idempotent substitution $\sigma = \{x_1 \mapsto t_1, \ldots, x_n \mapsto t_n\}$ (such that $\sigma\sigma = \sigma$), $\hat{\sigma}$ denotes the corresponding tree solved form. A set of equations is said to be in *dag solved* *form* if they can be arranged as a list $x_1 =^? t_1, \ldots, x_n =^? t_n$ where (a) each left-hand side x_i is a distinct variable, and (b) $\forall 1 \leq i \leq j \leq n$: x_i does not occur in t_j. A set of equations $\{x_1 =^? t_1, \ldots, x_n =^? t_n\}$ is a *cycle* if for any $i \in [1, n-1], x_{i+1} \in Var(t_i), x_1 \in Var(t_n)$, and there exists $j \in [1, n]$ such that t_j is not a variable. Given two variables x and y, $x = y$ is said to be *solved* in a set of equations G if x does not occur in $G \backslash \{x = y\}$. Then, x is said to be *solved* in G. Given two disjoint signatures Σ_1 and Σ_2 and any $i = 1, 2$, Σ_i-terms (including the variables) and Σ_i-equations (including the equations between variables) are called *i-pure*. For any $\Sigma_1 \cup \Sigma_2$-theory E, an E-unification problem is in *separate form* if it is a conjunction $G_1 \wedge G_2$, where G_i is a conjunction of Σ_i-equations for $i = 1, 2$. A term t is called a Σ_i-*rooted* term if its root symbol is in Σ_i. An *alien* subterm of a Σ_i-rooted term t is a Σ_j-rooted subterm s $(i \neq j)$ such that all superterms of s are Σ_i-rooted. A *term rewrite system* (TRS) is a pair (Σ, R), where Σ is a signature and R is a finite set of rewrite rules of the form $l \rightarrow r$ such that l, r are Σ-terms, l is not a variable and $Var(r) \subseteq Var(l)$. A term s *rewrites* to a term t w.r.t R, denoted by $s \rightarrow_R t$ (or simply $s \rightarrow t$), if there exist a position p of s, $l \rightarrow r \in R$, and substitution σ such that $s|_p = l\sigma$ and $t = s[r\sigma]_p$. A TRS R is *terminating* if there are no infinite reduction sequences with respect to $\rightarrow_R$. A TRS R is *confluent* if, whenever $t \rightarrow_R^* s_1$ and $t \rightarrow_R^* s_2$, there exists a term w such that $s_1 \rightarrow_R^* w$ and $s_2 \rightarrow_R^* w$. A confluent and terminating TRS is called *convergent*. In a convergent TRS R, we have the existence and the uniqueness of R-normal forms, denoted by $t\downarrow_R$ for any term t. A substitution σ is *normalized* if, for every variable x in the domain of σ, $x\sigma$ is a normal form. A convergent TRS R is said to be *subterm convergent* if for any $l \rightarrow r \in R$, r is either a strict subterm of l or a constant. To simplify the notation, we often use tuples of terms, say $\bar{u} = (u_1, \ldots, u_n)$, $\bar{v} = (v_1, \ldots, v_n)$. Applying a substitution σ to $\bar{u}$ is the tuple $\bar{u}\sigma = (u_1\sigma, \ldots, u_n\sigma)$. The tuples $\bar{u}$ and $\bar{v}$ are said E-equal, denoted by $\bar{u} =_E \bar{v}$, if $u_1 =_E v_1, \ldots, u_n =_E v_n$. Similarly, $\bar{u} \rightarrow_R^* \bar{v}$ if $u_1 \rightarrow_R^* v_1, \ldots, u_n \rightarrow_R^* v_n$, $\bar{u}$ is R-normalized if $u_1, \ldots, u_n$ are R-normalized, and $\bar{u} =^? \bar{v}$ is $u_1 =^? v_1 \wedge \cdots \wedge u_n =^? v_n$.

3   Forward Closure

In this section, we define the central notion of finite forward closure. To define the forward closure as in [6], let us first introduce the notion of redundancy. For a given convergent TRS R, assume a reduction ordering $<$ such that $r < l$ for any $l \rightarrow r \in R$ and $<$ is total on ground terms. Since (rewrite) rules are multisets of two terms, the multiset extension of $<$ leads to an ordering on rules, also denoted by $<$, which is total on ground instances of rules. A rule ρ is *strictly redundant in R* if any ground instance $\rho\sigma$ of ρ is a logical consequence of ground instances of R that are strictly smaller w.r.t $<$ than $\rho\sigma$. A rule ρ is *redundant in R* if ρ is strictly redundant in R or ρ is an instance of some rule in R. Given two rules $\rho_1 = (g \rightarrow d)$, $\rho_2 = (l \rightarrow r)$ and a non-variable position p of d such that $d|_p$ and l are unifiable, $Fwd(\rho_1, \rho_2, p)$ denotes the rule $(g \rightarrow d[r]_p)\sigma$ where $\sigma = mgu(d|_p, l)$. Forward closure steps are inductively defined as follows:

- $FC_0(R) = NR_0(R) = R$,
- $FC_{k+1}(R) = FC_k(R) \cup NR_{k+1}(R)$ where $NR_{k+1}(R)$ is the set of rules $\rho_3 = Fwd(\rho_1, \rho_2, p)$ such that $\rho_1 \in NR_k(R)$, $\rho_2 \in R$, p is a non-variable position of the right-hand side of ρ_1, and ρ_3 is not redundant in $FC_k(R)$.

The *forward closure* of R is $FC(R) = \bigcup_{k \geq 0} FC_k(R)$. A TRS R is *forward-closed* if $FC(R) = R$. A TRS is *forward-closed convergent* if it is both forward-closed and convergent.

Example 1. Any subterm convergent TRS has a finite forward closure. Subterm convergent TRSs are often used in the verification of security protocols [1], e.g., $\{dec(enc(x, y), y) \to x\}$ and $\{fst(pair(x, y)) \to x,\ snd(pair(x, y)) \to y\}$.

Example 2. The following TRSs are forward-closed convergent:
$\{f(x) + f(y) \to f(x * y)\}$ $\{g(h(x, y), z) \to h(x, y * z)\}$
$\{f(x) + y \to f(x * y)\}$ $\{d(e(x, a), a) \to x * a\}$
$\{exp(exp(a, x), y) \to exp(a, x * y)\}$ $\{pdt(pair(x, y)) \to x * y\}$

We will study the unification problem in a combination of any of these TRSs with an equational theory over $*$, such as $C = \{x * y = y * x\}$ (Commutativity) or $AC = \{x * (y * z) = (x * y) * z, x * y = y * x\}$ (Associativity-Commutativity).

    It has been shown in [6] that for any convergent TRS R, R has a finite forward closure if and only if R has the finite variant property (FVP, for short). When a TRS has the FVP, any R-unification problem G admits a computable finite complete set of R-variants of G, say $V_R(G)$, such that solving G reduces to solve the syntactic unification problems in $V_R(G)$. In many cases, the computation of $V_R(G)$ can be prohibitive even with an efficient implementation of folding variant narrowing [9, 14]. For these cases, it is interesting to have an alternative to this brute force method, possibly via a rule-based R-unification procedure that does not impose a full reduction to syntactic unification.

4   Rule-Based Unification in Forward-Closed Theories

To design a rule-based unification procedure for forward-closed theories, we basically reuse the *BSM* unification procedure initially developed for the class of theories saturated by paramodulation [17], where *BSM* stands for *Basic Syntactic Mutation*. The *BSM* procedure extends syntactic unification with some additional mutation rules applied in a *don't know* non-deterministic way. These mutation rules are parameterized by a finite set of axioms corresponding to a resolvent presentation (cf. Sect. 2). The resulting *BSM* unification procedure is similar to the mutation-based unification procedures designed for syntactic theories [16, 20] but with the additional property of being terminating. To get termination, it makes use of boxed terms. Variables can be considered as implicitly boxed, and terms are boxed according to the following rules:

- Subterms of boxed terms are also boxed.

- Terms boxed in the premises of an inference rule remain boxed in the conclusion.
- When the "box" status of a term is not explicitly given in an inference rule, it can be either boxed or unboxed. For instance, each occurrence of f in the premise of **Imit** rule (cf. Fig. 1) can be either boxed or unboxed.

Boxed terms allow us to focus on particular R-normalized solutions of a unification problem. Hence, we are interested in R-normalized solutions σ such that $t\sigma$ is R-normalized for each boxed term t occurring in the unification problem.

Definition 3. *Let G be a unification problem and σ be a substitution. We say that (G, σ) is R-normalized if σ is R-normalized, and for any term t in G, $t\sigma$ is R-normalized whenever t is boxed.*

Assuming a forward-closed convergent TRS R is sufficient to replay the correctness proofs of BSM originally stated for theories saturated by paramodulation. Thus, BSM can be rephrased by using directly a forward-closed convergent TRS R as input. In this setting, the equational theory of any forward-closed convergent TRS R is syntactic and a resolvent presentation is used as the parameter of BSM mutation rules. This leads to a BSM procedure providing a unification algorithm for forward-closed theories, detailed in [10].

In this paper, we are also interested in solving the unification problem in the union of a forward-closed theory R_1 and a non-disjoint theory E_2. For this more general problem we develop a new and simplified mutation-based unification algorithm called BSM'. The new algorithm simplifies conflicts and therefore we need only a single mutation rule and thus a simpler mutation algorithm overall. These changes in turn allow for simpler correctness proofs as there are fewer cases to check. The single mutation rule, called **MutConflict** in Fig. 1, aims at applying rewrite rules in R instead of equalities in the resolvent presentation. This restriction to R is sufficient if there is no equation between two non-variable terms. This form of equations can be easily avoided by splitting such equation $s = t$ into two equations $x = s$ and $x = t$ involving a common fresh variable x. Thanks to this additional transformation called **Split** in Fig. 1, the classical decomposition rule of syntactic unification is superfluous.

All the BSM' rules are given in Fig. 1. Let $\mathbf{B}'$ be the subset of BSM' that consists of rules with boxed terms, i.e., **Imit**, **MutConflict** and **ImitCycle**. BSM' rules are applied according to the following order of priority (from higher to lower): **Coalesce**, **Split** and $\mathbf{B}'$, where all $\mathbf{B}'$ rules are applied in a non-deterministic way (using a "don't know" non-determinism). The BSM' unification procedure consists in applying repeatedly the BSM' rules until reaching normal forms. The procedure then only returns those sets of equations which are in dag solved form. The BSM' unification can be used as an equivalent alternative to BSM. Compared to BSM, the BSM' alternative has the advantage of being easily combinable as shown in Sect. 6.

Theorem 4. *If R is a forward-closed convergent TRS, then the BSM' unification procedure provides an R-unification algorithm.*

Theorem 4 is subsumed by Theorem 11 that will be presented in Sect. 6.

Coalesce $\{x = y\} \cup G \vdash \{x = y\} \cup (G\{x \mapsto y\})$
where x and y are distinct variables occurring both in G.

Split $\{f(\bar{v}) = t\} \cup G \vdash \{x = f(\bar{v}), x = t\} \cup G$
where t is a non-variable term and x is a fresh variable.

Imit $\bigcup_i \{x = f(\bar{v}_i)\} \cup G \vdash \{x = \boxed{f(\bar{y})}\} \cup \bigcup_i \{\bar{y} = \bar{v}_i\} \cup G$
where $i > 1$, $\bar{y}$ are fresh variables and there are no more equations $x = f(\dots)$ in G.

MutConflict $\{x = f(\bar{v})\} \cup G \vdash \{x = \boxed{t}, \boxed{\bar{s}} = \bar{v}\} \cup G$
where a fresh instance $f(\bar{s}) \rightarrow t \in R$, $f(\bar{v})$ is unboxed, and (there is another equation $x = u$ in G with a non-variable term u or $x = f(\bar{v})$ occurs in a cycle).

ImitCycle $\{x = f(\bar{v})\} \cup G \vdash \{x = \boxed{f(\bar{y})}, \bar{y} = \bar{v}\} \cup G$
where $f(\bar{v})$ is unboxed, $\bar{y}$ are fresh variables and $x = f(\bar{v})$ occurs in a cycle.

Fig. 1. BSM' rules

5   Forward-Closed Combination

Along the lines of hierarchical combination [12], we study a form of non-disjoint combination defined as a convergent TRS R_1 combined with a base theory E_2. The TRS R_1 must satisfy some properties to ensure that $E = R_1 \cup E_2$ is a conservative extension of E_2. We focus here on cases where it is possible to reduce the E-equality between two terms into the E_2-equality of their R_1-normal forms. In addition, we assume that R_1 is forward-closed. The following definition clearly introduces the forward-closed combinations studied in the rest of the paper.

Definition 5. *Let Σ_1 and Σ_2 be two disjoint signatures. A forward-closed combination (FC-combination, for short) is a pair (E_1, E_2) such that*

- *E_1 is an equational $\Sigma_1 \cup \Sigma_2$-theory given by a forward-closed convergent TRS R_1 whose left-hand sides are Σ_1-terms;*
- *E_2 is a regular and collapse-free equational Σ_2-theory;*
- *for any terms s, t, we have (i) $s =_{E_1 \cup E_2} t$ iff $s \downarrow_{R_1} =_{E_2} t \downarrow_{R_1}$, and (ii) if $s =_{E_2} t$ then s is R_1-reducible iff t is R_1-reducible.*

Let us discuss the ingredients of the above definition. First of all, it is important to note that Σ_1 and Σ_2 are disjoint signatures. Thus, the TRS is a standard rewrite system defined on the signature $\Sigma_1 \cup \Sigma_2$ where Σ_2-symbols can occur only in right-hand sides. For this TRS, we do not have to rely on the notions of E_2-confluence and E_2-coherence introduced for class rewrite systems [15].

Proposition 6. *Assume Σ_1, Σ_2 and E_2 are given as in Definition 5. If E_1 is an equational $\Sigma_1 \cup \Sigma_2$-theory given by a forward-closed convergent TRS whose left-hand sides are linear Σ_1-terms, then (E_1, E_2) is an FC-combination.*

Example 7. Consider R_1 as any TRS mentioned in Example 2 and $\Sigma_2 = \{*\}$. An FC-combination is defined by R_1 together with any regular and collapse-free Σ_2-theory E_2, such as C or AC.

From now on, we assume $E = E_1 \cup E_2$ and (E_1, E_2) is an FC-combination given by a forward-closed convergent TRS R_1.

6 Unification in Forward-Closed Combinations

We now study how the BSM' unification procedure can be combined with an E_2-unification algorithm to solve the unification problem in $E = E_1 \cup E_2$.

VA $\{s = t[u]\} \cup G \vdash \{s = t[x], x = u\} \cup G$
where u is an alien subterm of t, x is a fresh variable, and u is boxed iff $t[u]$ is boxed.

Solve $G_1 \wedge G_2 \vdash \bigvee_{\sigma_2 \in CSU_{E_2}(G_2)} G_1 \wedge \hat{\sigma}_2$
if $G_1 \wedge G_2$ is a separate form, G_2 is E_2-unifiable and not in tree solved form, where
w.l.o.g $\forall \sigma_2 \in CSU_{E_2}(G_2) \ \forall x \in Dom(\sigma_2)$, $(x\sigma_2$ is a variable$) \Rightarrow x\sigma_2 \in Var(G_2)$.

Fig. 2. Additional rules for the combination with E_2

Consider the inference system for Basic Syntactic Combination, say BSC, given by **Coalesce**, **Split** and **B′** rules defining BSM' in Sect. 4, where f is now supposed to be a function symbol in Σ_1; plus the two additional rules given in Fig. 2, namely **VA** and **Solve**. The rule **VA** applies the classical Variable Abstraction transformation [3,11,21] to purify terms and so to get a separate form, while **Solve** calls an E_2-unification algorithm to solve the set of Σ_2-equations in a separate form. The repeated application of rules in $\{$**Coalesce**, **Split**, **VA**, **Solve**$\}$ computes particular separate forms defined as follows.

Definition 8. *A separate form $G_1 \wedge G_2$ is* mutable *if **Coalesce** does not apply on $G_1 \wedge G_2$, G_1 is a set of Σ_1-equations $x = t$ (where x is a variable), and G_2 is a set of Σ_2-equations in solved form. A* compound solved form *is a mutable separate form in dag solved form.*

BSC rules are applied according to the following order of priority (from higher to lower): **Coalesce**, **Split**, **VA**, **Solve**, and **B′** where **Solve** computes each solution of the subproblem G_2 in a separate form $G_1 \wedge G_2$ and **B′** rules are applied on a separate form $G_1 \wedge G_2$ in a non-deterministic way as in Sect. 4. Due to the order of priority, **Solve** applies only if **Coalesce**, **Split**, **VA** are not applicable and **B′** rules apply only if $G_1 \wedge G_2$ is a mutable separate form. Notice that any compound solved form is in normal form w.r.t BSC.

Lemma 9. *Given an E-unification problem G as input, the repeated application of BSC rules always terminates.*

Proof. To prove termination we use a complexity measure, similar to the one in [17], which decreases, according to the lexicographic ordering, with each application of a rule. This reduction is illustrated in the following table.

Rule	m	n	i_1	i_2	p	q
Imit	$\geq$	$>$				
MutConflict	$>$					
ImitCycle	$>$					
Coalesce	$\geq$	$\geq$	$\geq$	$\geq$	$>$	
VA	$\geq$	$\geq$	$>$			
Split	$\geq$	$\geq$	$\geq$	$>$		
Solve	$\geq$	$\geq$	$\geq$	$\geq$	$\geq$	$>$

Where m is the number of unboxed Σ_1-symbols; n is the number of Σ_1-symbols; i_1 is the sum of sizes of impure terms; i_2 is the number of equations $f(\bar{v}) = t$ such that $f \in \Sigma_1$ and t is a non-variable term; p is the number of variables in G that are unsolved and not generated by E_2-unification; and $q \in \{0, 1\}$ such that $q = 0$ iff G is a separate form $G_1 \wedge G_2$ and G_2 is in solved form. □

Given an E-unification problem G, $\mathrm{BSC}(G)$ denotes the normal forms of G w.r.t BSC, which correspond to the compound solved forms of G. Following Lemma 9, the BSC unification procedure works as follows: apply the BSC rules on a given E-unification problem G until reaching normal forms, and return all the dag solved forms in $\mathrm{BSC}(G)$. Below, we show that BSC rules are applied without loss of completeness. We denote by $G \xrightarrow[BSC]{} G'$ an application of a BSC rule to a unification problem G producing a modified problem G'.

Lemma 10. *If (G, σ) is R_1-normalized, $E \models G\sigma$ and G is not a compound solved form, then there exist some G' and a substitution σ' such that $G \xrightarrow[BSC]{} G'$, (G', σ') is R_1-normalized, $E \models G'\sigma'$ and $\sigma' \leq_E^{Var(G)} \sigma$.*

Hence BSC leads to a terminating and complete E-unification procedure.

Theorem 11. *Given any FC-combination (E_1, E_2) and an E_2-unification algorithm, BSC provides an $E_1 \cup E_2$-unification algorithm.*

According to Definition 5, an FC-combination can be obtained by considering an arbitrary forward-closed convergent TRS R_1 and the empty theory E_2 over the empty signature Σ_2. In that particular case, BSC reduces to BSM' and so the fact that BSC is both terminating and correct provides a proof for Theorem 4.

Example 12. Assume f, g, a are in Σ_1 and E_2 is the C theory for $*$. Consider the separate form $G = \{x = f(y), x = z * y\}$ and the following possible cases:

- If $R_1 = \{f(v) \to a\}$, then **MutConflict** can be applied on G and we get $\{x = \boxed{a}, x = z * y\}$, which is in normal form w.r.t BSC but not in dag solved form. So, it has no solution.

– If $R_1 = \{f(v) \to v*a\}$, then **MutConflict** can be applied on G and we obtain $\{x = \boxed{y * a}, x = z*y\}$. Then **Solve** leads to the solved form $\{x = y*a, z = a\}$.
– If $R_1 = \{g(v) \to v * a\}$, then G is in normal form w.r.t BSC but not in dag solved form. So, it has no solution.

Example 13. Let $R_1 = \{exp(exp(a, x), y) \to exp(a, x*y)\}$ and let E_2 be the AC theory for $*$. Consider the problem $G = \{exp(x_1, x_2) = exp(a, x_2 * x_3)\}$ and a run of BSC on G. Applying **VA** and **Split** leads to the mutable separate form $\{z_2 = exp(x_1, x_2), z_2 = exp(a, z_1), z_1 = x_2 * x_3\}$. At this point one possibility is to apply **MutConflict** (introducing z_3, z_4) followed by **Coalesce** (replacing z_4 by x_2), leading to $\{z_2 = \boxed{exp(a, z_3 * x_2)}, x_1 = \boxed{exp(a, z_3)}, x_2 = z_4, z_2 = exp(a, z_1), z_1 = x_2 * x_3\}$ Then **VA** applies, leading to $\{z_2 = \boxed{exp(a, z_5)}, x_1 = \boxed{exp(a, z_3)}, x_2 = z_4, z_2 = exp(a, z_1), z_1 = x_2 * x_3, z_5 = \boxed{z_3 * x_2}\}$, By applying **Imit** (introducing z_6, z_7) followed by **Coalesce** (replacing z_5, z_7 by z_1), we obtain $\{z_2 = \boxed{exp(z_6, z_1)}, z_6 = a, z_7 = z_5, z_7 = z_1, x_1 = \boxed{exp(a, z_3)}, x_2 = z_4, z_1 = x_2 * x_3, z_1 = \boxed{z_3 * x_2}\}$ At this point an AC-unification algorithm can be used to solve $\{z_1 = x_2 * x_3, z_1 = z_3 * x_2\}$. The AC-unifier $\{z_3 \mapsto x_3\}$ leads to an expected solution of G, which is $\{x_1 \mapsto exp(a, x_3)\}$.

To complete the family picture on unification in FC-combinations, it is also possible to develop brute force methods relying on a reduction to general E_2-unification. Unsurprisingly, a possible method is based on the computation of variants. Another variant-free method consists in the non-deterministic application of some mutation/imitation rules. These two methods are described in [10].

7   Implementation

When choosing to implement the algorithms developed in this paper, we have selected the Maude programming language[1]. Maude provides a nice environment for a number of reasons. First, it provides a more natural environment for expressing the rules of algorithms such as BSM'. Second, it has both variant generation and several unification algorithms, such as AC, built-in. Indeed, having both the variant-based unification and the rule-based unification developed here implemented in Maude is the best way to compare them in practice. In addition, having both approaches implemented offers alternatives for selecting the most suitable method for an application (for example, in cases when the number of variants is high). One can now easily switch between the most appropriate approach for their situation.

Implementation of the above procedures is ongoing[2]. Currently, the focus of the implementation is on the BSM' algorithm which in itself provides a new alternative method for solving the unification problem in forward closed theories.

---

[1] http://maude.cs.illinois.edu/w/index.php/The_Maude_System.
[2] https://github.com/ajayeeralla/BSM.

Significantly, once the forward closure of a system is computed, the implementation of BSM' provides a unification procedure for any problem in the theory. In other words, the computation of a forward closure can be reused for any unification problem for that theory. The implementation also takes advantage of the flexibility of Maude, allowing the rules of the BSM' procedure to be instantiated by a theory input to the algorithm via a Maude-module. This will also make the program easier to incorporate into a larger tools.

After the BSM' implementation the focus will be on the combination and experimentation. Due to the importance of AC in practical applications, we plan to focus on the case of forward-closed combinations with AC-symbols, for which it is possible to reuse the AC-unification algorithm implemented in Maude in a way similar to [11].

8   Conclusion

In this paper we develop a rule-based unification algorithm which can be easily combined, even for some non-disjoint unions of theories, and does not require the computation of variants. By applying this rule-based unification algorithm, we present, in addition, a new non-disjoint, terminating combination procedure for a base theory extended with a non-disjoint forward-closed TRS. The new combination allows for the addition of such often used theories as AC and C.

Until now, we assume that the TRS is defined in a simple way, by using syntactic matching for the rule application. A possible extension would be to consider an equational TRS defined modulo the base theory, where equational matching is required for the rule application. Considering an equational TRS instead of a classical one, two natural problems arise. First, the possible equivalence between the finite forward closure and the FVP is an open problem when the TRS is equational. Second, another problem is to highlight a combination algorithm for solving unification problems modulo an equational TRS having the FVP.

References

1. Abadi, M., Cortier, V.: Deciding knowledge in security protocols under equational theories. Theor. Comput. Sci. **367**(1–2), 2–32 (2006)
2. Baader, F., Nipkow, T.: Term Rewriting and All That. Cambridge University Press, New York (1998)
3. Baader, F., Schulz, K.U.: Unification in the union of disjoint equational theories: combining decision procedures. J. Symb. Comput. **21**(2), 211–243 (1996)
4. Basin, D., Mödersheim, S., Viganò, L.: An on-the-fly model-checker for security protocol analysis. In: Snekkenes, E., Gollmann, D. (eds.) ESORICS 2003. LNCS, vol. 2808, pp. 253–270. Springer, Heidelberg (2003). https://doi.org/10.1007/978-3-540-39650-5_15
5. Blanchet, B.: Modeling and verifying security protocols with the applied Pi calculus and ProVerif. Found. Trends Priv. Secur. **1**(1–2), 1–135 (2016)

6. Bouchard, C., Gero, K.A., Lynch, C., Narendran, P.: On forward closure and the finite variant property. In: Fontaine, P., Ringeissen, C., Schmidt, R.A. (eds.) FroCoS 2013. LNCS (LNAI), vol. 8152, pp. 327–342. Springer, Heidelberg (2013). https://doi.org/10.1007/978-3-642-40885-4_23

7. Ciobâcă, S., Delaune, S., Kremer, S.: Computing knowledge in security protocols under convergent equational theories. J. Autom. Reasoning **48**(2), 219–262 (2012)

8. Comon-Lundh, H., Delaune, S.: The finite variant property: how to get rid of some algebraic properties. In: Giesl, J. (ed.) RTA 2005. LNCS, vol. 3467, pp. 294–307. Springer, Heidelberg (2005). https://doi.org/10.1007/978-3-540-32033-3_22

9. Durán, F., Eker, S., Escobar, S., Martí-Oliet, N., Meseguer, J., Talcott, C.: Built-in variant generation and unification, and their applications in Maude 2.7. In: Olivetti, N., Tiwari, A. (eds.) IJCAR 2016. LNCS (LNAI), vol. 9706, pp. 183–192. Springer, Cham (2016). https://doi.org/10.1007/978-3-319-40229-1_13

10. Eeralla, A.K., Erbatur, S., Marshall, A.M., Ringeissen, C.: Unification in non-disjoint combinations with forward-closed theories. http://hal.inria.fr

11. Eeralla, A.K., Lynch, C.: Bounded ACh Unification. CoRR abs/1811.05602 (2018). http://arxiv.org/abs/1811.05602

12. Erbatur, S., Kapur, D., Marshall, A.M., Narendran, P., Ringeissen, C.: Hierarchical combination. In: Bonacina, M.P. (ed.) CADE 2013. LNCS (LNAI), vol. 7898, pp. 249–266. Springer, Heidelberg (2013). https://doi.org/10.1007/978-3-642-38574-2_17

13. Escobar, S., Meadows, C., Meseguer, J.: Maude-NPA: cryptographic protocol analysis modulo equational properties. In: Aldini, A., Barthe, G., Gorrieri, R. (eds.) FOSAD 2007-2009. LNCS, vol. 5705, pp. 1–50. Springer, Heidelberg (2009). https://doi.org/10.1007/978-3-642-03829-7_1

14. Escobar, S., Sasse, R., Meseguer, J.: Folding variant narrowing and optimal variant termination. J. Log. Algebr. Program. **81**(7–8), 898–928 (2012)

15. Jouannaud, J., Kirchner, H.: Completion of a set of rules modulo a set of equations. SIAM J. Comput. **15**(4), 1155–1194 (1986). https://doi.org/10.1137/0215084

16. Kirchner, C., Klay, F.: Syntactic theories and unification. In: Logic in Computer Science 1990 Proceedings of the Fifth Annual IEEE Symposium on Logic in Computer Science, LICS 1990, pp. 270–277, June 1990. https://doi.org/10.1109/LICS.1990.113753

17. Lynch, C., Morawska, B.: Basic syntactic mutation. In: Voronkov, A. (ed.) CADE 2002. LNCS (LNAI), vol. 2392, pp. 471–485. Springer, Heidelberg (2002). https://doi.org/10.1007/3-540-45620-1_37

18. Meier, S., Schmidt, B., Cremers, C., Basin, D.: The TAMARIN prover for the symbolic analysis of security protocols. In: Sharygina, N., Veith, H. (eds.) CAV 2013. LNCS, vol. 8044, pp. 696–701. Springer, Heidelberg (2013). https://doi.org/10.1007/978-3-642-39799-8_48

19. Meseguer, J.: Variant-based satisfiability in initial algebras. Sci. Comput. Program. **154**, 3–41 (2018)

20. Nipkow, T.: Proof transformations for equational theories. In: Logic in Computer Science 1990 Proceedings of the Fifth Annual IEEE Symposium on Logic in Computer Science, LICS 1990, pp. 278–288 June 1990

21. Schmidt-Schauß, M.: Unification in a combination of arbitrary disjoint equational theories. J. Symb. Comput. **8**, 51–99 (1989)

Extensions of the Caucal Hierarchy?

Paweł Parys[✉] [iD]

Institute of Informatics, University of Warsaw, Warsaw, Poland
parys@mimuw.edu.pl

Abstract. The Caucal hierarchy contains graphs that can be obtained from finite graphs by alternately applying the unfolding operation and inverse rational mappings. The goal of this work is to check whether the hierarchy is closed under interpretations in logics extending the monadic second-order logic by the unbounding quantifier U. We prove that by applying interpretations described in the MSO+U^{fin} logic (hence also in its fragment WMSO+U) to graphs of the Caucal hierarchy we can only obtain graphs on the same level of the hierarchy. Conversely, interpretations described in the more powerful MSO+U logic can give us graphs with undecidable MSO theory, hence outside of the Caucal hierarchy.

Keywords: Caucal hierarchy · Boundedness · WMSO+U logic

1 Introduction

This paper concerns the class of finitely describable infinite graphs introduced in Caucal [9], called a Caucal hierarchy. Graphs on consecutive levels of this hierarchy are obtained from finite graphs by alternately applying the unfolding operation [14] and inverse rational mappings [8]. Since both these operations preserve decidability of the monadic second-order (MSO) theory, graphs in the Caucal hierarchy have decidable MSO theory. It turns out that this class of graphs has also other definitions. It was shown [5,7] that the Caucal hierarchy contains exactly ε-closures of configuration graphs of all higher-order pushdown automata [15]; while generating trees, these automata are in turn equivalent to a subclass of higher-order recursion schemes called safe schemes [19]. Moreover, Carayol and Wöhrle [7] prove that the defined classes of graphs do not change if we replace the unfolding operation by the treegraph operation [26], and similarly, if we replace inverse rational mappings by the stronger operation of MSO-transductions [13]. One can also replace inverse rational mappings by the operation of FO-interpretations, assuming that the FO formulae have access to the descendant relation [10].

In this paper we try to replace inverse rational mappings or MSO-interpretations in the definition of the Caucal hierarchy by interpretations in

Work supported by the National Science Centre, Poland (grant no. 2016/22/E/ST6/00041).

C. Martín-Vide et al. (Eds.): LATA 2019, LNCS 11417, pp. 368–380, 2019.
https://doi.org/10.1007/978-3-030-13435-8_27

some extensions of the MSO logic. Namely, we investigate logics obtained from MSO by adding the unbounding quantifier U introduced by Bojańczyk [1]. The meaning of a formula U$X.\varphi$ is that φ holds for arbitrarily large finite sets X. In the MSO+U^{fin} logic we can write U$X.\varphi$ only for formulae φ whose free variables cannot represent infinite sets (this fragment subsumes the more known WMSO+U logic in which all monadic variables can only represent finite sets [2,4]). We prove that the Caucal hierarchy does not change if we use MSO+U^{fin}-interpretations in its definition. In other words, by applying MSO+U^{fin}-interpretations to graphs in the Caucal hierarchy, we only obtain graphs on the same level of the hierarchy.

This result shows robustness of the Caucal hierarchy, but is a bit disappointing (but rather not surprising): it would be nice to find a class of graphs with decidable properties, larger than the Caucal hierarchy. We remark that the class of trees generated by all (i.e., not necessarily safe) higher-order recursion schemes (equivalently, by collapsible pushdown automata [17]) is such a class: these trees have decidable MSO theory [20], and some of them are not contained in the Caucal hierarchy [24]. This class lacks a nice machine-independent definition (using logics, like for the Caucal hierarchy), though. For some other classes of graphs we only have decidability of first-order logics [12,25].

Going further, we also check the full MSO+U logic, where the use of the U quantifier is unrestricted. For this logic we obtain graphs outside of the Caucal hierarchy; among them there are graphs with undecidable MSO theory. This is very expected, since the MSO+U logic is undecidable itself [3].

2 Preliminaries

2.1 Logics

A *signature* $\varXi$ (of a relational structure) is a list of relation names, $R_1, \ldots, R_n$, together with an arity assigned to each of the names. A *(relational) structure* $\mathcal{S} = (U^{\mathcal{S}}, R_1^{\mathcal{S}}, \ldots, R_n^{\mathcal{S}})$ over such a signature $\varXi$ is a set $U^{\mathcal{S}}$, called the *universe*, together with *relations* $R_i^{\mathcal{S}}$ over $\mathcal{S}$, for all relation names in the signature; the arity of the relations is as specified in the signature. Following the literature on the Caucal hierarchy [5–9] we forbid the universe to have isolated elements: every element of $U^{\mathcal{S}}$ has to appear in at least one of the relations $R_1^{\mathcal{S}}, \ldots, R_n^{\mathcal{S}}$.

We assume three countable sets of variables: $\mathcal{V}^{FO}$ of first-order variables, $\mathcal{V}^{fin}$ of monadic variables representing finite sets, and $\mathcal{V}^{inf}$ of monadic variables representing arbitrary sets. First-order variables are denoted using lowercase letters $x, y, \ldots$, and monadic variables (of both kinds) are denoted using capital letters $X, Y, \ldots$. The atomic formulae are

- $R(x_1, \ldots, x_n)$, where R is a relation name of arity n (coming from a fixed signature $\varXi$), and $x_1, \ldots, x_n$ are first-order variables;
- $x = y$, where x, y are first-order variables;
- $x \in X$, where x is a first-order variable, and X a monadic variable.

Formulae of the *monadic second-order logic with the unbounding quantifier*, MSO+U, are built out of atomic formulae using the boolean connectives $\vee, \wedge, \neg$, the first-order quantifiers $\exists x$ and $\forall x$, the monadic quantifiers $\mathsf{U}X$, $\exists_{\mathsf{fin}}X$, and $\forall_{\mathsf{fin}}X$ for $X \in \mathcal{V}^{fin}$, and the monadic quantifiers $\exists X$ and $\forall X$ for $X \in \mathcal{V}^{inf}$.

We use the standard notion of *free variables*. In this paper, we also consider three syntactic fragments of MSO+U. Namely, in the *monadic second-order logic*, MSO, we are not allowed to use variables from $\mathcal{V}^{fin}$, and thus the quantifiers using them: $\mathsf{U}X$ (most importantly), $\exists_{\mathsf{fin}}X$, and $\forall_{\mathsf{fin}}X$. In the MSO+U^{fin} logic, the use of the unbounding quantifier is syntactically restricted: we can write $\mathsf{U}X.\varphi$ only when all free variables are from $\mathcal{V}^{FO} \cup \mathcal{V}^{fin}$ (i.e., φ has no free variables ranging over infinite sets). In the *weak fragment*, WMSO+U, we cannot use variables from $\mathcal{V}^{inf}$, together with the quantifiers $\exists X$ and $\forall X$.

In order to evaluate an MSO+U formula φ over a signature $\varXi$ in a relational structure $\mathcal{S}$ over the same signature, we also need a *valuation* ν, which is a partial function that maps

- variables $x \in \mathcal{V}^{FO}$ to elements of the universe of $\mathcal{S}$;
- variables $X \in \mathcal{V}^{fin}$ to finite subsets of the universe of $\mathcal{S}$;
- variables $X \in \mathcal{V}^{inf}$ to arbitrary subsets of the universe of $\mathcal{S}$.

The valuation should be defined at least for all free variables of φ. We write $\mathcal{S}, \nu \models \varphi$ when φ is *satisfied* in $\mathcal{S}$ with respect to the valuation ν; this is defined by induction on the structure of φ. For most constructs the definition is as expected, thus we made it explicit only for φ of the form $\mathsf{U}X.\psi$: we have $\mathcal{S}, \nu \models \mathsf{U}X.\psi$ if for every $n \in \mathbb{N}$ there exists a finite subset $X^{\mathcal{S}}$ of the universe of $\mathcal{S}$ having cardinality at least n, such that $\mathcal{S}, \nu[X \mapsto X^{\mathcal{S}}] \models \psi$ (in other words: $\mathsf{U}X.\psi$ says that ψ is satisfied for arbitrarily large finite sets X).

We write $\varphi(x_1, \ldots, x_n)$ to denote that the free variables of φ are among $x_1, \ldots, x_n$. Then given elements $u_1, \ldots, u_n$ in the universe of a structure $\mathcal{S}$, we say that $\varphi(u_1, \ldots, u_n)$ is satisfied in $\mathcal{S}$ if φ is satisfied in $\mathcal{S}$ under the valuation mapping x_i to u_i for all $i \in \{1, \ldots, n\}$.

For a logic $\mathcal{L}$, an *$\mathcal{L}$-interpretation* from $\varXi_1$ to $\varXi_2$ is a family I of $\mathcal{L}$-formulas $\varphi_R(x_1, \ldots, x_n)$ over $\varXi_1$, for every relation name R of $\varXi_2$, where n is the arity of R. Having such an $\mathcal{L}$-interpretation, we can apply it to a structure $\mathcal{S}$ over $\varXi_1$; we obtain a structure $I(\mathcal{S})$ over $\varXi_2$, where every relation $R^{I(\mathcal{S})}$ is given by the tuples $(v_1, \ldots, v_n)$ of elements of the universe of $\mathcal{S}$ for which $\varphi_R(v_1, \ldots, v_n)$ is satisfied in $\mathcal{S}$. The universe of $I(\mathcal{S})$ is given implicitly as the set of all elements occurring in the relations $R^{I(\mathcal{S})}$ (because isolated elements are disallowed by the definition of a structure, there is no need to have a separate formula defining the universe).

2.2   Graphs and the Caucal Hierarchy

We consider directed, edge-labeled graphs. Thus, for a finite set $\varSigma$, a *$\varSigma$-labeled graph* G is a relational structure over the signature $\varXi_\varSigma$ containing binary relation names E_a for all $a \in \varSigma$. In other words, $G = (V^G, (E_a^G)_{a \in \varSigma})$, where V^G is a set of vertices, and $E_a^G \subseteq V^G \times V^G$ is a set of a-labeled edges, for every $a \in \Sigma$ (and where we assume that there are no isolated vertices, i.e., for every $v \in V^G$ there is an edge (v, w) or (w, v) in E_a^G for some $w \in V^G$ and $a \in \Sigma$). A graph is *deterministic* if for every $v \in V^G$ and $a \in \Sigma$ there is at most one vertex $w \in V^G$ such that $(v, w) \in E_a^G$.

A *path* from a vertex u to a vertex v labeled by $w = a_1 \ldots a_n$ is a sequence $v_0 a_1 v_1 \ldots a_n v_n \in V^G(\Sigma V^G)^*$, where $v_0 = u$, and $v_n = v$, and $(v_{i-1}, v_i) \in E_{a_i}^G$ for all $i \in \{1, \ldots, n\}$. A graph is called an (edge-labeled) *tree* when it contains a vertex r, called the *root*, such that for every vertex $v \in V^G$ there exists a unique path from r to v. The *unfolding* $Unf(G, r)$ of a graph $G = (V^G, (E_a^G)_{a \in \Sigma})$ from a vertex $r \in V^G$ is the tree $T = (V^T, (E_a^T)_{a \in \Sigma})$, where V^T is the set of all paths in G starting from r, and E_a^T (for every $a \in \Sigma$) contains pairs (w, w') such that $w' = w \cdot a \cdot v$ for some $v \in V^G$.

The Caucal hierarchy is a sequence of classes of graphs and trees; we use here the characterization from Carayol and Wöhrle [7] as a definition. We define *Graph*(0) to be the class containing all finite Σ-labeled graphs, for all finite sets of labels Σ. For all $n \geq 0$, we let

$$Tree(n + 1) = \{ Unf(G, r) \mid G \in Graph(n), r \in V^G \}, \qquad \text{and}$$
$$Graph(n + 1) = \{ I(T) \mid T \in Tree(n + 1), I \text{ an MSO-interpretation} \}.$$

We do not distinguish between isomorphic graphs.

2.3 Higher-Order Recursion Schemes

The set of *sorts* (aka. simple types) is constructed from a unique ground sort o using a binary operation $\rightarrow$; namely o is a sort, and if α and β are sorts, so is $\alpha \rightarrow \beta$. By convention, $\rightarrow$ associates to the right, that is, $\alpha \rightarrow \beta \rightarrow \gamma$ is understood as $\alpha \rightarrow (\beta \rightarrow \gamma)$. The *order* of a sort α, denoted $ord(\alpha)$ is defined by induction: $ord(o) = 0$ and $ord(\alpha_1 \rightarrow \ldots \rightarrow \alpha_k \rightarrow o) = \max_i(ord(\alpha_i)) + 1$ for $k \geq 1$.

Having a finite set of symbols Σ (an alphabet), a finite set of sorted nonterminals $\mathcal{N}$, and a finite set of sorted variables V, *(applicative) terms* over $(\Sigma, \mathcal{N}, V)$ are defined by induction:

- every nonterminal $N \in \mathcal{N}$ of sort α is a term of sort α;
- every variable $x \in V$ of sort α is a term of sort α;
- if $K_1, \ldots, K_k$ are terms of sort o, and $a \in \Sigma$ is a symbol, then $a\langle K_1, \ldots, K_k \rangle$ is a term of sort o;
- if K is a term of sort $\alpha \rightarrow \beta$, and L is a term of sort α, then $K L$ is a term of sort β.

The order of a term K, written $ord(K)$, is defined as the order of its sort.

A *(higher-order) recursion scheme* is a tuple $\mathcal{G} = (\Sigma, \mathcal{N}, \mathcal{R}, S)$, where Σ is a finite set of symbols, $\mathcal{N}$ a finite set of sorted nonterminals, and $\mathcal{R}$ a function assigning to every nonterminal $N \in \mathcal{N}$ of sort $\alpha_1 \rightarrow \ldots \rightarrow \alpha_k \rightarrow o$ a rule of the form $N x_1 \ldots x_k \rightarrow K$, where the sorts of variables $x_1, \ldots, x_k$ are $\alpha_1, \ldots, \alpha_k$, respectively, and K is a term of sort o over $(\Sigma, \mathcal{N}, \{x_1, \ldots, x_k\})$; finally, $S \in \mathcal{N}$

is a starting nonterminal of sort o. The order of a recursion scheme is defined as the maximum of orders of its nonterminals.

Unlike trees in the Caucal hierarchy, trees generated by recursion schemes are node-labeled; actually, these are infinite terms. They are defined by coinduction: for a finite set Σ and for $r \in \mathbb{N}$, a Σ-node-labeled tree of maximal arity r is of the form $a\langle T_1, \ldots, T_k \rangle$, where $a \in \Sigma$, and $k \leq r$, and $T_1, \ldots, T_k$ are again Σ-node-labeled trees of maximal arity r. For a tree $T = a\langle T_1, \ldots, T_k \rangle$, its set of vertices is defined as the smallest set such that

- ε is a vertex of T, labeled by a, and
- if u is a vertex of T_i for some $i \in \{1, \ldots, k\}$, labeled by b, then iu is a vertex of T, also labeled by b.

Such a tree can be seen as a relational structure over signature $\Xi_{\Sigma, r}^{nlt}$ containing unary relation names L_a for all $a \in \Sigma$, and binary relation names Ch_i for all $i \in \{1, \ldots, r\}$. Its universe is the set of vertices of T; for $a \in \Sigma$ the relation L_a^T contains all vertices labeled by a; for $i \in \{1, \ldots, r\}$ the i-th child relation Ch_i contains pairs (u, ui) such that both u and ui are vertices of T.

Having a recursion scheme $\mathcal{G}$, we define a rewriting relation $\rightarrow_{\mathcal{G}}$ among terms of sort o over $(\Sigma, \mathcal{N}, \emptyset)$: we have $N\, L_1 \ldots L_k \rightarrow_{\mathcal{G}} K[L_1/x_1, \ldots, L_k/x_k]$, where N is a nonterminal such that the rule $\mathcal{R}(N)$ is $N\, x_1 \ldots x_k \rightarrow K$ (and where $K[L_1/x_1, \ldots, L_k/x_k]$ is the term obtained from K by substituting L_1 for x_1, L_2 for x_2, and so on). We then define a tree *generated* by $\mathcal{G}$ from a term K of sort o over $(\Sigma, \mathcal{N}, \emptyset)$, by coinduction:

- if there is a reduction sequence from K to a term of the form $a\langle L_1, \ldots, L_k \rangle$, then the tree equals $a\langle T_1, \ldots, T_k \rangle$, where $T_1, \ldots, T_k$ are the trees generated by $\mathcal{G}$ from $L_1, \ldots, L_k$, respectively;
- otherwise, the tree equals $\omega\langle\rangle$ (where ω is a distinguished symbol).

A tree generated by $\mathcal{G}$ (without a term specified) is the tree generated by $\mathcal{G}$ from the starting nonterminal S.

We define when a term is *safe*, by induction on its structure:

- all nonterminals and variables are safe,
- a term $a\langle K_1, \ldots, K_k \rangle$ is safe if the subterms $K_1, \ldots, K_k$ are safe,
- a term $M = K\, L_1 \ldots L_k$ is safe if K and $L_1, \ldots, L_k$ are safe, and if $ord(x) \geq ord(M)$ for all variables x appearing in M.

Notice that not all subterms of a safe term need to be safe. A recursion scheme is safe if right sides of all its rules are safe.

2.4   Higher-Order Pushdown Automata

We actually need to consider two models of higher-order pushdown automata: nondeterministic (non-branching) automata of Carayol and Wöhrle [7], where letters are read by transitions, and deterministic tree-generating automata of Knapik, Niwiński, and Urzyczyn [19], where there are special commands for creating labeled tree vertices. We use the name *edge-labeled pushdown automata* for the former model, and *node-labeled pushdown automata* for the latter model. We only recall those fragments of definitions of these automata that are relevant for us.

For every $n \in \mathbb{N}$, and every finite set Γ containing a distinguished initial symbol $\bot \in \Gamma$, there are defined

- a set $PD_n(\Gamma)$ of pushdowns of order n over the stack alphabet Γ,
- an initial pushdown $\bot_n \in PD_n(\Gamma)$,
- a finite set $Op_n(\Gamma)$ of operations on these pushdowns, where every $op \in Op_n(\Gamma)$ is a partial function from $PD_n(\Gamma)$ to $PD_n(\Gamma)$, and
- a function $top \colon PD_n(\Gamma) \to \Gamma$ (returning the topmost symbol of a pushdown).

We assume that $Op_n(\Gamma)$ contains the identity operation *id*, mapping every element of $PD_n(\Gamma)$ to itself.

Having the above, we define an *edge-labeled pushdown automaton of order* n as a tuple $\mathcal{A} = (Q, \Sigma, \Gamma, q_I, \Delta)$, where Q is a finite set of states, Σ is a finite input alphabet, Γ is a finite stack alphabet, $q_I \in Q$ is an initial state, and $\Delta \subseteq Q \times \Gamma \times (\Sigma \cup \{\varepsilon\}) \times Q \times Op_n(\Gamma)$ is a transition relation. It is assumed that for every pair (q, γ) either all tuples $(q, \gamma, a, q', op) \in \Delta$ have $a = \varepsilon$, or all have $a \in \Sigma$. The automaton is *deterministic* if for every pair (q, γ) there is either exactly one transition (q, γ, a, q', op), where $a = \varepsilon$, or there are $|\Sigma|$ such transitions, one for every $a \in \Sigma$. A *configuration* of $\mathcal{A}$ is a pair $(q, s) \in Q \times PD_n(\Gamma)$, and $(q_I, \bot_n)$ is the initial configuration. For $a \in \Sigma \cup \{\varepsilon\}$, there is an a-labeled *transition* from a configuration (p, s) to a configuration (q, t), written $(p, s) \xrightarrow{a}_{\mathcal{A}} (q, t)$, if in Δ there is a tuple $(p, top(s), a, q, op)$ such that $op(s) = t$. The *configuration graph* of $\mathcal{A}$ is the edge-labeled graph of all configurations of $\mathcal{A}$ reachable from the initial configuration, with an edge labeled by $a \in \Sigma \cup \{\varepsilon\}$ from c to d if there is a transition $c \xrightarrow{a}_{\mathcal{A}} d$. The ε-closure of such a graph G is the Σ-labeled graph obtained from G by removing all vertices with only outgoing ε-labeled edges and adding an a-labeled edge between v and w if in G there is a path from v to w labeled by a word in $a\varepsilon^*$. The graph *generated* by $\mathcal{A}$ is the ε-closure of the configuration graph of $\mathcal{A}$.

Next, we define a *node-labeled pushdown automaton of order* n as a tuple $\mathcal{A} = (Q, \Sigma, \Gamma, q_I, \delta)$, where Q, Σ, Γ, q_I (and configurations) are as previously, and $\delta \colon Q \times \Gamma \to (Q \times Op_n(\Gamma)) \uplus (\Sigma \times Q^*)$ is a transition function. This time transitions are not labeled by anything; we have $(p, s) \to_{\mathcal{A}} (q, t)$ when $\delta(p, top(s)) = (q, op)$ and $op(s) = t$. We define when a node-labeled tree over alphabet $\Sigma \cup \{\omega\}$ is generated by $\mathcal{A}$ from (p, s), by coinduction:

- if $(p, s) \to_{\mathcal{A}}^* (q, t)$, and $\delta(q, top(t)) = (a, q_1, \ldots, q_k) \in \Sigma \times Q^*$, and trees $T_1, \ldots, T_k$ are generated by $\mathcal{A}$ from $(q_1, t), \ldots, (q_k, t)$, respectively, then the tree $a\langle T_1, \ldots, T_k \rangle$ is generated by $\mathcal{A}$ from (p, s),
- if there is no (q, t) such that $(p, s) \to_{\mathcal{A}}^* (q, t)$ and $\delta(q, top(t)) \in \Sigma \times Q^*$, then $\omega\langle\rangle$ is generated by $\mathcal{A}$ from (p, s).

While talking about the tree generated by $\mathcal{A}$, without referring to a configuration, we mean generating from the initial configuration $(q_I, \bot_n)$.

3 Between Caucal Hierarchy and Safe Recursion Schemes

The Caucal hierarchy is closely related to safe recursion schemes. Indeed, we have the following two results, from Carayol and Wöhrle [7, Theorem 3] and Knapik et al. [19, Theorems 5.1 and 5.3].

Fact 1. *For every* $n \in \mathbb{N}$, *a graph* G *is generated by some edge-labeled pushdown automaton of order* n *if and only if* $G \in Graph(n)$.

Fact 2. *For every* $n \in \mathbb{N}$, *a tree* T *is generated by some node-labeled pushdown automaton of order* n *if and only if it is generated by some safe recursion scheme of order* n.

It looks like the connection between the Caucal hierarchy and safe recursion schemes is already established by these two facts, but the settings of edge-labeled and node-labeled pushdown automata are not immediately compatible. Indeed, beside of the superficial syntactical difference between edge-labeled graphs from Fact 1 (and trees being their unfoldings) and node-labeled trees from Fact 2 we have two problems. First, node-labeled trees are only finitely branching (and moreover deterministic), while edge-labeled trees may have infinite branching. To deal with this, we use a fact from Carayol and Wöhrle [7, Theorem 2].

Fact 3. *For every* $G \in Graph(n)$, *where* $n \geq 1$, *there exists a tree* T *that is an unfolding of a deterministic graph* $G_{n-1} \in Graph(n-1)$, *and an MSO-interpretation[1]* I *such that* $G = I(T)$.

A second problem is that an edge-labeled pushdown automaton of order n generating a deterministic graph need not to be deterministic itself (and only deterministic edge-labeled automata can be easily turned into node-labeled automata). We thus need a fact from Parys [21, Theorem 1.1] (proved also in the Carayol's Ph.D. thesis [6, Corollary 3.5.3]).

Fact 4. *If a deterministic graph is generated by some edge-labeled pushdown automaton of order* n, *then it is also generated by some deterministic edge-labeled pushdown automaton of order* n.

Having all the recalled facts, it is now easy to prove the following lemma.

Lemma 5. *For every* $n \geq 1$, *a graph* G *is in* $Graph(n)$ *if and only if it can be obtained by applying an MSO-interpretation to a tree generated by a safe recursion scheme of order* $n - 1$.

Proof (sketch). Suppose first that $G = I(T)$ for some MSO-interpretation I and for some safe recursion scheme $\mathcal{G}$ of order $n - 1$ generating a tree T. By Fact 2, T is generated by a node-labeled pushdown automaton $\mathcal{A}$ of order $n - 1$. It is a routine to switch to the formalism of edge-labeled pushdown automata, that is,

---

[1] Carayol and Wöhrle say about an inverse rational mapping, which is a special case of an MSO-interpretation.

- change the node-labeled tree T (which is a structure over the signature $\Xi_{\Sigma,r}^{nlt}$) to a "similar" edge-labeled tree T' (which is a structure over the signature $\Xi_{\Sigma \cup \{1,\dots,r\}}$), where every edge of T from a vertex to its i-th child becomes an i-labeled edge, and where below every a-labeled vertex u we create a fresh vertex v_u with an a-labeled edge from u to v_u;
- change the node-labeled pushdown automaton $\mathcal{A}$ to an edge-labeled pushdown automaton $\mathcal{A}'$ of the same order $n-1$ such that T' is the unfolding of the graph generated by $\mathcal{A}'$;
- change the MSO-interpretation I evaluated in T to an MSO-interpretation I' evaluated in T', such that $I'(T') = I(T)$.

Finally, we use Fact 1 to say that the graph generated by $\mathcal{A}'$ belongs to $Graph(n-1)$. In effect its unfolding T' belongs to $Tree(n)$, and hence $G = I'(T')$ belongs to $Graph(n)$.

For the opposite direction, consider some graph $G \in Graph(n)$. We first use Fact 3 to say that there exists a tree T that is an unfolding of a deterministic graph $G_{n-1} \in Graph(n-1)$, and an MSO-interpretation I such that $G = I(T)$. By Fact 1 we obtain that G_{n-1} is generated by some edge-labeled pushdown automaton $\mathcal{A}$ of order $n-1$. Because of Fact 4 we can assume that $\mathcal{A}$ is deterministic. By definition, the vertex r such that $T = Unf(G_{n-1}, r)$ can be arbitrary; let $\mathcal{A}'$ be a modification of $\mathcal{A}$ that first reaches configuration r using ε-transitions, and then operates as $\mathcal{A}$ from r.

We now change $\mathcal{A}'$ into a node-labeled pushdown automaton $\mathcal{A}''$. To this end, we fix some order on the letters in Σ: let $\Sigma = \{a_1, \dots, a_k\}$. Moreover, without loss of generality we assume that for all transitions (q, γ, a, q', op) of $\mathcal{A}$ with $a \neq \varepsilon$, the operation op is id. Then, if from a pair (q, γ) we have transitions $(q, \gamma, a_1, q_1, id), \dots, (q, \gamma, a_k, q_k, id)$, we define $\delta(q, \gamma) = (\diamond, q_1, \dots, q_k)$, and for pairs (q, γ) being a source of ε-transitions $(q, \gamma, \varepsilon, q', op)$ we define $\delta(q, \gamma) = (q', op)$. The $\{\diamond, \omega\}$-node-labeled tree T' generated by $\mathcal{A}'$, after removing all ω-labeled vertices, and while treating an edge leading to the i-th child as a_i-labeled, equals T. It is easy to modify the interpretation I into I' such that $I'(T') = I(T) = G$. Finally, we use Fact 2 to say that T' is generated by a safe recursion scheme of order $n-1$. □

4 Closure Under MSO+U^{fin}-Interpretations

We now present the main theorem of this paper.

Theorem 6. *For every $n \in \mathbb{N}$, if $G \in Graph(n)$ and if I is an MSO+U^{fin}-interpretation, then $I(G) \in Graph(n)$.*

This theorem can be deduced from our previous result, which we recall now. We say that a $\Sigma \times \Gamma$-node-labeled tree T' enriches a Σ-node-labeled tree T, if it has the same vertices, and every vertex u labeled in T by some a is labeled in T' by a pair in $\{a\} \times \Gamma$.

Lemma 7. *Let $n \in \mathbb{N}$. For every MSO+U^{fin} formula φ and every safe recursion scheme $\mathcal{G}$ of order n generating a tree T there exists a safe recursion scheme $\mathcal{G}_+$ of order n that generates a tree T' enriching T, and an MSO formula φ_{MSO} such that for every valuation ν in T (defined at least for all free variables of φ) it holds that $T', \nu \models \varphi_{MSO}$ if and only if $T, \nu \models \varphi$.*

Proof. This result was shown in Parys [22, Lemma 5.4], without observing that the resulting recursion scheme $\mathcal{G}_+$ is of the same order as $\mathcal{G}$, and that it is safe when $\mathcal{G}$ is safe. We thus need to inspect the proof, to see this. Although the proof is not so simple, it applies only two basic kinds of modifications to the recursion scheme $\mathcal{G}$, in order to obtain $\mathcal{G}_+$.

First, it uses a construction of Haddad [16, Sect. 4.2] (described also in Parys [23, Sect. B.1]) to compose a recursion scheme with a morphism into a finitary applicative structure. It is already observed in Parys [23, Lemma 10.2] that this construction preserves the order. It is not difficult to see that it preserves safety as well: when a subterm K is transformed into a subterm M, then their order is the same, and their sets of free variables are essentially also the same, up to the fact that every single free variable of K corresponds to multiple free variables of M, all being of the same order as the free variable of K.

The second basic kind of modifications applied to the recursion scheme is the composition with finite tree transducers. This is realized by converting the recursion scheme to a collapsible pushdown automaton generating the same tree [17], composing the automaton with the transducer, and then converting it back to a recursion scheme. When the original recursion scheme is safe, we can convert it to a higher-order pushdown automaton, which can be converted back to a safe recursion scheme; as stated in Fact 2, this preserves the order. Moreover composing a higher-order pushdown automaton with a finite tree transducer is as easy as for collapsible pushdown automata, and clearly preserves the order. $\square$

Corollary 8. *Let $n \in \mathbb{N}$. For every safe recursion scheme $\mathcal{G}$ of order n generating a tree T, and every MSO+U^{fin}-interpretation I evaluated in T, there exists a safe recursion scheme $\mathcal{G}_+$ of order n generating a tree T_+, and an MSO-interpretation I_{MSO} such that $I_{MSO}(T_+) = I(T)$.*

Proof. Suppose that $I = (\varphi_i)_{i \in \{1,\dots,k\}}$. Basically, we apply Lemma 7 consecutively for all the formulae of I. More precisely, after $i - 1$ steps (where $i \in \{1, \dots, k\}$) we have a recursion scheme $\mathcal{G}_{i-1}$ (assuming $\mathcal{G}_0 = \mathcal{G}$) that generates a tree T_{i-1} enriching T. We modify φ_i to φ'_i that evaluated in T_{i-1} behaves like φ_i evaluated in T, that is, ignores the part of labels of T_{i-1} that was not present in T. Using Lemma 7 for the recursion scheme $\mathcal{G}_{i-1}$ and for the formula φ'_i we obtain a recursion scheme $\mathcal{G}_i$ that generates a tree T_i enriching T_{i-1} (hence enriching T), and an MSO formula $\varphi'_{MSO,i}$ such that for every valuation ν in T (defined at least for free variables of φ_i) it holds that $T_i, \nu \models \varphi'_{MSO,i}$ if and only if $T_{i-1}, \nu \models \varphi'_i$, that is, if and only if $T, \nu \models \varphi_i$. At the very end, for every $i \in \{1, \dots, k\}$ we modify $\varphi'_{MSO,i}$ into $\varphi_{MSO,i}$ that ignores the part of T_k appended after step i; we then have $T_k, \nu \models \varphi_{MSO,i}$ if and only if $T_i, \nu \models \varphi'_{MSO,i}$, that is, if and only if $T, \nu \models \varphi_i$. Taking $\mathcal{G}_+ = \mathcal{G}_k$, $T_+ = T_k$, and $I_{MSO} = (\varphi_{MSO,i})_{i \in \{1,...,k\}}$ we have $I_{MSO}(T_+) = I(T)$, as required. All the created recursion schemes are safe and of order n. $\quad\square$

Proof (Thorem 6). The class $Graph(0)$ contains exactly all finite graphs, and while interpreting in a finite graph we can only obtain a finite graph; this establishes the theorem for $n = 0$. We thus assume below that $n \geq 1$. In this case Lemma 5 gives us a safe recursion scheme $\mathcal{G}$ of order $n - 1$ generating a tree T, and an MSO-interpretation I_2 such that $I_2(T) = G$.

Suppose that $I_2 = (\varphi_a(x_1, x_2))_{a \in \Lambda}$ and $I = (\psi_\alpha(x_1, x_2))_{\alpha \in \Sigma}$ We create an MSO-interpretation I_3 such that $I_3(T) = I(I_2(T)) = I(G)$. To this end, in every formula ψ_α of I we replace every atomic formula $a(y, z)$ by the corresponding formula $\varphi_a(y, z)$ of I_2. Moreover, quantification in ψ_α should be restricted to those vertices of T that are actually taken to G, that is, to vertices y satisfying $\varphi_a(y, z)$ or $\varphi_a(z, y)$ for some $a \in \Lambda$ and some vertex z of T.

Corollary 8 gives us then a safe recursion scheme $\mathcal{G}_+$ of order $n-1$ generating a tree T_+, and an MSO-interpretation I_{MSO} such that $I_{MSO}(T_+) = I_3(T) = I(G)$. We conclude that $I(G) \in Graph(n)$ by Lemma 5. $\quad\square$

5 MSO+U-Interpretations Lead to Difficult Graphs

In this section we consider the full MSO+U logic, for which we prove the following theorem.

Theorem 9. *There is a tree $T \in Tree(2)$ and an MSO+U-interpretation I such that $I(T)$ is a graph with undecidable MSO theory; in effect, $I(G) \notin Graph(n)$ for any $n \in \mathbb{N}$.*

One can expect such a result, since the MSO+U logic is undecidable over infinite words [3]. We remark, though, that undecidability of a logic does not automatically imply that the logic can define some complicated ("undecidable") sets. For example, over rational numbers the MSO logic with quantification over cuts (real numbers) defines the same sets as the standard MSO logic quantifying only over rational numbers, but the latter logic is decidable while the former is not [11]. However, using arguments from topological complexity one can easily see that MSO+U is more expressible than MSO+U^{fin}: it is known that MSO+U can define sets located arbitrarily high in the projective hierarchy [18], while the topological complexity of MSO+U^{fin} can be bounded using the automaton model given in Parys [22]. Nevertheless, expressivity of the logic itself does not imply anything in the matter of interpretations: as we have seen in previous sections, MSO+U^{fin} is more expressive than MSO, and MSO is more expressive than FO, but interpretations in these logics define the same hierarchy of graphs.

Proof (Theorem 9, sketch). Because of Lemma 5, as the source of the interpretation I we can take a node-labeled tree T generated by a safe recursion scheme of order 2. We define the *depth-k comb* as the tree C_k such that $C_k = a\langle C_{k-1}, C_k \rangle$, where $C_0 = a\langle\rangle$. We also consider a depth-2 comb with first

i vertices marked by b: $C_{2,0} = C_2$ and $C_{2,i} = b\langle C_1, C_{2,i-1}\rangle$ for $i \geq 1$; and a depth-k comb (where $k \geq 3$) with first i vertices of every depth-2 comb marked by b: $C_{k,i} = a\langle C_{k-1,i}, C_{k,i}\rangle$.

We base on the undecidability proof from Bojańczyk, Parys, and Toruńczyk [3]. This proof, given a Minsky machine M constructs an MSO+U sentence φ_M that is true in an infinite forest of finite trees of height 3 if and only if the forest encodes a (finite) accepting run of M. Such a forest is then encoded in an infinite word, but is even easier to encode it in the depth-4 comb: we just need a set (a monadic variable) X saying which vertices of the comb appear in the considered forest (where roots of depth-k combs attached below a depth-$(k+1)$ comb represent children of the root of the latter comb).

Moreover, the recalled encoding of a run of M in the forest (checked by φ_M) requires that the arity of the first child of all (except finitely many) trees in the forest contains the initial value of the first counter, that is zero. We remove the part of φ_M saying that the initial value of the first counter is zero, and instead we add a part saying that from the first depth-2 comb in every depth-3 comb we take to X exactly left children of all b-labeled vertices. This way we obtain a sentence φ'_M (of the form $\exists X.\varphi''_M$) which, for every $i \in \mathbb{N}$, is true in $C_{4,i}$ if and only if M has an accepting run from the configuration c_i with value i in the first counter, value 0 in the second counter, and initial state.

We now consider the tree T_0 consisting of an infinite branch, where below the $(i+1)$-th node of this branch we attach $C_{4,i}$; formally, we define T_0 by coinduction: $T_i = a\langle C_{4,i}, T_{i+1}\rangle$ for $i \in \mathbb{N}$. We also consider the interpretation I_M consisting of two formulae: $\psi_a(x_1, x_2)$ that is true if x_1 and x_2 are consecutive vertices on the main branch, and $\psi_b(x_1, x_2)$ that is true if x_2 is a root of a comb $C_{4,i}$ in which φ'_M is true, and x_1 is its parent. The effect is that $I_M(T_0)$ consists of an infinite path with a-labeled edges, where for $i \in \mathbb{N}$ such that M accepts from c_i, we additionally have a b-labeled edge starting in the $(i+1)$-th vertex of that path.

Take a Minsky machine M such that the problem "given i, does M accept from c_i?" is undecidable. For such a machine, the graph $I_M(T_0)$ has undecidable MSO theory. And such a machine clearly exists: one can take a Minsky machine simulating a universal Turing machine, where the input to the latter is encoded in the value of the first counter.

It remains to observe that T_0 is generated by the safe recursion scheme of order 2 with the following rules:

$$S \to T\,C_2 \qquad\qquad C_4\,x \to a\langle C_3\,x, C_4\,x\rangle \quad C_2 \to a\langle C_1, C_2\rangle$$
$$T\,x \to a\langle C_4\,x, T\,b\langle C_1, x\rangle\rangle \quad C_3\,x \to a\langle x, C_3\,x\rangle \qquad C_1 \to a\langle a\langle\rangle, C_1\rangle \qquad\qquad \square$$

Acknowledgements. We thank Mikołaj Bojańczyk, Szymon Toruńczyk, and Arnaud Carayol for discussions preceding the process of creating this paper.

References

1. Bojańczyk, M.: A bounding quantifier. In: Marcinkowski, J., Tarlecki, A. (eds.) CSL 2004. LNCS, vol. 3210, pp. 41–55. Springer, Heidelberg (2004). https://doi.org/10.1007/978-3-540-30124-0_7

2. Bojańczyk, M.: Weak MSO with the unbounding quantifier. Theory Comput. Syst. **48**(3), 554–576 (2011). https://doi.org/10.1007/s00224-010-9279-2

3. Bojańczyk, M., Parys, P., Toruńczyk, S.: The MSO+U theory of (N, <) is undecidable. In: STACS, pp. 21:1–21:8 (2016). https://doi.org/10.4230/LIPIcs.STACS.2016.21

4. Bojańczyk, M., Toruńczyk, S.: Weak MSO+U over infinite trees. In: STACS, pp. 648–660 (2012). https://doi.org/10.4230/LIPIcs.STACS.2012.648

5. Cachat, T.: Higher order pushdown automata, the Caucal hierarchy of graphs and parity games. In: Baeten, J.C.M., Lenstra, J.K., Parrow, J., Woeginger, G.J. (eds.) ICALP 2003. LNCS, vol. 2719, pp. 556–569. Springer, Heidelberg (2003). https://doi.org/10.1007/3-540-45061-0_45

6. Carayol, A.: Automates infinis, logiques et langages. Ph.D. thesis. Université de Rennes 1 (2006)

7. Carayol, A., Wöhrle, S.: The Caucal hierarchy of infinite graphs in terms of logic and higher-order pushdown automata. In: Pandya, P.K., Radhakrishnan, J. (eds.) FSTTCS 2003. LNCS, vol. 2914, pp. 112–123. Springer, Heidelberg (2003). https://doi.org/10.1007/978-3-540-24597-1_10

8. Caucal, D.: On infinite transition graphs having a decidable monadic theory. In: Meyer, F., Monien, B. (eds.) ICALP 1996. LNCS, vol. 1099, pp. 194–205. Springer, Heidelberg (1996). https://doi.org/10.1007/3-540-61440-0_128

9. Caucal, D.: On infinite terms having a decidable monadic theory. In: Diks, K., Rytter, W. (eds.) MFCS 2002. LNCS, vol. 2420, pp. 165–176. Springer, Heidelberg (2002). https://doi.org/10.1007/3-540-45687-2_13

10. Colcombet, T.: A combinatorial theorem for trees. In: Arge, L., Cachin, C., Jurdziński, T., Tarlecki, A. (eds.) ICALP 2007. LNCS, vol. 4596, pp. 901–912. Springer, Heidelberg (2007). https://doi.org/10.1007/978-3-540-73420-8_77

11. Colcombet, T.: Composition with algebra at the background - on a question by Gurevich and Rabinovich on the monadic theory of linear orderings. In: Bulatov, A.A., Shur, A.M. (eds.) CSR 2013. LNCS, vol. 7913, pp. 391–404. Springer, Heidelberg (2013). https://doi.org/10.1007/978-3-642-38536-0_34

12. Colcombet, T., Löding, C.: Transforming structures by set interpretations. Log. Methods Comput. Sci. **3**(2) (2007). https://doi.org/10.2168/LMCS-3(2:4)2007

13. Courcelle, B.: Monadic second-order definable graph transductions: a survey. Theoret. Comput. Sci. **126**(1), 53–75 (1994). https://doi.org/10.1016/0304-3975(94)90268-2

14. Courcelle, B., Walukiewicz, I.: Monadic second-order logic, graph coverings and unfoldings of transition systems. Ann. Pure Appl. Logic **92**(1), 35–62 (1998). https://doi.org/10.1016/S0168-0072(97)00048-1

15. Engelfriet, J.: Iterated stack automata and complexity classes. Inf. Comput. **95**(1), 21–75 (1991). https://doi.org/10.1016/0890-5401(91)90015-T

16. Haddad, A.: IO vs OI in higher-order recursion schemes. In: FICS, pp. 23–30 (2012). https://doi.org/10.4204/EPTCS.77.4

17. Hague, M., Murawski, A.S., Ong, C.L., Serre, O.: Collapsible pushdown automata and recursion schemes. In: LICS, pp. 452–461 (2008). https://doi.org/10.1109/LICS.2008.34

18. Hummel, S., Skrzypczak, M.: The topological complexity of MSO+U and related automata models. Fundam. Inform. **119**(1), 87–111 (2012). https://doi.org/10.3233/FI-2012-728

19. Knapik, T., Niwiński, D., Urzyczyn, P.: Higher-order pushdown trees are easy. In: Nielsen, M., Engberg, U. (eds.) FoSSaCS 2002. LNCS, vol. 2303, pp. 205–222. Springer, Heidelberg (2002). https://doi.org/10.1007/3-540-45931-6_15

20. Ong, C.L.: On model-checking trees generated by higher-order recursion schemes. In: LICS, pp. 81–90 (2006). https://doi.org/10.1109/LICS.2006.38

21. Parys, P.: Variants of collapsible pushdown systems. In: CSL, pp. 500–515 (2012). https://doi.org/10.4230/LIPIcs.CSL.2012.500

22. Parys, P.: Recursion schemes, the MSO logic, and the U quantifier (submitted). https://arxiv.org/abs/1810.04763

23. Parys, P.: A type system describing unboundedness (submitted). https://hal.archives-ouvertes.fr/hal-01850934

24. Parys, P.: On the significance of the collapse operation. In: LICS, pp. 521–530 (2012). https://doi.org/10.1109/LICS.2012.62

25. Penelle, V.: Rewriting higher-order stack trees. Theory Comput. Syst. **61**(2), 536–580 (2017). https://doi.org/10.1007/s00224-017-9769-6

26. Walukiewicz, I.: Monadic second-order logic on tree-like structures. Theoret. Comput. Sci. **275**(1–2), 311–346 (2002). https://doi.org/10.1016/S0304-3975(01)00185-2

Tight Bounds on the Minimum Size
of a Dynamic Monopoly

Ahad N. Zehmakan[✉]

Department of Computer Science, ETH Zürich, Zürich, Switzerland
abdolahad.noori@inf.ethz.ch

Abstract. Assume that you are given a graph $G = (V, E)$ with an initial coloring, where each node is black or white. Then, in discrete-time rounds all nodes simultaneously update their color following a predefined deterministic rule. This process is called two-way r-bootstrap percolation, for some integer r, if a node with at least r black neighbors gets black and white otherwise. Similarly, in two-way α-bootstrap percolation, for some $0 < \alpha < 1$, a node gets black if at least α fraction of its neighbors are black, and white otherwise. The two aforementioned processes are called respectively r-bootstrap and α-bootstrap percolation if we require that a black node stays black forever.

For each of these processes, we say a node set D is a dynamic monopoly whenever the following holds: If all nodes in D are black then the graph gets fully black eventually. We provide tight upper and lower bounds on the minimum size of a dynamic monopoly.

Keywords: Dynamic monopoly · Bootstrap percolation ·
Threshold model · Percolating set · Target set selection

1   Introduction

Suppose for a graph G by starting from an initial configuration (coloring), where each node is either black or white, in each round all nodes simultaneously update their color based on a predefined rule. This basic abstract model, which is commonly known as cellular automaton, has been studied extensively in different areas, like biology, statistical physics, and computer science to comprehend the behavior of various real-world phenomena.

In two-way r-bootstrap percolation (or shortly r-BP), for some positive integer r, in each round a node gets black if it has at least r black neighbors, and white otherwise. Furthermore, in two-way α-bootstrap percolation (α-BP), for some $0 < \alpha < 1$, a node gets black if at least α fraction of its neighbors are black, and white otherwise. (Notice that there should not be any confusion between two-way α-BP and two-way r-BP since r is an integer value larger than equal to 1 and $0 < \alpha < 1$.) These two processes are supposed to model social phenomena like opinion forming in a community, where black and white could represent respectively positive and negative opinion concerning a reform proposal or a new

© Springer Nature Switzerland AG 2019
C. Martín-Vide et al. (Eds.): LATA 2019, LNCS 11417, pp. 381–393, 2019.
https://doi.org/10.1007/978-3-030-13435-8_28

product. For instance in a social network, if a certain number/fraction of someone's connections have a positive opinion regarding a particular topic, s/he will adapt the same opinion, and negative otherwise.

r-BP and α-BP are defined analogously, except we require that a black node stays unchanged. The main idea behind these two variants is to model monotone processes like rumor spreading in a society, fire propagation in a forest, and infection spreading among cells. For example, an individual gets informed of a rumor if a certain number/fraction of his/her friends are aware of it and stays informed forever, or a tree starts burning if a fixed number/fraction of the adjacent tress are on fire.

If we can try to convince a group of individuals to adopt a new product or innovation, for instance by providing them with free samples, and the goal is to trigger a large cascade of further adoptions, which set of individuals should we target and how large this set should be? This natural question brings up the well-studied concept of a dynamic monopoly. For each of the above four models, we say a node set is a dynamic monopoly, or shortly dynamo, whenever the following holds: If all nodes in the set are black initially then all nodes become black eventually.

Although the concept of a dynamo was studied earlier, e.g. by Balogh and Pete [3], it was formally defined and studied in the seminal work by Kempe, Kleinberg, and Tardos [14] and independently by Peleg [15], respectively motivated from viral marketing and fault-local mending in distributed systems. There is a massive body of work concerning the minimum size of a dynamo in different classes of graphs, for instance hypercube [2], the binomial random graph [5,7,17], random regular graphs [4,11], and many others. Motivated from the literature of statistical physics, a substantial amount of attention has been devoted to the d-dimensional lattice, for instance see [1,3,8,10,13].

In the present paper, we do not limit ourselves to a particular class of graphs and aim to establish sharp lower and upper bounds on the minimum size of a dynamo in general case, in terms of the number of nodes and the maximum/minimum degree of the underlying graph. See Table 1 for a summary.

Some of the bounds are quite trivial. For example in (two-way) r-BP, r is an obvious lower bound on the minimum size of a dynamo and it is tight since the complete graph K_n has a dynamo of size r (see Lemma 1 for the proof). However, some of the bounds are much more involved. For instance, an interesting open problem in this literature is whether the minimum size of a dynamo in two-way α-BP for $\alpha > 1/2$ is bounded by $\Omega(\sqrt{n})$ or not. We prove that this is true for $\alpha > \frac{3}{4}$. The case of $\frac{1}{2} < \alpha \leq \frac{3}{4}$ is left for the future research.

The proof techniques utilized are fairly standard and straightforward (some new and some inspired from prior work), however they turn out to be very effective. The upper bounds are built on the probabilistic method. We introduce a simple greedy algorithm which always returns a dynamo. Then, we discuss if this algorithm visits the nodes in a random order, the expected size of the output dynamo matches our desired bound. For the lower bounds, we define a suitable potential function, like the number of edges whose endpoints have different colors in a configuration. Then, careful analysis of the behavior of such a potential function during the process allows us to establish lower bounds on the size of a dynamo. To prove the tightness of our results, we provide explicit graph constructions for which the minimum size of a dynamo matches our bounds.

A simple observation is that by adding an edge to a graph the minimum size of a dynamo in (two-way) r-BP does not increase (for a formal argument, please see Sect. 2.1). Thus, if one keeps adding edges to a graph, eventually it will have a dynamo of minimum possible size, i.e. r. Thus, it would be interesting to ask for the degree-based density conditions that ensure that a graph G has a dynamo of size r. This was studied for r-BP, in the terms of the minimum degree, by Freund, Poloczek, and Reichman [9]. They proved that if the minimum degree $\delta(G)$ is at least $\lceil \frac{r-1}{r}n \rceil$, then there is a dynamo of size r in G. Gunderson [12] showed that the statement holds even for $\delta(G) \geq \frac{n}{2}+r$, and this is tight up to an additive constant. We study the same question concerning the two-way variant and prove that if $\delta(G) \geq \frac{n}{2}+r$ then the graph includes $\Omega(n^r)$ dynamos of size r. Note that this statement is stronger than Gunderson's result. Firstly, we prove that there is a dynamo of size r in two-way r-BP, which implies there is a dynamo of size r in r-BP. Moreover, we show that there is not only one but also $\Omega(n^r)$ of such dynamos. It is worth to stress that our proof is substantially shorter and simpler.

We say a dynamo is monotone if it makes all nodes black monotonically, that is no black node ever becomes white during the process. In r-BP and α-BP, any dynamo is monotone by definition, but in the two-way variants this is not necessarily true. Monotone dynamos also have been studied in different classes of graphs, see e.g. [1,8,15]. We provide tight bounds on the minimum size of a monotone dynamo in general graphs and also the special case of trees. In particular in two-way α-BP for $\alpha > \frac{1}{2}$, we prove the tight lower bound of $\sqrt{\frac{\alpha}{1-\alpha}}n$ on the minimum size of a monotone dynamo in general case. Interestingly, this bound drastically increases if we limit the underlying graph to be a tree. A question which arises, is whether there is a relation among the minimum size of a monotone dynamo and the girth of the underlying graph or not? This is partially answered in [6] for r-BP.

If all nodes in a dynamo are black, then black color will occupy the whole graph eventually. What if we only require the black color to survive in all upcoming rounds, but not necessarily occupy the whole graph? To address this question, we introduce two concepts of a stable set and immortal set. A non-empty node set S is stable (analogously immortal) whenever the following hold: If initially S is fully black, it stays black forever (respectively, black color survives forever). (See Sect. 1.1 for formal definitions) Trivially, a stable set is also immortal, but not necessarily the other way around. Similar to dynamo, we provide tight bounds on the minimum size of a stable and an immortal set; see Table 2. In r-BP and α-BP, a black node stays unchanged; thus, the minimum size of a stable/immortal set is equal to one. However, the situation is a bit more involved in two-way variants. Surprisingly, it turns out that in two-way 2-BP the parity of n, the number of nodes in the underlying graph, plays a key role.

The layout of the paper is as follows. First, we set some basic definitions in Sect. 1.1. Then, the bounds for dynamos, monotone dynamos, and stable/immortal sets are presented respectively in Sects. 2.1, 2.2, and 2.3.

1.1 Preliminaries

Let $G = (V, E)$ be a graph that we keep fixed throughout. We always assume that G is connected. For a node $v \in V$, $\Gamma(v) := \{u \in V : \{u, v\} \in E\}$ is the *neighborhood* of v. For a set $S \subset V$, we define $\Gamma(S) := \bigcup_{v \in S} \Gamma(v)$ and $\Gamma_S(v) := \Gamma(v) \cap S$. Furthermore, $d(v) := |\Gamma(v)|$ is the *degree* of v and $d_S(v) := |\Gamma_S(v)|$. We also define $\Delta(G)$ and $\delta(G)$ to be respectively the maximum and the minimum degree in graph G. (To lighten the notation, we sometimes shortly write Δ and δ where G is clear form the context). In addition, for a node set $A \subset V$, we define the *edge boundary* of A to be $\partial(A) := \{\{u, v\} : v \in A \land u \in V \setminus A\}$.

A *configuration* is a function $\mathcal{C} : V \to \{b, w\}$, where b and w stand for black and white. For a node $v \in V$, the set $\Gamma_a^{\mathcal{C}}(v) := \{u \in \Gamma(v) : \mathcal{C}(u) = a\}$ includes the neighbors of v which have color $a \in \{b, w\}$ in configuration $\mathcal{C}$. We write $\mathcal{C}|_S = a$ for a set $S \subseteq V$ and color $a \in \{b, w\}$ if $\mathcal{C}(u) = a$ for every $u \in S$.

Assume that for a given initial configuration $\mathcal{C}_0$ and some integer $r \geq 1$, $\mathcal{C}_t(v)$, which is the color of node $v \in V$ in round $t \geq 1$, is equal to b if $|\Gamma_b^{\mathcal{C}_{t-1}}(v)| \geq r$, and $\mathcal{C}_t(v) = w$ otherwise. This process is called *two-way r-bootstrap percolation*. If we require that a black node to stay black forever, i.e., $\mathcal{C}_t(v) = w$ if and only if $|\Gamma_b^{\mathcal{C}_{t-1}}(v)| < r$ and $\mathcal{C}_{t-1}(v) = w$, then the process is called *r-bootstrap percolation*.

Assumptions. We assume that r is fixed while we let n, the number of nodes in the underlying graph, tend to infinity. Note that if $d(v) < r$ for a node v, it never gets black in (two-way) r-BP, except it is initially black; thus, we always assume that $r \leq \delta(G)$.

Furthermore, suppose that for a given initial configuration $\mathcal{C}_0$ and some fixed value $0 < \alpha < 1$, $\mathcal{C}_t(v) = b$ for $v \in V$ and $t \geq 1$ if $|\Gamma_b^{\mathcal{C}_{t-1}}(v)| \geq \alpha |\Gamma(v)|$ and $\mathcal{C}_t(v) = w$ otherwise. This process is called *two-way α-bootstrap percolation*. If again we require a black node to stay unchanged, i.e., $\mathcal{C}_t(v) = w$ if and only if $|\Gamma_b^{\mathcal{C}_{t-1}}(v)| < \alpha |\Gamma(v)|$ and $\mathcal{C}_{t-1}(v) = w$, then the process is called *α-bootstrap percolation*.

For any of the above processes on a connected graph $G = (V, E)$, we define a node set D to be a *dynamic monopoly*, or shortly *dynamo*, whenever the following holds: If $\mathcal{C}_t|_D = b$ for some $t \geq 0$, then $\mathcal{C}_{t'}|_V = b$ for some $t' \geq t$. Furthermore, assume that for a non-empty node set S, if $\mathcal{C}_t|_S = b$ for some $t \geq 0$, then $\mathcal{C}_{t'}|_S = b$ for any $t' \geq t$; then, we say S is a *stable set*. Finally, a node set I is an *immortal set* when the following is true: If $\mathcal{C}_t|_I = b$ for some $t \geq 0$, then for any $t' \geq t$ there exists a node $v \in V$ so that $\mathcal{C}_{t'}(v) = b$.

For a graph G we define the following notations:

- $MD_r(G) :=$ The minimum size of a dynamo in r-BP.
- $\overleftarrow{MD}_r(G) :=$ The minimum size of a dynamo in two-way r-BP.

- $MS_r(G) :=$ The minimum size of a stable set in r-BP.
- $\overleftarrow{MS}_r(G) :=$ The minimum size of a stable set in two-way r-BP.
- $MI_r(G) :=$ The minimum size of an immortal set in r-BP.
- $\overleftarrow{MI}_r(G) :=$ The minimum size of an immortal set in two-way r-BP.

We analogously define $MD_\alpha(G)$, $MS_\alpha(G)$, $MI_\alpha(G)$ for α-BP and $\overleftarrow{MD}_\alpha(G)$, $\overleftarrow{MS}_\alpha(G)$, and $\overleftarrow{MI}_\alpha(G)$ for two-way α-BP.

As a warm-up, let us compute some of these parameters for some specific class of graphs in Lemma 1, which actually come in handy several times later for arguing the tightness of our bounds.

Lemma 1. *For complete graph K_n, $MD_r(K_n) = \overleftarrow{MD}_r(K_n) = r$ and $MD_\alpha(K_n) \geq \lceil \alpha n \rceil - 1$. Furthermore, for an r-regular graph $G = (V,E)$ and $r \geq 2$, $\overleftarrow{MD}_r(G) = n$.*

Proof. Firstly, $r \leq MD_r(K_n)$ because by starting from a configuration with less than r black nodes in r-BP, clearly in the next round all nodes will be white. Secondly, $\overleftarrow{MD}_r(K_n) \leq r$ because from a configuration with r black nodes in two-way r-BP, in the next round all the $n - r$ white nodes turn black and after one more round all nodes will be black because $n - r$ is at least $r + 1$ (recall that we assume that r is fixed while n tends to infinity). By these two statements and the fact that $MD_r(K_n) \leq \overleftarrow{MD}_r(K_n)$ (this is true since a dynamo in two-way r-BP is also a dynamo in r-BP), we have $MD_r(K_n) = \overleftarrow{MD}_r(K_n) = r$. In two-way α-BP on K_n, by starting with less than $\lceil \alpha(n - 1) \rceil$ black nodes, the process gets fully white in one round, which implies that $MD_\alpha(K_n) \geq \lceil \alpha n \rceil - 1$. (The interested reader might try to find the exact value of $MD_\alpha(K_n)$ as a small exercise).

For two-way r-BP on an r-regular graph G, consider an arbitrary configuration with at least one white node, say v. Trivially, in the next round all nodes in $\Gamma(v)$ will be white. Thus, by starting from any configuration except the fully black configuration, the process never gets fully black. This implies that $\overleftarrow{MD}_r(G) = n$. (We exclude $r = 1$ because a 1-regular graph is disconnected for large n.) □

2   Lower and Upper Bounds

2.1   Dynamos

In this section, we provide lower and upper bounds on the minimum size of a dynamo in α-BP (Theorem 1), two-way α-BP (Theorem 2), r-BP (Theorem 3), and two-way r-BP (Theorem 4). See Table 1 for a summary. Furthermore in Theorem 5, we present sufficient minimum degree condition for a graph to have a dynamo of size r in two-way r-BP.

In Theorem 1, the lower bound is trivial and the upper bound is proven by applying an idea similar to the one from Theorem 2.1 in [16].

Table 1. The minimum size of a dynamo. All bounds are tight up to an additive constant, except some of the bounds for two-way α-BP.

Model	Lower bound	Upper bound
α-BP	1	$(\frac{\delta+\frac{1}{\alpha}}{\delta+1})\,\alpha n$
Two-way α-BP $\alpha > \frac{3}{4}$	$2\alpha\sqrt{n} - 1$	n
Two-way α-BP $\alpha \leq \frac{3}{4}$	1	n
r-BP	r	$(\frac{r}{1+\delta})\,n$
Two-way r-BP $r \geq 2$	r	n
Two-way r-BP $r = 1$	1	2

Theorem 1. *For a graph* $G = (V, E)$, $1 \leq MD_\alpha(G) \leq (\frac{\delta+\frac{1}{\alpha}}{\delta+1})\,\alpha n$.

Proof. We apply an probabilistic method argument. Consider an arbitrary labeling $L : V \to [n]$, which assigns a unique label from 1 to n to each node. Define the set $D_L := \{v \in V : |\{u \in \Gamma(v) : L(u) < L(v)\}| < \alpha d(v)\}$. We claim that D_L is a dynamo in α-BP, irrespective of L. More precisely, we show that by starting from a configuration where D_L is fully black, in the t-th round for $t \geq 1$ all nodes with label t or smaller are black. This immediately implies that D_L is a dynamo since in at most n rounds the graph gets fully black. The node with label 1 is in D_L by definition; thus, it is black in the first round. As the induction hypothesis, assume that all nodes with label t or smaller are black in the t-th round for some $t \geq 1$. If node v with label $t+1$ is in D_L then it is already black; otherwise, it has at least $\alpha d(v)$ neighbors with smaller labels, which are black by the induction hypothesis. Thus, v will be black in the $(t+1)$-th round and all nodes with smaller labels also will stay black.

Assume that we choose a labeling L uniformly at random among all $n!$ possible labellings. Let us compute the expected size of D_L.

$$\mathbb{E}[|D_L|] = \sum_{v \in V} Pr[\text{node } v \text{ is in } D_L] = \sum_{v \in V} \frac{\lceil \alpha d(v) \rceil}{d(v) + 1} \leq \sum_{v \in V} \frac{\alpha d(v) + 1}{d(v) + 1}$$

$$\leq \sum_{v \in V} \frac{\alpha \delta + 1}{\delta + 1} = (\frac{\delta + \frac{1}{\alpha}}{\delta + 1})\alpha n.$$

Therefore, there exists a labeling L with $|D_L| \leq (\frac{\delta+\frac{1}{\alpha}}{\delta+1})\alpha n$, which implies that there exists a dynamo of this size.   □

Tightness. The lower bound is tight since in the star S_n, a tree with one internal node and n leaves, the internal node is a dynamo, irrespective of $0 < \alpha < 1$. Furthermore for the complete graph K_n, $MD_\alpha(K_n) \geq \lceil \alpha n \rceil - 1$ (see Lemma 1) and our upper bound is equal to $\alpha n + 1 - \alpha$ (by plugging in the value $\delta = n - 1$). Thus, the upper bound is tight up to an additive constant.

Theorem 2. *For a graph* $G = (V, E)$,

(i) $2\alpha\sqrt{n} - 1 \leq \overleftarrow{MD}_\alpha(G) \leq n$ for $\alpha > \frac{3}{4}$
(ii) $1 \leq \overleftarrow{MD}_\alpha(G) \leq n$ for $\alpha \leq \frac{3}{4}$.

Proof. We prove the lower bound of $2\alpha\sqrt{n} - 1$; all other bounds are trivial. Let D be an arbitrary dynamo in two-way α-BP for $\alpha > \frac{3}{4}$. Consider the initial configuration C_0 where $C_0|_D = b$ and $C_0|_{V\setminus D} = w$. Furthermore, we define for $t \geq 1$ $B_t := \{v \in V : C_{t'-1}(v) = C_{t'}(v) = b \text{ for some } t' \leq t\}$ to be the set of nodes which are black in two consecutive rounds up to the t-th round.

Now, we define the potential function $\Phi_t := |B_t| + |\partial(B_t)|$ to be the number of nodes in B_t plus the number of edges with exactly one endpoint in B_t. Since D is a dynamo, there exists some $T \geq 1$ such that $C_T|_V = b$, which implies that $C_{T+1}|_V = b$. Thus, we have $\Phi_{T+1} = |B_{T+1}| + |\partial(B_{T+1})| = n + 0 = n$. We prove that $\Phi_1 \leq \frac{1}{4\alpha^2}(|D| + (2\alpha - 1))^2$ and $\Phi_{t+1} \leq \Phi_t$ for any $t \geq 1$. Therefore, $n = \Phi_{T+1} \leq \Phi_1 \leq \frac{1}{4\alpha^2}(|D| + (2\alpha - 1))^2$, which results in

$$4\alpha^2 n \leq (|D| + (2\alpha - 1))^2 \Rightarrow 2\alpha\sqrt{n} - 2\alpha + 1 \leq |D| \xRightarrow{2\alpha < 2} 2\alpha\sqrt{n} - 1 \leq |D|.$$

Firstly, we prove that $\Phi_{t+1} \leq \Phi_t$ for any $t \geq 1$. Define the set $B := B_{t+1} \setminus B_t$. If $B = \emptyset$, then $B_t = B_{t+1}$, because by definition $B_t \subseteq B_{t+1}$, which implies that $\Phi_{t+1} = \Phi_t$. Thus, assume that $B \neq \emptyset$. A node $v \in B$ is black in both rounds t and $t + 1$, which means it has at least $\alpha d(v)$ black neighbors in the $(t - 1)$-th round and $\alpha d(v)$ black neighbors in the t-th round. Thus by the pigeonhole principle, at least $2\alpha - 1$ fraction of its neighbors are black in both rounds $t - 1$ and t. In other words, at least $2\alpha - 1$ fraction of its neighbors are in B_t. Note that $2\alpha - 1 > \frac{1}{2}$ for $\alpha > \frac{3}{4}$. Therefore, for each node $v \in B$ more than half of its neighbors are in B_t. This implies that $|\partial(B_{t+1})| \leq |\partial(B_t)| - |B|$. Thus,

$$\Phi_{t+1} = |B_{t+1}| + |\partial(B_{t+1})| \leq |B_t| + |B| + |\partial(B_t)| - |B| = |B_t| + |\partial(B_t)| = \Phi_t.$$

It remains to show that $\Phi_1 \leq \frac{1}{4\alpha^2}(|D| + (2\alpha - 1))^2$. Recall that for a node $v \in B_1$, $d_{V\setminus D}(v)$ and $d_{D\setminus B_1}(v)$ are the number of edges that v shares with nodes in $V \setminus D$ and $D \setminus B_1$, respectively. In addition, note that $C_0(v) = C_1(v) = b$ by the definition of B_1. Since all nodes in $V \setminus D$ are white in C_0 and v must be black in C_1, $\frac{d_{V\setminus D}(v)}{d(v)} \leq (1 - \alpha)$. Furthermore, since v has at most $|D| - 1$ neighbors in D (this is true because $C_0(v) = b$, that is $v \in D$), we have that $|D| - 1 + d_{V\setminus D}(v) \geq d(v)$. Thus, $\frac{d_{V\setminus D}(v)}{|D|-1+d_{V\setminus D}(v)} \leq (1 - \alpha)$, which implies that $d_{V\setminus D}(v) \leq \frac{1-\alpha}{\alpha}(|D| - 1)$. Moreover, since $B_1 \subseteq D$, $d_{D\setminus B_1}(v) \leq |D| - |B_1|$. Putting the last two statements together outputs $d_{V\setminus B_1}(v) \leq |D| - |B_1| + \frac{1-\alpha}{\alpha}(|D| - 1) = \frac{1}{\alpha}|D| - |B_1| + \frac{\alpha-1}{\alpha}$. Therefore

$$\Phi_1 = |B_1| + |\partial(B_1)| \leq |B_1| + |B_1| \cdot (\frac{1}{\alpha}|D| - |B_1| + \frac{\alpha - 1}{\alpha}) =$$
$$(\frac{1}{\alpha}|D| + \frac{2\alpha - 1}{\alpha})|B_1| - |B_1|^2 = \frac{|D| + (2\alpha - 1)}{\alpha}|B_1| - |B_1|^2.$$

The upper bound is maximized for $|B_1| = \frac{1}{2}(\frac{|D|+(2\alpha-1)}{\alpha})$. Thus, $\Phi_1 \leq \frac{1}{4\alpha^2}(|D| + (2\alpha - 1))^2$. $\square$

Tightness. Let us first consider part (i). We show that there exist n-node graphs with dynamos of size $k = \sqrt{\frac{\alpha}{1-\alpha}}n$ for $\alpha > \frac{3}{4}$ (actually our construction works also for $\alpha \geq 1/2$), which demonstrates that our bound is asymptotically tight. Consider a clique of size k and attach $\frac{n}{k} - 1$ distinct leaves to each of its nodes. The resulting graph has $k + k(\frac{n}{k} - 1) = n$ nodes. Consider the initial configuration C_0 in which the clique is fully black and all other nodes are white. In C_1, all the leaves turn black because their neighborhood is fully black in C_0. Furthermore, each node v in the clique stays black since it has $k - 1$ black neighbors and $k - 1 \geq \alpha d(v)$, which we prove below. Thus, it has a dynamo of size k.

$$\alpha d(v) = \alpha(k - 1 + \frac{n}{k} - 1) = \alpha(\sqrt{\frac{\alpha}{1-\alpha}}n + \sqrt{\frac{1-\alpha}{\alpha}}n - 2) =$$

$$\sqrt{\frac{\alpha}{1-\alpha}}(\alpha\sqrt{n} + (1-\alpha)\sqrt{n}) - 2\alpha \overset{\alpha \geq 1/2}{\leq} \sqrt{\frac{\alpha}{1-\alpha}}n - 1 = k - 1.$$

For $\alpha > \frac{1}{2}$ and the cycle C_n, a dynamo must include all nodes. Let C be a configuration on C_n with at least one white node, in two-way α-BP for $\alpha > \frac{1}{2}$ in the next round both its neighbors will be white. Thus, a configuration with one or more white nodes never reaches the fully black configuration. This implies that our trivial upper bound of n is tight.

Now, we provide the following observation for the cycle C_n, which implies that the lower bound in part (ii) is tight for $\alpha \leq \frac{1}{2}$. However, for $\frac{1}{2} < \alpha \leq \frac{3}{4}$ we do not know whether this bound is tight or not.

Observation 1. $\overleftarrow{MD}_\alpha(C_n) = 1$ for $\alpha \leq \frac{1}{2}$ and odd n.

Proof. Consider an odd cycle $v_1, v_2, \cdots, v_{2k+1}, v_1$ for $n = 2k + 1$. Assume that in the initial configuration C_0 there is at least one black node, say v_1. In configuration C_1, nodes v_2 and v_{2k+1} are both black because $\alpha \leq \frac{1}{2}$. With a simple inductive argument, after k rounds two adjacent nodes v_{k+1} and v_{k+2} will be black. In the next round, they both stay black and nodes v_k and v_{k+3} get black as well. Again with an inductive argument, after at most k more rounds all nodes will be black. □

Theorem 3. [16] *For a graph G, $r \leq MD_r(G) \leq (\frac{r}{1+\delta})n$.*

The upper bound is known by prior work [16]. The lower bound is trivial since if a configuration includes less than r black nodes in r-BP, no white node will turn black in the next round. Furthermore, $MD_r(K_n) = r$ (see Lemma 1), which implies that the lower bound and upper bound are both tight (note for K_n, $(\frac{r}{1+\delta})n = r$).

Theorem 4. *In a graph G,*

(i) *if $r \geq 2$, $r \leq \overleftarrow{MD}_r(G) \leq n$*
(ii) *if $r = 1$, $\overleftarrow{MD}_r(G) = 2$ if G is bipartite and $\overleftarrow{MD}_r(G) = 1$ otherwise.*

The bounds in part (i) are trivial. Regarding part (ii), if G is not bipartite, it has an odd cycle. We show that any node on such a cycle is a dynamo for two-way 1-BP. For a bipartite graph, there is no dynamo of size one, but any two adjacent nodes are a dynamo. The formal proof will be provided in the extended version of the paper.

Tightness. The bounds in part (i) are tight because $\overleftarrow{MD}_r(K_n) = r$ and $\overleftrightarrow{MD}_r(G) = n$ for any r-regular graph G and $r \geq 2$ (see Lemma 1).

For graphs $G = (V, E)$ and $G' = (V, E')$, if $E \subset E'$ then $MD_r(G') \leq MD_r(G)$ and $\overleftrightarrow{MD}_r(G') \leq \overleftrightarrow{MD}_r(G)$. This is true because by a simple inductive argument any dynamo in G is also a dynamo in G'. Thus, if we keep adding edges to any graph, eventually it will have a dynamo of minimum possible size, namely r, in both r-BP and two-way r-BP. Thus, it would be interesting to ask for the degree-based density conditions that ensure that a graph has a dynamo of size r. Gunderson [12] proved that if $\delta \geq \frac{n}{2} + r$ for a graph G (r can be replaced by $r - 3$ for $r \geq 4$), then $MD_r(G) = r$. We provide similar results for the two-way variant.

Theorem 5. *If $\delta \geq \frac{n}{2} + r$ for a graph $G = (V, E)$, then it has $\Omega(n^r)$ dynamos of size r in two-way r-BP.*

Note that this statement is stronger than Gunderson's result. Firstly, we prove that there is a dynamo of size r in two-way r-BP, which immediately implies that there is a dynamo of such size in r-BP. In addition, we prove that actually there exist $\Omega(n^r)$ of such dynamos (this is asymptotically the best possible since there are $\binom{n}{r} = \mathcal{O}(n^r)$ sets of size r). It is worth to mention that our proof is substantially shorter. (The proof of Theorem 5 is given in the extended version of the paper.)

2.2 Monotone Dynamos

Let us first define a monotone dynamo formally. For a graph $G = (V, E)$, we say a node set D is a *monotone dynamo* whenever the following holds: If $\mathcal{C}_0|_D = b$ and $\mathcal{C}_0|_{V \setminus D} = w$, then for some $t \geq 1$ we have $\mathcal{C}_t|_V = b$ and $\mathcal{C}_{t'-1} \leq \mathcal{C}_{t'}$ for any $t' \leq t$, which means any black node in $\mathcal{C}_{t'-1}$ is also black in $\mathcal{C}_{t'}$. Now, we provide bounds on the minimum size of a monotone dynamo. Since a dynamo in r-BP and α-BP is monotone by definition, our bounds from Sect. 2.1 apply.

For a graph $G = (V, E)$, the minimum size of a monotone dynamo in two-way r-BP is lower-bounded by $r + 1$. Assume that there is a monotone dynamo D of size r or smaller. If $\mathcal{C}_0|_D = b$ and $\mathcal{C}_0|_{V \setminus D} = w$, then $\mathcal{C}_1|_D = w$; this is in contradiction with the monotonicity of D. This lower bound is tight because in K_n, a set of size $r + 1$ is a monotone dynamo. Furthermore, the trivial upper bound of n is tight for r-regular graphs with $r \geq 2$ (see Lemma 1). For $r = 1$, any two adjacent nodes are a monotone dynamo; thus, the minimum size of a monotone dynamo is two.

In two-way α-BP, for $\alpha \leq \frac{1}{2}$, on the cycle C_n any two adjacent nodes are a monotone dynamo, which provides the tight lower bound of 2. However, we are not aware of any non-trivial upper bound. For $\alpha > \frac{1}{2}$, the trivial upper bound of n is tight for C_n. We provide the lower bound of $\sqrt{\frac{\alpha}{1-\alpha}n} - 1$ in Theorem 6, which is tight since the construction given for the tightness of Theorem 2 provides a monotone dynamo whose size matches our lower bound, up to an additive constant. Furthermore, we show that if we restrict the underlying graph to be a tree, we get the stronger bound of $\frac{\alpha}{2-\alpha}n$.

Theorem 6. *For a graph $G = (V, E)$ and two-way α-BP with $\alpha > \frac{1}{2}$, the minimum size of a monotone dynamo is at least $\sqrt{\frac{\alpha}{1-\alpha}n} - 1$, and at least $\frac{\alpha}{2-\alpha}n$ if G is a tree.*

Proof. Let set $D \subseteq V$ be a monotone dynamo in G. Suppose the process starts from the configuration where only all nodes in D are black. Let D_t denote the set of black nodes in round t. Then, $D_0 = D$ and $D_t \subseteq D_{t+1}$ by the monotonicity of D. Furthermore, define the potential function $\Phi_t := \partial(D_t)$. We claim that $\Phi_{t+1} \leq \Phi_t - |D_t \setminus D_{t-1}|$ because for any newly added black node, i.e. any node in $D_t \setminus D_{t-1}$, the number of neighbors in D_t is strictly larger than $V \setminus D_t$ (note that $\alpha > \frac{1}{2}$). In addition, since D is a dynamo, $C_T|_V = b$ for some $T \geq 0$, which implies $\Phi_T = 0$. Thus,

$$\Phi_T = 0 \leq \Phi_0 - (n - |D|) \Rightarrow n \leq \Phi_0 + |D|. \tag{1}$$

For $v \in D$, $d_{V \setminus D}(v) \leq \frac{1-\alpha}{\alpha}d_D(v)$ because D is a monotone dynamo and at least α fraction of v's neighbors must be in D. Furthermore, $d_D(v) \leq |D| - 1$, which implies that $d_{V \setminus D}(v) \leq \frac{1-\alpha}{\alpha}(|D| - 1)$. Now, we have

$$\Phi_0 = \partial(D) = \sum_{v \in D} d_{V \setminus D}(v) \leq \frac{1-\alpha}{\alpha} \sum_{v \in D} (|D| - 1) = \frac{1-\alpha}{\alpha}|D|^2 - \frac{1-\alpha}{\alpha}|D|. \tag{2}$$

Putting Eqs. 1 and 2 in parallel, plus some small calculations, imply that

$$n \leq \frac{1-\alpha}{\alpha}|D|^2 + (1 - \frac{1-\alpha}{\alpha})|D| \Rightarrow \sqrt{\frac{\alpha}{1-\alpha}n} - 1 \leq |D|.$$

When G is a tree, we have

$$\Phi_0 = \partial(D) = \sum_{v \in D} d_{V \setminus D}(v) \leq \frac{1-\alpha}{\alpha} \sum_{v \in D} d_D(v) \leq \frac{1-\alpha}{\alpha} 2(|D| - 1)$$

because the induced subgraph by D is a forest with $|D|$ nodes, which thus has at most $|D| - 1$ edges. Combining this inequality and Eq. 1 yields

$$n \leq \frac{2(1-\alpha)}{\alpha}|D| - 2 + |D| \Rightarrow n \leq \frac{2-\alpha}{\alpha}|D| \Rightarrow \frac{\alpha}{2-\alpha}n \leq |D|.$$

$\square$

2.3   Stable and Immortal Sets

In this section, we provide tight bounds on the minimum size of a stable/immortal set. In α-BP and r-BP a black node stays unchanged, which simply implies that $MS_\alpha(G) = MI_\alpha(G) = MS_r(G) = MI_r(G) = 1$ for a graph G. Thus, we focus on the two-way variants in the rest of the section. The bounds are given in Table 2.

Table 2. The minimum size of a stable/immortal set. All our bounds are tight up to an additive constant. $x = 0$ and $x = 1$ respectively for odd and even n.

Model	Stable set		Immortal set	
	Lower bound	Upper bound	Lower bound	Upper bound
Two-way α-BP $\alpha \le \frac{1}{2}$	$\lceil \frac{1}{1-\alpha} \rceil$	αn	1	αn
Two-way α-BP $\alpha > \frac{1}{2}$	$\lceil \frac{1}{1-\alpha} \rceil$	n	1	n
Two-way r-BP $r = 1$	2	2	1	1
Two-way r-BP $r = 2$	$r + 1$	n	2	$\frac{n}{1+x}$
Two-way r-BP $r \ge 3$	$r + 1$	n	r	n

Stable Sets in Two-Way α-BP. We present tight bounds on $\overleftarrow{MS}_\alpha(G)$ in Theorem 7, whose proof is given in the extended version of the paper. The proof of the lower bound is built on the simple observation that a set S is stable in two-way α-BP if for each node $v \in S$, $d_S(v) \ge \alpha d(v)$. Furthermore, the main idea to prove the upper bound of $\alpha n + \mathcal{O}(1)$ is to consider a partitioning of the node set into subsets of certain sizes such that the number of edges among them is minimized and then show that one of the subsets is stable and has our desired size.

Theorem 7. *For a graph $G = (V, E)$,*

(i) $\lceil \frac{1}{1-\alpha} \rceil \le \overleftarrow{MS}_\alpha(G) \le n$ *for* $\alpha > \frac{1}{2}$

(ii) $2 = \lceil \frac{1}{1-\alpha} \rceil \le \overleftarrow{MS}_\alpha(G) \le \alpha n + \mathcal{O}(1)$ *for* $\alpha \le \frac{1}{2}$.

Furthermore, these bounds are tight up to an additive constant.

Stable Sets in Two-Way r-BP. For a graph G, $\overleftarrow{MS}_r(G) = 2$ for $r = 1$ because two adjacent nodes create a stable set. For $r \ge 2$, we have tight bounds of $r + 1 \le \overleftarrow{MS}_r(G) \le n$. Notice if in a configuration less than $r + 1$ nodes are black, in the next round all black nodes turn white. Furthermore, the lower bound of $r + 1$ is tight for K_n and the upper bound is tight for r-regular graphs.

Immortal Sets in Two-Way α-BP. For a graph G, $1 \le \overleftarrow{MI}_\alpha(G) \le n$ for $\alpha > \frac{1}{2}$ and $1 \le \overleftarrow{MI}_\alpha(G) \le \alpha n + \mathcal{O}(1)$ for $\alpha \le \frac{1}{2}$. All bounds are trivial except $\alpha n + \mathcal{O}(1)$, which is a corollary of Theorem 7 (note that stability implies immortality).

The lower bound of 1 is tight since the internal node of the star graph S_n is an immortal set of size 1. Regarding the tightness of the upper bounds, we have $\overleftarrow{MI}_\alpha(K_n) \geq \alpha(n-1)$ for $\alpha \leq 1/2$ and $\overleftarrow{MI}_\alpha(C_n) = n$ for $\alpha > 1/2$ and odd n (this basically follows from the proof of Observation 1 by replacing black with white and $\alpha \leq \frac{1}{2}$ with $\alpha > \frac{1}{2}$).

Immortal Sets in Two-Way r-BP. We provide tight bounds on $\overleftarrow{MI}_r(G)$ in Theorem 8, which is proven in the extended version of the paper. Interestingly, the parity of n plays a key role concerning the minimum size of an immortal set for $r = 2$.

Theorem 8. *For a graph $G = (V, E)$,*

(i) if $r = 1$, $\overleftarrow{MI}_r(G) = 1$

(ii) if $r = 2$, $2 \leq \overleftarrow{MI}_r(G) \leq \frac{n}{1+x}$ ($x = 0$ for odd n and $x = 1$ for even n).

(iii) if $r \geq 3$, $r \leq \overleftarrow{MI}_r(G) \leq n$. Furthermore, these bounds are tight.

References

1. Balister, P., Bollobás, B., Johnson, J.R., Walters, M.: Random majority percolation. Random Struct. Algorithms **36**(3), 315–340 (2010)
2. Balogh, J., Bollobás, B., Morris, R.: Bootstrap percolation in high dimensions. Comb. Probab. Comput. **19**(5–6), 643–692 (2010)
3. Balogh, J., Pete, G.: Random disease on the square grid. Random Struct. Algorithms **13**(3–4), 409–422 (1998)
4. Balogh, J., Pittel, B.G.: Bootstrap percolation on the random regular graph. Random Struct. Algorithms **30**(1–2), 257–286 (2007)
5. Chang, C.L., Lyuu, Y.D.: Triggering cascades on strongly connected directed graphs. In: 2012 Fifth International Symposium on Parallel Architectures, Algorithms and Programming (PAAP), pp. 95–99. IEEE (2012)
6. Coja-Oghlan, A., Feige, U., Krivelevich, M., Reichman, D.: Contagious sets in expanders. In: Proceedings of the Twenty-Sixth Annual ACM-SIAM Symposium on Discrete Algorithms, pp. 1953–1987. Society for Industrial and Applied Mathematics (2015)
7. Feige, U., Krivelevich, M., Reichman, D., et al.: Contagious sets in random graphs. Ann. Appl. Probab. **27**(5), 2675–2697 (2017)
8. Flocchini, P., Lodi, E., Luccio, F., Pagli, L., Santoro, N.: Dynamic monopolies in tori. Discret. Appl. Math. **137**(2), 197–212 (2004)
9. Freund, D., Poloczek, M., Reichman, D.: Contagious sets in dense graphs. Eur. J. Comb. **68**, 66–78 (2018)
10. Gärtner, B., N. Zehmakan, A.: Color War: cellular automata with majority-rule. In: Drewes, F., Martín-Vide, C., Truthe, B. (eds.) LATA 2017. LNCS, vol. 10168, pp. 393–404. Springer, Cham (2017). https://doi.org/10.1007/978-3-319-53733-7_29
11. Gärtner, B., Zehmakan, A.N.: Majority model on random regular graphs. In: Bender, M.A., Farach-Colton, M., Mosteiro, M.A. (eds.) LATIN 2018. LNCS, vol. 10807, pp. 572–583. Springer, Cham (2018). https://doi.org/10.1007/978-3-319-77404-6_42

12. Gunderson, K.: Minimum degree conditions for small percolating sets in bootstrap percolation. arXiv preprint arXiv:1703.10741 (2017)
13. Jeger, C., Zehmakan, A.N.: Dynamic monopolies in reversible bootstrap percolation. arXiv preprint arXiv:1805.07392 (2018)
14. Kempe, D., Kleinberg, J., Tardos, É.: Maximizing the spread of influence through a social network. In: Proceedings of the Ninth ACM SIGKDD International Conference on Knowledge Discovery and Data Mining, pp. 137–146. ACM (2003)
15. Peleg, D.: Size bounds for dynamic monopolies. Discret. Appl. Math. **86**(2–3), 263–273 (1998)
16. Reichman, D.: New bounds for contagious sets. Discret. Math. **312**(10), 1812–1814 (2012)
17. Zehmakan, A.N.: Opinion forming in binomial random graph and expanders. arXiv preprint arXiv:1805.12172 (2018)

10. Hinton on the Minimum Siner..., University of Minneapolis, 1994

12. Grigoreva, A. K.: Minimal distance unit transformation, periodic array in hoon range probabilities arrays prepared ... 1999,003; 117 & 4287?

13. Alexei, G. Z. Khmelo, A. S.: Divisors, nonlinear transformation boundary parallel ..., 1994, GXS: prepared by Utah, 1999, 18,3,4

14. Kelber, B.; Digument, A.; Davey, C.; Maclouring, the spread of influence through a social network 1999, Procedings, of the Ninth ACM SIGKDD International Con reference on Knowledge Discovery and Data Mining, 2000, 137-146, ACM (California)

16. Li, B., R.; Mashunda, Pes bounded propagation. Discrete Appl. Math, 2004, 91, 500, 177 & 920, 293

17. Derkadine, N.: arbitrary bounds and mudpits with Discrete Math. 313.1002 (197,1212) 420, 293

18. McLemonon, A. G.: Options functions with transition bipartite graph bounds, array X a provide, 2013, https://177,7010

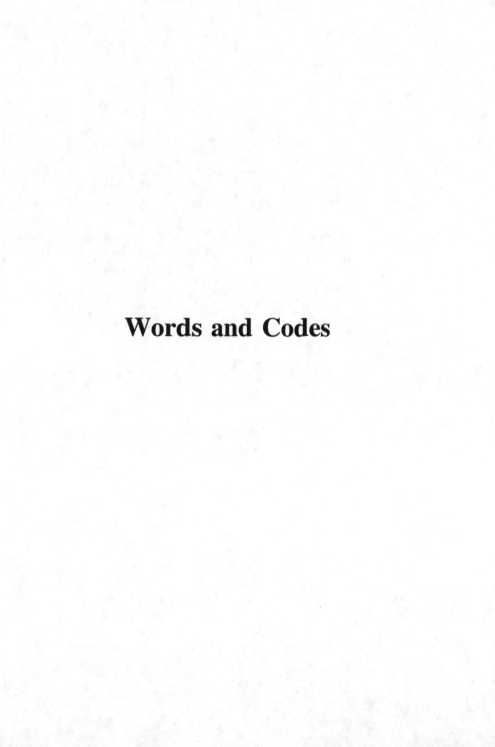

Words and Codes

Recurrence in Multidimensional Words

Émilie Charlier[1], Svetlana Puzynina[2,3]([✉]), and Élise Vandomme[4]

[1] University of Liege, Liège, Belgium
echarlier@uliege.be
[2] Saint Petersburg State University, Saint Petersburg, Russia
s.puzynina@gmail.com
[3] Sobolev Institute of Mathematics, Novosibirsk, Russia
[4] Czech Technical University in Prague, Prague, Czech Republic
elise.vandomme@fjfi.cvut.cz

Abstract. In this paper we study various modifications of the notion of uniform recurrence in multidimensional infinite words. A d-dimensional infinite word is said to be *uniformly recurrent* if for each $(n_1, \ldots, n_d) \in \mathbb{N}^d$ there exists $N \in \mathbb{N}$ such that each block of size $(N, \ldots, N)$ contains the prefix of size $(n_1, \ldots, n_d)$. We introduce and study a new notion of uniform recurrence of multidimensional infinite words: for each rational slope $(q_1, \ldots, q_d)$, each rectangular prefix must occur along this slope, that is in positions $\ell(q_1, \ldots, q_d)$, with bounded gaps. Such words are called *uniformly recurrent along all directions*. We provide several constructions of multidimensional infinite words satisfying this condition, and more generally, a series of three conditions on recurrence. We study general properties of these new notions and in particular we study the strong uniform recurrence of fixed points of square morphisms.

Keywords: Uniform recurrence · Multidimensional words · Multidimensional morphisms

Combinatorics on words in one dimension is a well-studied field of theoretical computer science with its origins in the early 20th century. The study of two-dimensional words is less developed, even though many concepts and results are naturally extendable from the unidimensional case (see, e.g., [1,2,4,7]). However, some words problems become much more difficult in dimensions higher than one. One of such questions is the connection between local complexity and periodicity. In dimension one, the classical theorem of Morse and Hedlund states that if for some n the number of distinct length-n blocks of an infinite word is less than or equal to n, then the word is periodic. In the two-dimensional case a similar assertion is known as Nivat's conjecture, and many efforts are made by

The second author is partially supported by Russian Foundation of Basic Research (grant 18-31-00118). The last author acknowledges financial support by the Ministry of Education, Youth and Sports of the Czech Republic (project no. CZ.02.1.01/0.0/0.0/16_019/0000778).

C. Martín-Vide et al. (Eds.): LATA 2019, LNCS 11417, pp. 397–408, 2019.
https://doi.org/10.1007/978-3-030-13435-8_29

scientists for checking this hypothesis [3,5,6]. In this paper, we are interested in two-dimensional uniform recurrence.

A first and natural attempt to generalize the notion of recurrence to the multidimensional setting turns out to be rather unsatisfying. Recall that an infinite word $w \colon \mathbb{N} \to A$ (where A is a finite alphabet) is said to be *recurrent* if every prefix occurs as least twice. A straightforward extension of this definition is to say that a bidimensional infinite word is recurrent if each rectangular prefix occurs at least twice. However, with such a definition of bidimensional recurrence, a binary bidimensional infinite word containing one row filled with 1 and the rest filled with 0 is considered as recurrent, even though any column is not recurrent in the unidimensional sense of recurrence.

In order to avoid this kind of undesirable phenomenon, a natural strengthening is to ask that every prefix occurs *uniformly*. In this paper, we investigate several notions of recurrence of multidimensional infinite words $w \colon \mathbb{N}^d \to A$, generalizing the usual notion of uniform recurrence of infinite words.

This paper is organized as follows. In Sect. 1, we define two new notions of uniform recurrence of multidimensional infinite words: the URD words and the SURD words. We also make some first observations in the bidimensional setting. We then show that these two new notions of recurrence along directions do not depend on the choice of the origin. In Sect. 2, we study fixed points of multidimensional square morphisms. In particular, we provide some infinite families of SURD multidimensional infinite words. We give a complete characterization of SURD bidimensional infinite words that are fixed points of square morphisms of size 2. Finally, in Sect. 3, we show how to build uncountably many SURD bidimensional infinite words. In particular, the family of bidimensional infinite words so-obtained contains uncountably many non-morphic SURD elements.

1   Uniform Recurrence Along Directions

Here and throughout the text, A designates an arbitrary finite alphabet and d is a positive integer. For $m, n \in \mathbb{N}$, the notation $[\![m, n]\!]$ designates the interval of integers $\{m, \ldots, n\}$ (which is considered empty for $n < m$). We write $(s_1, \ldots, s_d) \leq (t_1, \ldots, t_d)$ if $s_i \leq t_i$ for each $i \in [\![1, d]\!]$.

A *d-dimensional infinite word* over A is a map $w \colon \mathbb{N}^d \to A$. A *d-dimensional finite word* over A is a map $w \colon [\![0, s_1 - 1]\!] \times \cdots \times [\![0, s_d - 1]\!] \to A$, where $(s_1, \ldots, s_d) \in \mathbb{N}^d$ is the *size* of w. A *factor* of a d-dimensional infinite word w is a finite word f of some size $(s_1, \ldots, s_d)$ such that there exists $\mathbf{p} \in \mathbb{N}^d$ with $f(\mathbf{i}) = w(\mathbf{p} + \mathbf{i})$ for each $\mathbf{i} \in [\![0, s_1 - 1]\!] \times \cdots \times [\![0, s_d - 1]\!]$. A *factor* of a d-dimensional finite word is defined similarly. In both cases (infinite and finite), if $\mathbf{p} = \mathbf{0}$ then the factor f is said to be a *prefix* of w. Sometimes we will write $w_\mathbf{i}$ instead of $w(\mathbf{i})$ for brevity.

Note that d-dimensional words can be considered over $\mathbb{Z}^d$, i.e., $w \colon \mathbb{Z}^d \to A$. Although in our considerations it is more natural to consider one-way infinite words, since for example, we make use of fixed points of morphisms, most of our results and notions can be straightforwardly extended to words over $\mathbb{Z}^d$.

Definition 1 (UR). *A d-dimensional infinite word w is* uniformly recurrent *if for every prefix p of w, there exists a positive integer b such that every factor of w of size $(b, \ldots, b)$ contains p as a factor.*

Whenever $d = 1$, the previous definition corresponds to the usual notion of uniform recurrence of infinite words. Let us now introduce two new notions of uniform recurrence of multidimensional infinite words.

Throughout this text, when we talk about a *direction* $\mathbf{q} = (q_1, \ldots, q_d)$, we assume that $q_1, \ldots, q_d$ are coprime nonnegative integers. For the sake of conciseness, if $\mathbf{s} = (s_1, \ldots, s_d)$, we write $[\![0, \mathbf{s}-1]\!]$ in order to designate the d-dimensional interval $[\![0, s_1 - 1]\!] \times \cdots \times [\![0, s_d - 1]\!]$.

Let $w \colon \mathbb{N}^d \to A$ be a d-dimensional infinite word, $\mathbf{s} \in \mathbb{N}^d$ and $\mathbf{q} \in \mathbb{N}^d$ be a direction. The *word along the direction* $\mathbf{q}$ *with respect to the size* $\mathbf{s}$ *in* w is the infinite unidimensional word $w_{\mathbf{q}, \mathbf{s}} \colon \mathbb{N} \to A^{[\![0, \mathbf{s}-1]\!]}$, where elements of $A^{[\![0, \mathbf{s}-1]\!]}$ are considered as letters, defined by

$$\forall \ell \in \mathbb{N}, \forall \mathbf{i} \in [\![0, \mathbf{s}-1]\!], \ (w_{\mathbf{q}, \mathbf{s}}(\ell))(\mathbf{i}) = w(\mathbf{i} + \ell \mathbf{q}).$$

See Fig. 1 for an illustration in the bidimensional case.

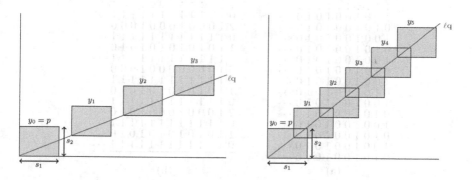

Fig. 1. The unidimensional word $w_{\mathbf{q}, \mathbf{s}}$ is built from the blocks of size $\mathbf{s}$ occurring at positions $\ell \mathbf{q}$ in w. Those blocks in $A^{[\![0, \mathbf{s}-1]\!]}$ may or may not overlap.

Definition 2 (URD). *A d-dimensional infinite word $w \colon \mathbb{N}^d \to A$ is* uniformly recurrent along all directions *(URD for short) if for all $\mathbf{s} \in \mathbb{N}^d$ and all directions $\mathbf{q} \in \mathbb{N}^d$, there exists $b \in \mathbb{N}$ such that, in $w_{\mathbf{q}, \mathbf{s}}$, two consecutive occurrences of the first letter $w_{\mathbf{q}, \mathbf{s}}(0)$ are situated at distance at most b.*

Definition 3 (SURD). *A d-dimensional infinite word $w \colon \mathbb{N}^d \to A$ is* strongly uniformly recurrent along all directions *(SURD for short) if for all $\mathbf{s} \in \mathbb{N}^d$, there exists $b \in \mathbb{N}$ such that, for each direction $\mathbf{q} \in \mathbb{N}^d$, in $w_{\mathbf{q}, \mathbf{s}}$, two consecutive occurrences of the first letter $w_{\mathbf{q}, \mathbf{s}}(0)$ are situated at distance at most b.*

We make some preliminary observations in the bidimensional setting. We choose the convention of representing a bidimensional infinite word $w \colon \mathbb{N}^2 \to A$ by placing the rows from bottom to top, and the columns from left to right (as for Cartesian coordinates):

$$
\begin{array}{cccc}
\vdots & \vdots & \vdots & \vdots \\
w(0,3) & w(1,3) & w(2,3) & w(3,3) \cdots \\
w(0,2) & w(1,2) & w(2,2) & w(3,2) \cdots \\
w(0,1) & w(1,1) & w(2,1) & w(3,1) \cdots \\
w(0,0) & w(1,0) & w(2,0) & w(3,0) \cdots
\end{array}
$$

The fact that all rows and columns of a bidimensional infinite word $w \colon \mathbb{N}^2 \to A$ are uniformly recurrent (in the unidimensional sense) does not imply that w is UR. Indeed, consider the word obtained by alternating two kind of rows as depicted in Fig. 2(a): $01F$ and $10F$ where $F = 01001010 \cdots$ is the Fibonacci word, that is, the fixed point of the morphism $0 \mapsto 01, 1 \mapsto 0$. Then the square prefix $\begin{bmatrix} 1 & 0 \\ 0 & 1 \end{bmatrix}$ only occurs within the first two columns.

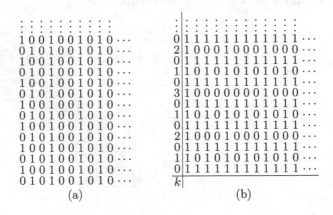

(a)   (b)

Fig. 2. (a) A non UR bidimensional infinite word having uniformly recurrent rows and columns. (b) A UR bidimensional infinite word with uniformly recurrent rows having different recurrence bounds.

Conversely, the fact that a bidimensional infinite word is uniformly recurrent does not imply that each of its row/column is uniformly recurrent with a shared uniform bound either. In a Toeplitz fashion, fill the rows indexed by k with the periodic word $u_k = (10^{2^k-1})^\omega$ as in Fig. 2(b). More precisely, the n-th row (with $n \in \mathbb{N}$) is indexed by k if $n \equiv 2^k - 1 \pmod{2^{k+1}}$. Such a bidimensional infinite word is UR. Indeed, consider a prefix p of size (m, n). Let $N = \max(\lceil \log_2(m) \rceil, \lceil \log_2(n) \rceil)$. The prefix p' of size $(2^N, 2^N)$ appears periodically according to the periods $(2^{N+1}, 0)$ and $(0, 2^{N+1})$. Therefore each factor of size $(2^{N+1} + 2^N - 1, 2^{N+1} + 2^N - 1)$ contains p'. So it contains p as well.

However, the distance between consecutive occurrences of 1 in a row can be arbitrarily large, even though the bidimensional infinite word is UR.

In the end of this section, we will discuss a variation the definition above – uniform recurrence along all directions from any origin. As a natural generalization of d-dimensional URD and SURD infinite words, we could ask that the recurrence property should not just be taken into account on the lines $\{\ell\mathbf{q}: \ell \in \mathbb{N}\}$ for all directions $\mathbf{q}$ but on all lines $\{\ell\mathbf{q} + \mathbf{p}: \ell \in \mathbb{N}\}$ for all origins $\mathbf{p}$ and directions $\mathbf{q}$. In fact, this would not be a real generalization; we will soon prove it.

Definition 4 (URDO). *A d-dimensional infinite word $w\colon \mathbb{N}^d \to A$ is uniformly recurrent along all directions from any origin (URDO for short) if for each $\mathbf{p} \in \mathbb{N}^d$, the translated d-dimensional infinite word $w^{(\mathbf{p})}\colon \mathbb{N}^d \to A$, $\mathbf{i} \mapsto w(\mathbf{i} + \mathbf{p})$ is URD.*

Definition 5 (SURDO). *A d-dimensional infinite word $w\colon \mathbb{N}^d \to A$ is strongly uniformly recurrent along all directions from any origin (SURDO for short) if for each $\mathbf{p} \in \mathbb{N}^d$, the translated d-dimensional infinite word $w^{(\mathbf{p})}\colon \mathbb{N}^d \to A$, $\mathbf{i} \mapsto w(\mathbf{i} + \mathbf{p})$ is SURD.*

Proposition 1. *A d-dimensional infinite word is URD (SURD, respectively) if and only if it is URDO (SURDO, respectively).*

Proof. Both conditions are clearly sufficient. Now we prove that they are necessary. Let $w\colon \mathbb{N}^d \to A$ be URD (SURD, respectively), let $\mathbf{p}, \mathbf{s} \in \mathbb{N}^d$ and let $f\colon [\![0, \mathbf{s} - 1]\!] \to A$ be the factor of w of size $\mathbf{s}$ at position $\mathbf{p}$: for all $\mathbf{i} \in [\![0, \mathbf{s} - 1]\!]$, $f(\mathbf{i}) = w(\mathbf{i} + \mathbf{p})$. We need to prove that for all directions $\mathbf{q}$, there exists $b \in \mathbb{N}$ such that (that there exists $b \in \mathbb{N}$ such that for all directions $\mathbf{q}$, respectively) consecutive occurrences of f at positions of the form $\ell\mathbf{q} + \mathbf{p}$ are situated at distance at most b. The situation is illustrated in Fig. 3. Consider the prefix p of size $\mathbf{p} + \mathbf{s}$ of w. Since the word is URD (SURD, respectively), for all directions $\mathbf{q}$, there exists b' such that (there exists b' such that for all directions $\mathbf{q}$, respectively) consecutive occurrences of p in positions $\ell\mathbf{q}$ are situated at distance at most b'. Since f occurs at position $\mathbf{p}$ in p, this implies the condition we need with $b = b'$.

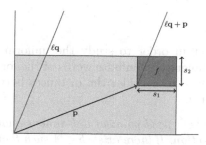

Fig. 3. Illustration of the proof of Proposition 1 in the bidimensional case.

2   Fixed Points of Multidimensional Square Morphisms

Similarly to unidimensional words, we can define morphisms and their fixed points in any dimension. For simplicity, we only consider constant length morphisms:

Definition 6. *A d-dimensional morphism of constant size* $\mathbf{s} = (s_1, \ldots, s_d) \in \mathbb{N}^d$ *is a map* $\varphi \colon A \to A^{[\![0, \mathbf{s}-1]\!]}$. *For each* $a \in A$ *and for each integer* $n \geq 2$, $\varphi^n(a)$ *is recursively defined as*

$$\varphi^n(a) \colon [\![0, \mathbf{s}^n - 1]\!] \to A, \ \mathbf{i} \mapsto \Big(\varphi\big((\varphi^{n-1}(a))(\mathbf{q}) \big) \Big)(\mathbf{r}),$$

where $\mathbf{i} = \mathbf{q}\mathbf{s} + \mathbf{r}$ *is the componentwise Euclidean division of* $\mathbf{i}$ *by* $\mathbf{s}$. *With these notation, the* preimage *of the letter* $(\varphi^n(a))(\mathbf{i})$ *is the letter* $(\varphi^{n-1}(a))(\mathbf{q})$. *In the case* $\mathbf{s} = (s, \ldots, s)$, *we say that* φ *is a d-dimensional square morphism of size s.*

If there exists $a \in A$ *with* $\varphi(a)_{0,0} = a$, *then the d-dimensional morphism* φ *is prolongable from a and, in that case, the limit* $\lim_{n \to \infty} \varphi^n(a)$ *is well defined. The limit d-dimensional infinite word obtained in this way is called the* fixed point *of φ beginning with a and it is denoted by* $\varphi^\omega(a)$.

Example 1. Figure 4 depicts the first five iterations of a bidimensional square morphism with the convention that a black (resp. white) cell represents the letter 1 (resp. 0).

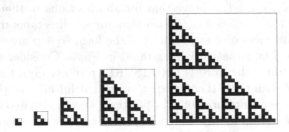

Fig. 4. The first five iterations of the 2D morphism. $0 \mapsto \begin{bmatrix} 0 & 0 \\ 0 & 0 \end{bmatrix}, 1 \mapsto \begin{bmatrix} 1 & 0 \\ 1 & 1 \end{bmatrix}$ starting from 1.

We first observe that in order to study the uniform recurrence along all directions (URD) of d-dimensional infinite words of the form $\varphi^\omega(a)$ for a square morphism φ, we only have to consider the distances between two consecutive occurrences of the letter a.

Proposition 2. *Let w be a fixed point of a d-dimensional square morphism of size s and let $\mathbf{q}$ be a direction. If there exists $b \in \mathbb{N}$ such that the distance between consecutive occurrences of $w(\mathbf{0})$ along $\mathbf{q}$ is at most b, then, for all $\mathbf{m} \in \mathbb{N}^d$, the distance between consecutive occurrences of the prefix of size $\mathbf{m}$ of w along $\mathbf{q}$ is at most $s^{\lceil \log_s (\max \mathbf{m}) \rceil} b$.*

Proof. Let $\mathbf{m} \in \mathbb{N}^d$ and let p be the prefix of size $\mathbf{m}$ of w. Let r be the integer defined by $s^{r-1} < \max \mathbf{m} \leq s^r$. In the d-dimensional infinite word w, the letter $w(\mathbf{0})$ occurs at position $\mathbf{i}$ if and only if the image $\varphi^r(w(\mathbf{0}))$ occurs at position $s^r \mathbf{i}$. Therefore, if the distance between consecutive occurrences of $w(\mathbf{0})$ along $\mathbf{q}$ is at most b, then the distance between consecutive occurrences of p along $\mathbf{q}$ is at most $s^r b$.

In order to provide a family of SURD d-dimensional infinite words, we introduce the following definition.

Definition 7. *For an integer $s \geq 2$ and $\mathbf{i} = (i_1, \ldots, i_d) \in (\mathbb{Z}/s\mathbb{Z})^d$ such that $i_1, \ldots, i_d$ are coprime, we define $\langle \mathbf{i} \rangle$ to be the additive subgroup of $(\mathbb{Z}/s\mathbb{Z})^d$ that is generated by $\mathbf{i}$:*

$$\langle \mathbf{i} \rangle = \{ k\mathbf{i} \colon k \in \mathbb{Z}/s\mathbb{Z} \}.$$

Then, we let $\mathcal{C}(s)$ be the family of all cyclic subgroups of $(\mathbb{Z}/s\mathbb{Z})^d$:

$$\mathcal{C}(s) = \{ \langle \mathbf{i} \rangle \colon \mathbf{i} \in (\mathbb{Z}/s\mathbb{Z})^d, \ \gcd(\mathbf{i}) = 1 \}.$$

If the alphabet A is binary (in which case we assume without loss of generality that $A = \{0, 1\}$), then we talk about *binary* morphism.

Proposition 3. *If φ is a d-dimensional square binary morphism of size s such that $\varphi(1)_{\mathbf{0}} = 1$ and if, for every $C \in \mathcal{C}(s)$, there exists $\mathbf{i} \in C$ such that $\varphi(0)_{\mathbf{i}} = \varphi(1)_{\mathbf{i}} = 1$, then its fixed point $\varphi^\omega(1)$ is SURD. More precisely, for all $\mathbf{m} \in \mathbb{N}^d$, the distance between consecutive occurrences of the prefix of size $\mathbf{m}$ of $\varphi^\omega(1)$ along any direction is at most $s^{\lceil \log_s(\max \mathbf{m}) \rceil + 1}$.*

Proof. Let $\mathbf{q} \in \mathbb{N}^d$ be a given direction. Let $\mathbf{r} = \mathbf{q} \bmod s$ (componentwise) and $d = \gcd(\mathbf{r})$. By hypothesis, there exists $\mathbf{i} \in \langle \frac{1}{d}\mathbf{r} \rangle$ such that $(\varphi(0))_{\mathbf{i}} = (\varphi(1))_{\mathbf{i}} = 1$. Let $k \in [\![0, s-1]\!]$ such that $\mathbf{i} = \frac{k}{d}\mathbf{r} \bmod s$. Then $k\mathbf{q} \equiv k\mathbf{r} \equiv d\mathbf{i} \bmod s$. Observe that $\gcd(d, s)$ divides $\mathbf{r}$ and s, hence also divides $\mathbf{q}$. This implies that $\gcd(d, s) = 1$. Let $\ell = d^{-1}k \bmod s$. Then $\ell\mathbf{q} \equiv \mathbf{i} \bmod s$. We obtain that for all $n \in \mathbb{N}$, $(\ell + ns)\mathbf{q} \equiv \mathbf{i} \bmod s$, hence $(\varphi^\omega(1))_{(\ell+ns)\mathbf{q}} = 1$. This proves that the distance between consecutive occurrences of the letter 1 in $\varphi^\omega(1)$ along the direction $\mathbf{q}$ is at most s.

Now let $\mathbf{m} \in \mathbb{N}^d$ and consider the prefix p of size $\mathbf{m}$ of $\varphi^\omega(1)$. From the first part of the proof and by Proposition 2, we obtain that the distance between consecutive occurrences of p along any direction is at most $s^{\lceil \log_s(\max \mathbf{m}) \rceil + 1}$.

Since each subgroup of $(\mathbb{Z}/s\mathbb{Z})^d$ contains $\mathbf{0}$, the following result is immediate.

Corollary 1. *Let φ be a d-dimensional square binary morphism of size s such that $(\varphi(1))_{\mathbf{0}} = (\varphi(0))_{\mathbf{0}} = 1$. Then the fixed point $\varphi^\omega(1)$ is SURD. More precisely, for all $\mathbf{m} \in \mathbb{N}^d$, the distance between consecutive occurrences of the prefix of size $\mathbf{m}$ of $\varphi^\omega(1)$ along any direction is at most $s^{\lceil \log_s(\max \mathbf{m}) \rceil + 1}$.*

Remark 1. For s prime, orbits of any two elements either coincide, or have only the element $\mathbf{0}$ in common. Therefore, for s prime, we have exactly $\frac{s^d-1}{s-1}$ distinct orbits. In particular, for $d = 2$, this gives $s + 1$ distinct orbits. Hence we can consider a partition of $(\mathbb{Z}/s\mathbb{Z})^2$ into $s+2$ sets: $s+1$ orbits without $\mathbf{0}$ and $\mathbf{0}$ itself. When s is not prime, the structure is a more complicated and we do not have such a nice partition. Examples below illustrate the two situations.

$$(a)\ s = 5 \qquad \begin{bmatrix} b & f & e & d & c \\ b & d & f & c & e \\ b & e & c & f & d \\ b & c & d & e & f \\ 0 & a & a & a & a \end{bmatrix}$$

b	j	h	f	d	c
b,f,k	i	d,j,e	k	c,h,i	e
b,d,h,ℓ	g	ℓ	c,f,j,g	ℓ	g
b,f,k	e	c,h,i	k	d,e,j	i
b	c	d	f	h	j
0	a	a,g,ℓ	a,k,i,e	a,g,ℓ	a

(b) $s = 6$

$$(c)\ s = 6 \qquad \begin{bmatrix} * & * & * & * & * & * \\ 1 & * & 1 & * & * & * \\ * & * & * & 1 & * & * \\ * & * & 1 & * & * & * \\ * & * & * & * & * & * \\ * & * & * & * & 1 & * \end{bmatrix}$$

Fig. 5. Partitions with $d = 2$ and an example on how to fill them.

Example 2. Partition for $s = 5$ and $d = 2$ is illustrated in Fig. 5(a) where each letter in $\{a, \dots, f\}$ represents an orbit. Due to Proposition 3, in order to obtain a SURD fixed point of a bidimensional square morphism, it is enough to have a 1 in the images of 0 and 1 in one of the coordinates marked by each letter. And by Corollary 1, having a 1 in the coordinate $(0,0)$ in both the images of 0 and 1 is enough.

Example 3. For $s = 6$ and $d = 2$, one has 12 orbits (which can be checked by considering the 36 possible cases of pairs of remainders of the Euclidean division by 6, out of which there are only 21 coprime pairs to consider). They are depicted using letters in $\{a, \dots, \ell\}$ in Fig. 5(b). We remark that here the orbits intersect. Due to Proposition 3, in order to obtain a SURD word, it suffices to have a 1 in both images of 0 and 1 in at least one of the elements of each orbit. For example, it is the case of the fixed point of any morphism with ones in the marked positions in the images of both 0 and 1 as in Fig. 5(c).

Since a power of a morphism φ has the same fixed points as φ, the following result is immediate.

Corollary 2. *If ψ is a d-dimensional square binary morphism of size s such that for some integer i, its power $\varphi = \psi^i$ satisfies the conditions of Proposition 3, then the fixed point $\psi^\omega(1)$ is SURD. More precisely, for each $\mathbf{m} \in \mathbb{N}^d$, the distance between consecutive occurrences of the prefix p of size $\mathbf{m}$ of $\psi^\omega(1)$ is at most $s^{i\lceil \log(\max \mathbf{m})\rceil + i}$.*

Example 4. The morphism

$$\psi: 0 \mapsto \begin{bmatrix} 0 & 0 & 0 \\ 1 & 1 & 1 \\ 0 & 1 & 0 \end{bmatrix}, \quad 1 \mapsto \begin{bmatrix} 0 & 1 & 0 \\ 1 & 0 & 1 \\ 1 & 1 & 0 \end{bmatrix}$$

satisfies the hypotheses of Corollary 2 for $s = 3$, $i = 2$. Indeed, it can be checked that for each $C \in C_9$, we can find a 1 in both images $\psi^2(0)$ and $\psi^2(1)$.

Now we give a family of examples of SURD d-dimensional words which do not satisfy the hypotheses of Corollary 2, showing that it does not give a necessary condition. We first need the following observation on unidimensional fixed points of morphisms.

Lemma 1. *Let s be a prime and let φ be a unidimensional binary morphism of constant length s such that $\varphi(1)_0 = 1$, $\varphi(0)_0 = 0$ and there exists $i \in [\![1, s-1]\!]$ such that $\varphi(0)_i = \varphi(1)_i = 1$. For all positive integers m, the maximal distance between consecutive occurrences of 1 in the arithmetic subsequence $k \mapsto \varphi^\omega(1)_{mk}$ of $\varphi^\omega(1)$ is at most s.*

Proof. Let $w = \varphi^\omega(1)$ and let m be a positive integer. Denote by d_m the maximal distance between consecutive occurrences of 1 in $k \mapsto w_{mk}$. We have to show that $d_m \leq s$. The integer m can be decomposed in a unique way as $m = s^e \ell$ with $e, \ell \in \mathbb{N}$ and $\ell \not\equiv 0 \bmod s$. We prove the result by induction on $e \in \mathbb{N}$. If $e = 0$ then $m \not\equiv 0 \bmod s$. Since there is a 1 in the i-th place of both the images of 0 and 1 and since $i \neq 0$, we obtain that $d_m \leq s$ in this case. Now suppose that $e > 0$ and that the result is correct for $e - 1$. Observe that, for every $k \in \mathbb{N}$, the preimage of the letter $w_{mk} = w_{s^e \ell k}$ is the letter $w_{\frac{m}{s}k} = w_{s^{e-1}\ell k}$. By definition of the morphism and since $m \equiv 0 \bmod s$, for each $k \in \mathbb{N}$, the letter w_{mk} is equal to 0 if its preimage is 0 and is equal to 1 if its preimage is 1. But by induction hypothesis, for all $k \in \mathbb{N}$, at least one of the s preimages $w_{\frac{m}{s}k}$, $w_{\frac{m}{s}(k+1)}, \ldots, w_{\frac{m}{s}(k+s-1)}$ is equal to 1. Therefore, we obtain that for all $k \in \mathbb{N}$, at least one of the s letters $w_{mk}, w_{m(k+1)}, \ldots, w_{m(k+s-1)}$ is equal to 1 as well, which shows that $d_m \leq s$.

Proposition 4. *If φ is a d-dimensional square binary morphism of a prime size s such that*

1. *$\forall i_2, \ldots, i_d \in [\![0, s-1]\!]$, $\varphi(1)_{0, i_2, \ldots, i_d} = 1$ and $\varphi(0)_{0, i_2, \ldots, i_d} = 0$*
2. *$\exists i_1 \in [\![0, s-1]\!]$, $\forall i_2, \ldots, i_d \in [\![0, s-1]\!]$, $\varphi(0)_{i_1, \ldots, i_d} = \varphi(1)_{i_1, \ldots, i_d} = 1$*

then $\varphi^\omega(1)$ is SURD.

Proof. By Proposition 2, we only have to show that there exists a uniform bound b such that the distance between consecutive occurrences of 1 along any direction of $\varphi^\omega(1)$ is at most b. It is sufficient to prove the result for the fixed point beginning with 1 of the morphism φ satisfying the hypotheses (1) and (2) and having 0 at any other coordinates in the images of both 0 and 1, since all other fixed points satisfying the hypotheses of the proposition differ from this one only by replacing some occurrences of 0 by 1. For example, for $d = 2$, the morphism φ is

$$\varphi: 0 \mapsto \begin{bmatrix} 0\,0\,\cdots\,0\,1\,0\,\cdots\,0 \\ 0\,0\,\cdots\,0\,1\,0\,\cdots\,0 \\ \vdots\;\vdots \qquad\quad \vdots \\ 0\,0\,\cdots\,0\,1\,0\,\cdots\,0 \end{bmatrix}, \quad 1 \mapsto \begin{bmatrix} 1\,0\,\cdots\,0\,1\,0\,\cdots\,0 \\ 1\,0\,\cdots\,0\,1\,0\,\cdots\,0 \\ \vdots\;\vdots \qquad\quad \vdots \\ 1\,0\,\cdots\,0\,1\,0\,\cdots\,0 \end{bmatrix}$$

(where the common columns of 1's are placed at position i_1 in both images). Each of the hyperplanes

$$H_k = \{\varphi^\omega(1)_{k,i_2\ldots,i_d} : i_2,\ldots,i_d \in \mathbb{N}\}, \quad \text{for } k \in \mathbb{N}$$

of $\varphi^\omega(1)$ contains either only 0's or only 1's. Therefore, for any direction $\mathbf{q} = (q_1,\ldots,q_d)$, we have $\varphi^\omega(1)_{\ell\mathbf{q}} = \varphi^\omega(1)_{\ell q_1,0,\ldots,0}$, hence the unidimensional infinite word $\mathbb{N} \to A, \ell \mapsto \varphi^\omega(1)_{\ell\mathbf{q}}$ is the fixed point of the unidimensional morphism

$$\sigma: 0 \mapsto \begin{bmatrix} 0\,0\,\cdots\,0\,1\,0\,\cdots\,0 \end{bmatrix}, \quad 1 \mapsto \begin{bmatrix} 1\,0\,\cdots\,0\,1\,0\,\cdots\,0 \end{bmatrix}$$

(where, again, the common 1's are placed at position i_1 in both images). By Lemma 1, we obtain that $\varphi^\omega(1)$ is SURD with the uniform bound $b = s$.

Now we give a sufficient condition for a d-dimensional word to be non URD.

Proposition 5. *Let φ be a d-dimensional square binary morphism of a prime size s. Let $\mathbf{q}$ be a direction and let $C = \langle \mathbf{q} \bmod s \rangle$. If $\varphi(1)_{\mathbf{0}} = 1$, $\varphi(0)_{\mathbf{0}} = 0$ and, for all $\mathbf{i} \in C \setminus \{\mathbf{0}\}$, $\varphi(1)_{\mathbf{i}} = \varphi(0)_{\mathbf{i}} = 0$, then $(\varphi^\omega(1)_{\ell\mathbf{q}})_{i \in \mathbb{N}} = 10^\omega$. In particular, $\varphi^\omega(1)$ is not recurrent along the direction $\mathbf{q}$.*

Proof. Suppose that the first occurrence of 1 after that in position $\mathbf{0}$ along the direction $\mathbf{q}$ occurs in position $\ell\mathbf{q}$. Since $\varphi(0)$ and $\varphi(1)$ have 0 on all places defined by $C \setminus \{\mathbf{0}\}$, the letter $\varphi^\omega(1)_{\ell\mathbf{q}}$ must be placed at the coordinate $\mathbf{0}$ of the image of 1. In particular, the preimage of $\varphi^\omega(1)_{\ell\mathbf{q}}$ must be 1. Because s is prime, ℓ must be divisible by s and the preimage of $\varphi^\omega(1)_{\ell\mathbf{q}}$ is $\varphi^\omega(1)_{\frac{\ell}{s}\mathbf{q}}$. But by the choice of ℓ and since $0 < \frac{\ell}{s} < \ell$, we must also have $\varphi^\omega(1)_{\frac{\ell}{s}\mathbf{q}} = 0$, a contradiction. $\square$

However, the next result shows that the condition of Proposition 5 is not necessary.

Proposition 6. *The fixed point $\varphi^\omega(1)$ of the morphism*

$$\varphi: 0 \mapsto \begin{bmatrix} 1\,1\,0 \\ 0\,0\,0 \\ 0\,0\,1 \end{bmatrix} \quad 1 \mapsto \begin{bmatrix} 1\,1\,1 \\ 0\,1\,0 \\ 1\,1\,0 \end{bmatrix}$$

is not recurrent in the direction $(1, 3)$.

The next theorem gives a characterization of SURD fixed points of square binary morphisms of size 2.

Theorem 1. *If a bidimensional binary square morphism φ of size 2 has a fixed point beginning with 1, then this fixed point is SURD if and only if either $\varphi(0)_{0,0} = 1$ or $\varphi(1) = \left[\begin{smallmatrix} 1 & 1 \\ 1 & 1 \end{smallmatrix}\right]$.*

The "if" part follows from Corollary 1. The "only if" part is proved with a rather technical argument involving a case study analysis and using certain properties of arithmetic progressions in the Thue-Morse word.

The previous theorem gives a characterization of strong uniform recurrence along all directions for fixed points of bidimensional square binary morphisms of size 2. For larger sizes of morphisms, we gave several conditions that are either necessary (given by the contraposition of Proposition 5) or sufficient (Propositions 3 and 4). An open problem is to find a condition that would be both necessary and sufficient in general.

Question 1. Find a characterization of strong uniform recurrence along all directions for bidimensional square binary morphisms of size bigger than 2.

3   Non-morphic Bidimensional SURD Words

In this section we provide a construction of non-morphic SURD words. To construct such a word $w: \mathbb{N}^2 \to \{0,1\}$, we proceed recursively:

Step 0. For each $(i,j) \in \mathbb{N}^2$, put $w(2i, 2j) = 1$.

Step 1. Fill anything you want in positions $(0,1)$, $(1,0)$ and $(1,1)$. For each $(i,j) \in \mathbb{N}^2$, put $w(4i, 4j+1) = w(0,1)$, $w(4i+1, 4j) = w(1,0)$, $w(4i+1, 4j+1) = w(1,1)$. Note that the filled positions are doubly periodic with period 4.

Step n. At step n, we have filled all the positions (i,j) for $i,j < 2^n$, and the positions with filled values are doubly periodic with period 2^{n+1}. Let S be a set of pairs (k, ℓ) with $k, \ell < 2^{n+1}$ which have not been yet filled in. Fill anything you want in the positions from S. Now for each (k, ℓ) and each $(k', \ell') \in S$, define $w(2^{n+2}k + k', 2^{n+2}\ell + \ell') = w(k', \ell')$. Note that the filled positions are doubly periodic with period 2^{n+2} (Fig. 6).

Proposition 7. *Any bidimensional infinite word w defined by the construction above is SURD. More precisely, for all $\mathbf{s} \in \mathbb{N}^2$, the distance between consecutive occurrences of the prefix of size $\mathbf{s}$ of w along any direction is at most $2^{\lceil \log_2(\max \mathbf{s}) \rceil}$.*

Proof. Let p be the prefix of w of size $\mathbf{s}$ and let $\mathbf{q}$ be a direction. We show that the square prefix p' of size $(2^k, 2^k)$ with $k = \lceil \log_2(\max \mathbf{s}) \rceil$ appears within any consecutive 2^{k+1} positions along $\mathbf{q}$, hence this is also true for p itself. By construction, at step k we have filled all the positions $\mathbf{i}$ for $\mathbf{i} < (2^k, 2^k)$, and the positions with filled values are doubly periodic with periods $(2^{k+1}, 0)$ and $(0, 2^{k+1})$. Therefore the factor of size $(2^k, 2^k)$ occurring at position $2^{k+1}\mathbf{q}$ in w is equal to p'. The claim follows.

$$\begin{array}{l}
\vdots \;\; \vdots \;\; \vdots \;\; \vdots \\
1 \cdot 1 \cdot 1 \cdot 1 \cdots \\
a \; b \cdot \cdot \; a \; b \cdot \cdots \\
1 \; c \; 1 \cdot 1 \; c \; 1 \cdots \\
\cdots\cdots\cdots \\
1 \cdot 1 \cdot 1 \cdot 1 \cdots \\
a \; b \cdot \cdot \; a \; b \cdot \cdots \\
1 \; c \; 1 \cdot 1 \; c \; 1 \cdots
\end{array}$$

Fig. 6. Construction of a non-morphic SURD bidimensional word.

Observe that the morphic words satisfying Corollary 1 for $s = 2$ can be obtained by this construction.

Proposition 8. *Among the bidimensional infinite words obtained by the construction above, there are words which are not morphic.*

Proof. The construction above provides uncountably many bidimensional infinite words. However, there exist only countably many morphic words.

Remark 2. This construction can be generalized for any $s \in \mathbb{N}$ instead of 2 and for an arbitrary alphabet. Moreover, on each step we can choose as a period any multiple of a previous period.

Acknowledgements. We are grateful to Mathieu Sablik for inspiring discussions.

References

1. Berthé, V., Vuillon, L.: Tilings and rotations on the torus: a two-dimensional generalization of Sturmian sequences. Discrete Math. **223**, 27–53 (2000)
2. Cassaigne, J.: Double sequences with complexity $mn + 1$. J. Autom. Lang. Combin. **4**, 153–170 (1999)
3. Cyr, V., Kra, B.: Nonexpansive $\mathbb{Z}^2$-subdynamics and Nivat's conjecture. Trans. Am. Math. Soc. **367**(9), 6487–6537 (2015)
4. Durand, F., Rigo, M.: Multidimensional extension of the Morse-Hedlund theorem. Eur. J. Combin. **34**, 391–409 (2013)
5. Kari, J., Szabados, M.: An algebraic geometric approach to Nivat's conjecture. In: Halldórsson, M.M., Iwama, K., Kobayashi, N., Speckmann, B. (eds.) ICALP 2015. LNCS, vol. 9135, pp. 273–285. Springer, Heidelberg (2015). https://doi.org/10.1007/978-3-662-47666-6_22
6. Nivat, M.: Invited talk at ICALP, Bologna (1997)
7. Vuillon, L.: Combinatoire des motifs d'une suite sturmienne bidimensionnelle. Theor. Comput. Sci **209**, 261–285 (1998)

A Note with Computer Exploration on the Triangle Conjecture

Christophe Cordero[(✉)]

Université Paris-Est, LIGM (UMR 8049), CNRS, ENPC, ESIEE Paris, UPEM,
77454 Marne-la-Vallée, France
christophe.cordero@u-pem.fr

Abstract. The triangle conjecture states that codes formed by words of the form $a^i b a^j$ are either commutatively equivalent to a prefix code or not included in a finite maximal code. Thanks to computer exploration, we exhibit new examples of such non-commutatively prefix codes. In particular, we improve a lower bound in a bounding due to Shor and Hansel. We discuss in the rest of the article the possibility of those codes to be included in a finite maximal code.

Keywords: Codes · Triangle conjecture ·
Commutative equivalence conjecture

General Notation: Let A be the alphabet $\{a, b\}$. For $n \geq 0$, let $A^{\leq n}$ be the set of words of A^* of length at most n. For any word $w \in A^*$, let $|w|_x$ be the number of occurrences of the letter $x \in A$ in w. For any integer n, let $[n]$ be the set $\{k \in \mathbb{N} : 1 \leq k \leq n\}$ and $[[n]]$ be the set $\{k \in \mathbb{N} : 0 \leq k \leq n - 1\}$. For a real number $x \in \mathbb{R}$, let $\lceil x \rceil$ be the least integer greater than or equal to x.

1 Introduction

Our introduction to the theory of codes follows the book [1]. We call a subset $X \subset A^*$ a *code* if for all $n, m \geq 0$ and $x_1, \ldots, x_n, y_1, \ldots, y_m \in X$ the condition

$$x_1 x_2 \cdots x_n = y_1 y_2 \cdots y_m$$

implies

$$n = m \text{ and } x_i = y_i \text{ for all } i \in [n].$$

For example, the set $\{aabb, abaaa, b, ba\}$ is not a code since

$$(b)(abaaa)(b)(b) = (ba)(ba)(aabb).$$

A code is *maximal* if it is not contained in any other code. A subset $X \subset A^*$ is *prefix* if no element of X is a proper prefix of another element in X. A prefix

© Springer Nature Switzerland AG 2019
C. Martín-Vide et al. (Eds.): LATA 2019, LNCS 11417, pp. 409–420, 2019.
https://doi.org/10.1007/978-3-030-13435-8_30

subset not containing the empty word is a code. A code X is *commutatively prefix* if there exists a prefix code P such that the multisets

$$\{(|x|_a, |x|_b) \ : \ x \in X\} \ \text{and} \ \{(|p|_a, |p|_b) \ : \ p \in P\}$$

are equal. In other words, it states that one can build a prefix code from X by allowing commutation between the letters. The *commutative equivalence conjecture* is one of the main open problem in the theory of codes. It states that all finite maximal code are commutatively prefix.

In this work, we study this conjecture for a particular case of codes called bayonet code. A code X is a *bayonet code* if $X \subset a^*ba^*$. This particular case of the conjecture is also called the *triangle conjecture*. It states that a non-commutatively prefix bayonet code is not included in a finite maximal code (see [9] for recent result). It is known that a bayonet code X is commutatively prefix if and only if

$$\left| X \cap A^{\leq n} \right| \leq n, \ \text{for all} \ n \geq 0. \tag{1}$$

In 1984, Shor [8] found the bayonet code

$$\{b, ba, ba^7, ba^{13}, ba^{14}, a^3b, a^3ba^2, a^3ba^4, a^3ba^6, a^8b, a^8ba^2, a^8ba^4, a^8ba^6,$$
$$a^{11}b, a^{11}ba, a^{11}ba^2\} \tag{2}$$

with 16 elements and included in $A^{\leq 15}$, thus it is a non-commutatively prefix code. It is the only known example of finite non-commutatively prefix code. It is still unknown if Shor's code (2) is included in a finite maximal code. If it is the case, then the commutative equivalence conjecture and a stronger conjecture called *factorisation conjecture* (see [2] for a recent note/summary) would be false.

It is known that for all finite maximal code X and for any letter $x \in A$, there exists k such that $x^k \in X$. We call the *order* of a letter x the smallest integer k such that x^k belongs to X. It has been showed that if Shor's code (2) is included in a finite maximal code then the order of the letter a is a multiple of 330.

In the first section, we mainly do some computer explorations of non-commutatively prefix bayonet codes. We exhibit new examples of such codes. In particular, we exhibit the smallest ones and deduce from these a better lower bound in a bounding due to by Shor [8] and Hansel [3]. We discuss in the rest of the article the possibility of those codes to be included in a finite maximal code. In the second section, we use factorisation of cyclic group theory to prove some lower bounds for the orders of the letter a. Finally, in the last section, we find the smallest known codes that are non-commutatively prefix and not included in a finite maximal code.

2   Non-commutatively Prefix Bayonet Codes

Given a bayonet code X, we call its *dual* the bayonet code

$$\delta(X) := \left\{ a^iba^j \mid a^jba^i \in X \right\}.$$

Of course, a bayonet code is commutatively prefix or included in a finite maximal code if and only if its dual is. Thus we consider in this work a bayonet code and its dual to be the same. Even if they cannot be equal in the case we are interested in.

Proposition 1. *If a bayonet code X is non-commutatively prefix then $X \neq \delta(X)$.*

Proof. Let X be an auto-dual bayonet code (i.e. $X = \delta(X)$) and n be an integer. Let E_i^n be the set

$$(a^i ba^* \cup a^* ba^i) \cap A^{\leq n},$$

for $i \geq 0$. Thus

$$X \cap A^{\leq n} = \bigsqcup_{0 \leq i \leq \lceil \frac{n}{2} \rceil} (X \cap E_i^n) \text{ and } |X \cap A^{\leq n}| = \sum_{0 \leq i \leq \lceil \frac{n}{2} \rceil} |X \cap E_i^n|.$$

It is enough to show that $|X \cap E_i^n| \leq 2$, for $i \geq 0$. Assume that $|X \cap E_i^n| > 2$. Then there exists $j_1 \geq i$ and $j_1 < j_2 < n - i$ such that

$$a^i ba^{j_1}, a^i ba^{j_2}, a^{j_1} ba^i, a^{j_2} ba^i \in X.$$

Thus

$$\left(a^i ba^{j_1}\right)\left(a^{j_2} ba^i\right) = \left(a^i ba^{j_2}\right)\left(a^{j_1} ba^i\right)$$

which contradicts the fact that X is a code. Thus $|X \cap A^{\leq n}| \leq n$ and thanks to (1), we conclude that X is commutatively equivalent to a prefix code. This concludes the proof by contraposition.   □

Remark 1. Notice that the auto-dual bayonet code $\{a^i ba^{n-1-i} : 0 \leq i < n\}$ reaches the bound (1).

2.1   Computer Exploration

We run an exhaustive search issuing the following algorithm directly deduced from the definition of a code. Given a set $X \subset a^* ba^* \cap A^{\leq n}$, we build the oriented graph $\mathcal{G}_{\text{abs}}(X)$ defined by the set of vertices $[[n]]$ and by the edges

$$\boxed{|i - k|} \longrightarrow \boxed{|j - \ell|},$$

for all $a^i ba^j, a^k ba^\ell \in X$ with $a^i ba^j \neq a^k ba^\ell$.

Example 1. Let X be the set $\{a^4 ba^3, a^2 ba^5, aba^5, b, ba^2\}$, the graph $\mathcal{G}_{\text{abs}}(X)$ is

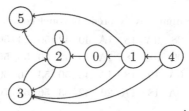

Thus X is a code (see Proposition 2).

Then we use the following proposition.

Proposition 2. *X is a code if and only if $\mathcal{G}_{abs}(X)$ does not contains a non-empty path from 0 to 0.*

Proof. Given $X \subset a^* b a^* \cap A^{\leq n}$, there is an edge from i to j in the graph $\mathcal{G}_{abs}(X)$ if and only if there exist $U, V \in X$ such that

$$a^i U = V a^j \text{ or } a^i U a^j = V.$$

By concatenation, there is a path from i to j in the graph if and only if there exist $U, V \in X^*$ such that

$$a^i U = V a^j \text{ or } a^i U a^j = V. \tag{3}$$

Assume that there is a non-empty path from 0 to 0 going through $k \neq 0$. Then by (3) there exist $U_1, U_2, V_1, V_2 \in X^*$ such that

$$U_1 = V_1 a^k \text{ and } U_2 = a^k V_2.$$

Hence $U_1 U_2 = V_1 U_2$ with $U_1 = V_1 a^k$, and thus X is not a code.

Conversely, if X is not a code then there exist $a^{i_1} b a^{j_1}, \ldots, a^{i_n} b a^{j_n}$, $a^{k_1} b a^{\ell_1}, \ldots, a^{k_n} b a^{\ell_n} \in X$ such that

$$\left(a^{i_1} b a^{j_1}\right)\left(a^{i_2} b a^{j_2}\right) \cdots \left(a^{i_n} b a^{j_n}\right) = \left(a^{k_1} b a^{\ell_1}\right)\left(a^{k_2} b a^{\ell_2}\right) \cdots \left(a^{k_n} b a^{\ell_n}\right).$$

Thus, the graph $\mathcal{G}_{abs}(X)$ contains the path

$$\boxed{0 = |i_1 - k_1|} \rightarrow \boxed{|j_1 - \ell_1| = |i_2 - k_2|} \rightarrow \cdots \rightarrow \boxed{|j_n - \ell_n| = 0}.$$

□

Remark 2. There already exist some algorithm to test in general if a given set is a code [7]. However, we noticed that for an exhaustive search of non-commutatively prefix bayonet code, our backtracking implementation of our algorithm (using mostly bitwise operation) runs faster.

We ran an exhaustive search of codes violating the condition (1), for $n \leq 15$. There is no such code for $n \leq 11$, $n = 13$, and $n = 14$. There are 4 codes for $n = 12$. We exhibit them below by representing the bayonet word $a^i b a^j$ by the two digits $x_i x_j$, where x_i is the i-th digit in base 17 $(0, \ldots, 9, A, \ldots, G)$.

ID	Non-commutatively prefix bayonet code											
X_1	00	02	08	0A	18	1A	40	42	50	53	56	90	92
X_2	01	03	09	0B	18	1A	40	42	50	53	56	90	92
X_3	02	08	0A	10	18	1A	42	50	53	56	60	92	A0
X_4	02	08	0A	18	1A	20	42	53	56	60	70	92	B0

Up to the knowledge of the author, these are the smallest (in cardinality and maximal word length) known non-commutatively prefix codes. In [8], Shor asked what is the maximal value of the ratio of the cardinality of a bayonet code divided by the length of its longest word. Hansel [3] proved an upper bound and Shor computed the lower bound $\frac{16}{15}$. Thanks to codes $(X_1\text{--}X_4)$, we improve Shor's lower bound to $\frac{13}{12}$.

There are 38 such codes for $n = 15$. They have in common the words

$$01 \ 07 \ 0D \ 0E \ 82 \ 84 \ 86 \ B1 \ B2 \tag{4}$$

Here follow the 38 codes, where for each code we only write the additional bayonets words.

ID	Code	ID	Code
Y_1	00 30 32 34 36 80 B0	Y_{20}	20 32 34 36 58 A0 D0
Y_2	00 30 32 34 36 80 B3	Y_{21}	20 34 36 3A 50 A0 D0
Y_3	00 30 34 36 3A 80 B0	Y_{22}	20 34 36 3A 58 A0 D0
Y_4	00 30 34 36 3A 80 B3	Y_{23}	30 32 34 36 60 B0 E0
Y_5	00 31 33 35 37 80 B0	Y_{24}	30 32 34 36 60 B3 E0
Y_6	00 31 33 35 37 80 B3	Y_{25}	30 34 36 3A 60 B0 E0
Y_7	00 31 35 37 3B 80 B0	Y_{26}	30 34 36 3A 60 B3 E0
Y_8	00 31 35 37 3B 80 B3	Y_{27}	31 33 35 37 60 B0 E0
Y_9	00 32 34 36 38 80 B0	Y_{28}	31 33 35 37 60 B3 E0
Y_{10}	00 32 34 36 38 80 B3	Y_{29}	31 35 37 3B 60 B0 E0
Y_{11}	00 33 35 37 39 80 B0	Y_{30}	31 35 37 3B 60 B3 E0
Y_{12}	00 33 35 37 39 80 B3	Y_{31}	32 34 36 38 60 B0 E0
Y_{13}	00 34 36 38 3A 80 B0	Y_{32}	32 34 36 38 60 B3 E0
Y_{14}	00 34 36 38 3A 80 B3	Y_{33}	33 35 37 39 60 B0 E0
Y_{15}	00 35 37 39 3B 80 B0	Y_{34}	33 35 37 39 60 B3 E0
Y_{16}	00 35 37 39 3B 80 B3	Y_{35}	34 36 38 3A 60 B0 E0
Y_{17}	10 32 34 36 40 90 C0	Y_{36}	34 36 38 3A 60 B3 E0
Y_{18}	10 34 36 3A 40 90 C0	Y_{37}	35 37 39 3B 60 B0 E0
Y_{19}	20 32 34 36 50 A0 D0	Y_{38}	35 37 39 3B 60 B3 E0

Notice that code (Y_1) is Shor's code. We also ran a partial search for $n = 16$.

ID	Non-commutatively prefix bayonet code
Z_1	00 01 02 0B 0C 3B 3C 50 51 52 80 82 84 86 D0 D1 D2
Z_2	00 01 02 0B 0C 3B 3C 50 51 52 81 83 85 87 D0 D1 D2
Z_3	00 01 02 0B 0C 3B 3C 50 51 5A 80 82 84 86 D0 D1 D2
Z_4	00 01 02 0B 0C 3B 3C 50 51 5A 81 83 85 87 D0 D1 D2
Z_5	00 01 0A 0B 0C 3B 3C 50 51 52 80 82 84 86 D0 D1 D2
Z_6	00 01 0A 0B 0C 3B 3C 50 51 52 81 83 85 87 D0 D1 D2
Z_7	00 01 0A 0B 0C 3B 3C 50 51 5A 80 82 84 86 D0 D1 D2
Z_8	00 01 0A 0B 0C 3B 3C 50 51 5A 81 83 85 87 D0 D1 D2
Z_9	00 02 0B 0C 11 3B 3C 50 52 61 83 85 87 91 D0 D2 E1
Z_{10}	00 02 0B 0C 11 3B 3C 50 5A 61 83 85 87 91 D0 D2 E1
Z_{11}	01 02 03 0C 0D 3B 3C 50 51 52 80 82 84 86 D0 D1 D2
Z_{12}	01 02 03 0C 0D 3B 3C 50 51 52 81 83 85 87 D0 D1 D2
Z_{13}	01 02 03 0C 0D 3B 3C 50 51 5A 80 82 84 86 D0 D1 D2
Z_{14}	01 02 03 0C 0D 3B 3C 50 51 5A 81 83 85 87 D0 D1 D2
Z_{15}	01 02 0B 0C 0D 3B 3C 50 51 52 80 82 84 86 D0 D1 D2
Z_{16}	01 02 0B 0C 0D 3B 3C 50 51 52 81 83 85 87 D0 D1 D2
Z_{17}	01 02 0B 0C 0D 3B 3C 50 51 5A 80 82 84 86 D0 D1 D2
Z_{18}	01 02 0B 0C 0D 3B 3C 50 51 5A 81 83 85 87 D0 D1 D2
Z_{19}	01 02 0B 0C 10 3B 3C 51 52 60 82 84 86 90 D1 D2 E0
Z_{20}	01 02 0B 0C 10 3B 3C 51 5A 60 82 84 86 90 D1 D2 E0
Z_{21}	01 02 0B 0C 1D 3B 3C 51 52 60 82 84 86 90 D1 D2 E0
Z_{22}	01 02 0B 0C 20 3B 3C 51 52 70 82 84 86 A0 D1 D2 F0
Z_{23}	01 02 0B 0C 20 3B 3C 51 5A 70 82 84 86 A0 D1 D2 F0
Z_{24}	01 02 0B 0C 2D 3B 3C 51 52 70 82 84 86 A0 D1 D2 F0
Z_{25}	01 02 0B 0C 2D 3B 3C 51 5A 70 82 84 86 A0 D1 D2 F0

There is no such code for $n = 17$ containing b. However, we found the following codes showing that there exist codes violating (1) even when n is prime.

$$01\ 02\ 03\ 0C\ 0D\ 3C\ 3D\ 51\ 52\ 5B\ 80\ 82\ 84\ 86\ D1\ D2\ D3\ G0 \quad (5a)$$
$$01\ 02\ 03\ 0C\ 0D\ 3C\ 3D\ 51\ 52\ 5B\ 81\ 83\ 85\ 87\ D1\ D2\ D3\ G0 \quad (5b)$$
$$01\ 02\ 03\ 0C\ 0D\ 3C\ 3D\ 51\ 52\ 5B\ 82\ 84\ 86\ 88\ D1\ D2\ D3\ G0 \quad (5c)$$

Let us recall that if one of these codes is included in a finite maximal code then the triangle conjecture is false. In the next sections, we try to complete each of these codes into a finite maximal one.

3   Factorisations of Cyclic Groups

In this section, we assume that the codes found in the previous section are included in some finite maximal code. Then we use factorisation of cyclic group theory to prove some lower bounds for the orders of the letter a in those finites codes.

Given $n \geq 1$, the ordered pair $(L, R) \subset [[n]]^2$ is a *factorisation* of $\mathbb{Z}/n\mathbb{Z}$ if for all $k \in [[n]]$ there exists a unique pair $(\ell, r) \in L \times R$ such that $k = \ell + r \mod n$.

Example 2. The ordered pair $(\{1, 3, 5\}, \{1, 2, 7, 8\})$ is a *factorisation* of $\mathbb{Z}/12\mathbb{Z}$.

In [5], Restivo, Salemi, and Sportelli showed the following link between factorisation and the theory of codes.

Theorem 1. *If X is a finite maximal code such that $b, a^n \in X$ then (L, R) is a factorisation of $\mathbb{Z}/n\mathbb{Z}$, where*

$$L = \{k \bmod n : a^k b^+ \in X\} \quad and \quad R = \{k \bmod n : b^+ a^k \in X\}.$$

Such a factorisation is called a factorisation associated to X.

In [6], Sands proved the following useful theorem.

Theorem 2. *If (L, R) is a factorisation of $\mathbb{Z}/n\mathbb{Z}$ and p is an integer relatively prime to $|L|$ then (pL, R) is a factorisation of $\mathbb{Z}/n\mathbb{Z}$.*

We call a *Sands factorisation* a factorisation (L, R) such that $p, q \in L$ and $1 \in R$ where p and q are relatively prime. We still do not know if there exists a factorisation associated to Shor's code, i.e. if there exists an integer n such that $(L \supseteq \{0, 3, 8, 11\}, R \supseteq \{0, 1, 7, 13, 14\})$ is a (Sands) factorisation of $\mathbb{Z}/n\mathbb{Z}$. We now study the factorisations associated to the other codes found in the previous section.

3.1   Known Factorisations

The reader can check that for $n \geq 2$,

$$\left(\{0, 4, 5, 9\}, \bigsqcup_{i \in [[n]]} \{8i, 8i + 2\} \right) \tag{6}$$

is a factorisation of $\mathbb{Z}/8n\mathbb{Z}$ associated to the code (X_1). In general, the integer n such that there exists a factorisation of $\mathbb{Z}/n\mathbb{Z}$ associated to the code (X_1) must be a multiple of 4. Indeed, $(\{0, 4, 5, 9\}, 2\{0, 2, 8, 10\})$ and $(2\{0, 4, 5, 9\}, \{0, 2, 8, 10\})$ are not factorisations because $0 + 2 \times 2 = 4 + 0$ and $2 \times 4 + 0 = 0 + 8$. Thus by Theorem 2 we have that $|L|$ and $|R|$ are multiples of 2, hence n is a multiple of 4. We did not find any factorisation associated to the code (X_1) where n is a multiple of 4 and not a multiple of 8.

The reader can check that for $n \geq 2$,

$$(\{0, 8, \cdots, 8(n - 1)\}, \{0, 1, 2, 3, 4, 7, 13, 14\}) \tag{7}$$

is a factorisation of $\mathbb{Z}/8n\mathbb{Z}$ associated to the codes $(Y_6, Y_8, Y_{10}, Y_{12}, Y_{14}, Y_{16})$. By a similar argument, we can show that in general a factorisation of $\mathbb{Z}/n\mathbb{Z}$ associated to those codes satisfies the fact that 4 divides n.

The others factorisations associated to codes found in the previous section are of Sands type or equivalent to a Sands factorisation.

3.2   Sands Factorisations

We did not find Sands factorisation but we can compute some constraints about their existence.

Assume that the code (Y_2) is included in a finite maximal code. Let (L, R) be a factorisation associated to this code, thus $L \supseteq \{0, 3, 8\}$ and $R \supseteq \{0, 1, 7, 13, 14\}$. Notice that $(L, 8R)$, $(L, 3R)$, and $(L, 5R)$ are not factorisations since $8 + 8 \times 0 = 0 + 8 \times 1$, $8 + 3 \times 0 = 0 + 3 \times 1$, and $8 + 5 \times 0 = 3 + 5 \times 1$, so that by Theorem 2, we have that $|R|$ is a multiple of $2 \times 3 \times 5$. Thus the order of the letter a is of the form $30 \times k$, where $k \geq 3$.

Following a similar argument we compute the following table.

Codes	Order of the letter a
Y_2, Y_4	$2 \times 3 \times 5 \times k = 30k$, with $k \geq 3$
$Y_5, Y_7, Y_9, Y_{11}, Y_{13}, Y_{15}$	$2 \times 3 \times 11 \times k = 66k$, with $k \geq 3$
Y_1, Y_3	$2 \times 3 \times 5 \times 11 \times k = 330k$, with $k \geq 4$
Z_1, Z_3, Z_5, Z_7	$2 \times 3 \times 5 \times 13 \times k = 390k$, with $k \geq 4$
Z_2, Z_4, Z_6, Z_8	$2 \times 3 \times 5 \times 13 \times k = 390k$, with $k \geq 3$
Z_9, Z_{10}	$2 \times 5 \times 13 \times k = 130k$, with $k \geq 3$

Remark 3. It is known [6] that (L, R) is a factorisation if and only if $(L, R - r)$ is a factorisation, where $r \in R$. Thus if $(L \supseteq \{0, 5, 13\}, R \supseteq \{0, 2, 11, 12\})$ is a factorisation associated to the codes (Z_9, Z_{10}) then $(L, R - 11)$ is a Sands factorisation.

The factorisations just take into account the words belonging to $ba^* \cup a^*b$. In the next section, we look for a more powerful tool.

4   Complete Modular Bayonet Code

In this section, we use a theorem by Perrin and Schützenberger to find the smallest known codes that are non-commutatively prefix and not included in a finite maximal code. Then, we propose a new approach of the triangle conjecture thanks to this theorem.

In [4], Perrin and Schützenberger proved the following theorem.

Theorem 3. *Let X be a finite maximal code. Let $x \in A$ be a letter and let n be the order of x. For all $\omega \in A^*$, the set*

$$C_x(\omega) := \{(i \bmod n, j \bmod n) : x^i \omega x^j \in X^*\}$$

has cardinal n.

We call a *n-modular bayonet code* a bayonet code X such that $\{a^n\} \cup X$ is a code and we said that it is *complete* if $|X| = n$. Thanks to Theorem 3, we know that to be included in a finite maximal code, a bayonet code must be included in a complete n-modular bayonet code.

Example 3. We call an n-permutation code a set of bayonet words $X \subseteq a^{<n}ba^{<n}$ such that the square binary matrix $\mathcal{M}$ of size n defined by

$$\mathcal{M}_{i,j} = 1 \text{ if and only if } a^i ba^j \in X$$

is a permutation matrix. An n-permutation code is a complete n-modular bayonet code.

We now try to find a complete n-modular bayonet code containing one of our codes that is non-commutatively equivalent to a prefix code.

4.1   Computer Exploration

We slightly modify the algorithm given in Sect. 2 to test whether or not a given set is an n-modular bayonet code. Given a set $X \in a^{<n}ba^{<n}$, we call $\mathcal{G}_{\mathrm{mod}}(X)$ the oriented graph defined by the set of vertices $[[n]]$ and by the edges

$$\boxed{i - k \bmod n} \longrightarrow \boxed{\ell - j \bmod n},$$

for all $a^i ba^j, a^k ba^\ell \in X$, with $a^i ba^j \neq a^k ba^\ell$. The set X is an n-modular bayonet code if and only if the graph $\mathcal{G}_{\mathrm{mod}}(X)$ does not contain a non-empty path from 0 to 0.

In the previous section, we show that the codes $(X_1, Y_6, Y_8, Y_{10}, Y_{12}, Y_{14}, Y_{16})$ might be included in a finite maximal code where the order of the letter a is of the form $4 \times k$ with $k \geq 4$. By an exhaustive computer search, we found that none of these codes is included in a complete n-modular bayonet code, where $n \leq 32$. Thus if (X_1) is included in a finite maximal code then the order of the letter a is of the form $4 \times k$, where $k \geq 10$ (there is no factorisation of $\mathbb{Z}/n\mathbb{Z}$ associated to this code, where $32 < n < 40$). If one of the codes $(Y_6, Y_8, Y_{10}, Y_{12}, Y_{14}, Y_{16})$ is included in a finite maximal code then the order of the letter a is of the form $4 \times k$, where $k \geq 9$ (there is no factorisation of $\mathbb{Z}/n\mathbb{Z}$ associated to those codes with $32 < n < 36$). In particular, the computer exploration implies the following proposition.

Proposition 3. *If X is one of the codes $(X_1\text{--}X_4)$, then $X \cup \{a^{16}\}$ is a code that is non-commutatively equivalent to a prefix code and not included in a finite maximal code.*

Proof. Let X be one of the codes $(X_1\text{--}X_4)$. Then $X \cup \{a^{16}\}$ is a code. Moreover, we checked by an exhaustive search that X is not included in a complete 16-modular bayonet code. We conclude the proof thanks to Theorem 3.   □

Up to the knowledge of the author, the four codes given in Proposition 3 are the smallest (in cardinality and maximal length word) known codes that are not commutatively equivalent to a prefix code and not included in a finite maximal code.

4.2   Transformations

In order to have a better understanding of the complete n-modular bayonet code, we look at some transformations.

Lemma 1. *If X is an n-modular code then for any $r \in [[n]]$, the set*

$$s_r(X) := \{a^i b a^j : a^i b a^p, a^q b a^j \in X \text{ and } p + q = r \text{ mod } n\}$$

is an n-modular code.

Proof. Given an integer $r \in [[n]]$, if $s_r(X)$ is not an n-modular bayonet code then there exists $a^{i_1} b a^{j_1}, \ldots, a^{i_m} b a^{j_m}, a^{k_1} b a^{\ell_1}, \ldots, a^{k_m} b a^{\ell_m} \in s_r(X)$, with $j_1 \neq \ell_1$ such that

$$\begin{cases} i_1 & = & k_1 \\ j_1 + i_2 & = & \ell_1 + k_2 \quad \text{mod } n \\ & \vdots & \\ j_{m-1} + i_m & = & \ell_{m-1} + k_m \text{ mod } n \\ j_m & = & \ell_m \end{cases}$$

By definition of $s_r(X)$, there exists $a^{i_1} b a^{p_1}, a^{q_1} b a^{j_1}, \ldots, a^{i_m} b a^{p_m}, a^{q_m} b a^{j_m} \in X$ and $a^{k_1} b a^{p'_1}, a^{q'_1} b a^{\ell_1}, \ldots, a^{k_m} b a^{p'_m}, a^{q'_m} b a^{\ell_m} \in X$ such that $p_t + q_t = p'_t + q'_t = r$ mod n, for all $t \in [m]$. Thus

$$\begin{cases} i_1 & = & k_1 \\ p_1 + q_1 & = & p'_1 + q'_1 \quad \text{mod } n \\ j_1 + i_2 & = & \ell_1 + k_2 \quad \text{mod } n \\ & \vdots & \\ j_{m-1} + i_m & = & \ell_{m-1} + k_m \text{ mod } n \\ p_m + q_m & = & p'_m + q'_m \quad \text{mod } n \\ j_m & = & \ell_m \end{cases}$$

Thus X is not an n-modular code, since it has a double factorisation. We conclude the proof by contraposition.   □

We use this lemma to prove the following theorem.

Theorem 4. *If X is an n-modular code then $|X| \leq n$.*

Proof. Assume that X is an n-modular code such that $|X| > n$. Thanks to Lemma 1, we know that for any $r \in [[n]]$, $s_r(X)$ is a code thus

$$\sum_{r \in [[n]]} |s_r(X)| = |X|^2$$

Thus there exists $r_1 \in [[n]]$ such that $|s_{r_1}(X)| \geq \left\lceil \frac{|X|^2}{n} \right\rceil \geq n + 1$. By iteration, there exists $r_2, \ldots, r_{n^2} \in [[n]]$ such that $\left| s_{r_{n^2}} (\cdots s_{r_2} (s_{r_1}(X)) \cdots) \right| > n^2$ which contradicts the fact that X belongs to $a^{<n} b a^{<n}$.   □

Thanks to this theorem, we now exhibit five transformations of an n-modular code that preserve the completeness.

Theorem 5. *If X is an n-modular code (respectively complete) then*

1. For all α and β, the set

$$\tau_{\alpha,\beta}(X) := \left\{ a^{i+\alpha \bmod n} b a^{j+\beta \bmod n} : a^i b a^j \in X \right\}$$

is an n-modular code (respectively complete).
2. For all q prime to n, the set

$$\rho_q(X) := \left\{ a^{qi \bmod n} b a^{qj \bmod n} : a^i b a^j \in X \right\}$$

is an n-modular code (respectively complete).
3. The set

$$\iota(X) := \left\{ a^{n-1-i} b a^j : a^i b a^j \in X \right\}$$

is an n-modular code (respectively complete).
4. The dual code $\delta(X)$ is an n-modular code (respectively complete).
5. For any $r \in [[n]]$, the set $s_r(X)$ is an n-modular code (respectively complete).

Proof

1. For any $\alpha, \beta \in [[n]]$, the graph $\mathcal{G}_{\mathrm{mod}}(\tau_{\alpha,\beta}(X))$ is equal to the graph $\mathcal{G}_{\mathrm{mod}}(X)$. Thus X is an n-modular code if and only if $\tau_{\alpha,\beta}(X)$ is an n-modular code.
2. For any q prime to n, the function that associates to $i \in [[n]]$ the integer qi mod n is a graph isomorphism from $\mathcal{G}_{\mathrm{mod}}(X)$ to $\mathcal{G}_{\mathrm{mod}}(\rho_q(X))$. Thus X is a code if and only if $\rho_q(X)$ is a code.
3. If $\iota(X)$ is not an n-modular bayonet code then the graph $\mathcal{G}_{\mathrm{mod}}(\iota(X))$ contains the paths

$$\boxed{0} \longrightarrow \boxed{i_1} \longrightarrow \boxed{i_2} \longrightarrow \cdots \boxed{i_m} \longrightarrow \boxed{0}$$

and

$$\boxed{0} \longrightarrow \boxed{-i_1 \bmod n} \longrightarrow \boxed{-i_2 \bmod n} \longrightarrow \cdots \boxed{-i_m \bmod n} \longrightarrow \boxed{0},$$

for $i_1, \ldots, i_m \in [[n]]$. Thus, the graph $\mathcal{G}_{\mathrm{mod}}(X)$ contains the path

$$\boxed{0} \longrightarrow \boxed{i_1} \longrightarrow \boxed{-i_2 \bmod n} \longrightarrow \boxed{i_3} \longrightarrow \boxed{-i_4 \bmod n} \longrightarrow \cdots \longrightarrow \boxed{0}$$

which contradicts the fact that X is an n-modular code. We conclude the proof by contraposition.
4. The graph $\mathcal{G}_{\mathrm{mod}}(\delta(X))$ is the graph $\mathcal{G}_{\mathrm{mod}}(X)$ with inverted arrows.
5. If X is an n-modular code then, by Lemma 1, $s_r(X)$ is an n-modular code. Let us prove that if $|X| = n$ then for any $r \in [[n]], |s_r(X)| = n$. Assume that $|s_r(X)| \neq n$ for $r \in [[n]]$ then there exists an $r' \in [[n]]$ such that $|s_{r'}(X)| > n$ which contradicts Theorem 4.

$\square$

The author wonders if the following Sands-like statement (a strong version of Theorem 5.2) is true.

Conjecture 1. *If X is a complete n-modular bayonet code then*

$$\varphi_q(X) := \left\{ a^{qi \bmod n} ba^j : a^i ba^j \in X \right\}$$

is a complete n-modular bayonet code, for all q prime to n.

If this conjecture is true, then one can compute some lower bound for the order of the letter a, using all the bayonet words. Notice that we already proved the case $q = n - 1$ of Conjecture 1 in Theorem 5, indeed $\varphi_q(X) = \tau_{1,0}(\iota(X))$.

5   Conclusion and Perspectives

We propose three main perspectives. Firstly, we would like to enumerate the bayonet codes that are non-commutatively equivalent to a prefix code. As we saw in the Sect. 2, the codes we found look closely related to each other. Secondly, we wonder if there exists a code non-commutatively equivalent to a prefix code smaller then the codes (X_1-X_4). Such a code would necessarily be a non-bayonet code. Finally, our main perspective is continuing our effort to find whether or not there exists a bayonet non-commutatively prefix code that is included in a complete modular bayonet code.

Acknowledgements. The author wants to thank Dominique Perrin for introducing him to the *commutatively prefix conjecture*, also his Ph.D. supervisors Samuele Giraudo and Jean-Christophe Novelli.

References

1. Berstel, J., Perrin, D., Reutenauer, C.: Codes and Automata, vol. 129. Cambridge University Press, Cambridge (2010)
2. De Felice, C.: A note on the factorization conjecture. Acta Informatica **50**(7–8), 381–402 (2013)
3. Hansel, G.: Baionnettes et cardinaux. Discrete Math. **39**(3), 331–335 (1982)
4. Perrin, D., Schützenberger, M.-P.: Codes et sous-monoïdes possédant des mots neutres. In: Theoretical Computer Science. LNCS, vol. 48, pp. 270–281. Springer, Heidelberg (1977). https://doi.org/10.1007/3-540-08138-0_23
5. Restivo, A., Salemi, S., Sportelli, T.: Completing codes. RAIRO-Theoret. Inf. Appl. **23**(2), 135–147 (1989)
6. Sands, A.D.: Replacement of factors by subgroups in the factorization of abelian groups. Bull. Lond. Math. Soc. **32**(3), 297–304 (2000)
7. Sardinas, A.A., Patterson, G.W.: A necessary and sufficient condition for unique decomposition of coded messages. In: Proceedings of the Institute of Radio Engineers, vol. 41, no. 3, pp. 425–425 (1953)
8. Shor, P.W.: A counterexample to the triangle conjecture. J. Comb. Theory Ser. A **38**(1), 110–112 (1985)
9. Zhang, L., Shum, K.P.: Finite maximal codes and triangle conjecture. Discrete Math. **340**(3), 541–549 (2017)

Efficient Representation and Counting of Antipower Factors in Words

Tomasz Kociumaka[ID], Jakub Radoszewski[ID], Wojciech Rytter[ID],
Juliusz Straszyński[ID], Tomasz Waleń[ID], and Wiktor Zuba[(✉)][ID]

Institute of Informatics, University of Warsaw, Warsaw, Poland
{kociumaka,jrad,rytter,jks,walen,w.zuba}@mimuw.edu.pl

Abstract. A k-antipower (for $k \geq 2$) is a concatenation of k pairwise distinct words of the same length. The study of antipower factors of a word was initiated by Fici et al. (ICALP 2016) and first algorithms for computing antipower factors were presented by Badkobeh et al. (Inf. Process. Lett., 2018). We address two open problems posed by Badkobeh et al. Our main results are algorithms for counting and reporting factors of a word which are k-antipowers. They work in $\mathcal{O}(nk \log k)$ time and $\mathcal{O}(nk \log k + C)$ time, respectively, where C is the number of reported factors. For $k = o(\sqrt{n/\log n})$, this improves the time complexity of $\mathcal{O}(n^2/k)$ of the solution by Badkobeh et al. Our main algorithmic tools are runs and gapped repeats. We also present an improved data structure that checks, for a given factor of a word and an integer k, if the factor is a k-antipower.

Keywords: Antipower · α-gapped repeat · Run (maximal repetition)

1 Introduction

Antipowers are a new type of regularity of words, based on diversity rather than on equality, that has been recently introduced by Fici et al. in [7,8]. Typical types of regular words are powers. If equality is replaced by inequality, other versions of powers are obtained.

Let us assume that $x = y_1 \cdots y_k$, where $k \geq 2$ and y_i are words of the same length d. We then say that:

- x is a k-*power* if all y_i's are the same;
- x is a k-*antipower* (or a (k,d)-antipower) if all y_i's are pairwise distinct;

T. Kociumaka and W. Rytter—Supported by the Polish National Science Center, grant no 214/13/B/ST6/00770.

J. Radoszewski and J. Straszyński—Supported by the "Algorithms for text processing with errors and uncertainties" project carried out within the HOMING program of the Foundation for Polish Science co-financed by the European Union under the European Regional Development Fund.

C. Martín-Vide et al. (Eds.): LATA 2019, LNCS 11417, pp. 421–433, 2019.
https://doi.org/10.1007/978-3-030-13435-8_31

– x is a *weak k-power* (or a weak (k,d)-power) if it is not a k-antipower, that is, if $y_i = y_j$ for some $i \neq j$;
– x is a *gapped (q,d)-square* if $y_1 = y_k$ and $q = k - 2$.

In the first three cases, the length d is called the *base* of the power or antipower x.

If w is a word, then by $w[i \mathinner{..} j]$ we denote a word composed of letters $w[i], \ldots, w[j]$ called a *factor* of w. A factor can be represented in $\mathcal{O}(1)$ space by the indices i and j. Badkobeh et al. [1] considered factors of a word that are antipowers and obtained the following result.

Fact 1 ([1]). *The maximum number of k-antipower factors in a word of length n is $\Theta(n^2/k)$, and they can all be reported in $\mathcal{O}(n^2/k)$ time. In particular, all k-antipower factors of a specified base d can be reported in $\mathcal{O}(n)$ time.*

Badkobeh et al. [1] asked for an output-sensitive algorithm that reports all k-antipower factors in a given word. We present such an algorithm. En route to enumerating k-antipowers, we (complementarily) find weak k-powers. Also gapped (q,d)-squares play an important role in our algorithm.

For a given word w, an *antipower query* (i,j,k) asks to check if a factor $w[i \mathinner{..} j]$ is a k-antipower. Badkobeh et al. [1] proposed the following solutions:

Fact 2 ([1]). *Antipower queries can be answered (a) in $\mathcal{O}(k)$ time with a data structure of size $\mathcal{O}(n)$; (b) in $\mathcal{O}(1)$ time with a data structure of size $\mathcal{O}(n^2)$.*

In either case, answering n antipower queries using Fact 2 requires $\Omega(n^2)$ time in the worst case. We show a trade-off between the data structure space and query time that allows answering any n antipower queries more efficiently.

Our Results. Our first main result is an algorithm that computes the number C of factors of a word of length n that are k-antipowers in $\mathcal{O}(nk \log k)$ time and reports all of them in $\mathcal{O}(nk \log k + C)$ time. We assume an integer alphabet $\{1, \ldots, n^{\mathcal{O}(1)}\}$.

Our second main result is a construction in $\mathcal{O}(n^2/r)$ time of a data structure of size $\mathcal{O}(n^2/r)$, for any $r \in \{1, \ldots, n\}$, which answers antipower queries in $\mathcal{O}(r)$ time. Thus, any n antipower queries can be answered in $\mathcal{O}(n\sqrt{n})$ time and space.

Structure of the Paper. Our algorithms are based on a relation between weak powers and two notions of periodicity of words: gapped repeats and runs. In Sect. 2, we recall important properties of these notions. Section 3 shows a simple algorithm that counts k-antipowers in a word of length n in $\mathcal{O}(nk^3)$ time. In Sect. 4, it is improved in three steps to an $\mathcal{O}(nk \log k)$-time algorithm. Finally, algorithms for reporting k-antipowers and answering antipower queries are presented in Sect. 5. Omitted proofs can be found in the full version [10].

2 Preliminaries

The length of a word w is denoted by $|w|$ and the letters of w are numbered 0 through $|w| - 1$, with $w[i]$ representing the ith letter. Let $[i \mathinner{..} j]$ denote the integer interval $\{i, i+1, \ldots, j\}$ and $[i \mathinner{.\,.} j)$ denote $[i \mathinner{.\,.} j-1]$. By $w[i \mathinner{.\,.} j]$ we denote the factor $w[i] \cdots w[j]$; if $i > j$, it denotes the empty word. Let us further denote $w[i \mathinner{.\,.} j) = w[i \mathinner{.\,.} j-1]$. We say that p is a *period* of the word w if $w[i] = w[i+p]$ holds for all $i \in [0 \mathinner{.\,.} |w| - p)$.

An α-gapped repeat γ (for $\alpha \geq 1$) in a word w is a factor uvu of w such that $|uv| \leq \alpha|u|$. The two occurrences of u are called *arms* of the α-gapped repeat and $|uv|$, denoted $\mathsf{per}(\gamma)$, is called *the period* of the α-gapped repeat. Note that an α-gapped repeat is also an α'-gapped repeat for every $\alpha' > \alpha$. An α-gapped repeat is called *maximal* if its arms can be extended simultaneously with the same character neither to the right nor to the left. In short, we call maximal α-gapped repeats α-*MGRs* and the set of α-MGRs in a word w is further denoted by $MGReps_\alpha(w)$. The first algorithm for computing α-MGRs was proposed by Kolpakov et al. [12]. It was improved by Crochemore et al. [6], Tanimura et al. [14], and finally Gawrychowski et al. [9], who showed the following result.

Fact 3 ([9]). *For a word w of length n and a parameter α, the set $MGReps_\alpha(w)$ satisfies $|MGReps_\alpha(w)| \leq 18\alpha n$, and it can be computed in $\mathcal{O}(n\alpha)$ time.*

A *run* (a maximal repetition) in a word w is a triple (i, j, p) such that $w[i \mathinner{.\,.} j]$ is a factor with the smallest period p, $2p \leq j-i+1$, that can be extended neither to the left nor to the right preserving the period p. Its *exponent* e is defined as $e = (j - i + 1)/p$. Kolpakov and Kucherov [11] showed that a word of length n has $\mathcal{O}(n)$ runs, with sum of exponents $\mathcal{O}(n)$, and that they can be computed in $\mathcal{O}(n)$ time. Bannai et al. [2] recently refined these combinatorial results.

Fact 4 ([2]). *A word of length n has at most n runs, and the sum of their exponents does not exceed $3n$. All these runs can be computed in $\mathcal{O}(n)$ time.*

A *generalized run* in a word w is a triple $\gamma = (i, j, p)$ such that $w[i \mathinner{.\,.} j]$ is a factor with a period p, not necessarily the shortest one, $2p \leq j - i + 1$, that can be extended neither to the left nor to the right preserving the period p. By $\mathsf{per}(\gamma)$ we denote p, called *the period* of the generalized run γ. The set of generalized runs in a word w is denoted by $GRuns(w)$.

A run (i, j, p) with exponent e corresponds to $\lfloor \frac{e}{2} \rfloor$ generalized runs (i, j, p), $(i, j, 2p)$, $(i, j, 3p)$, $\ldots$, $(i, j, \lfloor \frac{e}{2} \rfloor p)$. By Fact 4, we obtain the following

Corollary 5. *For a word w of length n, the set $GRuns(w)$ can be computed in $\mathcal{O}(n)$ time and it satisfies $|GRuns(w)| \leq 1.5n$.*

Our algorithm uses a relation between weak powers, α-MGRs, and generalized runs; see Fig. 1 for an example presenting the interplay of these notions.

An *interval representation* of a set X of integers is

$$X = [i_1 \mathinner{.\,.} j_1] \cup [i_2 \mathinner{.\,.} j_2] \cup \cdots \cup [i_t \mathinner{.\,.} j_t],$$

where $i_1 \leq j_1$, $j_1 + 1 < i_2$, $i_2 \leq j_2$, $\ldots$, $j_{t-1} + 1 < i_t$, $i_t \leq j_t$; the value t is called the *size* of the representation. The following simple lemma allows implementing unions on interval representations. Its proof can be found in the full version.

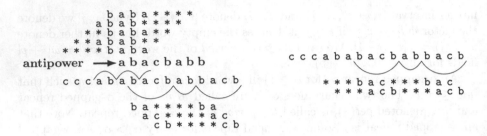

Fig. 1. To the left: all weak $(4,2)$-powers and one $(4,2)$-antipower in a word of length 16. An asterisk denotes any character. The first five weak $(4,2)$-powers are generated by the run ababa with period 2, and the last three are generated by the 1.5-MGR bacb ab bacb, whose period (6) is divisible by 2. To the right: all weak $(4,3)$-powers in the same word are generated by the same MGR because its period is a multiple of 3.

Lemma 6. *Assume that* $\mathcal{X}_1, \dots, \mathcal{X}_r$ *are non-empty families of subintervals of* $[0 \,.\,.\, n)$. *The interval representations of* $\bigcup \mathcal{X}_1, \bigcup \mathcal{X}_2, \dots, \bigcup \mathcal{X}_r$ *can be computed in* $\mathcal{O}(n + m)$ *time, where* m *is the total size of the families* $\mathcal{X}_i$.

Let $\mathcal{J}$ be a family of subintervals of $[0 \,.\,.\, m)$, initially empty. Let us consider the following operations on $\mathcal{J}$, where I is an interval: $\mathtt{insert}(I)$: $\mathcal{J} := \mathcal{J} \cup \{I\}$; $\mathtt{delete}(I)$: $\mathcal{J} := \mathcal{J} \setminus \{I\}$ for $I \in \mathcal{J}$; and $\mathtt{count}$, which returns $|\bigcup \mathcal{J}|$. It is folklore knowledge that all these operations can be performed efficiently using a static range tree (sometimes called a segment tree; see [13]). In the full version, we prove the following lemma for completeness.

Lemma 7. *There exists a data structure of size* $\mathcal{O}(m)$ *that, after* $\mathcal{O}(m)$*-time initialization, handles* $\mathtt{insert}$ *and* $\mathtt{delete}$ *in* $\mathcal{O}(\log m)$ *time and* $\mathtt{count}$ *in* $\mathcal{O}(1)$ *time.*

Let us introduce another operation $\mathtt{report}$ that returns all elements of the set $A = [0 \,.\,.\, m) \setminus \bigcup \mathcal{J}$. We also show in the full version that a static range tree can support this operation efficiently.

Lemma 8. *There exists a data structure of size* $\mathcal{O}(m)$ *that, after* $\mathcal{O}(m)$*-time initialization, handles* $\mathtt{insert}$ *and* $\mathtt{delete}$ *in* $\mathcal{O}(\log m)$ *time and* $\mathtt{report}$ *in* $\mathcal{O}(|A|)$ *time.*

3   Compact Representation of Weak k-powers

Let us denote by $Squares(q, d)$ the set of starting positions of occurrences of gapped (q, d)-squares in the input word w.

We say that an occurrence at position i of a gapped (q, d)-square is *generated* by a gapped repeat uvu if the gapped repeat has period $p = (q + 1)d$ and $w[i \,.\,.\, i + d), w[i + p \,.\,.\, i + p + d)$ are contained in the first arm and in the second arm of the gapped repeat, respectively; cf. Fig. 2. In other words, $u =$

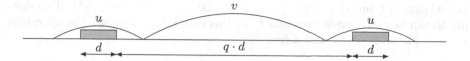

Fig. 2. An occurrence of a gapped (q, d)-square generated by a gapped repeat with period $(q + 1)d$. Gray rectangles represent equal words.

$u_1 u_2 u_3$, $|u_2| = d$, $|u_3 v u_1| = qd$, and uvu starts in the input word at position $i - |u_1|$.

An occurrence in w of a (q, d)-square is *generated* by a generalized run with period $p = (q + 1)d$ if it is fully contained in this generalized run. See Fig. 3 for a concrete example. The proof of the following lemma is in the full version.

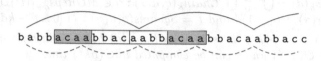

Fig. 3. An occurrence of a gapped $(2, 4)$-square acaa bbac aabb acaa generated by a generalized run with period 12. Note that the generalized run has its origin in a run with period 6 (depicted below) that itself *does not* generate this gapped square.

Lemma 9.

(a) *Every gapped (q, d)-square is generated by a $(q+1)$-MGR with period $(q+1)d$ or by a generalized run with period $(q + 1)d$.*

(b) *Each gapped repeat and each run γ with period $(q + 1)d$ generates a single interval of positions where gapped (q, d)-squares occur, denoted Squares (q, d, γ) (see Fig. 4). Moreover, this interval can be computed in constant time.*

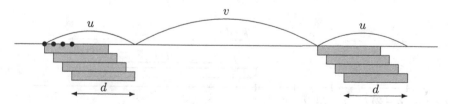

Fig. 4. An interval, represented as a sequence of four consecutive positions (black dots), of starting positions of occurrences of gapped (q, d)-squares generated by a gapped repeat with period $(q + 1)d$.

Let us denote $Chain_k(q, d, i) = \{ i,\, i - d,\, i - 2d,\, \ldots,\, i - (k - q - 2)d \}$. This definition can be extended to intervals I. To this end, let us introduce the operation $I \ominus r = \{ i - r : i \in I \}$ and define

$$Chain_k(q, d, I) \;=\; I \cup (I \ominus d) \cup (I \ominus 2d) \cup \cdots \cup (I \ominus (k - q - 2)d).$$

This set is further referred to as an *interval chain*; it can be stored in $\mathcal{O}(1)$ space.

We denote by $WeakPow_k(d)$ the set of starting positions in w of weak (k, d)-powers. A *chain representation* of a set of integers is its representation as a union of interval chains. The size of the chain representation is the number of chains. The following lemma shows how to compute small chain representations of the sets $WeakPow_k(d)$.

Lemma 10.

(a) $WeakPow_k(d) = \bigcup_{q=0}^{k-2} \bigcup_{i \in Squares(q,d)} Chain_k(q, d, i) \cap [0 .. n - kd]$.

(b) $WeakPow_k(d) = \bigcup_{q=0}^{k-2} \bigcup \{ Chain_k(q, d, I) : \gamma \in MGReps_{q+1}(w) \cup GRuns(w)$, *where* $\mathsf{per}(\gamma) = (q + 1)d$ *and* $I = Squares(q, d, \gamma) \} \cap [0 .. n - kd]$.

(c) *For* $d = 1, \ldots, \lfloor n/k \rfloor$, *the sets* $WeakPow_k(d)$ *have chain representations of total size* $\mathcal{O}(nk^2)$ *which can be computed in* $\mathcal{O}(nk^2)$ *time.*

Proof. As for point **(a)**, $x = y_1 \cdots y_k$ for $|y_1| = \cdots = |y_k| = d$ is a weak (k, d)-power if and only if $y_i \cdots y_j$ is a gapped $(j - i - 1, d)$-square for some $1 \leq i < j \leq k$. Conversely, a gapped (q, d)-square occurring at position i implies occurrences of weak (k, d)-powers at positions in the set $Chain_k(q, d, i)$, limited to the interval $[0 .. n - kd]$ due to the length constraint; see Fig. 5.

The formula in **(b)** follows from point **(a)** by Lemma 9. Indeed, Lemma 9**(a)** shows that every gapped (q, d)-square is generated by a $(q+1)$-MGR with period $(q+1)d$ or a generalized run with period $(q+1)d$. By Lemma 9**(b)**, the starting positions of all such gapped squares that are generated by an MGR or a generalized run γ form an interval $I = Squares(q, d, \gamma)$. Hence, it yields an interval chain $Chain_k(q, d, I)$ of starting positions of weak (k, d)-powers by point **(a)**.

Finally, we obtain point **(c)** by applying the formula from point **(b)** to compute the chain representations of sets $WeakPow_k(d)$ for all $d = 1, \ldots, \lfloor n/k \rfloor$. This is also shown in the first part of the SimpleCount algorithm, where the resulting chain representations are denoted as $\mathcal{C}_d$. The total number of interval

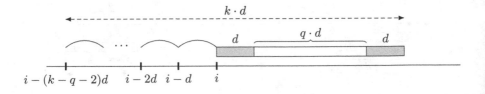

Fig. 5. The fact that $i \in Squares(q, d)$ is a witness of inclusion $(Chain_k(q, d, i) \cap [0 .. n - kd]) \subseteq WeakPow_k(d)$.

Algorithm 1. SimpleCount(w, n, k)

$(\mathcal{C}_d)_{d=1}^{\lfloor n/k \rfloor} := (\emptyset, \ldots, \emptyset)$
for $q := 0$ **to** $k - 2$ **do**
    foreach $(q + 1)$-*MGR or generalized run* γ *in* w **do**
        $p := \mathsf{per}(\gamma)$
        if $(q + 1) \mid p$ **then**
            $d := \frac{p}{q+1}$
            $I := Squares(q, d, \gamma)$
            $\mathcal{C}_d := \mathcal{C}_d \cup \{ Chain_k(q, d, I) \}$
$antipowers := 0$
for $d := 1$ **to** $\lfloor n/k \rfloor$ **do**
    $WeakPow_k(d) := (\bigcup \mathcal{C}_d) \cap [0 .. n - kd]$
    $antipowers := antipowers + (n - kd + 1) - |WeakPow_k(d)|$
return $antipowers$

---

chains in these representations is $\mathcal{O}(nk^2)$ because, for each $q \in [0 .. k - 2]$, the number of $(q + 1)$-MGRs and generalized runs γ is bounded by $\mathcal{O}(nk)$ due to Facts 3 and 4, respectively. □

Lemma 10 lets us count k-antipowers by computing the size of the complementary sets $WeakPow_k(d)$. Thus, we obtain the following preliminary result.

Proposition 11. *The number of k-antipower factors in a word of length n can be computed in $\mathcal{O}(nk^3)$ time.*

Proof. See Algorithm 1. We use Lemma 10, points **(b)** and **(c)**, to express the sets $WeakPow_k(d)$ for all $d = 1, \ldots, \lfloor n/k \rfloor$ as a union of $\mathcal{O}(nk^2)$ interval chains. That is, the total size of the sets $\mathcal{C}_d$ is $\mathcal{O}(nk^2)$. Each of the interval chains consists of at most k intervals. Hence, Lemma 6 can be applied to compute interval representations of the sets $WeakPow_k(d)$ in $\mathcal{O}(nk^3)$ total time. Finally, the size of the complement of the set $WeakPow_k(d)$ (in $[0 .. n - kd]$) is the number of (k, d)-antipowers. □

Next, we improve the time complexity of this algorithm to $\mathcal{O}(nk \log k)$.

4 Counting k-antipowers in $\mathcal{O}(nk \log k)$ Time

We improve the algorithm SimpleCount threefold. First, we show that the chain representation of weak k-powers actually consists of only $\mathcal{O}(nk)$ chains. Then, instead of processing the chains by their interval representations, we introduce a geometric interpretation that reduces the problem to computing the area of the union of $\mathcal{O}(nk)$ axis-aligned rectangles. This area could be computed directly in $\mathcal{O}(nk \log n)$ time, but we improve this complexity to $\mathcal{O}(nk \log k)$ by exploiting properties of the dimensions of the rectangles.

4.1 First Improvement of SimpleCount

First, we improve the $\mathcal{O}(nk^2)$ bounds of Lemma 10(c). By inspecting the structure of MGRs, we actually show that the formula from Lemma 10(b) generates only $\mathcal{O}(nk)$ interval chains. A careful implementation lets us compute such a chain representation in $\mathcal{O}(nk)$ time.

We say that an α-MGR for integer α with period p is *nice* if $\alpha \mid p$ and $p \geq 2\alpha^2$. Let $NMGReps_\alpha(w)$ denote the set of nice α-MGRs in the word w. The following lemma provides a combinatorial foundation of the improvement.

Lemma 12. *For a word w of length n and an integer $\alpha > 1$, $|NMGReps_\alpha(w)| \leq 54n$.*

Proof. Let us consider a partition of the word w into blocks of α letters (the final $n \bmod \alpha$ letters are not assigned to any block). Let uvu be a nice α-MGR in w. We know that $2\alpha^2 \leq |uv| \leq \alpha|u|$, so $|u| \geq 2\alpha$. Now, let us fit the considered α-MGR into the structure of blocks. Since $\alpha \mid |uv|$, the indices in w of the occurrences of the left and the right arm are equal modulo α. We shrink both arms to u' such that u' is the maximal inclusion-wise interval of blocks which is encompassed by each arm u. Then, let us expand v to v' so that it fills the space between the two occurrences of u'.

Let us notice that $|uv| = |u'v'|$. Moreover, $|u'| \geq \frac{1}{3}|u|$ since u encompasses at least one full block of w. Consequently, $|u'v'| \leq 3\alpha|u'|$.

Let t be a word whose letters correspond to whole blocks in w and u'', v'' be factors of t that correspond to u' and v', respectively. We have $|u''| = |u'|/\alpha$ and $|v''| = |v'|/\alpha$, so $u''v''u''$ is a 3α-gapped repeat in t. It is also a 3α-MGR because it can be expanded by one block neither to the left nor to the right, as it would contradict the maximality of the original nice α-MGR. This concludes that every nice α-MGR in w has a corresponding 3α-MGR in t. Also, every 3α-MGR in t corresponds to at most one nice α-MGR in w, as it can be translated into blocks of w and expanded in a single way to a 3α-MGR (that can happen to be a nice α-MGR).

We conclude that the number of nice α-MGRs in w is at most the number of 3α-MGRs in t. As $|t| \leq n/\alpha$, due to Fact 3 the latter is at most $54n$. $\square$

Lemma 13. *For $d = 1, \ldots, \lfloor n/k \rfloor$, the sets $WeakPow_k(d)$ have chain representations of total size $\mathcal{O}(nk)$ which can be computed in $\mathcal{O}(nk)$ time.*

Proof. The chain representations of sets $WeakPow_k(d)$ are computed for $d < 2k - 2$ and for $d \geq 2k - 2$ separately.

From Fact 1, we know that all (k, d)-antipowers can be found in $\mathcal{O}(n)$ time. This lets us compute the set $WeakPow_k(d)$ (and its trivial chain representation) in $\mathcal{O}(n)$ time. Across all $d < 2k - 2$, this gives $\mathcal{O}(nk)$ chains and $\mathcal{O}(nk)$ time.

Henceforth we consider the case that $d \geq 2k - 2$. Let us note that if a gapped (q, d)-square with $d \geq 2(q+1)$ is generated by a $(q+1)$-MGR, then this $(q+1)$-MGR is nice. Indeed, by Lemma 9(a) this $(q+1)$-MGR has period $p = (q+1)d \geq 2(q+1)^2$. This observation lets us express the formula of Lemma 10(b) for $d \geq 2k - 2$ equivalently using $NMGReps_{q+1}(w)$ instead of $MGReps_{q+1}(w)$.

By Fact 4 and Lemma 12, for every q we have only $|NMGReps_{q+1}(w) \cup GRuns(w)| = \mathcal{O}(n)$ MGRs and generalized runs to consider. Hence, the total size of chain representations of sets $WeakPow_k(d)$ for $d \geq 2k - 2$ is $\mathcal{O}(nk)$ as well. The last piece of the puzzle is the following claim, proved in the full version.

Claim. The sets $NMGReps_\alpha(w)$ for $\alpha \in [1 \mathinner{..} k - 1]$ can be built in $\mathcal{O}(nk)$ time.

This concludes the proof.   □

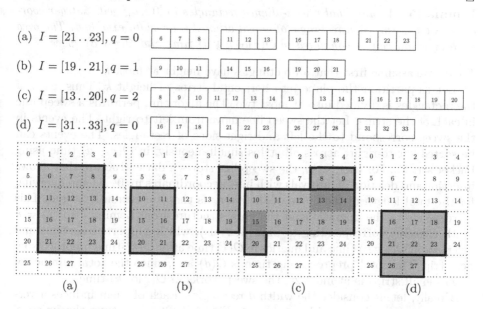

Fig. 6. Examples of decompositions of various interval chains $Chain_k(q, d, I)$ into orthogonal rectangles in the grid $\mathcal{G}_d$ for $d = 5$, $k = 5$, $n = 52$.

4.2 Second Improvement of SimpleCount

We reduce the problem to computing unions of sets of orthogonal rectangles with bounded integer coordinates.

For a given value of d, let us fit the integers from $[0 \mathinner{..} n - kd]$ into the cells of a grid of width d so that the first row consists of numbers 0 through $d - 1$, the second of numbers d to $2d - 1$, etc. Let us call this grid $\mathcal{G}_d$. A proof of the following lemma can be found in the full version. The main idea is shown in Fig. 6.

Lemma 14. *The set $Chain_k(q, d, I)$ is a union of $\mathcal{O}(1)$ orthogonal rectangles in $\mathcal{G}_d$, each of height at most k or width exactly d. The coordinates of the rectangles can be computed in $\mathcal{O}(1)$ time.*

Thus, by Lemma 13, our problem reduces to computing the area of unions of rectangles in subsequent grids $\mathcal{G}_d$. In total, the number of rectangles is $\mathcal{O}(nk)$.

4.3 Third Improvement of SimpleCount

Assume that r axis-aligned rectangles in the plane are given. The area of their union can be computed in $\mathcal{O}(r \log r)$ time using a classic sweep line algorithm (see Bentley [4]). This approach would yield an $\mathcal{O}(nk \log n)$-time algorithm for counting k-antipowers. We refine this approach in the case that the rectangles have bounded height or maximum width and their coordinates are bounded.

Lemma 15. *Assume that r axis-aligned rectangles in $[0 \mathinner{..} d]^2$ with integer coordinates are given, each rectangle of height at most k or width exactly d. The area of their union can be computed in $\mathcal{O}(r \log k + d)$ time and $\mathcal{O}(r + d)$ space.*

Proof. We assume first that all rectangles have height at most k.

Let us partition the plane into horizontal strips of height k. Thus, each of the rectangles is divided into at most two. The algorithm performs a sweep line in each of the strips. Let the sweep line move from left to right. The events in the sweep correspond to the left and right sides of rectangles. The events can be sorted left-to-right, across all strips simultaneously, in $\mathcal{O}(r + d)$ time using bucket sort [5]. For each strip, the sweep line stores a data structure that allows insertion and deletion of intervals with integer coordinates in $[0 \mathinner{..} k]$ and querying for the total length of the union of the intervals that are currently stored. This corresponds to the operations of the data structure from Lemma 7 for $m = k$ (with elements corresponding to unit intervals), which supports insertions and deletions in $\mathcal{O}(\log k)$ time and queries in $\mathcal{O}(1)$ time after $\mathcal{O}(k)$-time preprocessing per strip. The total preprocessing time is $\mathcal{O}(d)$ and, since the total number of events in all strips is at most $2r$, the sweep works in $\mathcal{O}(r \log k)$ time.

Finally, let us consider the width-d rectangles. Each of them induces a vertical interval on the second component. First, in $\mathcal{O}(r + d)$ time the union S of these intervals represented as a union of pairwise disjoint maximal intervals can be computed by bucket sorting the endpoints of the intervals. Then, each maximal interval in S is partitioned by the strips and the resulting subintervals are inserted into the data structures of the respective strips before the sweep. In total, at most $2r + d/k$ additional intervals are inserted so the time complexity is still $\mathcal{O}((r + d/k) \log k + d) = \mathcal{O}(r \log k + d)$. $\square$

We arrive at the main result of this section.

Theorem 16. *The number of k-antipower factors in a word of length n can be computed in $\mathcal{O}(nk \log k)$ time and $\mathcal{O}(nk)$ space.*

Proof. We use Lemma 13 to express the sets $WeakPow_k(d)$ for $d = 1, \ldots, \lfloor n/k \rfloor$ as sums of $\mathcal{O}(nk)$ interval chains. This takes $\mathcal{O}(nk)$ time. Each chain is represented on the corresponding grid $\mathcal{G}_d$ as the union of a constant number of rectangles using Lemma 14. This gives $\mathcal{O}(nk)$ rectangles in total on all the grids $\mathcal{G}_d$, each of height at most k or width exactly d, for the given d.

As the next step, we renumber the components in the grids by assigning consecutive numbers to the components that correspond to rectangle vertices. This can be done in $\mathcal{O}(nk)$ time, for all the grids simultaneously, using bucket sort [5]. The new components store the original values. After this transformation, rectangles with height at most k retain this property and rectangles with width d have maximal width. Let the maximum component in the grid $\mathcal{G}_d$ after renumbering be equal to M_d and the number of rectangles in $\mathcal{G}_d$ be R_d; then $\sum_d R_d = \mathcal{O}(nk)$ and $\sum_d M_d = \mathcal{O}(nk)$.

As the final step, we apply the algorithm of Lemma 15 to each grid to compute $|WeakPow_k(d)|$ as the area of the union of the rectangles in the grid. One can readily verify that it can be adapted to compute the areas of the rectangles in the original components. The algorithm works in $\mathcal{O}(\sum_d R_d \log k + \sum_d M_d) = \mathcal{O}(nk \log k)$ time. In the end, the number of (k,d)-antipower factors equals $n - kd + 1 - |WeakPow_k(d)|$. $\qquad\square$

5   Reporting Antipowers and Answering Queries

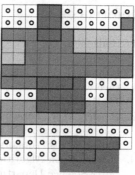

The same technique can be used to report all k-antipower factors. In the grid representation, they correspond to grid cells of $\mathcal{G}_d$ that are not covered by any rectangle, as shown in the figure to the right. Hence, in Lemma 15, instead of computing the area of the rectangles with the aid of Lemma 7, we need to report all grid cells excluded from rectangles using Lemma 8. The computation takes $\mathcal{O}(r \log k + d + C_d)$ time where C_d is the number of reported cells. By plugging this routine into the algorithm of Theorem 16, we obtain

Theorem 17. *All factors of a word of length n being k-antipowers can be computed in $\mathcal{O}(nk \log k + C)$ time and $\mathcal{O}(nk)$ space, where C is the output size.*

Finally, we present our data structure for answering antipower queries that introduces a smooth trade-off between the two data structures of Badkobeh et al. [1] (see Fact 2). Let us recall that an antipower query (i, j, k) asks to check if a factor $w[i .. j]$ of the word w is a k-antipower.

Theorem 18. *Assume that a word of length n is given. For every $r \in [1 .. n]$, there is a data structure of size $\mathcal{O}(n^2/r)$ that can be constructed in $\mathcal{O}(n^2/r)$ time and answers antipower queries in $\mathcal{O}(r)$ time.*

Proof. Let w be a word of length n and let $r \in [1 .. n]$. If an antipower query (i, j, k) satisfies $k \le r$, we answer it in $\mathcal{O}(k)$ time using Fact 2(a). This is always $\mathcal{O}(r)$ time, and the data structure requires $\mathcal{O}(n)$ space.

Otherwise, if $w[i .. j]$ is a k-antipower, then its base is at most n/r. Our data structure will let us answer antipower queries for every such base in $\mathcal{O}(1)$ time.

Let us consider a positive integer $b \le n/r$. We group the factors of w of length b by the remainder modulo b of their starting position. For a remainder $g \in [0 .. b-1]$ and index $i \in [0 .. \lfloor \frac{n-g}{b} \rfloor)$, we store, as $A_g^b[i]$, the smallest index $j > i$ such that $w[jb+g .. j(b+1)+g) = w[ib+g .. i(b+1)+g)$ ($j = \infty$ if it does not exist). We also store a data structure for range minimum queries over A_g^b for each group; it uses linear space, takes linear time to construct, and answers queries in constant time (see [3]). The tables take $\mathcal{O}(n)$ space for a given b, which gives $\mathcal{O}(n^2/r)$ in total. They can also be constructed in $\mathcal{O}(n^2/r)$ total time, as shown in the following claim (for a proof, see the full version).

Claim. The tables A_g^b for all $b \in [1 .. m]$ and $g \in [0 .. b - 1]$ can be constructed in $\mathcal{O}(nm)$ time.

Given an antipower query (i, j, k) such that $(j - i + 1)/k = b$, we set

$$g = i \bmod b, \quad i' = \lfloor \tfrac{i}{b} \rfloor, \quad j' = \lfloor \tfrac{i+1}{b} \rfloor - 2,$$

and ask a range minimum query on $A_g^b[i'], \ldots, A_g^b[j']$. Then, $w[i .. j]$ is a k-antipower if and only if the query returns a value that is at least $j' + 2$.   □

References

1. Badkobeh, G., Fici, G., Puglisi, S.J.: Algorithms for anti-powers in strings. Inf. Process. Lett. **137**, 57–60 (2018). https://doi.org/10.1016/j.ipl.2018.05.003
2. Bannai, H., I, T., Inenaga, S., Nakashima, Y., Takeda, M., Tsuruta, K.: The "runs" theorem. SIAM J. Comput. **46**(5), 1501–1514 (2017). https://doi.org/10.1137/15M1011032
3. Bender, M.A., Farach-Colton, M., Pemmasani, G., Skiena, S., Sumazin, P.: Lowest common ancestors in trees and directed acyclic graphs. J. Algorithms **57**(2), 75–94 (2005). https://doi.org/10.1016/j.jalgor.2005.08.001
4. Bentley, J.L.: Algorithms for Klee's rectangle problems. Unpublished notes, Computer Science Department, Carnegie Mellon University (1977)
5. Cormen, T.H., Leiserson, C.E., Rivest, R.L., Stein, C.: Introduction to Algorithms, 3rd edn. MIT Press, Cambridge (2009)
6. Crochemore, M., Kolpakov, R., Kucherov, G.: Optimal bounds for computing α-gapped repeats. In: Dediu, A.-H., Janoušek, J., Martín-Vide, C., Truthe, B. (eds.) LATA 2016. LNCS, vol. 9618, pp. 245–255. Springer, Cham (2016). https://doi.org/10.1007/978-3-319-30000-9_19
7. Fici, G., Restivo, A., Silva, M., Zamboni, L.Q.: Anti-powers in infinite words. In: Chatzigiannakis, I., Mitzenmacher, M., Rabani, Y., Sangiorgi, D. (eds.) Automata, Languages and Programming, ICALP 2016. LIPIcs, vol. 55, pp. 124:1–124:9. Schloss Dagstuhl-Leibniz-Zentrum für Informatik (2016). https://doi.org/10.4230/LIPIcs.ICALP.2016.124
8. Fici, G., Restivo, A., Silva, M., Zamboni, L.Q.: Anti-powers in infinite words. J. Comb. Theory Ser. A **157**, 109–119 (2018). https://doi.org/10.1016/j.jcta.2018.02.009
9. Gawrychowski, P., I, T., Inenaga, S., Köppl, D., Manea, F.: Tighter bounds and optimal algorithms for all maximal α-gapped repeats and palindromes: finding all maximal α-gapped repeats and palindromes in optimal worst case time on integer alphabets. Theory Comput. Syst. **62**(1), 162–191 (2018). https://doi.org/10.1007/s00224-017-9794-5
10. Kociumaka, T., Radoszewski, J., Rytter, W., Straszyński, J., Waleń, T., Zuba, W.: Efficient representation and counting of antipower factors in words. arXiv preprint arXiv:1812.08101 (2018)

11. Kolpakov, R., Kucherov, G.: Finding maximal repetitions in a word in linear time. In: 40th Annual Symposium on Foundations of Computer Science, FOCS 1999, pp. 596–604. IEEE Computer Society (1999). https://doi.org/10.1109/SFFCS.1999. 814634

12. Kolpakov, R., Podolskiy, M., Posypkin, M., Khrapov, N.: Searching of gapped repeats and subrepetitions in a word. J. Discrete Algorithms **46–47**, 1–15 (2017). https://doi.org/10.1016/j.jda.2017.10.004

13. Rubinchik, M., Shur, A.M.: Counting palindromes in substrings. In: Fici, G., Sciortino, M., Venturini, R. (eds.) SPIRE 2017. LNCS, vol. 10508, pp. 290–303. Springer, Cham (2017). https://doi.org/10.1007/978-3-319-67428-5_25

14. Tanimura, Y., Fujishige, Y., I, T., Inenaga, S., Bannai, H., Takeda, M.: A faster algorithm for computing maximal α-gapped repeats in a string. In: Iliopoulos, C., Puglisi, S., Yilmaz, E. (eds.) SPIRE 2015. LNCS, vol. 9309, pp. 124–136. Springer, Cham (2015). https://doi.org/10.1007/978-3-319-23826-5_13

On the Maximum Number of Distinct Palindromic Sub-arrays

Kalpana Mahalingam$^{(\boxtimes)}$ and Palak Pandoh

Department of Mathematics, Indian Institute of Technology Madras,
Guindy, Chennai 600036, India
kmahalingam@iitm.ac.in, palakpandohiitmadras@gmail.com

Abstract. We investigate the maximum number of distinct palindromic sub-arrays in a two-dimensional finite word over a finite alphabet Σ. For any finite array in $\Sigma^{m \times n}$, we find an upper bound for the number of distinct palindromic sub-arrays and improve it by giving a tight bound on the maximum number of distinct palindromes in an array in $\Sigma^{2 \times n}$ for $|\Sigma| = 2$. We then, propose a better upper bound for any finite array in $\Sigma^{m \times n}$.

Keywords: Combinatorics on words · Two-dimensional words · 2D palindromes · Maximum palindromes

1 Introduction

Identification of palindromes in sequences plays a major role in various fields like biology, modeling quasi-crystals, string matching, Diophantine approximation etc. due to their symmetrical structure. The study of palindromes includes rich words (containing the maximum number of distinct palindromes), palstars (product of even length palindromes), anti-palindromes etc. [8,14,21]. Palindrome-based languages were proved to be linear-time recognizable in [10,18].

Several authors have studied the total palindrome complexity (number of distinct palindromic factors) of finite 1D words and have given lower and upper bounds on the number of palindromes in finite and infinite (1D) words [4,8,9]. An upper bound for the palindrome complexity of a sequence in terms of its factor complexity was given in [1]. The factor complexity function is the number of distinct factors of a given length in a given sequence. The notion of complexity was extended to two-dimensional (2D) words over a finite alphabet in [2]. The complexity function $p_w(m, n)$ that counts the number of different rectangles in $\Sigma^{m \times n}$ that are factors of the two-dimensional sequence w is a two-dimensional analogue of the factor complexity.

In this paper, we give an upper bound on the number of distinct 2D palindromes in any given array. Two-dimensional palindromes were introduced by Berthé et al. [5] in order to characterize 2D Sturmian sequences in terms of 2D palindromes. It has applications in data compression, face recognition, pattern

© Springer Nature Switzerland AG 2019
C. Martín-Vide et al. (Eds.): LATA 2019, LNCS 11417, pp. 434–446, 2019.
https://doi.org/10.1007/978-3-030-13435-8_32

recognition to name a few [7,15,17]. A relation between 2D palindromes and 2D primitive words was studied in [19]. In [3], authors gave numerical values concerning palindrome complexity of two-row binary arrays of smaller size. An algorithm for finding the maximal 2D palindromes was given in [12].

In this paper, we investigate the maximum number of distinct 2D palindromic sub-arrays in any finite word in $\Sigma^{m \times n}$, m, $n \geq 2$. In Sect. 3, we find an upper bound for this number. In Sect. 4, we improve the bound by proving the fact that for $|\Sigma| = 2$, the maximum number of distinct palindromes in a word in $\Sigma^{2 \times n}$, $n \geq 2$, is equal to $2n + \lfloor \frac{n}{2} \rfloor - 1$. We later give an estimate of a better upper bound for a word in $\Sigma^{m \times n}$ in Sect. 5.

2 Basic Definitions and Notations

An alphabet Σ is a finite non-empty set of symbols. A 1-dimensional (1D) finite word $w = [a_i]_{1 \leq i \leq n}$ over the alphabet Σ is defined to be a finite string of letters where $w = a_1 a_2 a_3 \cdots a_n$ and $a_i \in \Sigma$. The length of a word w is the number of symbols in w and is denoted by $|w|$. Let Σ^* denote the set of all words over Σ including the empty word λ. Let Σ^+ be the set of all non-empty words over Σ. The reversal of $w = a_1 a_2 \cdots a_n$ is defined to be the string $w^R = a_n \cdots a_2 a_1$. We denote by $Alph(w)$, the set of all sub-strings of w of length 1. A word w is said to be a palindrome (or a 1D palindrome) if $w = w^R$. The palindrome complexity $P(w)$ is the number of distinct non-empty palindromic factors in the 1D word w and $PAL(w)$ is the set of all distinct non-empty palindromic factors of w. For all other concepts in formal language theory and combinatorics on words, the reader is referred to [16,20].

2.1 Two-Dimensional Arrays

We recall certain basic notions and some basic properties pertaining to two-dimensional word concepts. For more information, we refer the reader to [5,11, 13]. We denote by $\Sigma^{m \times n}$, the set of all $m \times n$ rectangular arrays of elements chosen from the alphabet Σ. A factor of w is a sub-array (or sub-word or sub-block) of w and if w is a finite 2D word, then factors of w are in $\Sigma^{s \times t}$ where $1 \leq s \leq m$, $1 \leq t \leq n$. In the case of 2D words, we use λ to denote the empty word. The set of all 2D (rectangular) words including the empty word λ over Σ is denoted by Σ^{**} whereas, Σ^{++} is the set of all non-empty 2D words over Σ. Note that, the words in $\Sigma^{m \times 0}$ and $\Sigma^{0 \times n}$ are not defined.

Definition 1. Let $u = [u_{i,j}]_{1 \leq i \leq m_1, 1 \leq j \leq n_1} \in \Sigma^{m_1 \times n_1}$ and $[v_{i,j}]_{1 \leq i \leq m_2, 1 \leq j \leq n_2} \in \Sigma^{m_2 \times n_2}$.

1. *The column concatenation of u and v (denoted by $\oplus$) is a partial operation, defined if $m_1 = m_2 = m$, and it is given by*

$$u \oplus v = \begin{matrix} u_{1,1} & \cdots & u_{1,n_1} & v_{1,1} & \cdots & v_{1,n_2} \\ \vdots & \ddots & \vdots & \vdots & \ddots & \vdots \\ u_{m,1} & \cdots & u_{m,n_1} & v_{m,1} & \cdots & v_{m,n_2} \end{matrix}$$

2. *The row concatenation of u and v (denoted by $\ominus$) is a partial operation defined if $n_1 = n_2 = n$, and it is given by*

$$u \ominus v = \begin{matrix} u_{1,1} & \cdots & u_{1,n} \\ \vdots & \ddots & \vdots \\ u_{m_1,1} & \cdots & u_{m_1,n} \\ v_{1,1} & \cdots & v_{1,n} \\ \vdots & \ddots & \vdots \\ v_{m_2,1} & \cdots & v_{m_2,n} \end{matrix}$$

It is clear that the operations of row and column concatenation are associative but not commutative. Moreover, the column and row concatenation of u and the empty word λ is always defined and λ is a neutral element for both the operations.

In [2], prefix of a 2D word w is defined to be a rectangular sub-array that contains one corner of w, whereas suffix of w is defined to be a rectangular sub-array that contains the diagonally opposite corner of w. However, in this paper, we consider prefix of a 2D word w (defined in [19]) to be a rectangular sub-array that contains the top left corner of w, and suffix of w to be a rectangular sub-array that contains the bottom right corner of w. Formally,

Definition 2. *Given $u \in \Sigma^{**}$, $v \in \Sigma^{**}$ is said to be a prefix of u (respectively, suffix of u), denoted by $v \leq_p u$ (respectively $v \leq_s u$) if $u = (v \ominus x) \oplus y$ or $u = (v \oplus x) \ominus y$ (respectively, $u = y \oplus (x \ominus v)$ or $u = y \ominus (x \oplus v)$) for $x, y \in \Sigma^{**}$.*

Definition 3. *Let $w = [w_{ij}]_{1 \leq i \leq m, 1 \leq j \leq n} \in \Sigma^{m \times n}$.*

1. *The reverse image of w, denoted by $w^R = [w_{m-i+1, n-j+1}]_{1 \leq i \leq m, 1 \leq j \leq n}$.*

$$w^R = \begin{matrix} w_{m,n} & w_{m,n-1} & \cdots & w_{m,1} \\ w_{m-1,n} & w_{m-1,n-1} & \cdots & w_{m-1,1} \\ \vdots & \vdots & \ddots & \vdots \\ w_{1,n} & w_{1,n-1} & \cdots & w_{1,1} \end{matrix}$$

2. *The transpose of w, denoted by $w^T = [u_{ij}]_{1 \leq i \leq n, 1 \leq j \leq m}$ such that $u_{ij} = w_{ji}$.*

$$w^T = \begin{matrix} w_{1,1} & w_{2,1} & \cdots & w_{m,1} \\ w_{1,2} & w_{2,2} & \cdots & w_{m,2} \\ \vdots & \vdots & \ddots & \vdots \\ w_{1,n} & w_{2,n} & \cdots & w_{m,n} \end{matrix}$$

If $w = w^R$, then w is said to be a *two-dimensional palindrome* [5, 12]. We call a palindrome in $\Sigma^{m \times n}$ to be an $m \times n$ palindrome. By $P_{2d}(w)$, we denote the number of all non-empty distinct 2D palindromic sub-arrays in w and $PAL_{2d}(w)$ is the set of all non-empty palindromic sub-arrays of the 2D word w. For example, if $\Sigma = \{a, b, c\}$, then $w = \begin{matrix} a\,b\,c\,a \\ b\,c\,c\,b \\ a\,c\,b\,a \end{matrix}$ is a 3×4 palindrome over Σ.

We use the following notion of horizontal and vertical palindromes.

Definition 4. *Let $w \in \Sigma^{m \times n}$.*

1. *The horizontal palindromes of w are the $1 \times i$ palindromic sub-arrays of w, where $1 \leq i \leq n$.*
2. *The vertical palindromes of w are the $j \times 1$ palindromic sub-arrays of w, where $2 \leq j \leq m$.*

Throughout the paper, by the number of palindromes, we simply refer to the number of non-empty distinct palindromic sub-arrays counted without repetition.

3 Maximum Number of Palindromes in 2D Words over an Arbitrary Alphabet

In this section, we find an upper bound for the number of palindromes in a word $w \in \Sigma^{m \times n}$ with the convention that m, $n \geq 2$.

Let $w \in \Sigma^{m \times n}$ such that $|Alph(w)| = 1$. Note that w is of the form $(a^{m\ominus})^{n\odot}$ and it has exactly mn palindromes. Therefore, we have the following observation.

Proposition 5. *A unary word in $\Sigma^{m \times n}$ where $m \geq 1$, $n \geq 1$, has exactly mn palindromes.*

Now, consider the 2D word $w = \begin{array}{cccc} a & b & b & a \\ a & b & a & b \end{array}$. Then, $P_{2d}(w) = 9$ as $PAL_{2d}(w) = \{a, b, bb, aba, bab, abba, a \ominus a, b \ominus b, ba \ominus ab\}$. Hence, we infer that in the case of a non-unary word, the number of palindromes in a word in $\Sigma^{m \times n}$ can be greater than mn. We find a loose upper bound for this using the following Lemma.

Lemma 6. *Let $w \in \Sigma^{m \times n}$, m, $n \geq 2$, and $y = (a_1 \ominus a_2 \cdots \ominus a_m)$, $a_i \in \Sigma$. Then, $w \odot y$ and $y \odot w$ create at most one extra distinct $m \times t$ palindrome for $t \geq 1$.*

Proof. Let $w \in \Sigma^{m \times n}$, m, $n \geq 2$ and $y = (a_1 \ominus a_2 \cdots \ominus a_m)$, $a_i \in \Sigma$ for each i. Assume, for the sake of contradiction that, a $m \times r$ and a $m \times s$ palindrome where $r < s$, $r \geq 1$ are created on the concatenation of $w = [w_{i,j}]$ and y. Then,

$$w \odot y = \begin{array}{ccccccccc} w_{1,1} & \cdots & w_{1,n-s+1} & a_m & \cdots & w_{1,n-r+1} & a_m & \cdots & w_{1,n} & a_1 \\ w_{2,1} & \cdots & w_{2,n-s+1} & a_{m-1} & \cdots & w_{2,n-r+1} & a_{m-1} & \cdots & w_{2,n} & a_2 \\ \vdots & & \vdots & & & \vdots & & & \vdots & \\ w_{m,1} & \cdots & w_{m,n-s+1} & a_1 & \cdots & w_{m,n-r+1} & a_1 & \cdots & w_{m,n} & a_m \end{array}$$

We only prove for the case when m is odd, as the case when m is even is similar. If m is odd, then the $m \times r$ palindrome created is

$$u_1 \ominus u_2 \cdots \ominus u_{\lfloor \frac{m}{2} \rfloor} \ominus (\alpha \alpha^R) \ominus u_{\lfloor \frac{m}{2} \rfloor}^R \cdots \ominus u_2^R \ominus u_1^R$$

where each $u_i, 1 \leq i \leq \lfloor \frac{m}{2} \rfloor$, and $\alpha\alpha^R$ are rows of the $m \times r$ palindromic suffix of $w \oplus y$. Now, let p_1 be the newly created $m \times s$ palindrome. Then,

$$p_1 = \beta_1 u_1 \ominus \cdots \ominus \beta_{\lfloor \frac{m}{2} \rfloor} u_{\lfloor \frac{m}{2} \rfloor} \ominus (\beta\alpha\alpha^R) \ominus \beta_{\lfloor \frac{m}{2} \rfloor}^R u_{\lfloor \frac{m}{2} \rfloor}^R \cdots \ominus \beta_1^R u_1^R$$

where each $\beta_i u_i, 1 \leq i \leq \lfloor \frac{m}{2} \rfloor$, and $\beta\alpha\alpha^R$ are rows of the $m \times s$ palindromic suffix of $w \oplus y$. Now, consider

$$p_1^R = u_1\beta_1 \cdots \ominus u_{\lfloor \frac{m}{2} \rfloor}\beta_{\lfloor \frac{m}{2} \rfloor} \ominus (\alpha\alpha^R\beta) \ominus u_{\lfloor \frac{m}{2} \rfloor}^R \beta_{\lfloor \frac{m}{2} \rfloor}^R \cdots \ominus u_1^R\beta_1^R.$$

It has $u_1 \ominus u_2 \cdots \ominus u_{\lfloor \frac{m}{2} \rfloor} \ominus (\alpha\alpha^R) \ominus u_{\lfloor \frac{m}{2} \rfloor}^R \cdots \ominus u_2^R \ominus u_1^R$ as its $m \times r$ prefix. Hence, we conclude that the $m \times r$ palindrome is present as a prefix of the $m \times s$ palindrome. Hence, concatenation of w and $(a_1 \ominus a_2 \cdots \ominus a_m)$, $a_i \in \Sigma$ for all i creates at most one extra distinct $m \times t$ palindrome for $t \geq 1$.   □

We recall the following result from [4].

Proposition 7. *The total palindrome complexity $P(w)$ of any finite 1D word w satisfies $P(w) \leq |w|$.*

Hence, one can conclude that concatenation of a letter to a 1D word can create at most one extra palindrome. We have an immediate Corollary to Lemma 6 and Proposition 7.

Corollary 8. *Let $w \in \Sigma^{2 \times n}$, $n \geq 1$. Then, the word $w \oplus (a_1 \ominus a_2)$ and $(a_1 \ominus a_2) \oplus w$ where a_1, $a_2 \in \Sigma$ create at most 3 extra palindromes i.e. at most 2 horizontal and at most one $2 \times t$ palindrome for $t \geq 1$.*

Remark 9. Based on Corollary 8, one can observe that the number of palindromes in a word in $\Sigma^{2 \times n}$, $n \geq 2$, is at most $3n$.

We show (Lemma 10) the existence of a word $w_n \in \Sigma^{2 \times n}$ such that $P_{2d}(w_n) > 2n$ for all $n \geq 4$. We recall the definition [6] of the fractional power of a 1D word u of length q denoted by $u^{(\frac{p}{q})}$ which is the prefix of length p of u^p. For example, for a word $w = aba$, $w^{(\frac{5}{3})} = abaab$.

Consider the word $x = (ab)^{(\frac{5}{2})} \in \Sigma^+$ and $u = xx^R$, $v = (ab)^k$. Let $w_{2k} = u \ominus v \in \Sigma^{2 \times 2k}$. Let w_{2k-1} be the word in $\Sigma^{2 \times (2k-1)}$ obtained by the removal of the last column from w_{2k}.

Lemma 10. *There exists a word w_n, $n \geq 3$ in $\Sigma^{2 \times n}$, with $P_{2d}(w_n) > 2n$.*

Examples of such words are $w_8 = \dfrac{ababbaba}{abababab}$ and $w_9 = \dfrac{ababaabab}{abababab a}$.

Remark 11. It can be easily observed that the word w_n, $n \geq 3$, constructed in Lemma 10, has exactly $2n + \lfloor \frac{n}{2} \rfloor - 1$ palindromes.

Consider a word $u = u_1 \oplus u_2 \oplus \cdots \oplus u_n \in \Sigma^{2 \times n}$, $n \geq 4$, such that $P_{2d}(u) \geq P_{2d}(w)$ for all words $w \in \Sigma^{2 \times n}$. Then, by Lemma 10, $P_{2d}(u) > 2n$. Now, u can be considered as a 1D word over the alphabet of columns A such that

$A = \{u_1, u_2, \cdots u_n\}$. For example, for the word $u = \dfrac{abbabb}{babbaa}$, $A = \left\{\begin{matrix} a & b & b \\ b' & a' & b \end{matrix}\right\}$. Then by Proposition 7, we have $P(u) \leq n$ over A. We show that $P(u) \leq n - 1$ over A. Note that all the $2 \times t$ palindromes of u for $t \geq 1$, are horizontal palindromes of u over A. If u has n palindromes over A, it can be observed by Corollary 8 and Proposition 7 that every alphabet of u over A must create one extra distinct palindrome. Hence, $A = \left\{\dfrac{a_1}{a_1} : a_1 \in \Sigma\right\}$. This implies that both the rows of u considered as a 2D word over Σ are same. Hence, there are at most n horizontal palindromes in u over Σ. Thus, over Σ,

$$P_{2d}(u) = [\# \text{ of horizontal palindromes of } u \text{ over } \Sigma]$$
$$+ [\# \text{ of horizontal palindromes of } u \text{ over } A] \leq 2n$$

which is a contradiction. Hence, the number of $2 \times t$ palindromes for $t \geq 1$ in u is at most $(n - 1)$. Now, consider the following Lemma.

Lemma 12. *Let* $w = w_1 \ominus w_2 \in \Sigma^{2 \times n}$, $n \geq 1$. *If* $|Alph(w_1)| = 1$ *or* $|Alph(w_2)| = 1$, *then* $P_{2d}(w) \leq 2n$.

Proof. We prove the result by induction on n. For $n = 1$, the word is of the form $a \ominus a_1$, $a_1 \in \Sigma$ and the statement holds true. Let the result be true for $n = k - 1$. Assume that $w \in \Sigma^{2 \times k}$ such that $w = w_1 \ominus w_2$ and $w_1 = a^k$. By induction hypothesis, there are at most $2(k - 1)$ palindromes in the $2 \times (k - 1)$ prefix of w. Now, the last column can be either $a \ominus b$ or $a \ominus a$ where a and b are distinct. If it is $a \ominus b$, then no extra $2 \times t$ palindrome for $t \geq 1$ is created. Hence, there are at most 2 extra horizontal palindromes. So, $P_{2d}(w) \leq 2k$. If the last column is $a \ominus a$ and a $2 \times t$ palindrome for $t \geq 1$ is created, then it is of the form $a^t \ominus a^t$. Suppose that, a horizontal palindrome is created in w_2, then it is of the form $a^t w_3 a^t$ for some 1D word w_3. But, this is a contradiction to the fact that $a^t \ominus a^t$ is a newly created palindrome. Hence, at most 2 extra palindromes can be created. So, $P_{2d}(w) \leq 2k$. Hence, by induction, the result holds for all n. □

It can be observed by Lemmas 10 and 12 that in a word $u \in \Sigma^{2 \times n}$ such that $P_{2d}(u) \geq P_{2d}(w)$ for all words $w \in \Sigma^{2 \times n}$, there are at most $2n - 2$ horizontal palindromes and the number of $2 \times t$ palindromes in u can be at most $n - 1$. Thus, $P_{2d}(u) \leq (2n - 2) + (n - 1) = 3n - 3$. The result also holds for $n = 3$. Hence, we conclude the following.

Theorem 13. *Let* $w \in \Sigma^{2 \times n}$, $n \geq 3$. *Then,* $P_{2d}(w) \leq 3n - 3$.

We now give an upper bound for the number of palindromes in a word in $\Sigma^{m \times n}$, $m, n \geq 3$.

Lemma 14. *For* $w \in \Sigma^{m \times n}$, $m, n \geq 3$, *we have,*

$$P_{2d}(w) \leq \begin{cases} \frac{mn(m+1) - 3m}{2}, & \text{if } m \text{ is even} \\ \frac{mn(m+1) - (3m-3)}{2}, & \text{if } m \text{ is odd.} \end{cases}$$

Proof. Let $w \in \Sigma^{m \times n}$, $m, n \geq 3$. We have the following.

1. If m is even, then $w = A_1 \ominus A_2 \ominus \cdots \ominus A_{\frac{m}{2}}$ where each $A_i \in \Sigma^{2 \times n}$. By Theorem 13, $P_{2d}(A_i) \leq (3n - 3)$. Note that, all the horizontal palindromes in w are counted. The $2 \times t$ palindromes for $t \geq 1$ that are still not counted are in the sub-arrays formed by taking the last row of A_i and first row of A_{i+1} for $1 \leq i \leq (\frac{m}{2} - 1)$. There are $(\frac{m}{2} - 1)$ such sub-arrays in total. By Lemma 6, the number of $2 \times t$ palindromes in each of them for $t \geq 1$ is at most n. Also, the number of $(i+1) \times n$ sub-words of w are $(m - i)$ for $2 \leq i \leq m - 1$, so the number of $i \times t$ palindromes for $i \geq 3$ and $t \geq 1$ is at most $n \sum_{i=2}^{m-1} (m - i)$. So, $P_{2d}(w) \leq (3n - 3)(\frac{m}{2}) + n(\frac{m}{2} - 1) + n \sum_{i=2}^{m-1} (m - i) = \frac{mn(m+1) - 3m}{2}$.

2. If m is odd, then by a similar calculation, we have, $P_{2d}(w) \leq (3n - 3)(\frac{m-1}{2}) + n + n(\frac{m-1}{2}) + n \sum_{i=2}^{m-1} (m - i) = \frac{mn(m+1) - (3m-3)}{2}$.

$\square$

We now obtain the maximum number of distinct non-empty palindromic sub-arrays in a word in $\Sigma^{2 \times n}$, where $|\Sigma| = 2$ along with the number of words that attain it by a computer program.

Table 1. Maximum number of palindromes in a word in $\Sigma^{2 \times n}$.

$m \times n$	Total words	Max (P_{2d})	# of words attaining maximum
2×2	16	4	14
2×3	64	6	56
2×4	256	9	12
2×5	1024	11	100
2×6	4096	14	24
2×7	16384	16	204
2×8	65536	19	8
2×9	262144	21	164
2×10	1048576	24	32

Clearly, the upper bound found in Lemma 14 is not a tight bound. So, in the next section, we prove the tight bound for a word in $\Sigma^{2 \times n}$, where $|\Sigma| = 2$ by a case by case analysis.

4 Binary Words in $\Sigma^{2 \times N}$

Throughout this section, we consider Σ to be a binary alphabet i.e. $|\Sigma| = 2$. From Table 1, we can observe the following.

1. The maximum number of distinct palindromic factors in a word in $\Sigma^{2 \times n}$, $n \geq 2$, is $2n + \lfloor \frac{n}{2} \rfloor - 1$.

2. The number of words in $\Sigma^{2\times n}$, $n \geq 2$, with the maximum number of palindromic factors is at least 8.

We denote $2n + \lfloor \frac{n}{2} \rfloor - 1$ by x_n, in the rest of the paper. From Lemma 10, we have the following result.

Proposition 15. *There exists a word in $\Sigma^{2\times n}, n \geq 3$ with x_n palindromes.*

We show that there are at least 8 words with exactly x_n palindromes for $n \geq 3$. For any word w over a binary alphabet $\Sigma = \{a, b\}$, we define the complement of w denoted by w^c to be the word $\phi(w)$ where ϕ is a morphism such that $\phi(a) = b$ and $\phi(b) = a$. For example, if $w = ababb$, then $w^c = babaa$.

Lemma 16. *There are at least 8 words with x_n palindromes in $\Sigma^{2\times n}$ for $n \geq 3$.*

Proof. Let n be even, say $n = 2k$ and $x = (ab)^{(\frac{k}{2})} \in \Sigma^+$. We have two cases:
1. For an even k, consider the sets

$$A_{2k} = \{(ab)^{\frac{n}{2}} \ominus xx^R, (ab)^{\frac{n}{2}} \ominus x^R x, (ba)^{\frac{n}{2}} \ominus xx^R, (ba)^{\frac{n}{2}} \ominus x^R x\},$$

$$B_{2k} = \{w^R : w \in A_{2k}\}$$

2. For an odd k, consider the sets

$$A_{2k} = \{(ab)^{\frac{n}{2}} \ominus xx^R, (ba)^{\frac{n}{2}} \ominus xx^R, xx^R \ominus (ab)^{\frac{n}{2}}, xx^R \ominus (ba)^{\frac{n}{2}}\},$$

$$B_{2k} = \{w^c : w \in A_{2k}\}$$

As $x \neq x^R$, then for all $w_1, w_2 \in S_{2k}$, $w_1 \neq w_2$. The required 8 words in $\Sigma^{2\times 2k}$ with x_{2k} palindromes are in the set S_{2k}, where $S_{2k} = \{w : w \in A_{2k} \cup B_{2k}\}$. Similarly, for n odd, say $n = 2k - 1$, the required 8 words in $\Sigma^{2\times(2k-1)}$ are in the set S_{2k-1}, where $S_{2k-1} = \{w : w$ is the $2 \times (2k - 1)$ prefix of the words in $S_{2k}\}$. $\square$

Remark 17. Let $w \in S_n$ for n even, where S_n is the set mentioned in Lemma 16. Let w' be a word obtained on removal of the $2 \times t$ prefix and $2 \times t$ suffix of w where $2t < n$. Then, $w' \in S_{n-2t}$.

We observe that, if $w = w_1 \ominus w_2 \in \Sigma^{2\times n}$ is a palindrome, then $w_2 = w_1^R$. Hence, we conclude the following.

Lemma 18. *If $w \in \Sigma^{2\times n}$, $n \geq 2$, is a 2D palindrome, then $P_{2d}(w) \leq 2n$.*

We use the following results to prove $P_{2d}(w) \leq x_n$ for $w \in \Sigma^{2\times n}$, $n \geq 2$.

Proposition 19. *Let T_n, $n \geq 6$ and n even, be the set of all the words $w \in \Sigma^{2\times n}$ that satisfies the following:*

1. $P_{2d}(w) = x_n$.
2. There is a $1 \times n$ palindromic sub-array in w.

3. *On the removal of the $2 \times k$ prefix and the $2 \times k$ suffix of w for all k such that $2k < n$, we get a word in $\Sigma^{2 \times (n-2k)}$ that contains x_{n-2k} palindromes.*

Then, $T_n = S_n$ where, S_n is the set mentioned in Lemma 16.

Proof. We prove $T_n \subseteq S_n$ by strong induction on n. For $n = 6$, let $w = w_1 \ominus w_2 \in T_6$. Suppose, w_1 is a palindrome. Let $w' = w_1' \ominus w_2'$ be the word obtained by removing the 2×1 prefix and suffix of w. Then, $P_{2d}(w') = x_4 = 9$ and by Lemma 12, w_1' is either $abba$ or $baab$. By Theorem 13, the maximum number of palindromes in a word in $\Sigma^{2 \times 3}$ is 6. Thus, 3 palindromes are removed by the removal of the first column of w'. If $w' = abba$, then by direct computation w' is either $abba \ominus baba$ or $abba \ominus abab$. By the structure of w, $P_{2d}(w) = 14$ and w_1 is a palindrome. So, w_1 is either $aabbaa$ or $babbab$.

If $w_1 = aabbaa$, then w has 5 more palindromes than w'. By Corollary 8, as there can be at most 4 new horizontal palindromes, a new $2 \times t$ palindrome is created for $t \geq 1$. The only $2 \times t$ palindrome that can be created in w is $aa \ominus aa$. Hence, w is either $aabbaa \ominus xbabaa$ or $aabbaa \ominus aababx$ where, $x \in \{a, b\}$. None of these have x_6 palindromes. If $w_1 = babbab$, then using a similar argument, one can verify that w is either $babbab \ominus ababab$ or $babbab \ominus bababa$.

Similarly, if $w_1' = baab$, then w is either $abaaba \ominus bababa$ or $abaaba \ominus ababab$. (Note that remaining 4 words in S_6 are obtained by considering w_2 as a palindrome.) This implies, $T_6 \subseteq S_6$. Assume that the result holds true for words in $\Sigma^{2 \times n}$, where n is even. Consider a word $w = w_1 \ominus w_2 \in T_{n+2}$. Let w' be the word obtained by removing the 2×1 prefix and suffix of w, then $w' \in T_n \subseteq S_n$. There are 8 such words. We show the result for one of the words in S_n and the result follows similarly for others. For $n = 2k$, let $x = (ab)^{(\frac{k}{2})} \in \Sigma^*$ and $w' = u \ominus v$, where $u = xx^R$ and $v = (ab)^k$. As $w \in T_{n+2}$, w_1 must be a palindrome. So, w_1 is either $axx^R a$ or $bxx^R b$.

Let $w_1 = axx^R a$. Since $P_{2d}(w) - P_{2d}(w') = 5$, by Corollary 8, a $2 \times t$ palindrome for $t \geq 1$, must be created. The only such palindrome in w can be $aa \ominus aa$ which can be present as a prefix of w. Two new horizontal palindromes are created in the first row: $aa, axx^R a$. Note that, at most one more distinct horizontal palindrome can be created in w_2 which is either $(ab)^{\frac{n}{2}} a$ or bb. Hence, no such word exists. Let $w_1 = bxx^R b$. Using a similar argument, we get the word $bub \ominus bva \in S_{n+2}$. Hence, $T_{n+2} \subseteq S_{n+2}$. Hence, by induction $T_n \subseteq S_n$, $n \geq 6$, and n even. The proof of $S_n \subseteq T_n$ for all $n \geq 6$, and n even follows from Remark 17 and the structure of words in S_n. Hence, $T_n = S_n$, $n \geq 6$, and n even. $\qquad\square$

By a similar argument, we can conclude the following.

Proposition 20. *Let T_n, $n \geq 7$, and n odd, be the set of all the words $w \in \Sigma^{2 \times n}$ that satisfies the following:*

1. $P_{2d}(w) = x_n$.
2. *There is a $1 \times n$. palindromic sub-array in w.*
3. *The recursive removal of the $2 \times k$ suffix and the $2 \times k$ prefix of w removes $3k$ and $2k$ palindromes respectively from w till we reach a word in $\Sigma^{2 \times 2}$ and at each stage, the resultant word in $\Sigma^{2 \times l}$ has x_l palindromes.*

Then, $T_n = \phi$, $n \geq 7$, and n odd.

We now show that for a word $w \in \Sigma^{2 \times n}$, $n \geq 2$, the number of distinct palindromic sub-arrays in w cannot exceed x_n.

Theorem 21. *Let $w \in \Sigma^{2 \times n}$, $n \geq 2$. Then, $P_{2d}(w) \leq 2n + \lfloor \frac{n}{2} \rfloor - 1$.*

Proof. We only give a sketch of the proof. The proof is by strong induction on n. The base case can be verified from Table 1, for $2 \leq n \leq 10$. Let $w = c_1 \oplus c_2 \cdots \oplus c_{n+1} \in \Sigma^{2 \times n+1}$ and let $u_{i \to j} = c_i \oplus c_{i+1} \oplus \cdots \oplus c_j$. We only need to prove for the Case when $P_{2d}(u_{1 \to n}) = x_n$ and n even as the other cases are direct. Let $y_{i,j}$ and $z_{i,j}$ be the number of palindromes removed from the word $u_{i \to j}$ on removing c_i and c_j respectively. We prove by contradiction i.e., we assume that $z_{1,n+1} = 3$, so, $P_{2d}(w) = x_n + 3 = x_{n+1} + 1$. By Corollary 8, $P_{2d}(u_{1 \to (n-1)}) = x_{n-1}$ and $P_{2d}(u_{1 \to (n-2)}) \leq x_{n-2}$. Thus, $z_{1,n-1}$ is either 2 or 3. If $z_{1,n-1} = 2$, then $y_{1,n-2} = 3$ and $z_{2,n-2}$ is either 2 *(Case 1)* or 3 *(Case 2)*.

Case 1: If $z_{2,n-2} = 2$, then $P_{2d}(u_{2 \to (n-3)}) = x_{n-4}$. Therefore $y_{2,n-3} \geq x_{n-4} - x_{n-5} = 3$ and hence, $P_{2d}(u_{3 \to (n-3)}) = x_{n-5}$. If there is no $1 \times (n-2)$ or $2 \times (n-2)$ palindrome in $u_{1 \to (n-2)}$, then $P_{2d}(u_{1 \to (n-3)}) = x_{n-5} + 3 + 3 = x_{n-3} + 1$, a contradiction. Otherwise, $u_{2 \to (n-3)}$ and $u_{1 \to (n-2)}$ have similar structure. The recursive removal of the first and the last column either results in *Case 1* or in *Case 2*. If it results in *Case 1* always, we have a contradiction by Lemma 18 and Proposition 19. If it results in *Case 2* at some stage i, then we discuss further.

Case 2: If $z_{2+i,n-2+i} = 3$ at same stage i, then $P_{2d}(u_{2+i \to (n-3-i)}) = x_{n-4-2i} - 1$. By the induction hypothesis, $y_{2+i,(n-3-i)} \geq 2$.

Case 2.1: If $y_{2+i,(n-3-i)} = 2$, then we have $P_{2d}(u_{3+i \to (n-3-i)}) = x_{n-5-2i}$ and $y_{3+i,n-3-i} \geq 2$.

Case 2.1.1: If $y_{3+i,n-3-i} = 2$, then $P_{2d}(u_{4+i \to (n-3-i)}) = x_{n-6-2i}$. If there is no $1 \times (n-3-2i)$ or $2 \times (n-3-2i)$ palindrome in $u_{2+i \to (n-2-i)}$, then $u_{3+i \to (n-2-i)}$ is similar to the word in *Case 1* and the proof follows, unless we end with a word of size $(2,2)$ in *Case 2.1.1*, which is not possible. If there is such a palindrome, consider $u_{3+i \to (n-3-i)}$ which is similar to $u_{2+i \to (n-2-i)}$ in the *Case 2.1*. The recursive removal of the first and the last column either results in *Case 2.1.1* or in *Case 2.1.2*. If the recursive removal of the first and the last column results in *2.1.1* always, we have a contradiction by Lemma 18 and Proposition 20. If the recursive removal of the first and the last column results in *Case 2.1.2* at some stage j.

Case 2.1.2: If $y_{3+i+j,n-3-i-j} = 2$ at some stage j, then $P_{2d}(u_{4+i+j \to (n-3-i-j)}) = x_{n-6-2i-2j} - 1$. If there is no $1 \times (n-3-2i-2j)$ or $2 \times (n-3-2i-2j)$ palindrome in $u_{2+i+j \to (n-2-i-j)}$, then $u_{3+i+j \to (n-2-i-j)}$ is similar to $u_{1 \to (n-2)}$ in *Case 2*. By considering words in $\Sigma^{2 \times 6}$, we can verify that there are no words that always results in *Case 2.1.2*. Hence, there exists such a palindrome. Note that, $P_{2d}(u_{3+i+j \to (n-2-i-j)}) = x_{n-4-2i-2j}$. Similar words are already discussed in *Case 2.1.1* and one can verify that these words do not result back to this case as such words in $\Sigma^{2 \times 2}$ have exactly 5 palindromes, which is a contradiction.

Case 2.2: If $y_{2+i,n-3-i} = 3$, then $P_{2d}(u_{3+i\rightarrow(n-3-i)}) = x_{n-5-2i}-1$. $z_{3+i,n-3-i} \geq 2$. If $z_{3+i,n-3-i} = 3$, then $u_{2+i\rightarrow(n-3-i)}$ is similar to the word in *Case 2* and the proof follows. If $z_{3+i,n-3-i} = 2$, then $P_{2d}(u_{3+i\rightarrow(n-4-i)}) = x_{n-6}$ and $u_{3+i\rightarrow(n-4-i)}$ is similar to $u_{2\rightarrow(n-3)}$ in *Case 2.1.1*. We cannot end in this case as we get a word in $\Sigma^{2\times3}$ with 7 palindromes which is a contradiction.

If $z_{2,n-2} = 3$, then $u_{1\rightarrow(n-1)}$ is similar to $u_{2\rightarrow(n-2)}$ in *Case 2*.   □

By direct computation on the words in $\Sigma^{2\times4}$ for arbitrary alphabet Σ, we observe that the tight bound of $x_4 = 9$ palindromes is never achieved over a non-binary alphabet. So, we propose that for a given word in $\Sigma^{2\times n}$, $n \geq 4$, the maximum number of palindromes occur over a binary alphabet. By Lemma 14 and Theorem 21, we now have a better upper bound for the number of palindromes in a word in $\Sigma^{m\times n}$.

Corollary 22. *Let* $w \in \Sigma^{m\times n}$, m, $n \geq 2$, *then*

$$P_{2d}(w) \leq \begin{cases} \frac{m(mn+\lfloor\frac{n}{2}\rfloor-1)}{2}, & \text{if } m \text{ is even} \\ \frac{m(mn+\lfloor\frac{n}{2}\rfloor-1)}{2} - \frac{n-1}{2}, & \text{if } m \text{ is odd.} \end{cases}$$

5 2D Words of Size (m, n)

The following table depicts the maximum number of distinct non-empty palindromic sub-arrays in any word in $\Sigma^{m\times n}$ for larger values of m and n obtained by a computer program.

Table 2. Maximum number of palindromes in a word in $\Sigma^{m\times n}$.

$m \times n$	Max (P_{2d})	$m \times n$	Max (P_{2d})
3×3	10	4×3	15
3×4	15	4×4	20
3×5	19	4×5	25
3×6	23	5×2	11
3×7	27	5×3	19
4×2	9	5×4	25

From Table 2, we propose the following upper bound for the number of palindromes in a word $w \in \Sigma^{m\times n}$, $m < n$, and m, $n \geq 3$.

$$P_{2d}(w) \leq mn + (m-1)\lfloor\frac{n}{2}\rfloor$$

One can observe that for any 2D word w, p is a palindromic sub-array of the word w iff p^T is a palindromic sub-array of the word w^T i.e., $P_{2d}(w) = P_{2d}(w^T)$. Hence, we conclude that the maximum number of palindromes in a word in $\Sigma^{m\times n}$ is same as that of the maximum number of palindromes in a word in $\Sigma^{n\times m}$, m, $n \geq 2$.

6   Conclusions

In this paper, we have investigated the upper bound on the number of palindromes in a finite 2D word over an arbitrary alphabet. We give the exact number for a binary word in $\Sigma^{2 \times n}$, $n \geq 2$. We then propose a better upper bound for the words in $\Sigma^{m \times n}$, m, $n \geq 3$. It will be interesting to study the relation between the number of distinct sub-arrays and the number of distinct palindromic sub-arrays in a given 2D word.

References

1. Allouche, J.P., Baake, M., Cassaigne, J., Damanik, D.: Palindrome complexity. Theor. Comput. Sci. **292**(1), 9–31 (2003)
2. Amir, A., Benson, G.: Two-dimensional periodicity in rectangular arrays. SIAM J. Comput. **27**(1), 90–106 (1998)
3. Anisiua, M.C., Anisiu, V.: Two-dimensional total palindrome complexity. Ann. Tiberiu Popoviciu Semin. Funct. Eqs. Approx. Convexity **6**, 3–12 (2008). ISSN 1584-4536
4. Anisiua, M.C., Anisiu, V., Kása, Z.: Total palindromic complexity of finite words. Discrete Math. **310**, 109–114 (2010)
5. Berthé, V., Vuillon, L.: Palindromes and two-dimensional Sturmian sequences. J. Autom. Lang. Combin. **6**(2), 121–138 (2001)
6. Brandenburg, F.: Uniformly growing k-th powerfree homomorphisms. Theor. Comput. Sci. **23**, 69–82 (1983)
7. De Natale, F., Giusto, D., Maccioni, F.: A symmetry-based approach to facial features extraction. In: Proceedings of 13th International Conference on Digital Signal Processing, vol. 2, pp. 521–525 (1997)
8. Droubay, X., Pirillo, G.: Palindromes and Sturmian words. Theor. Comput. Sci. **223**(1–2), 73–85 (1999)
9. Fici, G., Zamboni, L.Q.: On the least number of palindromes contained in an infinite word. Theor. Comput. Sci. **481**, 1–8 (2013)
10. Galil, Z., Seiferas, J.: A linear-time on-line recognition algorithm for "palstar". J. Assoc. Comput. Mach. **25**(1), 102–111 (1978)
11. Gamard, G., Richomme, G., Shallit, J., Smith, T.J.: Periodicity in rectangular arrays. Inf. Process. Lett. **118**, 58–63 (2017)
12. Geizhals, S., Sokol, D.: Finding maximal 2-dimensional palindromes. In: Grossi, R., Lewenstein, M. (eds.) 27th Annual Symposium on Combinatorial Pattern Matching (CPM 2016), vol. 54, pp. 19:1–19:12 (2016)
13. Giammarresi, D., Restivo, A.: Two-dimensional languages. In: Rozenberg, G., Salomaa, A. (eds.) Handbook of Formal Languages, pp. 215–267. Springer, Heidelberg (1997). https://doi.org/10.1007/978-3-642-59126-6_4
14. Glen, A., Justin, J., Widmer, S., Zamboni, L.Q.: Palindromic richness. Eur. J. Combin. **30**(2), 510–531 (2009)
15. Hooda, A., Bronstein, M.M., Bronstein, A.M., Horaud, R.P.: Shape palindromes: analysis of intrinsic symmetries in 2D articulated shapes. In: Bruckstein, A.M., ter Haar Romeny, B.M., Bronstein, A.M., Bronstein, M.M. (eds.) SSVM 2011. LNCS, vol. 6667, pp. 665–676. Springer, Heidelberg (2012). https://doi.org/10.1007/978-3-642-24785-9_56

16. Hopcroft, J.E., Ullman, J.D.: Formal Languages and Their Relation to Automata. Addison-Wesley Longman Inc., Boston (1969)
17. Kiryati, N., Gofman, Y.: Detecting symmetry in grey level images: the global optimization approach. Int. J. Comput. Vis. **29**(1), 29–45 (1998)
18. Knuth, D.E., Morris, Jr., J.H., Pratt, V.R.: Fast pattern matching in strings. In: Computer Algorithms, pp. 8–35. IEEE Computer Society Press, Los Alamitos (1994)
19. Kulkarni, M.S., Mahalingam, K.: Two-dimensional palindromes and their properties. In: Drewes, F., Martín-Vide, C., Truthe, B. (eds.) LATA 2017. LNCS, vol. 10168, pp. 155–167. Springer, Cham (2017). https://doi.org/10.1007/978-3-319-53733-7_11
20. Lothaire, M.: Combinatorics on Words. Cambridge University Press, Cambridge (1997)
21. Richmond, L.B., Shallit, J.: Counting the palstars. Electron. J. Combin. **21**(3), 6 (2014). Paper 3.25

Syntactic View of Sigma-Tau Generation of Permutations

Wojciech Rytter[ID] and Wiktor Zuba[✉][ID]

Institute of Informatics, University of Warsaw, Warsaw, Poland
{rytter,w.zuba}@mimuw.edu.pl

Abstract. We give a syntactic view of the Sawada-Williams (σ, τ)-generation of permutations. The corresponding sequence of $\sigma\tau$-operations, of length $n! - 1$ is shown to be highly compressible: it has $\mathcal{O}(n^2 \log n)$ bit description. Using this compact description we design fast algorithms for ranking and unranking permutations.

1 Introduction

We consider permutations of the set $\{1, 2, ..., n\}$, called here n-permutations. For an n-permutation $\pi = (a_1, ..., a_n)$ denote:

$$\sigma(\pi) = (a_2, a_3, ..., a_n, a_1), \tau(\pi) = (a_2, a_1, a_3, ..., a_n).$$

In their classical book on combinatorial algorithms Nijenhuis and Wilf asked in 1975 if all n-permutations can be generated, each exactly once, using in each iteration a single operation σ or τ. This difficult problem was open for more than 40 years. Very recently Sawada and Williams presented an algorithmic solution at the conference SODA'2018. In this paper we give new insights into their algorithm by looking at the generation from syntactic point of view.

Usually in a generation of combinatorial objects of size n we have a starting object and some set Σ of very local operations. Next object results by applying an operation from Σ, the generation is efficient iff each local operation uses small memory and time. Usually the sequence of generated objects is exponential w.r.t. n. From a syntactic point of view the generation globally can be seen as a very large word in the alphabet Σ describing the sequence of operations. It is called the *syntactic sequence* of the generation. Its textual properties can help to understand better the generation and to design efficient ranking and unranking. Such syntactic approach was used for example by Ruskey and Williams in generation of (n-1)-permutations of an n-set in [3].

Here we are interested whether the syntactic sequence is highly compressible. We consider compression in terms of Straight-Line Programs (*SLP*, in short), which represent large words by recurrences, see [4], using operations of concatenation. We construct SLP with $\mathcal{O}(n^2)$ recurrences, which has $\mathcal{O}(n^2 \log n)$ bit description.

The syntactic sequence for some generations is highly compressible and for others is not. For example in case of reflected binary Gray code of rank n each

© Springer Nature Switzerland AG 2019
C. Martín-Vide et al. (Eds.): LATA 2019, LNCS 11417, pp. 447–459, 2019.
https://doi.org/10.1007/978-3-030-13435-8_33

local operation is the position of the changed bit. Here $\Sigma = \{1, 2, ..., n\}$ and the syntactic sequence $T(n)$ is described by the short SLP of only $\mathcal{O}(n)$ size: $T_1 = 1$; $T(k) = T(k-1), k, T(k-1)$ for $2 \le k \le n$.

In case of de Bruijn words of length n each operation corresponds to a single letter appended at the end. However in this case the syntactic sequence is not highly compressible though the sequence can be iteratively computed in a very simple way, see [7]. In this paper we consider the syntactic sequence SEQ_n (over alphabet $\Sigma = \{\sigma, \tau\}$) of Sawada-Williams $\sigma\tau$-generation of permutations presented in [5,6]. An SLP of size $\mathcal{O}(n^2)$ describing SEQ_n is given in this paper. The $\sigma\tau$-generation of n-permutations by Sawada and Williams can be seen as a Hamiltonian path $\mathsf{SW}(n)$ in the Cayley graph $\mathcal{G}_n$. The nodes of this graph are permutations and the edges correspond to operations σ and τ.

We assume that (simple) arithmetic operations used in the paper are computable in constant time.

Our results. We show:

1. SEQ_n can be represented by the straight-line program of $\mathcal{O}(n^2)$ size:
   - $\mathbf{W}_0 = \sigma$, $\mathbf{W}_k = \tau \cdot \prod_{i=1}^{n-2} \sigma^i \mathbf{W}_{\Delta(k,i)} \gamma_{n-2-i}$
     for $1 \le k < n-3$;
   - $\mathbf{V}_n = \gamma_{n-3} \cdot \prod_{i=2}^{n-3} \sigma^i \mathbf{W}_{\Delta(n-3,i)} \gamma_{n-2-i} \cdot \sigma^{n-1}$;
   - $\mathsf{SEQ}_n = \gamma_1^{n-2} \sigma^2 (\mathbf{V}_n \tau)^{n-2} \mathbf{V}_n$.
   where $\Delta(k,i) = \min(k-1, n-2-i)$ and $\gamma_k = \sigma^k \tau$.

2. **Ranking:** using compact description of SEQ_n the number of steps (the rank of the permutation) needed to obtain a given permutation from a starting one can be computed in time $\mathcal{O}(n\sqrt{\log n})$ using inversion-vectors of permutations.

3. **Unranking:** again using SEQ_n the t-th permutation generated by SEQ_n can be computed in $\mathcal{O}(n \frac{\log n}{\log \log n})$ time.

2   Preliminaries

Denote by $cycle(\pi)$ all permutations cyclically equivalent to π. Sawada and Williams introduced an ingenious concept of a seed: a *shortened permutation* representing a group of $(n-1)$ cycles. Informally it represents a set of permutations which are cyclically equivalent *modulo* one fixed element, which can appear in any place.

Let $\oplus$ denote a modified addition modulo $n-1$, where $n-1 \oplus 1 = 1$. It gives a cyclic order of elements $\{1, ..., n-1\}$. We write $a \ominus 1 = b$ iff $b \oplus 1 = a$.

Formally a *seed* is a $(n-1)$ tuple of distinct elements of $\{1, 2, ..., n\}$ of the form $\psi = (a_1, a_2, ..., a_{n-1})$, such that $a_1 = n$ and $(a_1, a_2 \oplus 1, a_2, ..., a_{n-1})$ is a permutation. The element $x = mis(\psi) = a_2 \oplus 1$ is called a *missing* element.

Denote by $perms(\psi)$ the set of all n-permutations resulting by making a single insertion of x into any position in ψ, and making cyclic shifts. The sets $perms(\psi)$ are called *packages*, the seed ψ is the *identifier* of its package

$perms(\psi)$. One of the main tricks in the Sawada-Williams construction is the requirement that the missing element equals $a_2 \oplus 1$. In particular this implies the following:

Observation 1. *A given n-permutation belongs to one or two packages. We can find identifiers of these packages in linear time.*

The algorithm of Sawada and Williams starts with a construction of a large and a small cycle (covering together the whole graph). The graph consisting of these two cycle is denote here by $\mathcal{R}_n$. The small cycle is very simple. Once $\mathcal{R}_n$ is constructed the Hamiltonian path is very easy: In each cycle one τ-edge is removed (the cycles become simple paths), then the cycles are connected by adding one edge to $\mathcal{R}_n$. First we introduce seed-graphs. Define the *seed-graph* of the seed ψ, denoted here by $\mathsf{SeedGraph}(\psi)$ (denoted by $Ham(\psi)$ in [6]), as the graph consisting of edges *implied* by the seed ψ. The set of nodes consists of $perms(\psi)$, the set of edges consists of almost all σ-edges between these nodes (except the edges of the form $(*, x, *, ..., *) \rightarrow (x, *, ..., *, *)$), but the set of τ-edges consists only of the edges of the form $(*, x, *, ..., *) \rightarrow (x, *, *, ..., *)$, where x is the *missing* element. see Fig. 1.

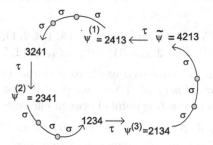

Fig. 1. Structure of $SeedGraph(\psi)$, where $\psi = (4, 1, 3)$, $mis(\psi) = 2$.

We say that an edge $u \rightarrow v$ conflicts with $u' \rightarrow v'$ iff $u = u', v \neq v'$. Non-disjoint packages ϕ, ψ can be joined into a simple cycle by removing two σ-edges conflicting with τ-edges.

By a union of graphs we mean set-theoretic union of nodes and set-theoretic union of all edges in these graphs.

Denote by $\mathcal{R}_n$ the graph $\bigcup_\psi \mathsf{SeedGraph}(\psi)$ in which we removed all σ-edges conflicting with τ-edges. The τ-edges have priority here. A version of the construction of a Hamiltonian path by Sawada-Williams, denoted by $\mathsf{SW}(n)$, can be written informally as:

Algorithm Compute PATH(n);

$P := \bigcup_{\psi \in SEEDS(n)} \text{SeedGraph}(\psi)$

remove from P all σ-edges conflicting with τ-edges in P

$\pi := (n, n-1, ..., 1)$; add to P the edge $\pi \to \sigma(\pi)$

remove edges $\pi \to \tau(\pi)$, $\tau(\sigma(\pi)) \to \sigma(\pi)$

return P {P is now a Hamiltonian path $\tau(\pi) \to^* \tau(\sigma(\pi))$ }

Lemma 2. PATH(n) = SW(n).

Our aim is to give a syntactic version of PATH(n): the sequence SEQ$_n$ of $\sigma\tau$-labels of PATH(n) represented compactly. We have to investigate more carefully the structure of seed-graphs and their interconnections.

2.1 Structure of Seed Graphs

For a seed $\psi = (a_1, a_2, ..., a_{n-1})$ with $mis(\psi) = x$, let

$$\psi^{(n-1)} = (x, a_2, ..., a_{n-1}, a_1), \quad \widetilde{\psi} = (a_1, x, a_2, a_3, ..., a_{n-1}).$$

For $1 \le i \le n-1$ denote $\psi^{(i)} = \gamma_{n-1}^i(\psi^{(n-1)})$. In other words $\psi^{(i)}$, for $n > i > 0$, is the word ψ right-shifted by $i-1$ and with x added at the beginning. Observe that: $\gamma_{n-1}(\psi^{(i)}) = \psi^{(i+1)}$ for $0 < i < n-1$.

Example 3. For $\psi = (5, 3, 2, 1)$ we have $\widetilde{\psi} = (5, 4, 3, 2, 1)$, $\psi^{(1)} = (4, 5, 3, 2, 1)$, $\psi^{(2)} = (4, 1, 5, 3, 2)$, $\psi^{(3)} = (4, 2, 1, 5, 3)$, $\psi^{(4)} = (4, 3, 2, 1, 5)$.

Each $perms(\psi)$ can be sequenced easily as a simple cycle in $\mathcal{G}_n$. Two seeds ϕ, ψ are called *neighbors* iff $perms(\phi) \cap perms(\psi) \ne \emptyset$. The permutations of type $\psi^{(i)}$ play crucial role as *connecting* points between packages of neighboring seeds.

Observation 4. *Two distinct seeds ϕ, ψ are neighbors iff $mis(\phi) = mis(\psi) \oplus 1$ or $mis(\psi) = mis(\phi) \oplus 1$, and after removing both $mis(\psi)$, $mis(\phi)$ from ϕ and ψ the sequences ϕ, ψ become identical.*

2.2 The Pseudo-tree ST$_n$ of Seeds

For a seed $\psi = a_1 a_2 ... a_{n-1}$ denote by $height(\psi)$ the maximal length k of a prefix of $a_2, a_3, ..., a_{n-1}$ such that $a_i = a_{i+1} \oplus 1$ for $i = 2, 3, ..., k$. For example $height(94326781) = 3$ (here the *missing* number is 5). For each two neighbors we distinguish one of them as a parent of the second one and obtain a tree-like structure called a *pseudo-tree* denoted by ST$_n$. If $height(\psi) > 1$ and $mis(\psi) = mis(\beta) \oplus 1$ we write $parent(\beta) = \psi$. Additionally if $\sigma^i(\psi^{(i)}) = \widetilde{\beta}$ we write $son(\psi, i) = \beta$ and we say that β is the i-th son of ψ.

The function *parent* gives the tree-like graph of the set of seeds, it is a cycle with hanging subtrees rooted at nodes of this cycle. The set of seeds on this cycle is denoted by Hub_n. For example

$$Hub_6 = \{(6, 5, 4, 3, 2), (6, 4, 3, 2, 1), (6, 3, 2, 1, 5), (6, 2, 1, 5, 4), (6, 1, 5, 4, 3)\}.$$

Due to Lemma 2 we have:

Observation 5. *If* $\psi \notin Hub_n$ *then* <u>*all*</u> τ-*edges of* SeedGraph(ψ) *are in* $PATH(n)$.

For $\psi \notin Hub_n$ let $Tree(\psi)$ be the subtree of ST_n rooted at ψ including ψ and nodes from which ψ is reachable by *parent*-links. For $\psi \notin Hub_n$ define $bunch(\psi) = \bigcup_{\beta \in Tree(\psi)} perms(\beta) - cycle(\widetilde{\psi}) \cup \{\widetilde{\psi}, \psi^{(n-1)}\}$.

In other words $cycle(\widetilde{\psi})$ connects $bunch(\psi)$ with the *"outside world"*, only through $\widetilde{\psi}, \psi^{(n-1)}$.

3   Compact Representation of Bunches

We start with properties of local interconnection between two packages.

Lemma 6. *Two seeds* $\phi \neq \psi$ *are neighbors iff one of them is the parent of another one. If* $\phi = parent(\psi)$ *then* $perms(\phi) \cap perms(\psi)$ *is the* σ-*cycle containing both* $\widetilde{\psi}$ *and* $\phi^{(i)}$, *for some* i, *and has a structure as shown in Fig. 2(A), where* ψ *is the* i-*th son of* ϕ. *If* $height(\phi) = k < n - 3$ *then* $height(\psi) = \Delta(k, i)$. *Furthermore* $son(\phi, i)$ *exists for all* $i \in \{1, ..., n - 3\}$.

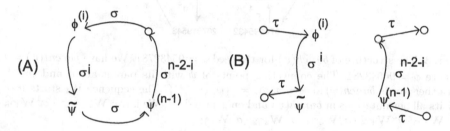

Fig. 2. (A) The anatomy of $perms(\phi) \cap perms(\psi)$: the graph SeedGraph($\psi$) $\cap$ SeedGraph(ϕ). (B) A part of the Hamiltonian path PATH(n) after removing two conflicting σ-edges, we have that ψ is the i-th son of ϕ.

For $k < n - 3$ and a seed ψ of height k we define $\mathbf{W}_k$ as the sequence of labels of a sub-path in PATH(n) starting in $\widetilde{\psi}$ and ending in $\psi^{(n-1)}$. In other words it is a $\sigma\tau$-sequence generating all n-permutations (each exactly once) of $bunch(\psi)$.

Observation 7. *By Lemma 6 every seed* ψ *such that* $1 < height(\psi) < n - 3$ *has exactly* $n - 3$ *sons whose heights depend only on height of* ψ. *Hence (by induction on heights) all trees* $Tree(\psi)$ *are isomorphic for seeds* ψ *of the same height. Consequently the definition of* $\mathbf{W}_k$ *is justified as it depends only on the height of* ψ.

For a permutation π and a sequence α of operations σ, τ denote by $GEN(\pi, \alpha)$ the set of all permutations generated from π by following α, including π.

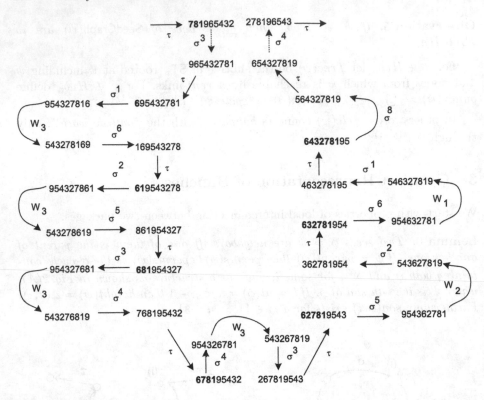

Fig. 3. The structure of $bunch(\psi)$ for the seed $\psi = 95432781$. We have $parent(\psi) = \phi$, where $\phi = 96543281$. The connecting points of ψ with its parent are $\widetilde{\psi}$ and $\psi^{(n-1)}$, in other words $bunch(\psi) \cap perms(\phi) = \{\widetilde{\psi}, \psi^{(n-1)}\}$. The sequence W_4 starts in $\widetilde{\psi}$, visits all permutations in $bunch(\psi)$ and ends in $\psi^{(n-1)}$. We have: $\mathbf{W}_4 = \tau \cdot \sigma^1 \mathbf{W}_3 \gamma_6 \cdot \sigma^2 \mathbf{W}_3 \gamma_5 \cdot \sigma^3 \mathbf{W}_3 \gamma_4 \cdot \sigma^4 \mathbf{W}_3 \gamma_3 \cdot \sigma^5 \mathbf{W}_2 \gamma_2 \cdot \sigma^6 \mathbf{W}_1 \gamma_1 \cdot \gamma_8$.

The word $\mathbf{W}_k$ satisfies:

$$GEN(\widetilde{\psi}, \mathbf{W}_k) = bunch(\psi) \text{ and } \mathbf{W}_k(\widetilde{\psi}) = \psi^{(n-1)}.$$

In this section we give compact representation of $\mathbf{W}_k$.

For example if $height(\psi) = 1$ then W_1 is a traversal of $perms(\psi)$ except $n-2$ cyclically equivalent permutations, common to $perms(\psi)$ and $perms(\phi)$, where $\phi = parent(\psi)$.

Recall that we denote $\gamma_k = \sigma^k \tau$

Theorem 8. *For $1 \le k < n-3$ we have the following recurrences:*

$$\mathbf{W}_0 = \sigma, \quad \mathbf{W}_k = \tau \cdot \prod_{i=1}^{n-2} \sigma^i \, \mathbf{W}_{\Delta(k,i)} \, \gamma_{n-2-i}$$

Proof. Assume $\psi \notin Hub_n$ is of height k, then by Lemma 6 the first, from left to right, $n - k - 1$ children of ψ in the subtree $Tree(\psi)$ are of height $k - 1$ and the next $k-2$ children are of heights $k-2, k-3, ..., 1$. The representative $\widetilde{\beta_i}$ of the i-th son β_i of ψ equals $\sigma^i(\psi^{(i)})$ (see Figs. 3 and 4).   □

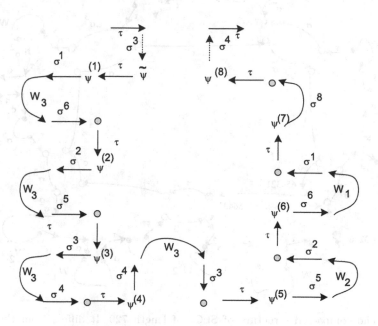

Fig. 4. Schematic view of structure from Fig. 3.

4   Compact Representation of the Whole Generation

We have the following fact:

Observation 9. *Assume two seeds ψ, β satisfy: $height(\psi) = k > 1$ and $\sigma^i(\psi^{(i)}) = \beta$. Then if $i = 1$ and $\psi \in Hub_n$ then $height(\beta) = height(\psi)$.*

Theorem 10. *The whole $\sigma\tau$-sequence SEQ_n starting at $\tau(n, n-1, ..., 1)$, ending at $\sigma\tau(n, n-1, ..., 1)$, and generating all n-permutations, has the following compact representation of $\mathcal{O}(n^2)$ size (together with recurrences for $\mathbf{W}_k$):*

$$\mathsf{SEQ}_n = \gamma_1^{n-2}\sigma^2 \left(\mathbf{V}_n\tau\right)^{n-2}\mathbf{V}_n, \quad where$$

$$\mathbf{V}_n = \gamma_{n-3} \cdot \prod_{i=2}^{n-3} \sigma^i \mathbf{W}_{\Delta(n-3,i)} \gamma_{n-2-i} \cdot \sigma^{n-1}.$$

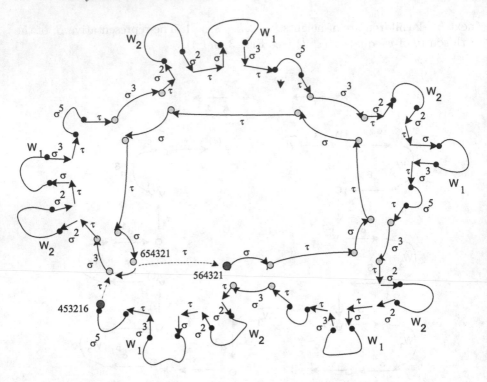

Fig. 5. The compacted structure of SEQ_6 of length 720. It differs from the structure of $\mathcal{R}_6$ by adding one σ-edge from 654321 and removing two (dotted) τ-edges to have Hamiltonian path. We have: $\mathsf{SEQ}_6 = (\sigma\tau)^4\sigma^2 \ (\mathbf{V}_6\tau)^4 \ \mathbf{V}_6$, where $\mathbf{V}_6 = \sigma^3\tau\,\sigma^2\mathbf{W}_2\sigma^2\tau\,\sigma\mathbf{W}_1\sigma^3\tau\,\sigma^5$. The structure is the union of graphs of 5 seeds in Hub_6 with *hanging* bunches. The *starting path* consists of permutations from 564321 to 654321.

Proof. For every non-hub seed ψ we had that $GEN(\widetilde{\psi}, \mathbf{W}_k) = bunch(\psi)$, where $k = height(\psi)$. The only difference for a hub seed ϕ is that $son(\phi, 1)$ cannot be considered as part of a tree rooted at ϕ (with already defined *parent*-links), since $son(\phi, 1) \in Hub_n$ and this would lead to a cycle ($son(\phi, 1)$ is reachable via parent-links from ϕ). Thus to prevent this problem we define V_n as W_{n-3} with the part corresponding to the first son removed (leaving only the γ_{n-2-1} part), and also delete the last symbol τ, as it does not appear at the end of the path (it corresponds to one of the τ-edges removed when joining two cycles into one path). Now Seq_n consists of $n - 1$ such segments V_n (corresponding to $n - 1$ hub seeds) joined by τ-edges (they are linked in the same way as if the previous V_n part was a son of the next one). Additionally it starts with γ_1^{n-1}-path representing the small path with the last τ-edge replaced by a σ-edge. $\qquad\square$

5   Ranking

We need some preprocessing to access later some values in constant time.

Observation 11. *All the values* $|W_k|$ *and* $\sum_{i=0}^{k}(|W_i| + n - 1)$ *for* $k \in \{0..n - 4\}$ *can be computed in* $\mathcal{O}(n)$ *total time and accessed in* $\mathcal{O}(1)$ *time afterwards.*

The ranks of representatives of hub seeds are easy to compute. For example for $n = 6$ we have (see Fig. 5): $rank(643215) = 1$, $rank(632154) = 3$, $rank(621543) = 5$, $rank(615432) = 7$, $rank(654321) = 9$.

Lemma 12. *For a given permutation* π *we can compute in time* $\mathcal{O}(n)$

(a) $rank(\pi) - rank(\widetilde{\psi})$ *if* $\pi \in perms(\psi)$,
(b) $rank(\pi)$ *if* $\pi \in perms(\psi)$ *for some* $\psi \in Hub_n$.

Hence we concentrate on ranking permutations of type $\widetilde{\psi}$ (representatives of seeds). We slightly abuse notation and for a seed ψ define $rank(\psi) = rank(\widetilde{\psi})$.

For a non-hub seed ψ denote by $anchor(\psi)$ the highest non-hub ancestor ϕ of ψ and let $hub(\psi) = parent(anchor(\psi))$. Observe that the anchor ϕ is the first contacting seed with the hub, it is the first ancestor of ψ such that $perms(\phi) \cap perms(\beta) \neq \emptyset$ for some $\beta \in Hub_n$, in fact for $\beta = hub(\psi)$.

$rank(\psi) - rank(anchor(\psi))$ for a non-hub seed ψ, can be treated as its *distance from* Hub_n. It happens that computing the rank of the anchor is much easier, since we have to deal only with the hub seeds. The bottleneck in ranking is computation of the distance of a seed representative from Hub_n. Define:

$$SUM(k, j) = |\tau \cdot \prod_{i=1}^{j-1} \sigma^i \mathbf{W}_{\Delta(k,i)} \gamma_{n-2-i}| + j.$$

Denote also by $ord(\psi)$ the position of $mis(\psi) + 1$ in ψ counting from the end of sequence ψ. For example for $\psi = (10\,6\,5\,9\,8\,4\,3\,1\,2)$ we have $ord(\psi) = 5$, since $mis(\psi) = 7$ and 8 is on the 5-th position from the right.

Observation 13. *If* $\phi = parent(\psi) \notin Hub_n$ *and* ψ *is the* i-th *son of* ϕ *then* $rank(\psi) - rank(\phi) = SUM(height(\phi), i)$.

Example 14. Let $\psi = 94326781$, then $parent(\psi) = \phi = 95432781$. The path from $\widetilde{\phi} = 965432781$ to $\widetilde{\psi} = 954326781$ is

$$\tau \sigma^1 W_3 \sigma^6 \tau \sigma^2 W_3 \sigma^5 \tau \sigma^3 W_3 \sigma^4 \tau \sigma^4,$$

see Fig. 3. Its length equals $SUM(4, 4)$, we have: $height(\phi) = 4$, $ord(\psi) = 4$.

Observation 15. $ord(\psi) = i$ *iff* ψ *is the* i-th *son of* $parent(\psi)$.

For the parent-sequence $\psi_0 = \psi, \psi_1, ..., \psi_m = anchor(\psi_i)$ denote

$$route(\psi) = (ord(\psi_0), ord(\psi_1), ..., ord(\psi_m)).$$

For a seed $\psi = a_1 a_2 ... a_{n-1}$ define the decreasing sequence of ψ, denoted by $dec\_seq(\psi)$, as the maximal sequence $a_{i_0} a_{i_1} ... a_{i_m}$, where $2 = i_0 < i_1 < i_2 < ... < i_m$ such that $i_{j-1} = i_j \oplus 1$ for $0 < j \le m$. Denote $level(\psi) = n - m - 3$. The length of the *parent-sequence* $\psi = \psi_0, \psi_1, \psi_2, ..., \psi_r = anchor(\psi)$ from ψ to its anchor is $r = level(\psi) - 1$.

Example 16. We have: $dec\_seq(96154238) = (6, 5, 4, 3)$. Hence the path from $\psi = (96154238)$ to $anchor(\psi) = (98765423)$ is of length $(9 - 3 - 3) - 1 = 2$. This path equals:

$$\psi_0 \to \psi_1 \to \psi_2 = 96154238 \to 97615423 \to 98765423.$$

We have: $ord(\psi_0) = 1$, $ord(\psi_1) = 5$, $ord(\psi_2) = 2$, $route(96154238) = (1, 5, 2)$.

The key point is that we do not need to deal with the whole parent-sequence, including explicitly seeds on the path, which is of quadratic size (in worst-case) but it is sufficient to deal with the sequence of orders of sons, which is an implicit representation of this path of only linear size.

Lemma 17. *For a non-hub seed ψ we can compute $route(\psi)$ and $anchor(\psi)$ in $\mathcal{O}(n\sqrt{\log n})$ time.*

Proof. We know the length of the parent sequence from ψ to its anchor, since we know $level(\psi)$. Now we use the following auxiliary problem

Inversion-Vector problem:
for a seed ψ compute for each element x the number $RightSm[x]$ of elements smaller than x which are to the right of x in ψ.

Assume $\psi = (a_1, a_2, ..., a_{n-1})$. We introduce a new linear order

$$a_2 \prec a_2 \ominus 1 \prec a_2 \ominus 2 \prec ... \prec a_2 \ominus (n-2).$$

Then we compute together the numbers $RightSm[z]$ w.r.t. linear order $\prec$ for each element z in ψ.

Now $ord(\psi_i)$ is computed separately for each i in the following way:

$$ord(\psi_i) := RightSm[x_i + 1] + 1, \text{where } x_i = mis(\psi_i)$$

The *Inversion-Vector* problem can be computed in $\mathcal{O}(n\sqrt{\log n})$ time, see [1]. Consequently the whole computation of numbers $ord(\psi_i)$ is of the same asymptotic complexity. We know that $hub(\psi) = (n, b, b \ominus 1, ..., b \ominus (n-3))$, where $b = a_2 \oplus level(\psi)$ and we know also which son of $hub(\psi)$ is $anchor(\psi)$. This knowledge allows to compute $anchor(\psi)$ within required complexity. This completes the proof. □

Corollary 18. *For a non-hub seed ψ the value $rank(\psi) - rank(anchor(\psi))$ can be computed in $\mathcal{O}(n\sqrt{\log n})$ time.*

Proof. Let the *parent-sequence* from ψ to its anchor be

$$\psi = \psi_0, \psi_1, \psi_2, ..., \psi_r = anchor(\psi), \text{where } r = level(\psi) - 1.$$

Then $rank(\psi_i) - rank(\psi_{i+1}) = SUM(height(\psi_{i+1}), ord(\psi_i))$, and $height(\psi_i) = \Delta(height(\psi_{i+1}), ord(\psi_i))$, which allows us to compute in $\mathcal{O}(n)$ time:

$$rank(\psi) - rank(anchor(\psi)) = \sum_{i=m-1}^{0} (rank(\psi_i) - rank(\psi_{i+1}))$$

Now the thesis is a consequence of Observations 11, 13 and Lemma 17. This completes the proof.   □

Example 19. (Continuation of Example 16) For ψ from Example 16 we have:

$$rank(\psi) - rank(anchor(\psi)) = SUM(5,5) + SUM(2,1)$$

The following result follows directly from Corollary 18, Lemma 12 and Observation 11.

Theorem 20. **[Ranking]** *For a given permutation π we can compute the rank of π in* SEQ_n *in time* $\mathcal{O}(n\sqrt{\log n})$.

6   Unranking

Denote by $Perm(t)$ the t-th permutation in SEQ_n, and for $t < |bunch(\psi)|$ let $Perm(\psi, t) = Perm(t + rank(\widetilde{\psi}))$ (it is the t-th permutation in $bunch(\psi)$, counting from the beginning of this bunch). The following case is an easy one.

Lemma 21. *If we know a seed ψ together with its rank, such that $Perm(t) \in perms(\psi)$, then we can recover $Perm(t)$ in linear time.*

We say that a permutation π is a *hub-permutation* if $\pi \in perms(\psi)$ for some $\psi \in Hub_n$.

Lemma 22. *We can test in $\mathcal{O}(n)$ time if $Perm(t)$ is a hub-permutation.*

(a) *If "yes" then we can recover $Perm(t)$ in $\mathcal{O}(n)$ time.*
(b) *Otherwise we can find in $\mathcal{O}(n)$ time an anchor-seed ψ together with $rank(\psi)$ such that $Perm(t) \in bunch(\psi)$.*

For a sequence $\mathbf{b} = (b_1, b_2, ..., b_m)$ of positive integers denote

$$MaxFrac(\mathbf{b}) = \max_i \frac{b_{i+1}}{b_i}, \quad MinFrac(\mathbf{b}) = \min_i \frac{b_{i+1}}{b_i}.$$

The sequence $\mathbf{b}$ is called here $D(m)$–*stably increasing* iff

$$MinFrac(\mathbf{b}) \geq 2, \text{ and } MaxFrac(\mathbf{b}) \leq D(m).$$

Lemma 23.

(a) *Assume we have a $D(m)$–stably increasing sequence* **b** *of length $\mathcal{O}(m)$. Then after linear preprocessing we can locate any integer t in the sequence* **b** *in $\mathcal{O}(\log \log D(m))$ time.*
(b) *The sequence* **b** $= (b_0, b_1, ..., b_{n-5})$, *where $b_k = \sum_{i=0}^{k}(|W_i| + n - 1)$ is n–stably increasing.*

Lemma 24. *After linear preprocessing if we are given a height of a non-hub seed ψ, and a number $t \le |bunch(\psi)|$ we can find the number j and $height(\beta)$ of the seed-son β of ψ such that $Perm(\psi, t) \in bunch(\beta)$ in $\mathcal{O}(\log \log n)$ time if $Perm(\psi, t) \notin perms(\psi)$.*

Proof. Let $k = height(\psi)$. We need j such that $SUM(k, j) - j \le t < SUM(k, j + 1) - (j + 1)$. For $j \le n - k$ we have $SUM(k, j) - j = (j - 1) \cdot (|W_{k-1}| + n - 1)$, hence if $t < SUM(k, n - k) - n + k$ the simple division by $|W_{k-1}| + n - 1$ suffices to find the appropriate j. Otherwise we look for j such that

$$|W_k| - SUM(k, j + 1) + j + 1 < s \le |W_k| - SUM(k, j) + j, \text{ where } s = |W_k| - t.$$

Let $b_i = |W_k| - SUM(k, n - 2 - i) + n - 2 - i = (\sum_{j=0}^{i} |W_j| + n - 1)$. By Lemma 23(b) $(b_0, ..., b_{k-2})$ is n–stably increasing (it is a prefix of $(b_0, ..., b_{n-5})$ for which we made the linear preprocessing). Hence by Lemma 23(a) we can find the required j in $\mathcal{O}(\log \log n)$ time.

Moreover if $SUM(k, j) < t < SUM(k, j) + |W_{\Delta(k,j)}|$, then $Perm(\psi, t) = Perm(\beta, t - SUM(k, j))$, where $\beta = son(\psi, j)$ has height $\Delta(k, j)$. Otherwise $Perm(\psi, t) \in perms(\psi)$.   □

Theorem 25. [Unranking] *For a given number t we can compute the t-th permutation in Sawada-Williams generation in $\mathcal{O}(n \frac{\log n}{\log \log n})$.*

Proof. From Lemma 22 we either obtain the required permutation (if it is a hub-permutation) or obtain its anchor-seed ϕ and $rank(\phi)$. In the second case we know that $Perm(t) \in bunch(\phi)$ and it equals $Perm(\phi, t - rank(\widetilde{\phi}))$. Now after the linear preprocessing we apply Lemma 24 exhaustively to obtain $route(\psi)$ for a seed ψ such that $Perm(t) \in perms(\psi)$. However we do not know ψ and have to compute it.

Claim. If we know $anchor(\psi)$ and $route(\psi)$ then ψ can be computed in $\mathcal{O}(n \frac{\log n}{\log \log n})$ time.

Proof. We can compute the second element a_2 of ψ as $a_2' \ominus m$ and $dec\_seq(\psi)$ as $(a_2, a_2 \ominus 1, ..., a_2 \ominus (n - m - 3))$ where a_2' is the second element of $anchor(\psi)$, and $m = |route(\psi)| - 1$. Then we use the order:

$$a_2 \prec a_2 \ominus 1 \prec a_2 \ominus 2 \prec ... \prec a_2 \ominus (n - 2).$$

We produce a linked list initialized with $dec\_seq(\psi)$. For $i \in \{0, ..., m - 1\}$ we want to insert $a_2 \oplus (m + 1 - i)$ after $ord(\psi_{m-1})$ position from the end of the current list (all the smaller elements are already in the list and we know, that after $a_2 \oplus (m + 1 - i)$ there are $ord(\psi_{m-1}) - 1$ such elements). ψ is composed of n and consecutive elements of the final list. The data structure from [2] allows us to achieve that in $\mathcal{O}(n \frac{\log n}{\log \log n})$ time.

Finally we use this claim and Lemma 21 to obtain the required permutation $Perm(t)$.   □

References

1. Chan, T.M., Patrascu, M.: Counting inversions, offline orthogonal range counting, and related problems. In: Charikar, M. (ed.) Proceedings of the Twenty-First Annual ACM-SIAM Symposium on Discrete Algorithms, SODA 2010, Austin, Texas, USA, 17–19 January 2010, pp. 161–173. SIAM (2010). https://doi.org/10.1137/1.9781611973075.15
2. Dietz, P.F.: Optimal algorithms for list indexing and subset rank. In: Dehne, F., Sack, J.-R., Santoro, N. (eds.) WADS 1989. LNCS, vol. 382, pp. 39–46. Springer, Heidelberg (1989). https://doi.org/10.1007/3-540-51542-9_5
3. Ruskey, F., Williams, A.: An explicit universal cycle for the $(n-1)$-permutations of an n-set. ACM Trans. Algorithms 6(3), 45:1–45:12 (2010). https://doi.org/10.1145/1798596.1798598
4. Rytter, W.: Grammar compression, LZ-encodings, and string algorithms with implicit input. In: Díaz, J., Karhumäki, J., Lepistö, A., Sannella, D. (eds.) ICALP 2004. LNCS, vol. 3142, pp. 15–27. Springer, Heidelberg (2004). https://doi.org/10.1007/978-3-540-27836-8_5
5. Sawada, J., Williams, A.: Solving the sigma-tau problem. http://socs.uoguelph.ca/~sawada/papers/sigmaTauCycle.pdf
6. Sawada, J., Williams, A.: A Hamilton path for the sigma-tau problem. In: Proceedings of the Twenty-Ninth Annual ACM-SIAM Symposium on Discrete Algorithms, SODA 2018, New Orleans, LA, USA, 7–10 January 2018, pp. 568–575 (2018). https://doi.org/10.1137/1.9781611975031.37
7. Sawada, J., Williams, A., Wong, D.: A surprisingly simple de Bruijn sequence construction. Discret. Math. 339(1), 127–131 (2016). https://doi.org/10.1016/j.disc.2015.08.002

Palindromic Subsequences
in Finite Words

Clemens Müllner[1] and Andrew Ryzhikov[2(✉)]

[1] CNRS, Université Claude Bernard - Lyon 1, Villeurbanne, France
mullner@math.univ-lyon1.fr
[2] LIGM, Université Paris-Est, Marne-la-Vallée, France
ryzhikov.andrew@gmail.com

Abstract. In 1999 Lyngsø and Pedersen proposed a conjecture stating that every binary circular word of length n with equal number of zeros and ones has an antipalindromic linear subsequence of length at least $\frac{2}{3}n$. No progress over a trivial $\frac{1}{2}n$ bound has been achieved since then. We suggest a palindromic counterpart to this conjecture and provide a nontrivial infinite series of circular words which prove the upper bound of $\frac{2}{3}n$ for both conjectures at the same time. The construction also works for words over an alphabet of size k and gives rise to a generalization of the conjecture by Lyngsø and Pedersen. Moreover, we discuss some possible strengthenings and weakenings of the named conjectures. We also propose two similar conjectures for linear words and provide some evidences for them.

Keywords: Palindrome · Antipalindrome · Circular words · Subsequences

1 Introduction

Investigation of subsequences in words is an important part of string algorithms and combinatorics, with applications to string processing, bioinformatics, error-correcting codes. A lot of research has been done in algorithms and complexity of finding longest common subsequences [1,4], their expected length in random words [9], codes with bounded lengths of pairwise longest common subsequences [10], etc. An important type of subsequences is a longest palindromic subsequence, which is in fact a longest common subsequence of a word and its reversal. Despite a lot of research in algorithms and statistics of longest common subsequences, the combinatorics of palindromic subsequences is not very well understood. We mention [2,5–7] as some results in this direction. In this note we recall some known conjectures on this topic and provide a number of new ones.

C. Müllner—This research was suported by the European Research Council (ERC) under the European Union's Horizon 2020 research and innovation programme under the Grant Agreement No 648132.

C. Martín-Vide et al. (Eds.): LATA 2019, LNCS 11417, pp. 460–468, 2019.
https://doi.org/10.1007/978-3-030-13435-8_34

The main topic of this note are finite words. A *linear word* (or just a *word*) is a finite sequence of symbols over some alphabet. A *subsequence* of a linear word $w = a_1 \ldots a_n$ is a word $w' = a_{i_1} \ldots a_{i_m}$ with $i_1 < \ldots < i_m$. A *circular word* is an equivalence class of linear words under rotations. Informally, a circular word is a linear word written on a circle, without any marked beginning or ending. A linear word is a *subsequence* of a circular word if it is a subsequence of some linear word from the corresponding equivalence class (such linear word is called a *linear representation*).

A word $w = a_1 \ldots a_n$ is a *palindrome* if $a_i = a_{n-i+1}$ for every $1 \leq i \leq \frac{n}{2}$. A word is called *binary* if its alphabet is of size two (in this case we usually assume that the alphabet is $\{0,1\}$). A binary word $w = a_1 \ldots a_n$ is an *antipalindrome* if $a_i \neq a_{n-i+1}$ for every $1 \leq i \leq \frac{n}{2}$. The *reversal* w^R of a word $w = a_1 \ldots a_n$ is the word $a_n \ldots a_1$.

In 1999 Lyngsø and Pedersen formulated the following conjecture motivated by analysis of an approximation algorithm for a 2D protein folding problem [8].

Conjecture 1 (Lyngsø and Pedersen, 1999). *Every binary circular word of length n divisible by 6 with equal number of zeros and ones has an antipalindromic subsequence of length at least $\frac{2}{3}n$.*

To the best of our knowledge, no progress has been achieved in proving this conjecture, even though it has drawn substantial attention from the combinatorics of words community. However, it is a source of other interesting conjectures.

In the mentioned conjecture, the position of a longest antipalindromic subsequence on the circle is arbitrary. A strengthening is to require the two halves of the subsequence to lie on different halves of the circle according to some partition of the circle into two parts of equal length. Surprisingly, experiments show that this does not change the bound.

Conjecture 2 (Brevier, Preissmann and Sebő, [3]). *Let w be a binary circular word of length n divisible by 6 with equal number of zeros and ones. Then w can be partitioned into two linear words w_1, w_2 of equal length, $w = w_1 w_2$, having subsequences s_1, s_2 such that $s_1 s_2$ is an antipalindrome and $|s_1| = |s_2| = \frac{1}{3}|w|$.*

We checked this conjecture up to $n = 30$ by computer. The worst known case for the both conjectures is provided by the word $w = 0^i 1^i (01)^i 1^i 0^i$ showing the tightness of the conjectured bound (by tightness everywhere in this note we understand the existence of a lower bound different from the conjectured bound by at most a small additive constant). The bound $\frac{1}{2}n$ instead of $\frac{2}{3}n$ can be easily proved, but no better bound is known.

Proposition 3 (Brevier, Preissmann and Sebő, [3]). *Conjecture 2 is true when replacing $|s_1| = |s_2| = \frac{1}{3}|w|$ by $|s_1| = |s_2| = \frac{1}{4}|w|$.*

Proof. Consider an arbitrary partition of w into two linear words w_1, w_2 of equal length, $w = w_1 w_2$. Assume that w_1 has less ones than w_2 zeros. By changing the partition by one letter each time (by adding a subsequent letter to the end of w_1 and removing one from the beginning), we get an opposite situation in $\frac{1}{2}n$ steps. That means that there exists a partition $w = w_1'w_2'$, $|w_1'| = |w_2'|$, such that the number of zeros in w_1' is the same as the number of ones in w_2' and vice versa. Thus, we can pick an antipalindromic subsequence 0^k1^k or 1^k0^k with $k = \frac{1}{4}n$ having the required properties.   □

2   Circular Words

A natural idea is to look at palindromic subsequences instead of antipalindromic ones. This leads to a number of interesting conjectures which we describe in this section. First, we formulate palindromic counterparts to Conjectures 1 and 2.

Conjecture 4. *Every binary circular word of length n has a palindromic subsequence of length at least $\frac{2}{3}n$.*

Conjecture 5. *Let w be a binary circular word of length n divisible by 6. Then w can be partitioned into 2 linear words w_1, w_2 of equal length, $w = w_1w_2$, having subsequences s_1, s_2 such that $s_1 = s_2^R$ (that is, s_1s_2 is a palindrome) and $|s_1s_2| = \frac{2}{3}|w|$.*

We checked both conjectures up to $n = 30$ by computer. The worst known case for Conjecture 5 is provided by the word $0^{2i}(10)^i1^{2i}$, showing the tightness of the conjectured bound. The word $0^i(10)^i1^i$ provides an upper bound of $\frac{3}{4}n$ for Conjecture 4. A better bound is discussed in Sect. 3.

In Conjecture 4 it is enough to pick the subsequence consisting of all appearances of the letter with the largest frequency to get the $\frac{1}{2}n$ lower bound. Using the same idea as in the proof of Proposition 3, it is also easy to prove the $\frac{1}{2}n$ bound for Conjecture 5. No better bounds are known to be proved.

Proposition 6. *Conjecture 5 is true when replacing $|s_1s_2| = \frac{2}{3}|w|$ by $|s_1s_2| = \frac{1}{2}|w|$.*

Conjecture 5 is about a palindromic subsequence aligned with some cut of the circular word into two equal halves. There are $\frac{n}{2}$ such cuts, so one attempt to simplify the conjecture is to look at only two cuts which are "orthogonal". This way we attempt to switch from the circular case to something close to the linear case, which is often easier to deal with.

Let w be a circular word of length n divisible by 4. Let $w_1w_2w_3w_4$ be some partition of w into four linear words of equal length. Let p_1p_1' and p_2p_2', $|p_1| = |p_1'|$, $|p_2| = |p_2'|$, be the longest palindromic subsequences of w such that p_1, p_1', p_2, p_2' are subsequences of w_1w_2, w_3w_4, w_2w_3, w_4w_1 respectively. Informally, these two palindromes are aligned to two orthogonal cuts of the word w into two linear words of equal length. The partitions w_1w_2, w_3w_4 and w_2w_3, w_4w_1 are two particular partitions (made by two orthogonal cuts) considered among all $\frac{n}{2}$ partitions in Conjecture 5.

Conjecture 7. *For every word w of length n divisible by 4 and its every linear representation $w = w_1 w_2 w_3 w_4$, the maximum of the lengths of $p_1 p'_1$ and $p_2 p'_2$ defined above is at least $\frac{1}{2}n$.*

We checked this conjecture up to $n = 30$ by computer. The worst known case is provided by the already appeared word $0^i (10)^i 1^i$ showing the tightness of the conjectured bound. The bound $\frac{1}{3}n$ can be proved as follows.

Proposition 8. *For every word w of length n divisible by 4 and its every linear representation $w = w_1 w_2 w_3 w_4$, the maximum of the lengths of $p_1 p'_1$ and $p_2 p'_2$ is at least $\frac{1}{3}n$.*

Proof. Suppose that $|p_1 p'_1| < \frac{1}{3}n$. Then without loss of generality we can assume that the number of zeros in $w_1 w_2$ and the number of ones in $w_3 w_4$ is less than $\frac{1}{6}n$. Then by the pigeonhole principle the number of ones in both w_1 and w_2, and the number of zeros in both w_3 and w_4 is at least $\frac{1}{12}n$. It means that we can pick a subsequence of $\frac{1}{12}n$ zeros and then $\frac{1}{12}n$ ones from $w_4 w_1$ and a symmetrical subsequence from $w_2 w_3$. Thus we get $|p_2 p'_2| \geq \frac{1}{3}n$. □

In fact, a slightly stronger statement that the total length of $p_1 p'_1$ and $p_2 p'_2$ is $\frac{2}{3}n$ can be proved this way. We conjecture the optimal bound for this value to be equal to n.

Even being proved, the bound of $\frac{1}{2}n$ in this conjecture would not improve the known bound for Conjecture 5. However, Conjecture 7 deals with palindromic subsequences of only two linear words, and thus seems to be easier to handle. Considering four regular cuts instead of two should already improve the bound for Conjecture 5.

3   Showing Asymptotic Tightness of Conjecture 4

In this section we present the main technical contribution of this paper, which is an infinite family of words providing a better upper bound for Conjecture 4. In fact, we show a stronger result for words over an arbitrary alphabet. Below we consider words over the alphabet $\{0, \ldots, k-1\}$, i.e. $w \in \{0, \ldots, k-1\}^*$.

Definition 9. *We say that w' is a* consecutive subword *of a word w if there exist words u, v with $w = uw'v$.*

We call a word $w \in \{0, \ldots, k-1\}^*$ of type n *if it is a consecutive subword of $(0^n 1^n \ldots (k-1)^n)^*$ or a consecutive subword of $((k-1)^n \ldots 1^n 0^n)^*$. In the first case we write $w \in S'_n$, in the second case we write $w \in S''_n$.*

Furthermore, we define $S_n = S'_n \cup S''_n$.

Thus $w \in S'_n$ if it is a concatenation of blocks $(0^n 1^n \ldots (k-1)^n)$, where the first and the last blocks may be shorter, and analogously for $w \in S''_n$.

We denote by $\overline{w}$ the word we get when exchanging every letter ℓ by $(k-1-\ell)$, e.g. $01 \ldots (k-1) = (k-1)(k-2) \ldots 0$. We see directly that $w \in S_n$ if and only if $\overline{w} \in S_n$. Furthermore, we have that $w \in S_n$ if and only if $w^R \in S_n$.

Lemma 10. *Let $w_1 \in S_{n_1}$ be a word of length ℓ_1 and $w_2 \in S_{n_2}$ be a word of length ℓ_2, where $n_1 > n_2$. Then, the length of the longest common subsequence of w_1 and w_2 is at most $\frac{\ell_1 + \ell_2}{k+1} + \ell_1 \frac{n_2}{n_1} + 2n_2$.*

Proof. Let w be a common subsequence of w_1 and w_2 of length ℓ. We see that w is of the form $a_1^{p_1} a_2^{p_2} \ldots a_s^{p_s}$, where all $a_j \in \{0, \ldots, k-1\}$, all p_j are positive and $a_j \neq a_{j+1}$. We find directly that for $i = 1, 2$:

$$s \leq \left\lceil \frac{\ell_i - 1}{n_i} \right\rceil + 1 \leq \frac{\ell_i - 1 + n_i - 1}{n_i} + 1 \leq \frac{\ell_i}{n_i} + 2. \tag{1}$$

We consider now the minimal length of a consecutive subword of w_i that contains $a_j^{p_j}$, where $p_j > n_i$. Thus, $a_j^{p_j}$ cannot be contained in one block of the form $(0^{n_i} 1^{n_i} \ldots (k-1)^{n_i})$. This shows that the minimal length of a consecutive subword of w_i that contains $a_j^{p_j}$ is at least kn_i.

This generalizes for $p_j > n_i r$ and we find that each $a_j^{p_j}$ spans a subsequence of length at least $kn_i(\lceil \frac{p_j}{n_i} \rceil - 1) \geq k(p_j - n_i)$ in w_i. Thus, we find $\ell_i \geq \sum_{j=1}^{s} k(p_j - n_i)$. This gives in total

$$\ell = \sum_{j=1}^{s} p_j \leq \frac{\ell_i}{k} + sn_i. \tag{2}$$

By combining (1) and (2) we find

$$\ell \leq \frac{\ell_2}{k} + \left(\frac{\ell_1}{n_1} + 2\right)n_2. \tag{3}$$

Furthermore, we find directly that $\ell \leq \ell_1$. This gives in total

$$\frac{k}{k+1}\ell \leq \frac{\ell_2}{k+1} + \ell_1 \frac{kn_2}{(k+1)n_1} + \frac{2k}{k+1}n_2 \leq \frac{\ell_2}{k+1} + \ell_1 \frac{n_2}{n_1} + 2n_2$$

$$\frac{1}{k+1}\ell \leq \frac{\ell_1}{k+1},$$

and by adding these inequalities, we find

$$\ell \leq \frac{\ell_1 + \ell_2}{k+1} + \ell_1 \frac{n_2}{n_1} + 2n_2.$$

$\square$

We think of $\frac{\ell_1 + \ell_2}{k+1}$ in the bound above as the "main term". Therefore, we need that $\frac{n_2}{n_1}$ is small. The remaining term origins from boundary phenomena due to incomplete blocks. We note that this "main term" is indeed sharp for large ℓ_1, ℓ_2, when $\frac{n_1}{n_2}$ is integer and $k\ell_1 = \ell_2$ as the following example shows.

Example 11. *We consider $n_1 = pn_2$, with p integer, and $w_1 = (0^{n_1} 1^{n_1} \ldots (k-1)^{n_1})^{\ell n_2}$, $w_2 = (0^{n_2} 1^{n_2} \ldots (k-1)^{n_2})^{k\ell n_1} = ((0^{n_2} 1^{n_2} \ldots (k-1)^{n_2})^{kp})^{\ell n_2}$. One finds that i^{n_1} is a subsequence of $(0^{n_2} 1^{n_2})^p$ and thus, w_1 is a subsequence of w_2. This gives directly $|w_1| = kn_1 \ell n_2, |w_2| = kn_2 k\ell n_1 = k|w_1|$ and $|w| = |w_1| = \frac{|w_1| + |w_2|}{k+1}$.*

For the following considerations we will need a generalization of the notion of antipalindromes to the case of non-binary alphabet. One natural version would be to say that w is an antipalindrome if w and w^R differ at every position. However, we work with a stronger notion, which still provides an interesting bound.

Definition 12. *We call a word $w \in \{0, \ldots, k-1\}^*$ a strong antipalindrome if $w = \overline{w}^R$.*

Theorem 13. *For every $\varepsilon > 0$ there exists a circular word over the alphabet $\{0, \ldots, k-1\}$ with equal number of 0's, 1's, $\ldots$, $(k-1)$'s (n occurences of each letter) such that any palindromic and any strongly antipalindromic subsequence of it is of length at most $(\frac{2}{k+1} + \varepsilon)kn$.*

Proof. Let us consider a circular word with a linear representation $w_1 w_2 \ldots w_r = w$, where $w_j = (0^{p^j} 1^{p^j} \ldots (k-1)^{p^j})^{p^{r-j}}$. We see directly that $|w_j| = kp^r$ and, thus, $kn := |w| = krp^r$. Furthermore, we have $w_j \in S_{p^j}$.

We only work in the palindromic case from now on, but the same reasoning also holds in the case of strong antipalindromes.

Let vv^R be a palindromic subsequence of even length. Thus, we find that v is a subsequence of the linear word $u_1' w_{i_1} w_{i_2} \ldots w_{i_a} u_2$ and v^R is a subsequence of the linear word $u_2' w_{j_1} w_{j_2} \ldots w_{j_b} u_1$, where $u_1 u_1' = w_{i_0}$, $u_2 u_2' = w_{j_0}$ and $i_k \neq j_\ell$ for all $0 \leq i \leq a, 0 \leq \ell \leq b$.

This shows that v is a common subsequence of $u_1' w_{i_1} w_{i_2} \ldots w_{i_a} u_2$ and $u_1^R w_{j_b}^R \ldots w_{j_1}^R u_2'^R$. By removing the parts of v that belong to the boundary blocks u_i we get v that is a common subsequence of $w_{i_1} w_{i_2} \ldots w_{i_a}$ and $w_{j_b}^R \ldots w_{j_1}^R$, where

$$|v| - |v'| \leq |w_{i_0}| + |w_{j_0}| = 2kp^r.$$

From now on, we only work with v'. We can rewrite v' as a concatenation of at most $(a + b - 1)$ blocks v_i, where each v_i is a common subsequence of some $w_1^{(i)} \in S_{p^{j_1(i)}}$ and $w_2^{(i)} \in S_{p^{j_2(i)}}$ where $j_1(i) \neq j_2(i)$. Furthermore, we have $a + b \leq r$ and

$$\sum_i |w_1^{(i)}| = akp^r$$

$$\sum_i |w_2^{(i)}| = bkp^r.$$

By using Lemma 10 we find that

$$|v'| = \sum_i |v_i|$$

$$\leq \sum_i \left(\frac{|w_1^{(i)}| + |w_2^{(i)}|}{k+1} + \frac{(|w_1^{(i)}| + |w_2^{(i)}|)}{p} + 2p^{r-1} \right)$$

$$\leq \frac{|w|}{k+1} + \frac{|w|}{p} + \frac{2|w|}{kp}.$$

This gives in total (together with the bound on $|v| - |v'|$)

$$|vv^R| \leq \frac{2|w|}{k+1} + |w| \left(\frac{4}{p} + \frac{4}{r} \right).$$

Thus, choosing $p = r \geq \frac{8}{\varepsilon}$ finishes the proof.   □

The trivial lower bound is $\frac{1}{k}$. For palindromes, this can be seen immediately. For strong antipalindromes the case for k odd works very similarly: We see that $\overline{(k-1)/2} = (k-1)/2$ and the word $((k-1)/2)^{|w|/k}$ is a strongly antipalindromic subsequence of length $|w|/k$. The case k is even slightly more complicated but can be dealt with in the same way as $k = 2$.

Theorem 13 deserves some remarks. First, it is interesting that the family of words constructed in the theorem provides the same bound for both palindromic and strongly antipalindromic subsequences. Second, it provides a generalization of the palindromic and strongly antipalindromic conjectures to the case of an alphabet of more than two letters. These conjectures also remain open.

Finally, for any $\varepsilon > 0$, we find that the bound $\frac{2n}{k+1+\varepsilon}$ holds almost surely for large n in the case when we choose every letter independently and uniformly in $\{0, \ldots, k-1\}$.

To see this, we fix a subsequence of length $\frac{n}{k+1+\varepsilon}$ and call it w_0. Then we try to find $w_0, \overline{w_0}, w_0^R$ or $\overline{w_0}^R$ as a subsequence of the remaining word w_1. However, any letter in w_1 is chosen independently and uniformly. Therefore, it takes on average k letters until one finds one specific letter. By the law of large numbers, the number of letters we have to read in a string of independent and uniformly chosen letters to find a specific subsequence of length ℓ is asymptotically normal distributed with mean ℓk and variance $\alpha \ell$ for some $\alpha > 0$. By the Chebyshev inequality, we find that w_0 (or any of the mentioned forms above) appears in w_1 almost surely for large n as $|w_1| = (k + \varepsilon)|w_0|$.

4   Linear Words

The minimum length of the longest palindromic/antipalindromic subsequence in the class of all linear binary words with n letters can be easily computed. However, for some restricted classes of words their behavior is more complicated. One of the simplest restrictions is to forbid some number of consecutive equal letters. The following proposition is then not hard to prove. It suggests some progress for Conjectures 1 and 4 for binary words without three consecutive equal letters.

Proposition 14. *Every binary word of length n without three consecutive equal letters has a palindromic subsequence of length at least $\frac{2}{3}(n-2)$. The same is true for an antipalindromic subsequence.*

Proof. Let w be a binary word without three consecutive equal letters. Consider the representation $w = w_1 w_2 \ldots w_m$ such that each w_i is composed of only zeros or only ones, and two consecutive words w_i and w_{i+1} consist of different letters. Then the length of each w_i is at most 2. Assume that m is even (otherwise remove w_m). Then one can pick at least one letter from each pair w_i, w_{m-i+1} (or two letters if both w_i, w_{m-i+1} are of the same length) and all the letters from $w_{\frac{m+1}{2}}$ in such a way that the resulting subsequence is a palindrome. This way we get a palindromic subsequence of length at least $\frac{2}{3}(n-2)$. The same proof can be done for antipalindromic subsequences. □

For the antipalindromic part, one can take the word $(001)^i$ to see tightness (we conjecture the bound $\frac{2}{3}n$ to be tight for words with equal number of zeros and ones, but we could not find an example providing tightness). For palindromic subsequences we conjecture a stronger bound.

Conjecture 15. *Every binary word of length n without three consecutive equal letters has a palindromic subsequence of length at least $\frac{3}{4}(n-2)$.*

We checked this conjecture up to $n = 30$. The worst known cases are provided by the word $(001)^i(011)^i$, showing the tightness of the conjectured bound.

Note that every binary word without two consecutive equal letters is a sequence of alternating zeros and ones, and thus has a palindromic subsequence of length $n - 1$, where n is the length of the word. For a three-letter alphabet it is not hard to prove the following.

Proposition 16. *Let w be a word of length n over a three-letter alphabet. If w has no two consecutive equal letters, then it has a palindromic subsequence of length at least $\frac{1}{2}(n-1)$.*

Proof. Assume that the number of letters in w is even (otherwise, remove the last letter). Let $w = w_1 w_2 \ldots w_m$ where w_i is a word of length 2. Each such word contains two different letters. Then for each pair w_i, w_{m-i+1} there exists a letter present in both words. By taking such a letter from every pair, we get a palindrome of length $m = \frac{1}{2}(n-1)$. □

Based on these observations and computer experiments, we formulate the following conjecture.

Conjecture 17. *Let w be a word of length n over an alphabet of size k, $k \geq 2$. If w has no two consecutive equal letters, then it has a palindromic subsequence of length at least $\frac{1}{k-1}(n-1)$.*

We checked this conjecture up to $n = 21$ for $k = 4$ and $n = 18$ for $k = 5$ by computer. A critical example for this conjecture is provided by a word which is a concatenation of the word $(a_1 a_2)^i$ and words $(a_{\ell+1} a_\ell)^{i-1} a_{\ell+1}$ for $1 < \ell < k - 1$. This word shows that the conjectured bound is tight.

5 Further Work

There are some questions besides the conjectures above that are worth mentioning. First, there is no known reduction between the palindromic and antipalindromic conjectures. Thus, it is interesting to know whether a bound for one of them implies some bound for the other one. Second, no non-trivial relation is known for the bounds for the same conjecture but different size of alphabets.

Acknowledgements. We thank anonymous reviewers for their comments on the presentation of the paper. The second author is also grateful to András Sebő, Michel Rigo and Dominique Perrin for many useful discussions during the course of the work.

References

1. Abboud, A., Backurs, A., Williams, V.V.: Tight hardness results for LCS and other sequence similarity measures. In: 2015 IEEE 56th Annual Symposium on Foundations of Computer Science, pp. 59–78 (2015). https://doi.org/10.1109/FOCS.2015.14
2. Axenovich, M., Person, Y., Puzynina, S.: A regularity lemma and twins in words. J. Comb. Theory Ser. A **120**(4), 733–743 (2013). https://doi.org/10.1016/j.jcta.2013.01.001
3. Brevier, G., Preissmann, M., Sebő, A.: Private communication
4. Bringmann, K., Kunnemann, M.: Quadratic conditional lower bounds for string problems and dynamic time warping. In: 2015 IEEE 56th Annual Symposium on Foundations of Computer Science, pp. 79–97 (2015). https://doi.org/10.1109/FOCS.2015.15
5. Bukh, B., Ma, J.: Longest common subsequences in sets of words. SIAM J. Discret. Math. **28**(4), 2042–2049 (2014). https://doi.org/10.1137/140975000
6. Bukh, B., Zhou, L.: Twins in words and long common subsequences in permutations. Isr. J. Math. **213**(1), 183–209 (2016). https://doi.org/10.1007/s11856-016-1323-8
7. Holub, Š., Saari, K.: On highly palindromic words. Discret. Appl. Math. **157**(5), 953–959 (2009). https://doi.org/10.1016/j.dam.2008.03.039
8. Lyngsø, R.B., Pedersen, C.N.: Protein folding in the 2D HP model. Technical report, University of Aarhus (1999)
9. Paterson, M., Dančík, V.: Longest common subsequences. In: Prívara, I., Rovan, B., Ruzička, P. (eds.) MFCS 1994. LNCS, vol. 841, pp. 127–142. Springer, Heidelberg (1994). https://doi.org/10.1007/3-540-58338-6_63
10. Sloane, N.: On single-deletion-correcting codes. In: Codes and Designs, vol. 10, pp. 273–291 (2000)

Author Index

Printed in the United States
By Bookmasters